Seismic Waves in Soil Layers

Olga V. Pavlenko

Seismic Waves in Soil Layers

Soil Behaviour During Recent Strong Earthquakes

Springer

Olga V. Pavlenko
Schmidt Institute of Physics of the Earth
Russian Academy of Sciences
Moscow, Russia

ISBN 978-3-031-52337-3 ISBN 978-3-031-52338-0 (eBook)
https://doi.org/10.1007/978-3-031-52338-0

This Springer imprint is published by the registered company Springer Nature Switzerland AG
The registered company address is: Gewerbestrasse 11, 6330 Cham, Switzerland

Paper in this product is recyclable.

Preface

The distribution of damage caused by earthquakes is largely dependent on the response of subsurface soils. Local site effects associated with soil response have always received great attention from engineering seismologists, because the majority of urbanized areas in the world are located in river valleys, on soft sedimentary deposits.

In addition to topographic effects, the following mechanisms of seismic wave transformations in soil layers are known: amplification of seismic waves in subsurface layers possessing lower values of seismic velocities and densities, resonant phenomena in softer layers, and the nonlinearity of stress–strain relations in soils. In conditions of strong motion, soil behavior can be nonlinear, and then, the evaluation of soil response becomes rather complicated, because it depends not only on the composition of soil layers and their saturation with water, but also on the intensity and spectral content of seismic motion.

The nonlinearity of soil response leads to changes (sometimes, significant) in the waveforms and spectra of seismic waves propagating in soil layers; resonant frequencies of soft layers become dependent on the intensity of motion and can noticeably decrease relative to the values estimated in microzonation studies based on the records of seismic noise and weak earthquakes. Rheological properties of soil layers change during strong motion, which is often due to the movements of the underground water.

The 2011 mega-thrust Tohoku earthquake revealed new features of soil behavior in conditions of long-lasting strong motion that involve soil hardening and generation of abnormally high peak ground accelerations (PGA).

Nowadays, estimation of soil response in strong ground motion is one of the most important and problematic issues in engineering seismology. Knowledge of parameters of ground motion during future strong earthquakes is necessary for the estimation of the response spectra, dynamic stresses and strains that can cause soil instability and damage of constructions, for the assessment of soil liquefaction potential.

Soil behavior in conditions of dynamic loading is calculated based on the dynamic characteristics of soils, and to date, various field and laboratory methods have been developed to estimate soil parameters in small- and large-strain ranges. However, they provide only limited information, because there remain differences between in situ conditions of soil loading and their laboratory and field analogues. The most promising approach is studying soil behavior in situ, i.e., interpretation of records of vertical (borehole) seismic arrays.

Since the 1990s, networks of seismic vertical arrays are developed, such as, KiK-net (Japan), DART and SMART (Taiwan), SCEC (Los Angeles), Bay Bridges (San Francisco), etc., and strong motion databases are growing. Vertical arrays are designed for a detailed study of the response of the upper meters of soils, which are most important from the viewpoint of earthquake engineering.

In the early 2000s, a method for processing records of seismic vertical arrays was developed by the author in collaboration with Prof. Kojiro Irikura (DPRI, Kyoto, Japan). The method allows the reconstruction of nonlinear stress–strain relations in soil layers, from the surface down to the location of the deepest device of the vertical array. The stresses and strains are estimated in successive small time intervals, so we can trace their changes during strong ground motion. Since the estimates are based only on real measurements, they are free of theoretical approximations and physical assumptions concerning mechanisms of processes arising in the medium during strong motion. These "models of soil behavior" allow us to visualize the behavior of different layers and transformations of seismic waves propagating in soils.

Since that time, the method was successfully applied to studying soil behavior during recent strong earthquakes, such as, the 1995 Kobe and 2000 Tottori earthquakes (Japan), the 1999 Chi-Chi (Taiwan) earthquake, the 2003 Tokachi-oki, and 2011 Tohoku earthquakes (Japan).

The catastrophic Tokachi-oki and mega-thrust Tohoku earthquakes provided records of strong ground motion at a large number of stations in the near-fault zones. The models of soil behavior during these earthquakes help us to understand the mechanisms of generation of abnormally high PGA in the near-fault zones.

The constructed models of soil behavior can be used in numerical experiments showing us transformations of intense seismic waves in soil layers. The types and quantitative characteristics of nonlinearity of soil response are defined based on the nonlinear identification methods developed in system analysis, i.e., the Volterra–Wiener theory and methodology. We discuss the quasi-stationary spectrum of seismic motion on the surface, modulation instability of seismic waves, application of high-order spectra and coherences to the identification of nonlinear-bounded components in seismic motion, and other nonlinear wave effects.

Based on the obtained results and conclusions, recommendations are given concerning accounting for nonlinearity of soil response in seismic microzonation studies.

The main results were obtained in my research works at the Schmidt Institute of Physics of the Earth of Russian Academy of Sciences (Moscow), in Strong Motion Laboratory of the Disaster Prevention Research Institute (DPRI), Kyoto University (Japan) in collaboration with Prof. Kojiro Irikura (the research programs of Japanese

Society for the Promotion of Science), and in the National Center for Research in Earthquake Engineering (NCREE) in Taiwan.

Moscow, Russia Olga V. Pavlenko

Acknowledgements I thank Prof. Kojiro Irikura, who invited me to his laboratory at DPRI, Kyoto University, for his kind attention and long-term cooperation, without which this book would not have been written. I am grateful to my colleagues and teachers in Russia Alexander Gusev (Institute of Volcanology and Seismology, Petropavlovsk-Kamchatsky), Alexei Nikolaev, Gennady Sobolev and Valentin Ulomov (Schmidt Institute of Physics of the Earth, Moscow) for their help and support, and to my son Vasily Pavlenko, the first reader of the manuscript, for his comments, which enabled me to improve the text.

Contents

Chapter 1
Dynamic Properties of Soils

Soil response to seismic motion (cyclic dynamic loading) depends on the mechanical properties of the soil. Soil behavior under various conditions of loading is defined by one or the other of its mechanical properties. In conditions of heavy loading, soils can exhibit dynamic instability, when their deformability and the probability of destruction increase.

Dynamic properties of soils have been investigated since the 1900s and addressed in a number of renowned studies of K. Terzaghi, N.M. Gersevanov, V.A. Florin, Ya.I. Frenkel, M. Biot, and others. A qualitatively higher level of the research has been reached in the late 1960s and early 1970s after the catastrophic earthquakes of 1964 in Anchorage (Alaska) and Niigata (Japan) and the 1971 earthquake in California. Severe damage and economic losses caused by these earthquakes were associated with numerous cases of liquefaction of sandy soils. Since in those years the construction of nuclear power plants began in the USA and Japan, investigation of dynamic properties of soils has become a research priority. The milestone developments were related to the studies of Hardin and Drnevich (1972), Casagrande (1975), Castro (1975), Castro and Poulos (1977), Ishihara et al. (1975, 1980, 1992, 1998, 1982, 1985), Youd et al. (1978, 1985), Seed (1968, 1978), Seed and Chan (1957), Seed and Lee (1966), Seed et al. (1976, 1992), Seed and Idriss (1970, 1971, 1982), Idriss and Seed (1967, 1968), Idriss (1990), Finn et al. (1977, 1978, 1986), Finn (2000), Lee (1976), Lee and Finn (1978) and others. To increase the reliability of estimation of the response of constructions on soil foundations in strong ground motion, elastic, filtering, damping characteristics of soils were studied, and the dynamic stability and changes in the rheological properties of soils under dynamic loading were evaluated, such as, the deformability and strength degradation. In laboratory experiments, nonlinear hysteretic stress–strain relations were obtained for various types of soils (Hardin and Drnevich 1972).

In Russia, dynamic properties of soils were studied theoretically and experimentally by Pokrovsky et al. (1938), Kompaneets (1956), Grigoryan (1960), Shemyakin (1963), Rakhmatullin et al. (1964), Melnikov and Rykov (1965), Stavnitser (1966),

O. V. Pavlenko, *Seismic Waves in Soil Layers*,
https://doi.org/10.1007/978-3-031-52338-0_1

Lyakhov (1966, 1974), Sagomonyan (1967), Gvozdev and Kuznetsov (1967), Nikolaev (1967), Vasil'ev et al. (1969, 1977), Krasnikov (1970), Rodionov et al. (1971), Zvolinsky (1982), Nikolaevsky (1984), Mishel and Shulman (1999), Voznesensky (1999) and others.

This chapter describes the main types of soils and their physical properties, methods for measuring their dynamic parameters, the most important features of soil behavior related to earthquake engineering, and models of soil behavior in conditions of their dynamic loading. The description of physical and dynamic properties of soils is taken from many authors, and the soil classification corresponds to the ideas of Mishel and Shulman (1999) and Voznesensky (1999).

1.1 Soils as Multiphase Dispersible Media; Types of Soils

Soils are materials originated from igneous, sedimentary, and metamorphic rocks that lie on the Earth's surface and are used in engineering practice as a basis for structures. According to their properties and classification accepted in Russia, soils are divided into rocky and non-rocky (Mishel and Shulman 1999). Non-rocky soils are mainly composed of Quaternary sediments: loose non-cohesive (coarse-grained, clastic, gravelly, sandy), weakly-cohesive soft, cohesive poorly lithified (silty clays) and biogenic ones. All the soils, in spite of the differences in their physical conditions, complexity, and variety of properties, can be considered as dispersible heterogeneous systems consisting of solid particles and pore space between them filled with water and air. In other words, real soils are multiphase systems generally consisting of solid, liquid and gaseous phases.

The solid phase is formed by mineral particles of various shapes, with various granulometric and mineralogical compositions. It is a porous matrix that is stable due to mutual particle-to-particle contacts, friction and inter-particle cementation bonds.

The mineral content strongly influences physical properties of a soil. Some minerals (quartz, feldspar) interact less actively with water, while others (like clays) more actively, and the mechanism of the interaction is different. The smaller the soil particles, the larger their specific surface areas, and the more centers of their interaction with water and contacts between the particles themselves. According to the mineralogical composition, soil particles are classified into four groups:

- primary materials—quartz, feldspar, mica, etc.;
- clayey (secondary) minerals formed by weathering of igneous and metamorphic rocks;
- salts—sulfates (gypsum, anhydride, etc.), carbonates (calcite, dolomite, etc.), halides;
- organic substances.

The properties of soils substantially depend on the particle sizes. Various soil classifications by the particle sizes were suggested. According to the classification by Mishel and Schulman (1999), soils are divided into:

- pebbly soils with particles > 20 mm (coarse/very coarse gravel (pebble) to cobble (64–256 mm) and boulder (>256 mm) on Wentworth scale (WS));
- gravelly soils with particles 20–2 mm (very fine (granule) to medium/coarse gravel (pebble) on WS);
- sandy soils with particles 2–0.05 mm (coarse silt to very coarse sand on WS);
- silty soils with particles 0.05–0.005 mm (same as on WS);
- clayey soils with particles < 0.005 mm (same as on WS).

Tsytovich (1968) recommended a simplified soil classification by the granulometric composition depending on the content of clay particles:

Soil	Weight ratio of clay particles to all solid particles, wt%
Clay	>30
Loam	30–10
Sandy ioam	10–3
Sand	<3

Sands are classified into five categories by the particle size:

- gravelly sands—more than 25 wt% of particles larger than 2 mm;
- coarse (medium) sands—more than 50 wt% of particles larger than 0.5 mm;
- medium sands—more than 50 wt% of particles larger than 0.25 mm;
- medium fine sands—more than 75 wt% of particles larger than 0.5 mm;
- fine sands—less than 75 wt% of particles larger than 0.5 mm.

Particle size analysis of aggregate materials includes another component—a coefficient of uniformity, which is a measure of particle gradation and defined as the ratio of the numbers of particles with different grain sizes in sands.

Clayey soils are very diverse in the grain sizes, shapes and mineralogy, and therefore, are not divided into groups. Clay particles are thin flakes whose thickness is 0.02–0.1 of their largest size; they can have rectangular (kaolinite) or acicular (attapulgite) shape. Soils consisting of clay particles are virtually waterproof and highly plastic when water-saturated.

Loesses are soils largely composed of silt-sized particles and characterized by high porosity (determining the uniqueness of their properties).

Arrangements of mineral particles, their interaction, and the way they are assembled define the *soil structure*, which can be:

- separate-grained (particles are virtually not connected to each other by cohesion);
- bridged-grained (some particles are connected to each other),
- aggregative (grains and aggregates are assembled into a system),
- fused (solid, dense, visually homogeneous aggregated clay mass).

Within this classification, types, kinds, and varieties of structures that differ from each other by the density, phase ratio, porosity and granulometric composition are distinguished. The diversity of soil structures is due to the diversity of diagenetic, constantly changing physical, geographical, and geological conditions of their formation and existence.

The voids between solid particles are filled with liquids and gases. *The liquid phase* in soils mainly consists of pore water, which is distinguished into free and bound water. Free water exists in gravitational and capillary states. Gravitational water experiences only the effect of gravity; it moves due to the pressure differences, while capillary water experiences the effects of gravity and capillary forces. Capillary water concentrates around the contact points of soil particles. Its mass is limited by the curved surfaces of particles and menisci. Gravity and surface tension pull particles together, forming a cohesion between them due to capillary moisture. The cohesion is stronger at smaller particle sizes and smaller distances between the particles.

Bound water is water affected by electric fields on the surfaces of the mineral soil particles; as a result, the molecules of water are oriented relative to the particle verges and lose the properties of free water. Bound water can be tightly bound (firm water, according to Gersevanov (1937)) and weakly bound, or film water, i.e., a polymolecular layer (several tens of molecular diameters thick) surrounding the particle surfaces covered by tightly bound water. Thus, in conditions in situ, soil particles cannot be in direct contacts with each other, but are always separated by thin films of molecular water, which has partially lost its liquid properties. According to Gersevanov (1937), if the soil pores are completely filled with water, the soil is called water-saturated, while dry soils are typically considered as soils with low natural humidity (up to 10–15%).

Weakly bound water only slightly affects the properties of coarse-grained and sandy soils, but this influence is significant for the properties of clayey soils, stipulating their low permeability. The larger the soil particles, the fewer the points of their contacts with each other and the stronger the forces acting on each contact, while capillary effects are insignificant. The mechanical properties of such a soil are defined by the arrangement of particles, their shapes, friction, and cohesion at the contact points. The friction here is not only the Coulomb dry friction, but also viscous friction, due to the viscosity of film water.

The smaller the particles, the more contact points and the higher viscous resistance. Thus, in finely dispersed clayey soils, electric molecular interaction forces are especially strong at the contacts of the particles and film water. When drying, soils first lose capillary water, then film water. Soils consisting of pebbly and gravelly particles are highly permeable for water; sands are water-permeable.

The gaseous phase in soils appears in the following states:

- trapped: gases are localized in voids between solid mineral particles, surrounded by films of bound water,
- free: gases (air) are connected to the atmosphere,
- dissolved in pore water.

It was established in experiments (e.g., (Lyakhov 1966)) that dynamic properties of soils, whose pores contain water and only a small amount of air, are close to the properties of completely water-saturated soils. At the same time, the presence of gas bubbles, trapped and dissolved in the pore water, can significantly affect soil deformability, causing compressibility of pore water and increasing elasticity of the soil. At the same time, the presence of free air in the soil does not really matter in soil mechanics, because gases do not participate in the distribution of pressure between soil particles.

Analysis of soil samples shows that in floodplains, the content of air in water-saturated sandy soils significantly changes with depth (Lyakhov 1974). In the upper layers that are annually dried, the content of trapped air is 2–5% of the total volume. This value continuously decreases with depth, and at 2–3 m below the water level, it drops to almost zero. As mentioned above, trapped air can exist in small amounts in deep non-drained layers in floodplains. The presence of air in these layers is explained by fluctuations in temperature and pressure.

Pore water contains dissolved air. As temperature increases or pressure decreases, the solubility of air in water decreases, causing it to be released in a free state. Increase of temperature from 0 to 10 °C leads to formation of bubbles with total volume of about 5.4 cm^2 in 1 L of water; i.e., at typical fluctuations of temperature in natural conditions, changes of the content of trapped air in a soil do not exceed tenths of a percent of soil volume. Therefore, since dynamic properties of soils substantially depend on the air content, different estimates can be obtained in the same soil in the morning and in the evening (Lyakhov 1974).

Trapped air can be located in the capillary water rise zone. In sandy soils, the height of this area above the groundwater level is several centimeters; in fine-grained soils with small pore sizes, it is tens of centimeters or even meters. Clayey soils in the area of capillary rise of water are water-saturated. Clayey soils at depths of 15–20 m, but above the groundwater level, can also be water-saturated (Lyakhov 1974). Thus, clayey soils can be water-saturated after a long stay above the groundwater level, which is impossible for sandy soils.

Such features of soil structure as the phase ratio and structural bonds between individual particles determine strength properties of soils and their dynamic stability. The genesis of structural bonds is complicated; it depends on the energy fields acting in the soils, which are related to molecular electromagnetic forces. These forces are influenced by the phase interface surfaces and by the composition, structure, and properties of substances filling interparticle spaces.

According to Mishel and Schulman (1999), structural bonds in soils are classified into crystallization bonds (water-resistant and non-water-resistant), which are brittle, rigid, and irreversible, and water-colloidal bonds (coagulation and condensation-types), which are viscous-plastic, soft, and reversible.

Physical properties of noncohesive soils. This soil category includes sands, gravels, finer-grained soils (silt), and clastic masses. Crystallization bonds, typical for noncohesive soils, arise due to the chemical forces; at the contact points of mineral particles, they form new polycrystalline compounds, which are strong, but brittle and non-regenerating when destroyed. The strength of these bonds depends on the mineral composition of the soils (Mishel and Schulman 1999). In coarse-grained soils with particle sizes > 2 mm, adhesion between the particles is absent at any humidity (Lyakhov 1974). In sandy soils, where particle sizes are relatively large, adhesion, caused by molecular attraction between solid grains and cementing action of salts, is weak. Capillary forces are of primary importance (sometimes this is called false coupling; it results from surface tension of menisci of capillary water in the intergranular space). These forces disappear when drying and no water remains in the pores, or at high humidity, when there is no air. Capillary cohesion is most pronounced at some intermediate humidity.

Density ρ *and porosity* n (the ratio of pore volume to the total volume) are of crucial importance, because any change in them causes a qualitative change in the soil behavior. For an idealized soil consisting of identical spherical grains, $n = 26\%$ at the densest packing of grains and $n = 48\%$ at the loosest packing. Loosely packed masses (like loose sands) are easily destroyed by shaking; large subsidence (tens percent of the volume) and residual deformations are possible (Tsytovich 1968). When a soil is below the groundwater level, the water filling the pores is squeezed out as the porosity decreases, and pore pressure built-up occurs. In this case, grains of sand become separated by water films, and spontaneous liquefaction occurs. Low water permeability of a sediment (e.g., due to the presence of clay particles, as in weakly cohesive soils) serves as a factor increasing the pore pressure. Therefore, the dynamic stability of loose sand deteriorates if it contains impurities of clay particles, and pore pressure in such a soil may not decrease for a long time after unloading. Boundary porosity values separating dense and loose sands are ~38–40% (Tsytovich 1968); based on these estimates, density and pressure values are established, at which the dilatancy (expansion during deformation) disappears.

The granulometric composition influences such soil parameters as density and dilatancy value observed in soils during shear deformations (dilatancy is much higher in coarse-grained than in fine-grained sandy soils).

Strength of granular masses. Selective weathering of unstable soil grains reduces the soil strength, turning it into a compressible mass; as the grains are further destroyed, large subsidence occurs in the soil. Weathering products are removed by water, and subsidence increases.

Particle shape. Soils composed of rounded isometric grains possess lower shear resistance than soils composed of angular and elongated grains.

Macrostructure. Structural features are of great importance: for example, thin layers of loams in sands and gravels reduce vertical water permeability of the soil column.

Humidity. Studying the humidity of homogeneous noncohesive soils is not necessary. If located below the water level, such a soil is fully water saturated, and if located above, its humidity varies depending on the environment. If the soil becomes saturated due to the rise of the groundwater, the water occupies ~85% of the pore volume in sands and ~96% in gravels. After a sharp decrease of the groundwater level, water saturation in sands and gravels drops to ~20 and ~10%, respectively (Tsytovich 1968).

In the practice of earthquake engineering, shear strains are the most widespread and critical type of strains in soils. During the propagation of shear strains, internal friction between the grains is important; it is defined by the laws of friction occurring when bodies shift relative to each other (Zaruba and Mencl 1976):

$$T = f \cdot N,$$

where T is the maximum shear resistance, f is the coefficient of friction, N is the normal force causing shear. In soil mechanics, f is commonly replaced by $tg\varphi$, where φ is the angle between the resultant force R and its normal component N, the so-called angle of internal friction (Fig. 1.1).

Due to the roughness of the contact surfaces, the angle can be large. For example, in sands it is 25–30°, up to 35°, whereas typical values for clays are about 20° (Zaruba and Mencl 1976).

Physical properties of cohesive and weakly-cohesive soils. Water-colloidal bonds, typical for clayey (cohesive) soils, are created by electric-molecular interaction between the soil particles, water films, and colloidal shells. Molecular bonds in clays are mainly caused by residual and not fully balanced charges, which is due to irregular arrangements of molecules on the surfaces of soil particles. The edges of the particles are positively charged, but the flat verges are negatively charged, which results in intergranular bonds. In addition, the cations adsorbed on the surfaces of the particles serve as connecting links between the particles. Whereas, molecules of water, involved in the inter-cation space by osmotic forces, separate the particles and reduce the soil strength. Thus, the cohesion of clayey soils decreases with increasing the water content. When water is squeezed out of the pore space, clay particles approach each other (hardening under pressure), and the cohesion increases. Therefore, clayey soils are characterized by the strength increase with increasing pressure (Mishel and Shulman 1999).

Thus, in clayey soils particles are held together by the cohesion, resulting from the molecular attraction between the grains, cementing action of salts, and the capillary forces. The most important factor is cohesion due to cementation of solid particles by salts. As humidity decreases, it increases, and capillary forces decrease. The increase of cohesion due to the increase of cementing effect of salts substantially exceeds the decrease of the capillary forces; therefore, when clayey soils dry out, the cohesion between the particles significantly increases. The deformability of clayey soils depends not only on the nature of the structural bonds, but also on the orientation of the structural elements (texture).

The cohesion of loess (weakly cohesive) soils is also due to the action of all these factors. Because of the higher porosity and larger (compared to clays) sizes of particles, the molecular forces are less important. Cohesion due to the cementing action of salts is the most important. With increasing humidity, the cohesion between the particles decreases faster in loesses than in clayey soils.

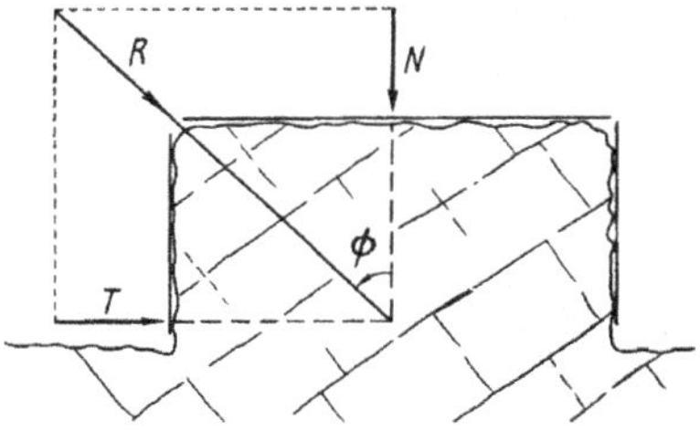

Fig. 1.1 The angle of internal friction in the analysis of shear resistance (from Zaruba and Mencl 1976)

Humidity is the most important physical parameter of cohesive soils, because it defines their strength and density. Depending on the humidity, soils can be solid, semi-solid or viscous, plastic and liquid. Since the influence of water is different in soils containing different amounts of clay particles, soils are also characterized by consistency. To describe the consistency of a soil, three parameters are used, such as, humidity w, plasticity limit w_p, and yield stress w_l. Humidity, at which the soil turns from a semi-solid state (w_p) to a liquid state (w_l), indicates how much water this soil can adsorb. Since this ability is defined by the quantity and petrographic composition of clay particles and by the content of cations, the difference $PI = w_l - w_p$ (plasticity index) is an important characteristic of the soil. Soils with PI below 15 may consolidate rather quickly at the foundations of structures (Tsytovich 1968). However, they are prone to liquefaction and become mud-like at the bottoms of pits, for example. In earth-fill structures they are more easily compacted than fat clays with higher plasticity index. Two factors reduce PI, such as, low content of clay particles and high content of particles that poorly retain water. The plasticity of wet clays is due to the secondary molecular bonds between particles.

The content of particles with sizes < 0.002 mm is another important characteristic of clayey soils: for example, residual strength of clays depends on their quantity. However, we cannot say that the size 0.002 mm represents an absolute boundary between clayey and dusty particles; it is only the most suitable arbitrarily chosen limit (Tsytovich 1968).

Macrostructure. The macrostructure of clays, permeated with a network of hairline cracks oriented parallel or perpendicular to the stratification or sectioning layers at different angles, is of great practical importance. The origin of these cracks has not been satisfactorily explained. The denser the clay, the more noticeable this is. Such cracks occur not only near the surface, but also at depth, but in this case, they are closed.

1.2 Measurements of Dynamic Parameters of Soils

Modern field and laboratory methods for estimating dynamic parameters of soils are described in detail, for example, by Kramer (1996) and Voznesensky (1999). Testing models, studying soil-structure interactions and interpretation of the observed in situ soil response in earthquakes also allow us to evaluate some of the dynamic parameters of soils. Below is a brief overview of these methods and some conclusions about the parameters and the accuracy of their estimation in experiments.

Field and laboratory methods for measuring dynamic parameters of soils can be divided into two groups: methods inducing low or high strains in soils.

Field Methods for Measuring Soil Parameters

Field methods for dynamic testing allow estimation of soil parameters in situ, when external and internal stresses, as well as chemical, thermal and structural environmental conditions are reflected in the measured parameters of soils. The advantage of field methods over laboratory methods is studying the response of large volumes, where induced strains are close to those induced in soils during earthquakes.

Field methods for measuring dynamic parameters of soils in the low-strain range (i.e., at strains < 0.001%, when soil behavior can be considered linear). These methods are based on the theory of seismic wave propagation in linear media.

Seismic reflection and refraction methods allow us to estimate average velocities of propagation of *P*- and *S*- waves in subsurface sediments, thicknesses of the layers, and the angles of incidence based on measurements on the surface, provided that the acoustic rigidity of each underlying layer is higher than that of the previous one (which is typically the case). Low-velocity layers located between high-velocity layers and thin high-velocity layers are not identified (Redpath 1973), and the resolution of the method is rather low. To study parameters of soils lying below the groundwater level, *S*- waves are preferable, since *P*- wave velocities strongly depend on water saturation, which complicates the interpretation of the results. The methods are widely used in studying large-scale or deep structures, but are rarely used to study parameters of thin soil layers.

Suspension logging method. A 5–6 m long suspension-type probe is lowered into a borehole filled with water or drilling fluid. A horizontal solenoid with alternating polarity is installed on the surface near the wellhead; it radiates a pulsed pressure wave into the liquid. Having reached the borehole wall, the pressure wave generates *P*- and *S*-waves in the surrounding soils, which propagate in the soil and transmit energy through the borehole fluid to two biaxial geophones located at the tip of the probe and spaced ~1 m apart. To better identify *P*- and *S*-waves, the procedure is repeated with a reverse polarity pulse. Based on the difference in the arrival times of the waves, *P*- and *S*-wave velocities in soils between the geophones are calculated. The method allows measuring seismic wave propagation velocities with one borehole, but the frequencies (500–2000 Hz for *S*-waves and 1000–3000 Hz for *P*-waves) are much higher than those used in geotechnical earthquake engineering. The method is especially effective at depths of ~2 km and more, where a resolution of 1 m depth can be achieved.

The method of standing (Rayleigh) waves avoids problems associated with picking wave arrivals. A vibrator radiating harmonic signals forms standing Rayleigh waves and wavy deformations of the Earth's surface. A seismic receiver moves along the profile and locating the nodes and antinodes, estimates the wavelength and the propagation velocity. Based on the Rayleigh wave velocities, *S*-wave velocities are estimated: usually the ratio $V_S = 1.09\ V_R$ is valid. If elastic moduli of the soil change with depth, the phase velocity of the Rayleigh wave also changes with depth; the estimated velocity refers to the depth range $\lambda_R/3$–$\lambda_R/2$ (where λ_R is the wavelength). By changing the radiation frequency, we can study changes of the velocity with depth. The method allows estimation of seismic velocities in subsurface layers, but does not provide high resolution. In engineering practice, it is often supplemented by spectral analysis of surface waves.

Spectral analysis of surface waves. Last decades, with the appearance of portable multichannel waveform analyzers, methods of spectral analysis of surface waves (SASW) have been developed. The methods provide information about the structure of the soil strata, including thicknesses of the layers and dynamic shear moduli. With modern equipment, we can obtain the dispersion curves for transmitted pulse or noise signals. The shape of the dispersion curve at a site is associated with seismic body wave velocities changing with depth. Standing wave method allows us to obtain a dispersion curve by performing measurements at various radiation frequencies. A source of pulse or noise signals and two receivers are installed on the surface along one line. Records of the two receivers are converted in frequency domain using fast Fourier transform (FFT), then the phase difference φ at each frequency is calculated, and the travel time is estimated by the formula

$$\Delta t(f) = \varphi(f)/2\pi f.$$

Since the distance between the receivers is known, the phase velocity and the wavelength are calculated. The calculations are performed in the field during the measurements. Measurements are carried out several times at different distances between the receivers. The estimations of the layer thicknesses and seismic wave velocities in the layers involve iterative procedures for comparing theoretical and experimental dispersion curves. To obtain theoretical dispersion curves, Thomas-Haskell method (Haskell 1953) is used that calculates the propagation of seismic waves in a system of homogeneous linear elastic soil layers of infinite horizontal extent. The initial estimates of the layers' thicknesses and velocities are clarified until an agreement is achieved between the theoretical and the experimental curves. Nonregular variations of the dispersion curve with depth can be associated with the influence of higher modes of Rayleigh waves. The method allows detection of low-velocity layers; it can be used down to depths of ~100 m and more, in areas where drilling or sampling are difficult (for example, on gravelly and coarse clastic soils). However, its application is limited to areas where Thomas-Haskell method is valid at least approximately.

Seismic cross-hole testing is considered as the most reliable method for estimating dynamic shear moduli in situ. At least two boreholes are required, in one of which a seismic source is placed, and in the other (or several others)—a receiver (receivers) at the same depth. Seismic wave velocities in the medium between the boreholes are measured. By moving the source and the receiver along the depth, we measure velocity as a function of depth. Thus, we can detect high-velocity layers that are undetectable by seismic reflection and refraction methods. The use of several boreholes with receivers located along one line answers a number of questions, such as, on the possible anisotropy of the medium and on reliable identification of the arrivals of refracted waves; also, it allows us to determine attenuation of seismic waves, if the medium is homogeneous. The distance between the boreholes should be as short as possible (~2–3 m) to avoid distortions of the initial parts of records by early arrivals of refracted waves (Woods 1978). Seismic waves are excited by mechanical shock sources (when estimating seismic velocities at depths of ~30–60 m) and by explosive sources (for depths > 60 m). Most sources generate seismic motion with shear strain amplitudes below ~0.001%, but some sources (like impulse shock shear wave generator developed in Seattle by Shannon and Wilson) achieve strains of ~0.1%, which allow estimation of soil liquefaction potential.

Seismic logging is carried out in one borehole, where seismic receivers are installed at certain depths, or one receiver moves along the wellbore. The source is located on the surface near the wellhead, so that the waves propagate vertically. One reference receiver is located near the source. The method allows detecting low-velocity layers enclosed between high-velocity layers. The equipment is the same as for seismic cross-hole testing. The reverse configuration is also possible, when the receiver is installed on the surface of a soil column, and the source moves in the borehole. Possible problems and difficulties of the method are related to: disturbances of the soil structure by drilling a well; the presence of water in the well; insufficiently low or too high energy of the source radiation; background seismic noise. Identification of *S*-wave arrivals may be difficult at depths below 30–60 m because of seismic wave attenuation. The method is also used to estimate seismic wave attenuation in the soil (Redpath 1973). Currently, seismic logging is often combined with seismic sounding, which does not require boreholes. In this case, a standard probe-cone embedded into the ground is equipped with a seismic mini-receiver that can measure *S*-wave velocities. Seismic waves are excited on the surface, usually by a blow. The use of two seismic cones allows testing soils in a cross-cone configuration, with a source installed in one cone. Measurement schemes are also proposed, in which the source and receiver are installed in the upper and lower parts of the same probe-cone.

Field methods for measuring dynamic parameters of soils in the high-strain range have been actively developed in recent decades, largely due to the appearance of new measuring equipment. Their results correlate with similar parameters obtained by the methods applied in the low-strain range. For geotechnical earthquake engineering, the most interesting methods are: standard penetration test, cone penetration methods, dilatometry and pressure measurements in soils, which allow estimation of the bearing capacity or the strength of a soil in conditions of its dynamic loading.

Standard penetration test is the most commonly used test in geotechnical engineering almost worldwide. The test counts the number of repeated blows needed to drive a standard soil sampler (length 81.3 cm, diameter 5.18 cm) to a depth of 30 cm by a slide hammer (63.6 kg) released from a height of 76.2 cm. The approach was developed by professor Seed's group (Seed 1979; Seed et al. 1983, 1985) based on testing results at sites previously affected by earthquakes. The method is based on empirical relationships between the number of blows N and the critical ratio of tangential and normal effective cyclic stresses (CSR_{cr}) that cause soil liquefaction during an earthquake with a magnitude $M_w = 7.5$. The number of blows N is a function of the soil type, confining pressure and soil density, but it is also influenced by the equipment and the applied procedures. Thus, the frequency of blows 30 to 40 per minute and the use of pre-drilled round wells of 10–13 cm diameter are recommended. Correlation relationships are obtained for soils with various contents of clay particles, which allow a reliable assessment of the liquefaction potential of the soil with known dispersibility under certain dynamic loading, based on the results of standard penetration. To increase the capabilities of the method, coefficients for earthquakes of various magnitudes were determined, and the procedures for correcting the results to various engineering, geological, and stress conditions were proposed.

For gravelly and pebble soils, a special *Becker-penetration test* based on the Becker drill system was developed. To assess soil liquefaction potential based on the results of Becker penetration test (converted to the equivalent parameters of standard penetration test), similar diagrams are used as for standard penetration. The following factors contribute to the uncertainty of the assessment of soil liquefaction potential: (1) the energy of a hammer blow depends on the conditions of fuel combustion, so the energy should be measured whenever possible (Finn 1993); (2) friction against side surfaces of the borehole is not taken into account; (3) the reliability of the relation between the results of standard penetration and Becker penetration tests remains unclear (Voznesensky 1999).

Cone penetration methods are developing rapidly recent years. A standard cone penetrometer with a conical tip of 10 cm^2, opening angle of 60°, and a cylindrical sleeve with a surface area of 150 cm^2 is pushed into the ground at a constant rate of 2 cm/s, while sensitive sensors mounted on the tip and on the sleeve measure the resistance of the soil. The resistance correlates with many soil parameters, including those estimated by standard penetration test and with the soil type. The result is a continuous profile of soil resistance, which allows detection of thin layers and seams undetectable by standard penetration. The capabilities of the method can be extended by installing additional sensors, e.g., to measure pore pressure (as in *piezo-cone penetration*) or seismic wave velocities (as in *seismic sounding*). However, the method cannot be applied on very hard rock, and its application is limited on gravelly-pebble materials.

Dilatometry. A flat steel dilatometer with a thin circular movable membrane on one side is inserted into the ground, and measurements are carried out at penetration intervals of 10–20 cm. The reservoir behind the membrane is filled with gas under pressure, and the pressures required to shift the center of the membrane by 0.05 mm and by 1.1 mm are recorded. These values are then used to calculate soil parameters.

Pressure measurements in the soil is a method that allows us to measure stress–strain relations in soils in situ. A cylindrical tensometer has a movable membrane that applies uniform pressure on the walls of the well. Strains are measured either using a fluid injected into a volume bounded by a membrane, or by a sensor device using compressed gas. The measured values are corrected for the operation of the system, the depth of penetration, and membrane parameters; then pressure–volume relations are constructed and recalculated to stress–strain relations of the soil. Devices that minimally disturb the soil structure are being developed.

Other field methods represent variations of the methods described above and new methods.

Geotomography uses algorithms developed in medicine for constructing 3D images for various parameters of the medium. Using several emitters and receivers, the matrix of travel times of seismic waves (with different travel paths) is constructed and compared with the predictions of the ray theory that accounts for reflections and refractions at internal inhomogeneities and boundaries. Item-by-item examination is applied to find the number and locations of the inhomogeneities and the slopes of the internal boundaries that provide the best agreement between the observed and calculated travel-time matrices. The possibilities of the method are great: two- and three-dimensional internal structures are mapped in this way.

Measurement of electrical conductivity of water-saturated soils with a special equipment is used to estimate in situ density, elastic low-strain moduli and the liquefaction potential. A *torsion system* is also used, consisting of two concentric cylindrical tubes installed at the bottom of a well; the angle of their rotation relative to each other is measured, depending on the torque applied to the inner tube. As a result, we obtain a relation similar to the stress–strain relation (Kramer 1996).

Vibration methods are based on studying the dynamic interaction of structures with the foundation soils by measuring parameters of various vibration modes of the structures. Analysis of these data allows the estimation of parameters of the soils.

Vibro-penetration test is based on the comparison of the soil resistance to penetration of the probe into it in static and dynamic regimes. Tokimatsu (1988) used a standard cone penetrometer as a probe, and estimated soil liquefaction potential based on the ratio of the corresponding values of the resistance to penetration.

Laboratory Methods for Measuring Soil Parameters

Laboratory tests are usually carried out on relatively small soil samples, and laboratory methods are diverse and numerous. Dynamic properties of soils are described by soil density, confining pressure, soil structure, soil age, its stress–strain history, and thixotropy. Many of these factors are losing in a sample, so we should take samples without disturbing the soil structure when possible. The results of laboratory measurements often depend on the experimental conditions. For example, to prepare sand samples, more than 10 different methods are currently used, such as, wet tamping, vibration packing, precipitation in a water column, pre-freezing, etc. At the same time, variations in the dynamic shear stresses that cause initial liquefaction in samples of the same sand with the same density, but prepared by different methods, reach 110% (Voznesensky 1999).

Laboratory methods for measuring dynamic parameters of soils in the low-strain range. There are few methods, such as, low-amplitude resonant column testing, ultrasonic pulse testing, and piezoelectric bender element testing.

Low–amplitude resonant column testing is the most common type of laboratory tests for measuring shear moduli, Young's moduli, and attenuation in soils in the low-strain range. A soil sample is placed to a triaxial compression chamber and subjected to longitudinal and torsional loading at variable frequencies until the resonance is achieved (noise- or impulse- loadings are also used). Based on the fundamental frequency, size and mass of the sample, elastic moduli are calculated; attenuation is estimated based on the decrease of the amplitudes of free oscillations with time.

Ultrasonic pulse testing allows measuring wave propagation velocities and wave amplitudes; the method is more often used in soft materials, like bottom sediments. A piezoelectric source radiating high-frequency pulses and an ultrasonic receiver are attached to the two ends of the sample; the wave propagation velocity is estimated based on the travel time of the impulse and the distance between the source and receiver.

Piezoelectric bender element testing also allows measurements of *S*-wave velocity. A bender element consists of two plates made of piezoelectric materials and connected to each other, so that the applied voltage causes one material to stretch and the other material to simultaneously compress, which makes the plates to oscillate. Any deformations of the bender element generate an electrical voltage in it, so that such elements are used both as sources and receivers. The source radiates a pulse signal, and *S*-wave is generated in the sample. When it reaches the other end of the sample, another pulse is generated in the receiver. *S*-wave velocity is estimated from the travel time of the pulse.

Laboratory methods for measuring dynamic parameters of soils in the high-strain range usually involve the control of drainage conditions and careful measurements of the volume and/or pore pressure. Loading of soil samples in drained conditions is accompanied by volumetric strains, while in undrained conditions, pore pressure and effective stresses in the sample increase.

Dynamic triaxial compression is most widely used in the world practice. Cylindrical samples are placed between the compressing upper and lower plates of the press and wrapped in a rubber membrane on the sides; thus, radial and axial stresses are applied to the sample. In this testing scheme, the principal stress axes are vertical and horizontal, and the deviatoric stresses (the difference of the axial and radial stresses) are sinusoidal both in stress-controlled and strain-controlled tests. Most often, constant radial stresses and axial stresses periodically changing with a frequency of ~1 Hz are used. The tests can be carried out in conditions of isotropic or anisotropic consolidation in a soil sample. Stresses created in the sample are close to those induced in the horizontal section of an elementary volume of a soil column by vertically incident *S*-waves from an earthquake. Measured stresses and strains are used to calculate the elastic moduli and attenuation of the soil; however, we should remember that in such tests, it is usually impossible to create conditions adequate to in situ conditions of seismic wave propagation in soil columns.

Dynamic simple shear tests rather adequately simulate loading conditions of an elementary volume of a soil column in seismic motion and are often used to assess soil liquefaction potential. A short cylindrical soil sample is firmly confined by a reinforced rubber membrane to prevent lateral strains. Cyclic shear stresses are applied to the upper or lower end of a sample and induce *S*-waves in it. For more adequate modeling of seismic motion, a shaking table can be used; to minimize nonuniformity of stress distribution in the sample, the ratio diameter/height should be ≥ 8.

Dynamic torsional shear test avoids many difficulties encountered during triaxial compression and simple shear tests. These tests allow us to create isotropic or anisotropic initial loading conditions and apply cyclic shear stresses to the horizontal faces of the sample, providing rotation of the principal stress axes. Thus, the seismic effect on the soil element is simulated more adequately. This method is most often used to measure stiffness and attenuation in soils in wide strain ranges. In cylindrical or annular hollow samples (for a more uniform distribution of strains in the cross section), shear strains are created, which parameters are measured and used to evaluate soil parameters.

Fatigue testing of soils is dynamic testing of soils with rigid structural bonds. Though the phenomenon of soil fatigue is almost unexplored and only few methods of fatigue testing of soils (mainly soils used as building materials) are developed, interest in this subject is increasing. Usually, the fatigue diagram is constructed based on test results of 10 or more samples at various loadings, providing the identity of frequency, shape, and asymmetry coefficient of the loading cycles. Usually, three types of loading patterns are applied, such as, compression–expansion, bending, and torsion. Quasi–static tests (at frequencies < 0.1 Hz), low–frequency tests (0.1–10 Hz), medium-frequency tests (10–100 Hz), high-frequency tests (0.1–10 kHz) and ultrasonic tests (>10 kHz) are performed. Testing results are processed with statistical analysis to find relations between the acting stresses, numbers of loading cycles, and the corresponding probability of failure. At stresses exceeding the dynamic yield stress, the range of low-cycle fatigue is defined; when exceeding $5\ 10^4$ loading cycles, there is the range of limited endurance, referring to multi-cycle fatigue. A diagram break was found at the junction of the ranges of the low-cycle and multi-cycle fatigue, which indicates changes in the mechanisms of fatigue failure and the stress state of the sample (Fracture Mechanics... 1990).

Model testing is a type of laboratory tests aimed at reproducing boundary loading conditions characteristic for particular geotechnical problems with the help of models. Models simulating geological objects (slopes, foundations of structures, etc.), engineering structures (dams, embankments, etc.), and "soil–structure" interacting systems are tested. A small-scale physical model of a large-scale prototype is constructed and subjected to cyclic loading. Models are used either to evaluate the behavior of the prototypes in conditions of loading, or to study the influence of various parameters on the geotechnical solution overall. In model testing, similarity ratios are used, and we should account for the fact that soil behavior is sensitive to the stress level, i.e., some soils experience compression under high stresses and expansion under lower stresses. Model tests can be divided into two groups, such as, tests in the gravity field 1 g (usually carried out on shaking tables) and tests at higher accelerations (carried out on geotechnical centrifuges).

Shaking table tests were the main type of model tests since late 1950s and provided us a basic understanding of the liquefaction, foundation settlements and shear stresses. On a horizontally shifting shaking table, soils experience simple shear loading. It was believed that the conditions created in model testing on shaking tables are more adequate to in situ conditions than conditions of laboratory tests, because of the absence of edge effects of stress concentration, nonuniform stress distribution and porosity variations, which are typical for small-scale simple shear tests on samples.

Most of shaking tables possess one degree of freedom and can move in one direction horizontally, but there exist shaking tables with several degrees of freedom. The frequency ranges of their shaking cover average and high seismic frequencies.

There exist large shaking tables that allow testing models of a size of several meters, but shaking tables cannot provide high accelerations > >1 g.

Centrifuge tests. A model is located at some distance from the axis of the centrifuge rotating at a certain speed; thus, accelerations several times higher than 1 g are created in the model. Measures are taken to ensure that accelerations experienced by each point of the model were similar to the accelerations acting in its full-scale prototype; thus, the behavior of the model will be similar to the behavior of the prototype. Therefore, when planning an experiment and interpreting the results, similarity ratios are very important: the scaling coefficients indicate how much dynamic loads should be increased in the centrifuge testing.

For example, stresses and strains induced by seismic waves in a 30-m earthen dam can be simulated in a 30-cm model on a centrifuge, where accelerations up to 100 g are created (however, the sizes of soil particles in the model correspond to the particle sizes of the prototype that are 100 times larger than in reality). Harmonic vibrations with a frequency of 1 Hz lasting 10 s for the prototype correspond to vibrations with a frequency of 100 Hz lasting 0.1 s for the model. However, pore pressure dissipation will proceed 10,000 times faster in the model than in field conditions; therefore, viscous liquids, like glycerin or silicone oil are often used as pore fluids in centrifuge tests.

The behavior of the prototypes can be evaluated more accurately by means of improving techniques and comparing the responses to certain seismic loadings of the models of different sizes, but with the same similarity ratios. Since the similarity ratios are applied to all parts of the models, then miniature sources, receivers, and connecting cables are required to minimize their influence on the model response.

Other tests on models. Recent years, hydraulic gradient tests are also used to measure dynamic soil properties (Zelikson 1969), in which the model is subjected to controlled hydraulic gradient loadings. However, the application of such models is limited to a rather narrow range of problems, because the acting body forces must be oriented along the field lines and flows, but the presence of obstacles and inhomogeneities reorient the forces, making them act in different directions at different points of the medium, which significantly complicates the interpretation of the results (Kramer 1996).

Studying soil-structure interactions (or soil-model interactions) also allows the assessment of dynamic characteristics of soils.

Resonant foundation method is used to estimate dynamic shear moduli of soils. The calculations require the values of fundamental frequencies and peak displacements of the "soil–foundation" systems in conditions of resonance. A model of the foundation of a given mass with a vibrator is installed on the ground, and the vibration frequency changes smoothly until the resonance is achieved. Usually, torsional vibrations or vertical vibrations are used (Srinivasulu and Vaidyanathan 1976). The thickness of the investigated soil layer is related to the weight of the foundation.

A water cannon is a tube installed vertically on a massive base. In its lower part, there is a chamber for explosive charge, separated from the main barrel by a membrane. The tube is filled with water ejected when the explosive is detonated in the lower chamber. The dynamic rigidity of the soil, on which the cannon is mounted, is estimated by comparing the momentum of the ejected water with the vertical response of the entire system. The method is also used to evaluate the dynamic response of bridges and underground structures.

Cylindrical in situ test is proposed to evaluate the response of soils and structures in conditions of high (up to 10%) shear strains (Bratton and Higgins 1978). Vertical and horizontal accelerometers are installed in the studied soil massive with some step in depth and in radial directions. The explosion in the central borehole produces pressures up to ~40 MN/m^2 on its walls, causing soil destructions near the cavity and significant shear strains in the vicinities. The soil response is measured, which allows us to estimate the dynamic properties of the soil and derive its state equation (parameters of the state equations allow calculation of the probable soil response to a dynamic excitation at any point of the soil massive). Calculations are carried out by the iteration method, in order to obtain velocigrams of the ground motion showing the best agreement with the recorded ones. In the explosion tests SIMQUAKE and MINISIMQUAKE (that are similar in the methodology), horizontally propagating plane waves are excited in the ground by a series of successive explosions in the boreholes located along one line (Higgins 1978); in multi-cycle tests, pressure at the soil-structure contacts is studied.

Studies of free and forced oscillations of foundations. Free oscillations are usually excited by shocks. Forced oscillations are used to control the behavior of existing structures; the emitters are installed on the foundations (Krasnikov 1970). Based on such tests, we can calculate dissipative characteristics of the soils, but the complexity and low reliability of the results should be mentioned.

Interpretation of in situ soil response. Instrumental observations of the response of real geological structures to seismic motion provide us valuable information about dynamic properties of soils and some geotechnical engineering parameters. The approach requires well-equipped observation sites, such as, seismic vertical arrays or dense surface arrays installed on sediments or rocks. Simultaneous records of motions on the crystalline basement and on sediments can be used to estimate the dynamic parameters of the sediments. Obviously, this approach requires earthquake records, and the strain levels, at which dynamic properties of soils are evaluated, are determined by the strains induced by the earthquake. The examples of such observations of soil response are given in Chap. 2, Sects. 2.2 and 2.4, Chaps. 3 and 4. As seismic observations are developing and improving, a growing number of records of soil response become available for processing and analysis.

1.3 Strength Properties and Dynamic Stability of Soils

Strength properties of soils characterize their resistance to external loads and generally depend on the composition of the soil, the ratios of solid, liquid and gaseous phases and structural bonds between the particles.

Soil resistance to compressive and shear loadings is defined by their strength parameters. Breakdown of structural bonds in conditions of soil loading is accompanied by violations of the soil structure, i.e., displacements of the mineral particles, rearrangement of their contacts, and volume deformations. The thickness of cementing films of salts and water between the grains is much smaller than the grain sizes, and the cohesion between the films and the grains is much lower than the strength of grain material; therefore, in conditions of loading, the films become zones of stress concentration, and the destruction of the skeleton begins from ruptures of cementing bonds (Zaruba and Mencl 1976). In sandy soils, stress concentration occurs near individual asperities, at interparticle contacts, and leads to compression of the grains themselves. Further loading causes mutual displacements of solid particles and their repacking to achieve higher density. This is accompanied by friction between solid particles, depending on the particle shape, roundness, and initial density of the soil. As a result of particle rearrangement under loading, the old structure is destroyed, a new structure is formed, and residual deformations appear.

Under small loads, rearrangement may not occur, and residual deformations are not formed. Static and dynamic loads of moderate intensity (shocks, vibrations, small explosions, etc.) usually cause small mutual displacements of particles, so that soil compaction occurs (Ivanov 1983). The increase of static loading, e.g., during the construction of a structure, causes both compaction of the skeleton of the soil massive at the basement of the structure, and redistribution of water and air in the pores of the skeleton, with their partial squeezing and removal. In sandy soils, squeezing water and air with increasing static loading takes minutes to hours, whereas, in clayey (cohesive) and loess (weakly cohesive) soils it takes days, weeks and even months. Under static and dynamic loads of long durations, soils behave as three-component media with varying water and air contents in the pores (Lyakhov 1974).

The occurrence of shear strains in soils is accompanied by the phenomenon of dilatancy, i.e. increase in soil volume during its deformation. When shear strains occur in a soil, shear stresses also occur, the cohesion between the particles is breaking, and sliding surfaces may form. In coarse-grained dense sandstones dilatancy and cracks up to 5 mm thick occur. The physical mechanism of dilatancy is the destruction of cohesion areas and rotation of particles along sliding surfaces (Zaruba and Mencl 1976).

When dilatancy occurs, sliding surfaces develop in the soil as thin as possible, with the thicknesses of fractions of a millimeter in clays and a few grains in sands. In soils showing dilatancy, the angle of the internal friction is larger than in materials where this phenomenon is absent. For dense and loose sands, the difference in the angle is about 10° (Zaruba and Mencl 1976).

The studies have shown that soils, except loose deposits, are dilative under low shear stresses, but contractive (negatively dilative) under high shear stresses (~4 kgf/cm^2 for dense clays). In the latter cases, plastic behavior of soils is observed, associated with volume decrease during shear deformation. In contrast to thin sliding surfaces observed in dense deposits with dilatant behavior, larger sliding zones are formed; this behavior is typical for loose sands and clastic deposits on mountain slopes (Zaruba and Mencl 1976). Dilatancy indicates the beginning of soil destruction.

Therefore, the orientation of sliding surfaces formed as a result of the destruction of soil structure during dilatancy, is partially predetermined before the strength limit is reached. At each stage of load increment, internal structural weakening occurs, and induced anisotropy develops. The geometry of large slip deformations is analyzed by plastic flow theory (Zaruba and Mencl 1976).

Soil strength is the maximum resistance that the soil provides to a particular loading. The strength of a soil element is usually defined as the shear stress that must be applied at some point to cause a rupture. Two measures of the soil strength are often considered (Kramer 1996), such as, dynamic strength characterizing the soil under cyclic loading and defined by the limiting cyclic strain during cyclic loading, and static strength, i.e., the limiting strength under static loading, which the soil possesses at the end of cyclic loading. Changes in static strength can be caused by changes of the soil structure during its cyclic loading. The degree of structural disturbance depends on the relationship between the amplitude of cyclic loading and strain, at which destruction occurs in conditions of static loading (Thiers and Seed 1969).

In field conditions, failure is usually associated with strains exceeding a certain limiting level, and the breakpoint is often defined in terms of limiting strains. The relationship between dynamic loading and the ultimate soil strength is one of the most important issues in geotechnical earthquake engineering. Slope stability, behavior of foundations and structures in conditions of their dynamic loading largely depend on the strength (dynamic resistance or stability) of soils under high strains (Kramer 1996; Voznesensky 1999).

Discussing soil stability, we usually consider cases of the behavior of coarse-grained non-cohesive and fine-grained cohesive soils in drained and undrained conditions.

Dynamic dilatancy of non-cohesive soils. Most studies on soil dynamics, both experimental and theoretical, deal with sands, which is due to painful experience of the liquefaction of sandy soils during past strong earthquakes (Coulter and Migliaccio 1966; Ishihara 1985; Youd et al. 1985, etc.). Typical forms of the response of sandy soils to dynamic loadings are (Voznesensky 1999):

- compaction of loose sands of any humidity,
- liquefaction of water-saturated sands due to the rapid built-up of the pore pressure, on the background of their compaction,
- degradation of their bearing capacity, i.e., cyclic mobility of dense sands (observed at lower shear strains),
- dilatancy and softening of dense wet sands.

Dynamic response of dry non-cohesive soils. In conditions of both static and dynamic loading, the solid skeleton counteracts to compression, and the compressibility of dry non-cohesive soils is defined by the compressibility of the skeleton, i.e. by deformations of solid particles at the contact points and their displacements relative to each other. The contents of small amounts of water and air in the pores determine the degree of cohesion of solid particles. With increasing dynamic loading, deformations of the skeleton grow, while compression of water and air also grow, and their resistance to the compression increases, and finally, at some pressure value, it becomes comparable to the resistance of the skeleton. Starting from a certain stress value σ_{κ}, a non-water-saturated soil can be considered as a three-component medium, neglecting the resistance from the skeleton (Lyakhov 1974). The σ_{κ} value depends on the porosity, humidity and granulometric composition of the soil. In a completely water-saturated soil, σ_{κ} is equal to the atmospheric pressure. With increasing the air content in the pores, σ_{κ} increases to hundreds, thousands and tens of thousands atmospheres.

Low-amplitude cyclic loading of dry non-cohesive soils causes slow accumulation of axial strains with amplitudes $\leq 0.04\%$ per cycle, with gradual stabilization of strains (Voznesensky 1999). With increasing the loading amplitudes, the stabilization can be replaced by slow progressive increase of strains. Sudden failure of the sample may be caused by slow formation of weakened zones in the soil. The critical level of the amplitude of repeated cyclic loadings (*CLRL*) increases with increasing soil plasticity. *CLRL* ranges from ~0.05 for non-plastic soils to ~0.55 for soils of high plasticity, such as clays (muds) in San Francisco area (Kramer 1996).

Dynamic response of loose water-saturated non-cohesive soils is associated with the liquefaction phenomenon. Liquefaction attracted the attention of seismologists after the catastrophic 1964 earthquakes in Anchorage (Alaska) and Niigata (Japan). Since the 1970s, the mechanisms of liquefaction and soil behavior before and after liquefaction have been intensively studied. Recent earthquakes have shown that a wide range of soils, from silty sandy to gravelly soils, are susceptible to liquefaction (Yoshida and Iai 1998).

Let's briefly consider physical mechanisms of liquefaction. In conditions of dynamic loading of a water-saturated soil, the skeleton is deformed; however, since compressibility of water and air (a small amount) is many times smaller than compressibility of the skeleton (as experiments show: e.g., Lyakhov 1974), soil compression is mostly resisted by the water and air, not the skeleton. Thus, the compressibility of a water-saturated soil is defined by the compressibility and the content of the water and air in the pores. With increasing normal compressive stresses σ, the soil (skeleton) strength increases, because of its denser repacking. However, pore water pressure provides the opposite effect. If we denote the pore pressure u, the remaining effective stress is $\sigma_{eff} = \sigma - u$. Since the soil strength changes depending on σ_{eff}, the soil resistance to friction and dilatancy change in water-saturated soils. For non-cohesive permeable soils, u is equal to the hydrostatic pressure of the groundwater. Destruction of the soil structure and soil compaction lead to a complete or partial transfer of the soil's weight and external load to the pore water. This increases the pore pressure, which then decreases, approaching its initial value as the water is squeezed out and the soil is compacted.

Under short-term dynamic loadings, non-cohesive water-saturated soils virtually do not exhibit skeleton creep effects (Lyakhov 1974). As experiments show (Ivanov 1983), the soil skeleton tends to be deformed instantly, and at the moment of loading, the pressure in the pore fluid increases to its maximum value almost instantly; therefore, water-saturated soils (in a wide range of initial densities) pass into completely liquefied states. Immediately after the complete liquefaction, soil consolidation begins. First, rearrangement of particles occurs in the lower parts of the layer, then it gradually spreads upwards.

In conditions of cyclic loading on the surface, layer-by-layer destruction of the soil structure and spreading the liquefied zone down is clearly observed. During the first vibrations, the upper layer is liquefied, which leads to unloading of the underlying layers and their transition to the liquefied state during subsequent vibrations. Layer-by-layer liquefaction and vibration lead to slow built-up of pressure in the fluid and a gradual development of pore pressure to deeper layers. The presence of trapped air in the pores of non-cohesive water–saturated soil changes the size of the zone of complete liquefaction, because instantaneous reduction of gas bubbles causes volume deformation, i.e., soil compaction, followed by squeezing water, which still continues after the end of soil compaction (Lyakhov 1974; Mishel and Shulman 1999).

In laboratory experiments, relative pore pressure sharply increases even in the first cycle of loading of loose water-saturated sands (Voznesensky 1999). Then a rapid accumulation of strains to ~2.5% occurs and almost instantaneous rearrangement of grains with a slight increase of the axial strain. Then the strain accumulation slows down, and the soil behavior can be described as cyclic mobility. Thus, in loose sands and sands of medium density, a rapid accumulation of strains occurs, abruptly turning into cyclic mobility. Liquefaction is possible at the inversion of the stress sign, which always happens during the propagation of shear waves (Voznesensky 1999). Voznesensky also notes that soil liquefaction during strong motion can occur in the following cases:

- in soils with loose structure and a high content of dusty-sandy material, i.e., when at least 32–40% of particles are smaller than 0.02 mm;
- if shallow sandy-gravelly soils are overlaid by a surface layer of significantly lower permeability, which prevents rapid dissipation of excess pore pressure during seismic motion. In this case, intrusions of gravelly soils into the overlying layers may occur.

Pure gravel-pebble soils without a significant amount of fine-grained fractions can be considered as dynamically stable, due to their low compressibility and high water permeability, which exclude the development of significant pore pressure.

There are two important features of soil behavior before and after liquefaction: (1) effective confining stresses in a soil decrease during cyclic loading at constant stress amplitude because of compaction (negative dilatancy), which results in the increase of strains (soil softening); (2) in triaxial tests with cyclic loading of water-saturated sands in undrained conditions, the samples in each cycle pass through the point of hydrostatic compression, which means disappearance of all tangential stresses, or liquefaction, when the phases of compression and tension change; after the loading path crossed the phase transformation line, dilatancy occurs, resulting in hardening of soil behavior, when shear strains increase: the hysteretic loop becomes inverse-*S* shaped (Yoshida and Iai 1998).

Dynamic response of dense water-saturated sands. Undrained tests on dense water-saturated sands indicate an extremely important role of the strain amplitude. Long-lasting loading at low strain amplitudes without inversion of the stress sign does not allow dense water-saturated sand to mobilize significant resistance to shear and leads to slow repacking of the grains and sudden liquefaction (Voznesensky 1999). Inversion of the stress sign in dense sands leads to accumulation of significant strains and disappearance of effective stresses at some moments of the cycle. A hysteretic loop is developed in the form of inverse *S*; this state corresponds to cyclic mobility. The failure is conditionally fixed by the achieved strain amplitude (usually, 2.5 or 5% per cycle).

Thus, dynamic instability of loose (with relative density up to 45%) sands is characterized by liquefaction phenomena, and dynamic instability of dense (with relative density above 60%) sands—by cyclic mobility (Voznesensky 1999).

Soil strength after an earthquake reflects the influence of cyclic loading during strong motion on the soil characteristics. As Castro and Poulos (1977) noted, the ultimate residual shear strength of water-saturated soils in undrained conditions is defined by their porosity and structure. If the soil structure was not changed after seismic motion, its loading history and the achieved strain level have little effect on the soil strength in undrained conditions, i.e., the soil strength after cyclic loading is the same as before loading. However, if structural changes have occurred, the stress–strain relation of such a soil changes, and its strength may noticeably decrease.

Since cyclic loading usually causes an increase of the pore pressure, effective stresses acting on an elementary soil volume, shifted monotonically after it has been subjected to cyclic loading, are less than in case of an identical elementary soil volume shifted monotonically without prior cyclic loading. Therefore, we can expect that an element that has been subjected to cyclic loading will exhibit higher dilatancy, but will have lower strength at early stages of monotonous loading in undrained conditions than an element that has not experienced cyclic loading.

In drained tests on dense water-saturated non-cohesive soils, a closed hysteretic loop is formed as early as the second loading cycle. Since the degree of pore pressure dissipation is controlled by the frequency of loading, at frequencies of the order of several hertz, accumulation of a large amount of excess pore pressure can occur, which will affect the soil strength (Voznesensky 1999).

Significant steady-state strains can develop, if the total (static plus cyclic) shear stresses exceed the dynamic shear strength (Kramer 1996). Estimates of dynamic limiting (destructive) shear stresses in soils can be obtained by measuring strains caused by earthquakes in dams and earth fences. These stresses were ~80% of the limiting soil strength in undrained conditions.

Thus, the behavior of non-cohesive soils under dynamic loading substantially depends on their density and loading conditions. Chapter 3 describes the behavior of sands (near the surface and at depth) during liquefaction. Records of the 1995 Kobe earthquake in the near-fault zones illustrate the loss of strength and liquefaction in subsurface layers and the behavior of water-saturated sands, described by a hysteretic curve in the form of inverse *S*.

Thixotropy and quasi-thixotropy of cohesive soils. Static loads on cohesive clayey soils lead to their compaction (consolidation), due to compression of water-colloidal shells of mineral particles, extrusion of water, and creep of soil skeleton. Under dynamic loading, clayey soils are compacted only slightly because of their cohesion, and their dynamic strength is 1.5–2 times higher than the static strength. The destruction of bonds occurs under uniaxial compaction by 4% and more (Voznesensky 1999). In cohesive soils, pore pressure increases with decrease in volume, caused by compression or compaction, and a long time period is required before excess water is squeezed out of the pores. Soil consolidation process is slow, and the longest consolidation is observed in the middle of the layer (Zaruba and Mencl 1976).

In conditions of dynamic loading of cohesive and weakly cohesive soils, thixotropy can be observed, i.e., a physical–chemical phenomenon typical for coagulated dispersible systems. It is expressed in the complete or partial destruction of structural bonds under dynamic loading and their subsequent spontaneous restoration at the same temperature and humidity (constant humidity means constant porosity, i.e. undrained conditions). An important property of thixotropic systems is their complete reversibility, i.e., recovery of strength to the initial level, regardless of the number of loading cycles. However, cohesive soils in situ usually do not possess this property: their final strength either does not reach the initial level after recovery, or exceeds it; therefore, we usually talk about quasi-thixotropic properties of cohesive soils (Voznesensky 1999).

Softening of a quasi-thixotropic soil can be partial or complete; in the latter case, soil liquefaction occurs (in weakly cohesive soils). Reduction and recovery of strength occurs due to cohesion. The degree of quasi-thixotropy of a soil depends on its humidity, chemical and mineralogical composition, dispersibility, the presence of organic material, and some other features; in the whole, it is characterized by the value of specific surface of the soil (Voznesensky 1999). An increase of humidity accelerates reduction of strength under dynamic loading (strength is maximal at humidity close to the yield point). Voznesensky also notes that the highest decrease in strength is observed at loads corresponding to accelerations of ~0.5–0.7 g.

The final strength of a thixotropic system after its hardening does not depend on the characteristics of the dynamic loading. In contrast, the recovered strength of a quasi-thixotropic soil is a function of the degree of destruction, i.e. of the loading intensity (Voznesensky 1999). With increasing loading, the final strength decreases or increases. For example, Thiers and Seed (1969) note that the ultimate strength of three types of clays decreased by less than 10% after cyclic loadings with a strain amplitude smaller than half of the fracture strain, determined in tests with monotonous loading; at larger cyclic amplitudes, larger strength losses were observed. Similar results were obtained in the experiments by other authors (Kramer 1996).

According to various researchers, the duration of strength recovery in clays varies depending on the composition, dispersibility and humidity, from several hours to several days. The largest increase in strength is observed during the first 3–10 min, when the structural lattice is formed; then the lattice hardens (Voznesensky 1999).

Dilatant-thixotropic phenomena in weakly cohesive soils. A rather large group of soils, ranging in their granulometric composition from dusty sands to sandy loams and loams, possess intermediate properties between non-cohesive dilatant and cohesive quasi-thixotropic soils. Summarizing data of various authors, Voznesensky (1999) highlights the following features of the dynamic response of weakly cohesive soils.

1. Even relatively weak dynamic loading causes an abrupt loss of strength and rapid liquefaction of weakly cohesive water-saturated soils (Ishihara et al. 1992), which is due to the features of their structure: the Coulomb friction is lower than in sands, and the coagulation lattice is weak and intermittent. Relatively low permeability hampers pore pressure dissipation and facilitates liquefaction. With increasing the number of fine particles (<0.074 mm) to more than 5%, the liquefaction potential noticeably increases (Seed et al. 1984).
2. Gradual and very slow hardening after termination of dynamic loading consists of a slight compaction (with release of water in water–saturated soils; the duration of the process is defined by the soil filtration capacity and pore pressure dissipation rate) and thixotropic hardening of the coagulation lattice, which is completed in a few hours to days. The hardening often occurs on the background of high pore pressure, which indicates the simultaneity of dilatant and thixotropic processes in weakly cohesive soils. Thixotropic hardening in a dry weakly cohesive soil appears only at a certain soil humidity and intensifies with increasing humidity.
3. The dynamic response substantially depends even on small changes of humidity, and at various humidity levels, positive or negative dilatancy can be observed.
4. Weakly cohesive soils possess selective sensitivity to loadings at certain frequencies in the range 15–45 Hz (at some frequencies, resonantly large softening is observed and sometimes, sudden liquefaction). The "resonant" frequencies are different for soils of different dispersibility, and as the dispersibility decreases, they shift to lower frequency ranges (for example, spectral maxima corresponding to softening of sandy loams and dusty sands decrease from 27–35 Hz to 10–12 and 21–25 Hz), and the resonant peaks become smoother (Voznesensky 1999).

As known, wet and especially water–saturated weakly cohesive soils are the most dynamically unstable. Loesses are among them. In conditions of dynamic loading, wet loesses can show compaction, which is usually referred to as seismic subsidence, and softening (Musaelyan 1964; Tankaeva 1964; Saifiddinov 1987, etc.).

Following various researchers, Voznesensky (1999) identifies the features of the dynamic response of loesses:

1. Sharp (2–3 times) drop in strength down to liquefaction with increasing humidity by 1–3% near its critical value, which is close to the yield point. Such behavior is explained by the key role of free water in the avalanche-like destruction of structural bonds in the soil, as well as the low total area of the interparticle contacts along with high porosity and relatively low water permeability, which slow down pore pressure dissipation.
2. At subcritical humidity values softening is usually small, about 15–20%.

3. Separate loess layers often noticeably differ in their dynamic response, while their sensitivity to loading decreases with depth. In most cases, at depths below 7.5–8 m softening of loesses is not observed even at accelerations of ~5 g.
4. Softening of wet and water-saturated loesses is accompanied by the breakdown of their structure, and it is noticeably increased with the increase of the loading amplitude. The maximum softening effect was observed at loadings at a frequency of about 25 Hz.
5. Thixotropic recovery of the strength of loesses is characterized by the following features:

   - strength recovery proceeds on the background of compaction,
   - a complete recovery is not achieved because of irreversible destruction of the cementation bonds,
   - the recovery process is completed within approximately two days, and the final strength can reach 85–95% of the initial value. The highest recovery rate is observed during the first hour after the end of loading.

Chapter 3 describes dilatant and thixotropic phenomena, reduction and recovery of shear moduli in subsurface layers of weakly cohesive soils during strong 1995 Kobe and 2000 Tottori earthquakes.

The effect of loading frequency on the dynamic response of soils is one of the insufficiently studied issues in soil dynamics (Voznesensky 1999).

The influence of loading frequency on the behavior of sands is controlled by the rates of accumulation or dissipation of the pore pressure. In undrained conditions, the rate of strain accumulation decreases with increasing frequency; in drained conditions, it decreases to a critical frequency, which depends on the soil density and water permeability, and then it increases due to a sharp decrease of the pore pressure dissipation rate. The response of dry and wet sands can be considered independent of the loading frequency.

Hardin (1965) has shown that shear moduli and damping in sands do not depend on the loading frequency. His results were confirmed by other researchers (Wilson 1985) in the low-strain range. Independence of stiffness and damping of dry sands on the loading frequency in the range 0.001–120 Hz was proved by resonant column tests at strains < 0.07% (Bolton and Wilson 1989).

It was shown (Kramer et al. 1992) that the influence of the loading frequency in the range 0.008–15 Hz on the stiffness and damping of clayey soils is rather small. Minimal damping was observed at intermediate frequencies 0.3–0.8 Hz, and the stiffness increased with increasing frequency. Similar results were obtained by a research group of the University of Texas (Kim et al. 1991): a noticeable increase in damping (almost twice) was observed at frequencies >5 Hz. However, these regularities are more clearly expressed at low shear strains (< 0.116%) and almost absent when shear strains increase to 1.5–2.5%. In cohesive soils, damping is considered independent of the loading frequency at f > 0.1 Hz and independent of shear strains at strains > 0.001% (Voznesensky 1999). However, soil strength and strain accumulation rate during its dynamic loading depend on the loading frequency, because, for example, increase of the loading frequency from 1 to 2 Hz substantially changes the number of loading cycles required for the destruction of silt samples from San Francisco Bay (Thiers and Seed 1969).

Based on the published data and experimental results, the following possible causes of the frequency dependence of soil response can be identified (Voznesensky 1999):

- different durations of loading and unloading phases in cycles at different frequencies,
- different rates of pore pressure dissipation,
- increase of the energy of dynamic loading with frequency,
- resonant phenomena,
- creep phenomena,
- thixotropic phenomena.

For weakly cohesive soils, the difference in the duration of loading and unloading phases at different frequencies leads to the fact that the strain accumulation rate grows with decreasing the loading frequency in undrained conditions, while in an open system, an inverse relationship can also be observed, because the rate of pore pressure built-up strongly depends on frequency (sometimes, the excess pore pressure does not have time to completely dissipate during the cycle due to low permeability of the soil) (Voznesensky 1999). According to Voznesensky (1999), a clear dependence of the dynamic response of weakly cohesive soils on the loading frequency at low (<1 Hz) and high (15–45 Hz) frequencies, where resonant effects occur, is due to their low permeability and general high sensitivity to dynamic loadings.

For frequencies in the range from fractions of hertz to the first hertzs and for frequencies >5 Hz, manifestations of these effects are different. If the frequency is rather low, and soil permeability is sufficient for complete dissipation of the excess pore pressure during the unloading phase in each cycle (which is possible in sandy and weakly cohesive soils), then the effective stresses in the soil are not reduced. This is an example of alternating static loading. Whereas, in undrained tests of sands and clays, even at very low loading frequencies (for example, 1 cycle in 10 h), the loading acts as dynamic one, because accumulation of the excess pore pressure occurs (Sangrey et al. 1969) and, therefore, either destruction (at stresses above critical) or equilibrium state with the formation of a closed hysteresis loop (at lower stresses) is observed.

At frequencies $f > 5$ Hz, with increasing the loading frequency, the energy of loading on the soil increases, causing enhanced destruction of the structural bonds and a progressive decrease of soil strength. However, this process is complicated by resonant effects. Most often, resonant "amplification" of softening is observed for clayey soils in the range 15–22 Hz (Voznesensky 1999); at frequencies < 1 Hz resonant phenomena do not occur.

Accumulation of strains and strength degradation can slow down with decreasing frequency due to partial thixotropic recovery of clayey soil during the periods between the loading phases (Seed and Chan 1957). Obviously, the soil humidity should play a key role in this process: along with decreasing the mobility of particles with decreasing humidity, the role of partial thixotropic recovery is noticeably reduced. Similar effect of partial thixotropic structure formation can be observed at loading frequencies of ~0.1 Hz and lower, including cases on the background of excess pore pressure, because the recovery of quasi-thixotropic soils begins about 5 s after the end of the dynamic loading (Voznesensky 1999). With increasing the period of loading to the values close to the relaxation time of the soil, creep effects appear even in undrained conditions. On the other hand, the strength of such a soil increases due to thixotropic recovery of the broken bonds.

Thus, the combined action of the resonant effects, partial thixotropic recovery, and creep leads to the fact that the soil strength usually slightly decreases with decreasing frequency at $f < 5$ Hz and with its increasing at $f > 5$ Hz. The higher the soil humidity, the higher the influence of the loading frequency on the soil behavior (Voznesensky 1999).

In conditions of strong seismic motion, the dependence of shear moduli and damping on frequency for cohesive and non-cohesive soils is related to the problem of nonlinear soil behavior, which is frequency-dependent, as shown in Chapter 5. During strong motion, shear moduli decrease, and nonlinear damping mechanisms are activated, leading to significant weakening of high-frequency vibrations. Because of nonlinearity of soil behavior in strong ground motion, attenuation of seismic waves in soil layers depends on frequency and on the intensity of motion, and these dependencies are complicated by resonant effects, which occur not only in soil profiles with clear seismic boundaries, but also in the profiles with seismic velocities smoothly increasing with depth. How strong such dependencies are for different types of soils can be investigated in numerical experiments with models of soil behavior, described in Chapter 3.

1.4 Models of Soil Behavior in Conditions of Dynamic Loading

Soils are multiphase media, and the propagation of seismic waves in such media can be studied by two ways. In the first approach, soils are considered as combinations of continuous media, each with their own parameters: velocities and pressures, and the propagation of seismic waves in the soils causes interpenetrating and interconnected movements of these media. In the second approach, a multiphase medium is considered as a single continuum described by the compression and shear diagrams that account for the contents and properties of its components. Elementary microvolumes of the medium are characterized by a single velocity value, but the pressure in the components can be different. The second approach is used in engineering seismology, where soil behavior is described by the compression and shear diagrams depending on the composition of the soil and its saturation with water. Thus, the type of stress–strain relation represents the model of soil behavior in condition of its dynamic loading.

As shown in the previous section, soil behavior can be quite complicated even under single static loadings. For practical applications, simple rational models are needed that characterize the most important aspects of the soil behavior with a sufficient accuracy. The optimal balance of the conflicting requirements of simplicity and accuracy depends on many factors, and various combinations have been proposed.

The simplest model used to describe soil behavior is the linear elastic model. It provides simple analytical solutions to wave propagation problems and it is widely used in soil mechanics to study static deformations of non-water-saturated soils under loads up to 3–5 kg/cm^2 and cyclic loads on soils from machine foundation vibrations. Linear rheological models, such as, Maxwell body model, Kelvin-Voigt model, and Boltzmann-Volterra linear hereditary creep model are used to solve similar problems, in which the effects of creep, aftereffect and heredity are important.

In Russia, a bilinear or bimodular model of a loose soil medium was proposed by Nikolaev (1967) to explain data obtained in field experiments with weak explosions. In the epicenter of an explosion, an upward rebound of soil was recorded, and it was shown that multiple breaks occur in loose weakly cohesive upper soil layers, i.e. the soil structure was disturbed and soil softened (Gvozdev and Kuznetsov 1967). According to the bilinear model, elastic moduli of subsurface soils possess different values for tensile and compressive strains, and characteristic distortions of the waveform occur in seismic fields of sufficiently high intensity, where the amplitudes of compressive strains are noticeably weaker than the amplitudes of tensile strains.

Maslov et al. (1985) solved the problem of the incidence of a finite-amplitude seismic wave on the free boundary of a bimodular medium. Lyakhovsky applied a bimodular model to seismic ray tracing and analysis of the stress state of fractured subsurface rocks (Lyakhovsky and Myasnikov 1986; Lyakhovsky 1990). He found that, since seismic wave velocities are related to the stress state of the medium and the direction of wave propagation, changes in the stress state can lead to significant changes in the ray paths, despite the smallness of the introduced corrections to Hook's law in the medium. The bilinear model is still used now to explain the mechanisms of soil nonlinearity.

A more complicated is a model of nonlinear-elastic medium, in which stress–strain relation is nonlinear, convex towards the strain axis, and the same for loading and unloading paths. This model is applicable to media, where the loading and unloading curves are only slightly different, and no residual deformations occur. Such models were used to describe the behavior of water-saturated soils during the propagation of shock waves in them; i.e., in cases of strong nonlinearity (Lyakhov 1974).

To describe the behavior of dry (non-water-saturated) soils, models of elastic–plastic media were used, in which the loading curves were assumed to be convex to the stress axis. The loading and unloading curves were approximated by three straight lines, corresponding to elastic and plastic deformations and to unloading. Rakhmatullin and Demyanov (1961) proposed a model of a soil as a plastic gas, i.e. an ideal medium, in which the loading part of the stress–strain relation is convex to the strain axis, while unloading part goes along a straight line parallel to the stress axis. This assumption greatly facilitates the solution of wave propagation problems, and the model was widely used for solving plane one-dimensional problems in cases, when the basic equations of motion in solid and liquid media are identical. Later, Rakhmatullin et al. (1964) proposed a model representing a deformation scheme for both volumetric and shear strains.

In the most general form, the model of a soil as an elastic–plastic medium (Fig. 1.2a) was proposed by Grigoryan (1960). The model allows descriptions of arbitrary ground motions. Grigoryan's model accounts for the main most significant features of the soil behavior, such as, the nonlinearity and irreversibility of volumetric strains, elastic–plastic shear, dependence of the shear yield stress on the average normal stress. In the low-strain range, the stress–strain relation corresponds to a linear elastic medium, and in the limiting state, it corresponds to Prandtl–Reis scheme with the Mises-Schleicher-Botkin plasticity condition.

Grigoryan also proposed the application of the plasticity condition to soils and rocks (Grigoryan 1959). This condition assumes that in the plastic state of a medium, the intensity of tangential stresses defined by the second invariant of the stress tensor deviator is a function of the average normal stress. Experimental studies on various soils have shown that this function can be considered linear (Lyakhov 1974). Grigoryan's model was widely used in solving problems of wave propagation in Russia and abroad.

Elastic–plastic models gave explanations to a number of experimental results, such as, decrease of the maximum stress and propagation velocity of explosive waves with distance along with increasing their duration, conversion of shock waves into low-amplitude compression waves, residual deformations after the passage of the waves, etc. However, some other experimental data did not fully fit these models and indicated the importance of accounting for the viscosity of the medium (Lyakhov 1974): sometimes, maximum strains were not reached at maximum stresses, but at their decrease, residual deformations depended on the duration of the wave, in some cases wave attenuation was higher than predicted by calculations based on elastic–plastic models, etc.

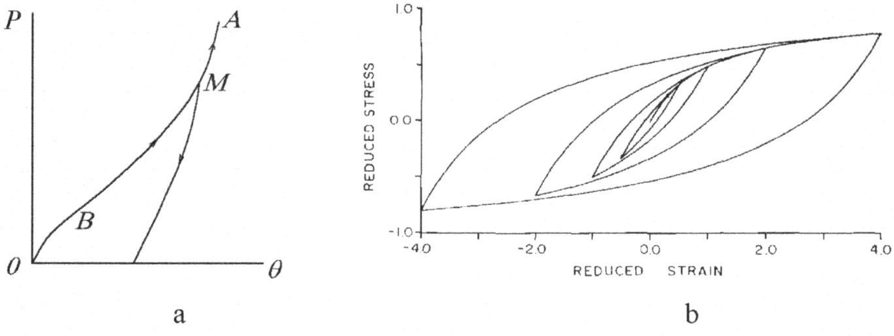

Fig. 1.2 **a**—Model of a soil as an elastic–plastic medium; **b**—initial loading curve and hysteretic stress–strain curves for the model of elastic–plastic soil behavior (from Hardin and Drnevich 1972): stresses and strains are in relative units, i.e. normalized so that shear stress in failure and shear modulus at low strains are equal to 1

Viscous properties of soils that result in the dependence of the stress–strain diagrams on the rates of loading, are known for a long time. Model of viscous media account for the fact that the rearrangement of grains is not instant, but takes a finite time period. In Russia, viscous properties of soils were studied experimentally by Pokrovsky et al. (1938), Lyakhov (1974), Melnikov and Rykov (1965), Stavnitzer (1966), etc. The experiments indicated that for many soils under loadings of the order of tens of atmospheres, the diagrams of dynamic and static compression are close to linear ones, and the soil behavior can be approximately described by linear viscoelastic models (Lyakhov 1974). Residual deformations do not occur in this case (in contrast to the viscous-plastic model), because the loading and unloading curves of a viscous-elastic medium are described by the same equation. However, accounting for viscosity provides only few percent corrections to ground motion parameters; therefore, the use of elastic–plastic models is acceptable (Lyakhov 1974).

Viscous-elastic, or so-called "equivalent linear" models are currently used in the calculations of the soil response. The approach is implemented in many computer programs; one of the most famous is SHAKE. The advantages and limitations of this approach are described in Chap. 2.

Hardin and Drnevich (1972) were the first who performed a series of laboratory experiments and obtained hysteretic stress–strain relations for cohesive and non-cohesive soils; also, they proposed a method for normalizing the relations (Fig. 1.2b).

Similar relations for positive stresses and strains were obtained in field experiments with small explosions by the researchers from Schmidt Institute of Physics of the Earth of the USSR Academy of Sciences Vasil'ev, Gvozdev, Ivanova, Shcherbo, etc. (Vasil'ev et al. 1969, 1977), who measured stresses and strains in explosive waves propagating in subsurface loams and loesses above the groundwater level (Fig. 1.3).

These studies revealed the features of the behavior of sediments in conditions of dynamic loading, such as, different loading and unloading paths, residual deformations, saturation of the loading curve.

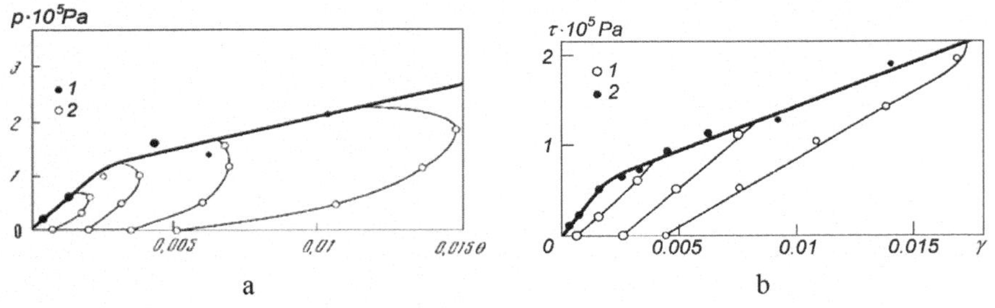

Fig. 1.3 Typical compression (**a**) and shear (**b**) diagrams in clays during loading (1) and unloading (2) (natural humidity 10–15%, $V_P = 300$–600 m/s, $V_S = 200$–250 m/s) obtained in field experiments with small explosions. The principal stress and strain components were measured in subsurface soils using small membrane sensors and borehole strainmeters with bases 250–300 mm. The areas of the circles are proportional to the numbers of measurements (from Vasil'ev et al. 1977)

The compression diagrams obtained in field experiments are discussed by Zvolinsky (1982). There are a wide diversity of soils, which differ in their granulometric composition, humidity, depth, etc., and a significant scatter in the measured stresses and strains; so we cannot indicate a unified reliable compression or shear diagram for all the soils. A pattern close to Grigoryan's model can serve as some general scheme (Fig. 1.2a); its characteristic feature is the presence of two parts: OB—convex upward and BA—concave upward; B is the point of inflection. Depending on the physical properties of a soil, the position of point B changes significantly. In some cases, point B shifts to the origin, leaving only a concave part (a "hard" diagram), or goes beyond the working section of the diagram (a "soft" diagram).

A number of experimental diagrams of this kind obtained in field experiments are available in (Rykov and Skobeev 1978). It was found that the convex upward part is weakly expressed in sands (in experiments with limiting stresses of a few kgf/cm^2) and is completely lost with increasing sand humidity, when the diagrams become hard (experiments were conducted up to 50–80 kgf/cm^2). Hard-type diagrams are especially typical for clays. For loess-like loams, the soft part is more pronounced and observed throughout the experiment up to stresses of ~30–40 kgf/cm^2; the convex area reduces with increasing humidity. In some experiments conducted up to stresses of ~50–100 kgf/cm^2, a convex part was not recorded, but this does not mean that it is really absent—it may be beyond the sensitivity of the equipment, configured for heavy loads, notes Zvolinsky (1982). Thus, the dependence shown in Fig. 1.2a describes the main features of the uniaxial compression diagrams. The unloading branch of the diagram has, on average, less curvature and a steeper slope.

Gvozdev et al. (1982) discuss the application of various models of soil behavior in engineering seismology. To describe soil behavior, many models are proposed, characterized by their own features and details; however, the results of field experiments do not allow us to compare these models with the established facts and rules. There is no doubt that attenuation of seismic waves increases as the soil passes to the plastic stage. We should study quantitative regularities of soil behavior, and the models being developed should be as simple as possible, so that they can be verified in practice (Gvozdev et al. 1982).

Nowadays, three broad classes of models of soil behavior are used in the world practice, such as, equivalent linear (viscous-elastic) models, cyclic nonlinear models, and advanced constitutive models (Kramer 1996). The simplest are equivalent linear models, but they are limited in their ability to represent some features of soil behavior in conditions of cyclic loading. On the other hand, advanced constitutive models reproduce various features of the dynamic behavior of soils, but their complexity and calibration difficulties currently make them inapplicable for solving problems of engineering seismology. All the models are based on hysteretic stress–strain relations, which are considered typical for soils deposited at some depth and far from crystalline structures (Kramer 1996).

The following is a brief overview of the results of studies aimed at finding some general patterns of soil behavior and describing the behavior of different soils by similar relations.

Hysteretic curves can be defined in two ways: directly by setting the loading paths $\tau(\gamma)$ (we consider shear strains, the most important in engineering seismology), or by parameters describing their shapes. In the low-strain range, two main parameters that describe a hysteretic curve are its slope and width. The slope of the curve depends on the rigidity, defined at each point by the tangential shear modulus G_{tan} (Fig. 1.4). G_{tan} changes during loading; its average value G_{sec} characterizes the entire loading curve and is defined by the ratio of shear stress τ_c to shear strain γ_c (Fig. 1.4):

$$G_{\mathrm{sec}} = \frac{\tau_c}{\gamma_c} \tag{1.4.1}$$

The width of the hysteretic curve is related to the area bounded by this curve, which is defined by the energy dissipation and is usually described by a dimensionless relative damping coefficient (Kramer 1996):

$$h = \frac{W_D}{4\pi W_S} = \frac{1}{2\pi}\frac{A_{loop}}{G_{\mathrm{sec}}\gamma_c^2}, \tag{1.4.2}$$

where W_D is the damped energy, W_S is the maximum strain energy, A_{loop} is the area inside the hysteretic curve.

Shear modulus can be measured in situ and in laboratory experiments (Sect. 1.2), however, the laboratory estimates of elastic moduli usually do not coincide with the values measured in situ (Yasuda and Yamaguchi 1984). This can be due to violations of the soil structure when taking a sample, because samples taken using the local soil freezing method provide estimates of elastic moduli close to the in situ values (Tokimatsu 1989).

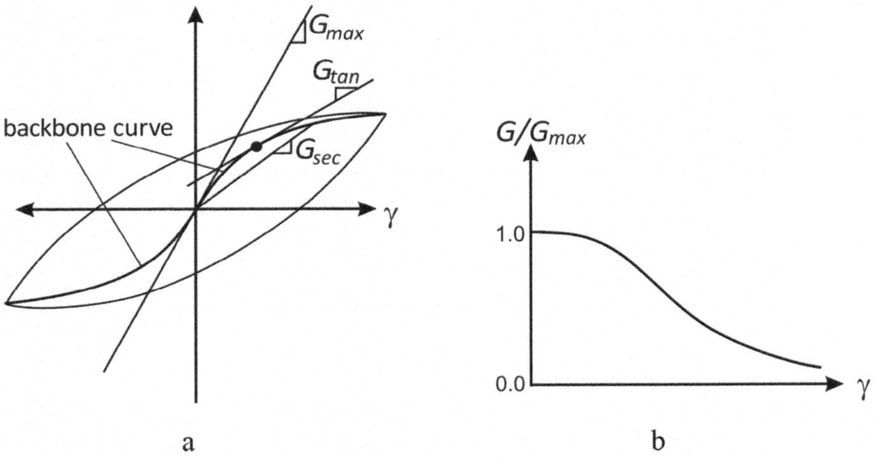

Fig. 1.4 Tangential G_{tan} and average G_{sec} shear moduli and a backbone curve (**a**) showing a typical case of reduction of the shear modulus (**b**) with increasing strains

The shear and compression moduli are calculated based on the velocities of *S*- and *P*-waves according to the following formulas:

$$G = \rho V_s^2, \quad K = \rho\left(V_p^2 - \frac{4}{3}V_s^2\right). \tag{1.4.3}$$

It is assumed that the compression modulus characterizes not only the soil skeleton, but the entire volume of water-saturated soil, and therefore it can be used in total, rather than effective, stress analysis. There are virtually no simple and reliable methods for measuring the compression modulus in situ, and in engineering practice, Poisson's ratio is often used for its estimation. Usually, the influence of the compression modulus on soil response in earthquakes is rather small (Yoshida and Iai 1998).

The locus of points connecting the tips of hysteretic loops at various loading levels is called a characteristic, skeleton, or backbone curve (Fig. 1.4); its slope at the origin is equal to the maximum value of the shear modulus $G_{\max}$.

Maximum shear modulus. Since most laboratory and field tests on soils induce shear strains lower than 0.0003% (Kramer 1996), measured *S*-wave velocities can be used to calculate $G_{\max}$ by the formula

$$G_{\max} = \rho\, V_S^2. \tag{1.4.4}$$

Measurement of *S*-wave velocities is often the most reliable way to estimate $G_{\max}$ in the soil in situ; field and laboratory measurements of dynamic soil parameters described in Sect. 1.2 are also used to estimate $G_{\max}$.

Figure 1.5 shows *S*-wave velocities measured by two methods, such as, downhole seismic logging and suspension logging method.

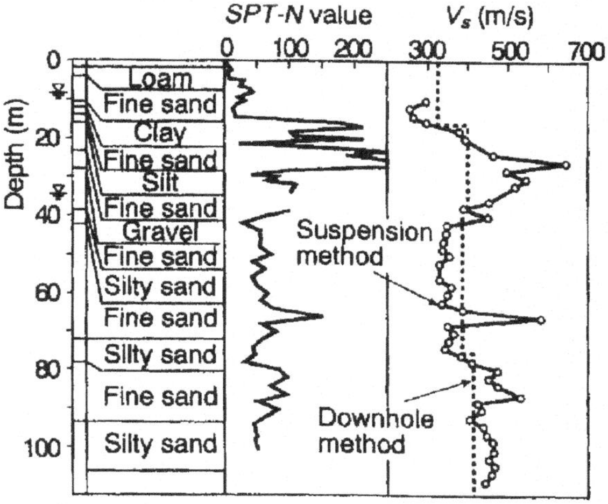

Fig. 1.5 Comparison of *S*-wave velocities measured by two methods: downhole seismic logging and suspension logging method, with the results of standard penetration test (from Kokusho 1992)

It is concluded that downhole seismic logging tends to average the velocity estimates, and depth distribution of *S*-wave velocities obtained by suspension logging well agrees with the velocity estimates obtained by standard penetration test. These observations are confirmed by further studies. The importance of knowing the exact values of *S*-wave velocities in soil response analysis is discussed in the next chapter.

In the absence of *S*-wave velocity measurements, $G_{\max}$ can be estimated by some other methods. Laboratory tests have shown that the maximum shear modulus obeys the empirical relation

$$G_{\max} = 625F(e)(OCR)^k p_a^{1-n}(\sigma'_m)^n, \tag{1.4.5}$$

where $F(e)$ is a function of porosity; OCR is overconsolidation ratio; k is overconsolidation ratio exponent; σ'_m is the average effective stress $(\sigma'_m=(\sigma'_1+\sigma'_2+\sigma'_3)/3)$; n is the stress exponent (usually assumed 0.5, can be estimated in laboratory tests); p_a is the atmospheric pressure.

Hardin (1978) proposed the formula $F(e) = 1/(0.3 + 0.7e^2)$, while Jamiolkowski et al. (1991) suggested that $F(e) = 1/e^{1.3}$. The overconsolidation ratio exponent k increases with increasing the plasticity index as shown in Table 1.1.

In addition to Eq. (1.4.5), other empirical relationships are known, for example:

$$G_{\max} = 1000\, K_{2,\max}(\sigma'_m)^{0.5}, \tag{1.4.6}$$

where $K_{2,\max}$ is a function of porosity and density (Table 1.2), and effective stresses σ'_m are expressed in lb/ft^2.

Table 1.1 Dependence of the overconsolidation ratio exponent k on the plasticity index (from Hardin and Drnevich 1972)

Plasticity index	k
0	0.00
20	0.18
40	0.30
60	0.41
90	0.48
>100	0.50

Table 1.2 Dependence of $K_{2,\max}$ on the porosity e and relative density D_r (from Seed and Idriss 1982)

e	$K_{2,\max}$	D_r,%	$K_{2,\max}$
0.4	70	30	34
0.5	60	40	40
0.6	51	45	43
0.7	44	60	52
0.8	39	75	59
0.9	34	90	70

Field experiments indicate that S-wave velocities and, consequently, G_{max} in gravels are noticeably higher than in sands. For gravels $K_{2,max}$ is typically in the range 80–180 (Seed and Idriss 1982).

For fine-grained soils, preliminary estimates of G_{max} can be obtained based on parameters measured in laboratory tests, such as, the plasticity index, overconsolidation ratio, and stress in undrained conditions s_u (Table 1.3).

Maximum shear modulus can be estimated based on the measurements of other parameters. Empirical relations linking G_{max} and some parameters measured in situ are established; the most famous of them are given in (Kramer 1996). The fact that we need to evaluate G_{max} parameter characterizing the range of low strains, based on parameters related to high strains, causes errors and large statistical scattering of the estimates obtained by various methods and various authors. Thus, such estimates are used only as preliminary ones.

Estimation of the shear modulus is complicated by the fact that the estimates depend on the rate of loading of a soil and on the total time of loading. Shear modulus can grow by several percent with increasing the loading rate by dozens of times; this is more expressed for soils with high plasticity and low *OCR*. Such effects can be significant when comparing G_{max} calculated from field measurements of S-wave velocities (usually using high-frequency pulse signals) and obtained in laboratory tests. Some seismologists explain the differences in G_{max} estimates obtained in field and laboratory experiments by different duration of loading, and suggest methods for correcting laboratory results to better match in situ parameters (Kramer 1996).

Reduction of shear moduli during soil loading. In the 1970s, reduction of shear moduli in coarse-grained and fine-grained soils under dynamic loading was treated as related to different mechanisms (Seed and Idriss 1982). However recently, a connection between these processes was found, so we can trace a smooth transition of the mechanisms for shear moduli reduction in non-plastic coarse-grained soils to similar mechanisms in plastic fine-grained soils. Kokusho (1982) was the first who noted the influence of soil plasticity on the shape of the shear modulus reduction curve: shear moduli of soils with high plasticity decrease more slowly than those of low-plasticity soils. Dobry and Vucetic (1987) and Sun et al. (1988) after analyzing a large amount of experimental data concluded that the shape of the shear modulus reduction curve is more influenced by the plasticity index than by the void ratio (Table 1.4) and constructed families of parametric relations with various indices *PI*. According to these curves, the area of "linear" soil behavior, i.e. the linearity threshold for strains

Table 1.3 Dependence of G_{max}/s_u on the overconsolidation ratio *OCR* and the plasticity index (from Kramer 1996)

Plasticity index *PI*	Overconsolidation ratio *OCR*		
	1	2	3
15–20	1100	900	600
20–25	700	600	500
35–45	450	380	300

turned out to be noticeably higher for soils with high plasticity, which is an important point in assessing soil response in earthquakes (Fig. 1.6a).

Reduction of shear moduli also depends on the effective confining pressure, especially for low-plasticity soils (Iwasaki et al. 1977): the linearity threshold is higher for larger confining pressures (Fig. 1.7a, b).

Damping ratio. Theoretically, the energy dissipation does not occur at strains that do not exceed the linearity threshold for a given soil. However, experiments show that a certain amount of energy is lost even at low strains. Thus, damping is always nonzero. In conditions of cyclic loading and at strains exceeding the linearity threshold, the widths of the hysteretic curves and damping increase with increasing the level of strains.

Table. 1.4 Influence of the environmental and loading conditions on G/G_{max} in normally consolidated and moderately overconsolidated soils (from Kramer 1996)

Factor	G/G_{max}
Confining pressure, σ'_m	Increases with increasing σ'_m; the effect decreases with increasing *PI*
Void ratio, *e*	Increases with increasing *e*
Geological age, *t*	May increase with increasing *t*
Cementation, *c*	May increase with increasing *c*
Overconsolidation ratio, *OCR*	Not affected
Plasticity index *PI*	Increases with increasing *PI*
Cyclic strain γ_c	Decreases with increasing γ_c
Strain rate	*G* increases with increasing, but G/G_{max} perhaps does not depend on
Number of loading cycles, *N*	Decreases after *N* cycles at large γ_c (G_{max} is measured before loading cycles) for clays; for sands, may increase in drained conditions or decrease in undrained conditions

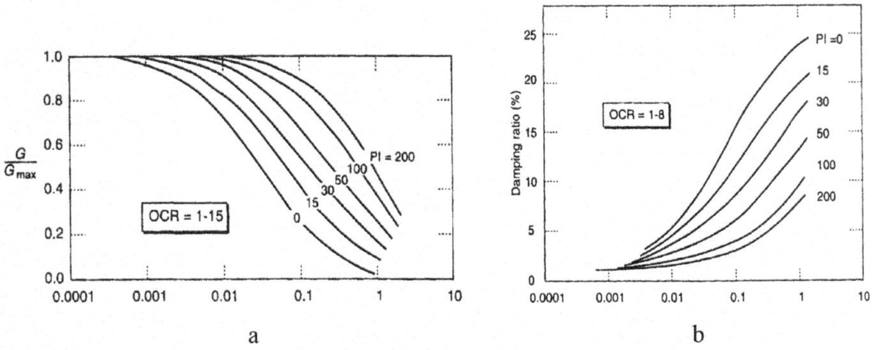

Fig. 1.6 Reduction of shear moduli (**a**) and damping (**b**) with increasing shear strains in fine-grained sands with various plasticity indexes (from Vucetic and Dobry 1991)

The soil plasticity affects both shear modulus reduction and damping. Damping ratio is lower in high-plasticity soils than in non-plastic soils at the same levels of strains (Fig. 1.6b). For soils with zero plasticity index $PI = 0$, the dependence of damping on the strain amplitude approximately coincides with a similar dependence for coarse-grained soils. For gravelly soils, damping mechanisms are similar to those in sandy soils (Seed and Idriss 1982).

Damping is also influenced by the confining pressure, especially in soils of low plasticity (Table 1.5, Fig. 1.7). Ishibashi and Zang derived an empirical relationship linking the damping ratio of a soil with its shear modulus and plasticity:

$$h = 0.333\frac{1 + \exp(-0.0145PI^{1.3})}{2}\left[0.586\left(\frac{G}{G_{\max}}\right)^2 - 1.547\frac{G}{G_{\max}} + 1\right] \quad (1.4.7)$$

Table. 1.5 Influence of the environmental and loading conditions on damping in normally consolidated and moderately overconsolidated soils (from Kramer 1996)

Factor	Damping ratio, h
Confining pressure, σ'_m	Decreases with increasing σ'_m ; the effect decreases with increasing PI
Void ratio, e	Decreases with increasing e
Geological age, t	Decreases with increasing t
Cementation, c	May decrease with increasing c
Overconsolidation ratio, OCR	Not affected
Plasticity index PI	Decreases with increasing PI
Cyclic strain γ_c	Increases with increasing γ_c
Strain rate	Does not depend or may increase with strain rate
Number of loading cycles N	Does not depend for average moderate γ_c and N

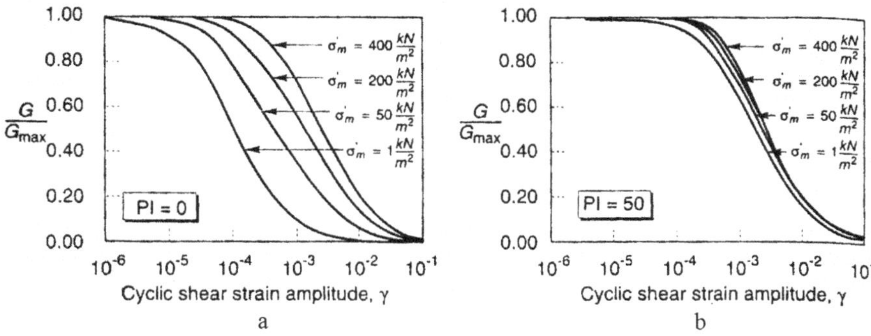

Fig. 1.7 Influence of the average effective confining pressure on the behavior of shear moduli reduction curves for non-plastic (**a**) and plastic (**b**) soils (from Ishibashi and Zhang 1993)

Seed and Idriss (1970) and Hardin and Drnevich (1972) have studied the dependencies of shear modulus G and damping ratio h on shear strain level γ. Figure 1.8 schematically shows the results of the laboratory tests. Usually, a hysteretic curve at the 10-th loading cycle is used to estimate such dependencies.

The dependencies $G(\gamma)$ and $h(\gamma)$ are affected by drainage conditions. Figure 1.9 shows stress–strain relations for soils in drained (a) and undrained (b) conditions. Soil behavior is different in these cases, which is explained by different mechanisms of changes of the soil volume when shear stresses are applied and by different mechanisms of dilatancy described above (Sect. 1.3). Under cyclic loading in drained conditions, soil samples are compacted, which results in hardening of the stress–strain curves (Fig. 1.9a). In undrained conditions, average effective stresses decrease, and elastic moduli degrade (Fig. 1.9b) down to liquefaction (stress–strain relations in the form of inverse S).

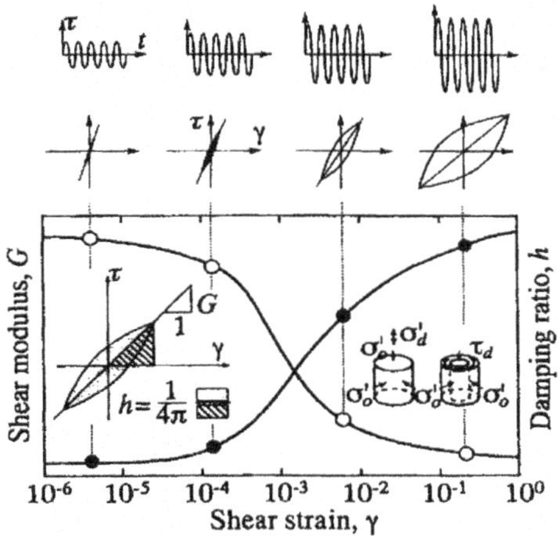

Fig. 1.8 A diagram explaining processing of laboratory test data to obtain estimates of the shear modulus and damping ratio (from Yoshida and Iai 1998)

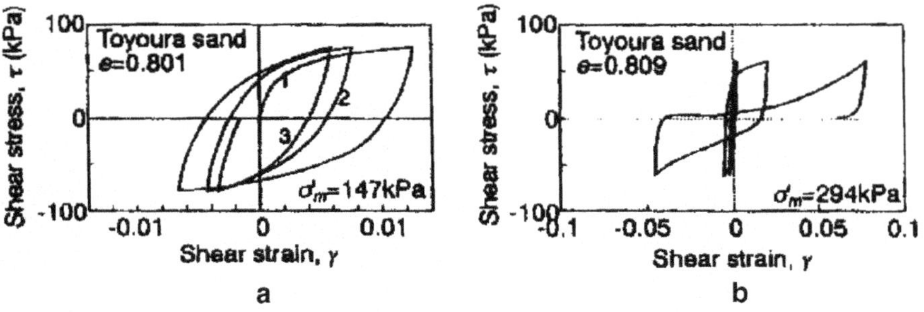

Fig. 1.9 Stress–strain relations obtained in laboratory simple shear tests in drained **a**— (1, 2, 3 are successive positions of the curve) and undrained **b**— conditions (from Towhata 1989)

Yoshida and Iai note that, despite the obvious differences in the stress–strain relations, these differences are not too noticeable in the soil behavior in drained and undrained conditions and in seismic records at strains below 10^{-3} (Yoshida and Iai 1998). With increasing the strain level, dilatancy effects appear, and the differences in soil behavior in the two cases become evident. Degradation of stress–strain relations in undrained conditions was observed even in clayey soils (Hyodo et al. 1988), but the effect was small.

Maximum strains, for which the dependencies $G(\gamma)$ and $h(\gamma)$ can be reliably determined in laboratory tests, are typically of the order of 10^{-2}. There are virtually no test data to determine $G(\gamma)$ and $h(\gamma)$ at strains above this level, except laboratory test data, where soil liquefaction is studied. In such tests, strains reach several percent, but it is difficult to obtain stress–strain relations. Kiku and Yoshida carried out cyclic tests on soils at high strains in conditions, where the hysteretic curves were not stabilized and strains increased rapidly. They recorded a closed hysteretic loop at the third loading cycle and found that damping h decreases, as well as strains in the soil, due to dilatancy (Kiku and Yoshida 1998).

Yoshida (1995) studied the dependencies $G(\gamma)$ and $h(\gamma)$ at high strains and concluded that:

1. The dependencies $G(\gamma)$ and $h(\gamma)$ are changed significantly at the beginning of loading and stabilized after several loading cycles. However, since peak amplitudes are often observed at the beginning of strong motion during an earthquake, it is important to investigate soil behavior in the first loading cycles and not so important to know the stabilized soil behavior.
2. Stress—strain relations in a soil can be different even if the dependencies $G(\gamma)$ and $h(\gamma)$ are identical, because in the calculations of $G(\gamma)$ and $h(\gamma)$ only the shape of the hysteretic curve is used, while its position in the plane $\tau - \gamma$ is not defined.

Thus, the dependencies $G(\gamma)$ and $h(\gamma)$ can be a good indicator of nonlinear soil behavior, but obviously are not well suited for calculating stress–strain relations, which could be used to estimate the soil response in strong ground motion (Yoshida 1995).

Dependencies $G(\gamma)$ and $h(\gamma)$ are widely used in engineering practice, in particular, to describe ***equivalent linear models*** of soil behavior (Chapter 2). In equivalent linear models, soils are considered as linear viscoelastic materials; the nonlinear properties of soils are taken into account by introducing the dependencies of their elastic moduli and damping on the level of strains. The reviews of studies using these dependencies and equivalent linear models are made by Richart (1978), Ishihara (1982), Woods (1991), etc. In some approaches, the dependencies $G(\gamma)$ and $h(\gamma)$ are used directly to describe soil behavior, while in others, knowledge of the entire hysteretic curves is necessary. A limitation of the equivalent linear model is the assumption that strains become zero when loading stops, and since linear-elastic material has no limits of elasticity, plasticity, etc., no ruptures and destructions occur in it.

Cyclic nonlinear models more accurately describe soil behavior under dynamic loading than equivalent linear models (Kramer 1996). Nonlinear models take into account shear stresses existing in the soil, pore pressure generation mechanisms, and changes of the effective stresses under cyclic loading in undrained conditions. Many cyclic nonlinear models have been developed; they are characterized by (1) a backbone curve and (2) a set of rules that describe soil behavior during unloading and repeated loading and define degradation of elastic moduli and other effects. In the simplest nonlinear models, the backbone curves have a simple form, and the number of rules is small. More complicated models include additional rules allowing them to better describe the effects of soil loading and pore pressure generation (Kramer 1996).

Calculations of soil response with nonlinear models can be illustrated by a simple case, when the shape of the backbone curve is described by the function:

$$\tau = F_{bb}(\gamma). \tag{1.4.8}$$

The shape of the curve is defined by two parameters, such as, stiffness in the origin (shear modulus at low strains) and the ultimate strength of the soil (shear stress in failure, in the high-strain range). For example, the characteristic curve can be a hyperbola:

$$F_{bb}(\gamma) = \frac{G_{\max}\gamma}{1 + (G_{\max}/\tau_{\max})|\gamma|}. \tag{1.4.9}$$

Hyperbolic backbone curve is shown in Fig. 1.4. Other types of stress–strain relations can also be used (for example, Ramberg–Osgood model (Ramberg and Osgood 1943). The backbone curve can be constructed based on the given shear modulus degradation curve *G(γ)* that describes the decrease of shear modulus with increasing loading, while $G_{\max}$ and $\tau_{\max}$ values are measured directly, calculated, or estimated by empirical relations.

For the hyperbolic model, soil response to cyclic loading obeys the following four rules (Kramer 1996):

1. During the first loading, the stress–strain relation coincides with the backbone curve.
2. If the stress or strain changes its sign at some point (γ_r, τ_r), then unloading follows the curve

$$\frac{\tau - \tau_r}{2} = F_{bb}\left(\frac{\gamma - \gamma_r}{2}\right). \tag{1.4.10}$$

In other words, the unloading and reloading curves have the same shape as the initial loading curve, with the origin shifted to the point, where the sign of the stress or the strain changed, and the scale expanded by a factor of two both vertically and horizontally.

However, these two rules called Masing's criterion (Masing 1926), are insufficient to describe the soil response under loading in general. Therefore, two additional rules are introduced:

3. If the unloading and reloading curves exceed maximum stress achieved in the previous loading cycles and cross the backbone curve, then after crossing they will follow the backbone curve until the next stress reversal.
4. If the unloading and reloading curves cross similar curves of the previous cycle, then after crossing they will follow the corresponding curves of the previous cycle.

Models that obey these four rules are called the extended Masing models (Kramer 1996).

Though the described hyperbolic model is simple and defined only in terms of effective stresses, it accounts for the hysteretic nature of damping and the dependence of shear modulus and damping on the strain amplitude. Alternative models describing unloading and reloading are suggested by Iwan (1967), Finn et al. (1977), Pyke (1979), Vucetic (1990), etc. In some cyclic nonlinear models, a slight attenuation is introduced to avoid uncertainty in assessing the soil response at low strains. Note that cyclic nonlinear models do not require zero shear strains at zero stresses. The ability of nonlinear models to describe the development of the deformation process is one of the most important advantages of these models compared to linear models.

A simple hyperbolic model does not allow us to determine changes in the volumetric strains caused by shear, which can lead to soil hardening in drained conditions, or to pore pressure increase accompanied by strength losses in undrained conditions. However, these factors are taken into account in most cyclic nonlinear models commonly used in engineering practice (Finn et al. 1977; Pyke 1979). The ability of nonlinear models to estimate pore pressure changes and consequently, effective stress changes, is another important advantage of cyclic nonlinear models over equivalent linear models. As pore pressure increases, effective stresses decrease, hence G_{max} and τ_{max} also decrease. Since the shape of the backbone curve depends on G_{max} and τ_{max}, the characteristic backbone curve "degrades" with increasing pore pressure. The stiffness of real soils, determined by their stress–strain relations, depends not only on the amplitude of cyclic loading, as assumed in equivalent linear models, but also on the history of soil loading. If we take this dependence into account in the calculations of soil response, such a cyclic nonlinear model will allow us to predict the generation, redistribution and dissipation of the pore pressure during and after the earthquake. These model capabilities are very useful for estimations of the soil liquefaction potential (Kramer 1996).

Advanced constitutive models use the most accurate and general methods based on the principles of mechanics to describe soil behavior in conditions of loading. They describe soil behavior (a) under initial loading conditions of the general type, (b) for a wide variety of loading paths, (c) with possible rotation of the principal stress axes, (d) in conditions of cyclic or monotonic loading, (e) at high and low loading rates, (e) in drained and undrained conditions. Such models usually include:

(1) some yield surface corresponding to the limiting stress conditions, at which linear soil behavior is still observed, (2) the law of hardening that describes changes in the size and shape of this surface when plastic deformations occur, (3) plastic flow rules linking increments of plastic strains to increments of stresses (Kramer 1996).

The first models of this type appeared back in the 1960s. Improvements of these models for more accurate description of shear strains can be achieved by introducing several initial yield centers on the sliding surface and several bounded yield surfaces and by accounting for a smooth transition from elastic to plastic soil behavior (Kramer 1996).

Though advanced constitutive models allow some flexibility in modeling soil response to cyclic loading, their description often requires far more parameters than equivalent linear models or cyclic nonlinear models. The estimation of these parameters can be difficult, and parameters obtained in some experiments may differ from those obtained in other experiments. Though such models are increasingly being used, practical difficulties noticeably limit their application in engineering practice (Kramer 1996).

Thus, models of various levels of complexity are developed to describe soil behavior in conditions of cyclic loading. The models considerably differ in the accuracy of describing the soil response, so that a model suitable for solving one engineering problem may not be applicable to another problem. The choice of the models requires careful consideration of the range of conditions of a particular engineering problem, analysis of the assumptions and limitations of the models, and full understanding of how a particular model is used in a given solution (Kramer 1996).

Alternative approach: representation of a soil as a collection of solid particles. A number of Russian and foreign studies attempt to describe soil behavior based on the characteristics of their internal structure. A soil is considered as a kind of granular medium consisting of a set of solid particles, the connections between which are carried out by compression and friction at their contact points. If we imagine a soil as an ensemble of identical discrete perfectly elastic particles, we can find some regularities in its behavior under small loads. For example, according to the theory, in conditions of uniaxial compression of a soil, its elastic shear modulus E_{tan} is proportional to the cubic root of the applied axial stress (Kramer 1996):

$$E_{\tan} = \frac{d\sigma}{d\varepsilon} = \frac{dN}{d\delta_N}\frac{2R}{4R^2} = \frac{1}{2R}\frac{dN}{d\delta_N} = \frac{3}{2}\left[\frac{2G}{3(1-\nu)}\right]^{2/3}\sigma^{1/3}, \qquad (1.4.11)$$

where G and ν are the elastic moduli of spherical elastic particles, σ is the average normal stress, δ_N is the change in distance between the centers of the particles, R is the radius of the particles, and N is the compressive force.

If a tangential stress T is applied, the elastic forces cause the centers of elastic particles to be displaced in the direction perpendicular to their original axis (Kramer 1996). With the increase of the tangential stress, sliding appears along the particle contacts throughout the volume and reorientation of the particles; volume changes (in drained conditions) and pore pressure build-up (in undrained conditions) cannot occur until mass sliding along the contacts has occurred. The shear strain corresponding to the onset of the mass sliding is called the volumetric threshold shear strain γ_{tv}; it is expressed by the formula

$$\gamma_{tv} = \frac{\delta_T}{2R} = 2.08\frac{(2-\nu)(1+\nu)f}{(1-\nu^2)^{1/2}E^{2/3}}\sigma^{2/3} \tag{1.4.12}$$

For confining pressures 25–200 kPa, which are rather common in practice, the volumetric threshold shear strain is 0.01–0.04% (Kramer 1996). The existence of the volumetric threshold shear strain, very close to the value predicted by theory, has been observed experimentally for sands in both drained and undrained conditions. The experiments also show that the threshold values of shear strains increase with increasing the plasticity index: the threshold strains for clays with plasticity index $PI = 50$ are approximately an order of magnitude higher than for sands, whose plasticity index is $PI = 0$. It was experimentally obtained that soils behave linearly at strains below the linear cyclic threshold of shear strains γ_{tl}, which is approximately 30 times lower than γ_{tv} (Vucetic 1994; Kramer 1996).

Analysis of interactions between individual soil particles helps us to understand the mechanical behavior of the entire soil column, and recent years, great success has been achieved in simulations of micromechanical processes in soils. Micromechanical models consider kinematics and contact interactions of individual soil particles. Numerical models have been developed with various limitations on the sample sizes, distributions of the particle shapes and sizes, kinematics of particle motions, contact interactions, etc. These calculations, where we could trace movements of individual particles, are sometimes very cumbersome, but they help us to understand some features of soil behavior.

However, we should note that the results of these calculations are confirmed experimentally mainly in tests with dry sandy soils and granular materials, like glass balls, shot, disks of photoelastic materials. Whereas, when solving practical problems of soil mechanics and soil dynamics, a phenomenological approach to determination of mechanical characteristics of soils prevails, in which soils are considered as continuous media, in some sense homogeneous and devoid of internal structure.

******

Thus, soils are dispersible multiphase systems consisting of solid, liquid and gaseous phases. Their physical properties are defined by the mineralogical composition, sizes and shapes of the constituent particles and by the soil structure, i.e., by the proportion of phases and structural bonds between particles; they significantly differ for the groups of non-cohesive, weakly cohesive and cohesive soils.

Measurements of dynamic characteristics of soils represent an important part of engineering seismology, and to date, many field and laboratory methods have been developed for measuring soil parameters in the low- and high-strain ranges. However, they provide only limited information about the behavior and properties of soils.

In field experiments, soil behavior is evaluated only in conditions prescribed in the experiments, and these conditions cannot be easily changed or controlled; also, we cannot create conditions for seismic wave propagation similar to the conditions of their propagation from an earthquake source and obtain strains close to those induced by strong earthquakes; the movement of the underground water cannot be controlled.

In laboratory tests, scattering of the measured soil parameters is due to the inherent place-to-place variability of soil properties, resulting from variable geological environments during soil formation, to anisotropy of soil properties related to the soil structure or texture, anisotropic loading conditions, structural disturbances from drilling and sampling, limitations introduced by the laboratory equipment, measurement and interpretation errors. As a consequence, the results of measurements of the same parameters of the same soils obtained by laboratory and field methods usually do not coincide. Therefore, when estimating dynamic properties of soils, we should use all the accumulated experience of laboratory and field studies and established empirical laws, and the most promising approach is interpretation of the observed in situ soil behavior.

Soils are multiphase media; however, when studying seismic wave propagation, soils are considered as continuous media described by the compression and shear diagrams that account for the contents and properties of the soil's components. The stress–strain relation is a model of soil behavior in conditions of its dynamic loading. In various periods, simple rational models were created that characterize the most important features of soil behavior, such as, linear elastic models, rheological models, bilinear or multi-modulus models, nonlinear elastic, elastic–plastic and viscous-elastic models.

Currently, three broad classes of models of soil behavior are used in engineering practice worldwide, such as, equivalent linear (viscous-elastic) models, cyclic nonlinear models, and advanced constitutive models. All the models are based on hysteretic stress–strain relations, and the research is aimed at identifying some general patterns of soil behavior and describing the behavior of various classes of soils by similar relations. Hysteretic stress–strain relations are defined by maximum shear moduli of soils estimated in the experiments and by the dependencies of the shear moduli and damping on the strain level.

An alternative approach is searching for analytical solutions to problems of engineering seismology. However, in most cases we cannot obtain them, because mathematical description of soils as multiphase media is extremely difficult: in multiphase media, it is more difficult than in single-phase media to describe processes in individual phases, while the effects of phase structures, changes in the structures, and interphase interactions are rather important. If we describe soils as granular media, analytical solutions are also almost impossible to obtain due to the complexity of the mechanical connections between the particles. If we consider soils as continuous media, then accounting for the viscosity, plasticity, multilayering, hysteresis nature of stress–strain relations in layers, etc. leads to mathematical difficulties that do not allow us to obtain solutions.

Thus, we should develop numerical methods for calculating soil behavior in conditions of dynamic loading, and when choosing a model, the preference should obviously be given to cyclic nonlinear models possessing optimal ratios of simplicity and accuracy. Calculation of soil behavior in conditions of loading should be based on the fullest possible consideration of all factors affecting the stress–strain state of the soil. When developing calculation methods, it is important to account for the main effects observed in dynamic loading of soils and involve, where possible, known theoretical positions and where necessary, empirical relations and parameters.

References

Bolton, M.D., Wilson, J.M.R.: An experimental and theoretical comparison between static and dynamic torsional soil tests. Geotechnique **39**(4), 585–599 (1989)

Bratton, J.L., Higgins, C.J.: Measuring dynamic in situ geotechnical properties, pp. 272–289. Proceeding ASCE Geotechnical Engineering Division Special Conference Pasadena, California (1978)

Casagrande, A.: Liquefaction and cyclic deformation of sands: a critical review. Lect. 5th Panam. Conf. Soil Mech. Found. Eng. Buenos Aires 80–133 (1975)

Castro, G.: Liquefaction and cyclic mobility of saturated sands. Proc. ASCE **101**(GT6), 551–569 (1975)

Castro, G., Poulos, S.J.: Factor affecting liquefaction and cyclic mobility. J. Geotech. Eng. Divis. ASCE **103**(GT6), 501–516 (1977)

Coulter, H.W., Migliaccio, R.R.: Effects of the earthquake of March, 27, 1964 at Valdez, Alaska. The Alaska earthquake, March 27 1964: effects on communities. Professional Paper 542-C (1966)

Dobry, R., Vucetic, M.: State-of-the-Art Report: Dynamic properties and seismic response of soft clay deposits. Proc. Int. Symp. Ceotech. Eng. Soft Soils. Mexico City **2**, 51–87 (1987)

Finn, W.D.L., Martin, G.R., Lee, M.K.W.: Comparison of dynamic analysis of saturated sand. Proc. ASCE Geotech. Eng. Div. Spec. Conf. Earthq. Eng. Soil Dyn. Pasadena 472–491 (1978)

Finn, W.D.L.: Evaluation of liquefaction potential. Soil Dyn. Lectures. Univ. Brit, Columbia (1993)

Finn, W.D.L.: State-of-the-art of geotechnical earthquake engineering practice. Soil Dyn. Earthq. Eng. **20**(1–4), 1–15 (2000)

Finn, W.D.L., Lee, K.W., Martin, G.R.: An effective stress model for liquefaction. GED, ASCE **103**(GT6), 517–533 (1977)

Finn, W.D.L., Yogendrakumar, M., Yoshida, N., Yoshida, M.: TARA-3: A Program for Nonlinear Static and Dynamic Effective Stress Analysis. Soil Dynamics Group University British Columbia, Vancouver (1986)

Fracture mechanics and material strength. Fatigue and cyclic crack resistance of structural materials. Kiev, Naukova Dumka (1990) (in Russian)

Gersevanov, N.M.: Fundamentals of Soil Mass Dynamics. ONTI, Moscow (1937). (in Russian)

Grigoryan, S.S.: On General Equations of soil dynamics. Doklady Acad. Sci. USSR **124**(2) (1959) (in Russian)

Grigoryan, S.S.: On basic concepts of soil dynamics. Appl. Math. Mechan. **24**(6), (1960) (in Russian)

Gvozdev, A.A., Kuznetsov, V.V.: Splitting phenomena in soils observed in seismic exploration. Izvestiya Acad. Sci. USSR, Phys. of the Earth **5** (1967) (in Russian)

Gvozdev, A.A., Zvolinsky, N.V., Kovshov, A.N.: Seismic vibrations of the Earth's surface depending on local conditions. Izvestiya Acad. Sci. USSR, Phys. Earth **9**, 10–29 (1982) (in Russian)

Hardin, B.O., Drnevich, V.P.: Shear modulus and damping in soils: Measurement and parameter effects. Proc. Am. Soc. Civil Eng. J. Soil Mech. Found. Div. **98**, 603–624 (1972)

Hardin, B.O.: The nature of damping in sands. Proc. Am. Soc. Civil Eng., J. Soil Mech. Found. Div. **91**(SM1), 63–97 (1965)
Hardin, B.O.: The nature of stress-strain behaviour for soils. Proc. Earthq. Eng. Soil Dy-nam. ASCE, Pasadena, 19–21 June, 3–89 (1978)
Haskell, N.A.: The dispersion of surface waves on multilayered media. Bull. Seism. Soc. Am. **43**, 17–34 (1953)
Higgins, C.J.: Ground motion induced interface pressures. In: Proceeding ASCE Special Conference Earthquake English Soil Dynamics Geotechnical English Division Pasadena (1978)
Hyodo, M., Yasuhara, K., Murata, H., Hirao, K.: Behaviour of soft clay deposit subjected to long-term cyclic shear stresses. Technol. Rep. Yamaguchi Univ. **4**(2), 171–191 (1988)
Idriss, I.M., Seed, H.B.: Seismic response of horizontal soil layers. Proc. Am. Soc. Civil Eng. J. Soil Mech. Found. Div. **94**, 1003–1031 (1968)
Idriss, I.M., Seed, H.B.: Response of Horizontal Soil Layers During Earthquakes. Department Civil Engineering University Berkeley, California (1967)
Idriss, I.M.: Response of soft soil sites during earthquakes. In: Proceeding H.B. Seed Memorial Symposium, Berkeley, California (1990)
Ishibashi, I., Zhang, X.: Unified dynamic shear moduli and damping ratios of sand and clay. Soils Found. **33**(1), 182–191 (1993)
Ishihara, K., Tatsuoka, F., Yasuda, S.: Undrained deformation and liquefaction of sand under cyclic stress. Soils Found. **15**(1), 29–44 (1975)
Ishihara, K., Kokusho, T., Silver, M.L.: Recent developments in evaluating liquefaction characteristics of local soils. Proc. 12th Int. Conf. Soil Mech. Found. Eng. Rio de Janeiro 1989, pp 2719–2734. Rotterdam, Brookfield (1992)
Ishihara, K., Cubrinovski, M., Takeshi, N.: Characterization of undrained behavior of soils in the reclaimed area of Kobe. Soils Found. Spec. Issue on Geotechnical Aspects of the January 17 1995 Hyogoken-Nambu Earthquake **2,** 33–46 (1998)
Ishihara, K.: Evaluation of soil properties for use in earthquake response analysis. Proc. Int. Symp. Numeric. Mod. Geomechanics. Zurich 237–259 (1982)
Ishihara, K.: Stability of natural deposits during earthquakes. Proc. 11th Int. Conf. Soil Mech. Found. Eng. San Francisco **2**, 321–376 (1985)
Ishihara, K. et al.: Effective stress analyses of ground and soil structures. Proc. Symp. Behav. Ground Soil Struct. Earthq. JSSMFE 50–136 (1980)
Ivanov, P.L.: Compaction of weakly-cohesive soils by explosions. Nedra, Moscow (1983). (in Russian)
Iwan, W.D.: On a class of models for the yielding behavior of continuous and composite systems. J. Appl. Mech. **34**, 612–617 (1967)
Iwasaki, T., Tatsuoka, F., Takagi, Y.: Shear moduli of sands under cyclic torsional shear loading. Techn. Mem. Public Works Research Inst. Ministry Constr. Chiba-Shi, Japan **1264** (1977)
Jamiolkowski, M., Lerueil, S., Lopresti, D.K.: Design parameters from theory to practice. In: Procceding geo-Coast '91, Yokohama, Japan, pp. 1–41 (1991)
Kiku, H., Yoshida, N.: Dynamic deformation property tests of sand at large strains. In: Proceeding 33rd Japan National Conference Geotechology Engineering, pp. 869–870 (1998) (in Japanese)
Kim, D.-S., Stokoe, K.H., Roesset, J.M.: Characterization of material damping of soils using resonant column and torsional shear tests. In: Proceeding 5th International Conference Soil Dynamics and Earthquake Engineering, Karlsruhe, Germany. CMP and Elsevier, N.Y., London 189–200 (1991)
Kokusho, T.: Dynamic deformation characteristics of soil and nonlinear response of ground. Rep. 301. Central Electric Res. Inst. 207–240 (1982) (in Japanese)
Kokusho, T.: Dynamic characteristics of ground. Lect. Analytic. Meth. Interact. Ground Struct. Tsuchito-Kiso **40**(4), 76–84 (1992)
Kompaneets, A.S.: Shock waves in a plastic compacting medium. Doklady Acad. Sci. USSR **109**(1), 49–52 (1956) (in Russian)

Kramer, S.L.: Geotechnical Earthquake Engineering, pp. 230–238. Prentice Hall, Upper Saddle River, N.J. (1996)
Kramer, S.L., von Laun, F.-Y., Sivaneswaran, N.: Strain-controlled, variable frequency cyclic loading system for soft soils. Geotechn. Testing J. **15**(3), 264–270 (1992)
Krasnikov, N.D.: Dynamic Properties of Soils and Methods for their Determination. Stroyizdat, Leningrad (1970). (in Russian)
Lee, M.K.W., Finn, W.D.L.: DESRA-2, Dynamic Effective Stress Response Analysis of soil deposits with energy transmitting boundary including assessment of liquefaction potential. Soil Mech. Dep. Civil Eng. Univ. of British Columbia, Vancouver (1978)
Lee, K.L.: Influence of end restraint in cyclic triaxial tests. WES CR S-76–1, USAE Waterways Experiment Station, Corps of Engineers, Vicksburg, Miss (1976)
Lyakhov, G.M.: Fundamentals of the Dynamics of Blast Waves in soils and rocks. Nedra, Moscow (1974). (in Russian)
Lyakhov, G.M.: Determination of dynamic compressibility of soils. Basements, foundations and soil mechanics **3** (1966) (in Russian)
Lyakhovsky, V.A., Myasnikov, V.P.: Heterogeneity, anisotropy and reflecting boundaries. Izvestiya Acad. Sci. USSR Phy. Earth **11** (1986) (in Russian)
Lyakhovsky, V.A.: Application of a heterogeneous model to the analysis of the stress state of rocks. Izvestiya Acad. Sci. USSR Phy. Earth **2** (1990) (in Russian)
Masing, G.: Eigenspannungen und Verfestigung Beim Messing. Proc. Int. Congr. Appl. Mech. 332–335 (1926) (in German)
Maslov, V.P., Mosolov, P.P., Sosnina, E.V.: On the types of discontinuities of solutions of the equation for longitudinal, free, one-dimensional motions in a multi-modulus medium. Questions of nonlinear continuum mechanics. Valgus, Tallinn (1985) (in Russian)
Melnikov, V.V., Rykov, G.V.: On the influence of the deformation rate on the compressibility of loess soils. Prikladnaya Mekhanika i Tekhnicheskaya Fizika **2** (1965) (in Russian)
Mishel, A.G., Shulman, S.G.: Dynamics of Multiphase Soil Media. VNIIG im. Vedeneeva, St. Petersburg (1999)
Musaelyan, A.A.: Preliminary results of laboratory studies of soils under dynamic loadings, Doklady Acad. Sci. Taj. SSR **8**, (1964) (in Russian)
Nikolaev, A.V.: Seismic properties of a loose medium. Izvestiya Acad. Sci. USSR, Physics of the Earth **2** (1967) (in Russian)
Nikolaevsky, V.N.: Mechanics of Porous and Fractured Media. Nedra, Moscow (1984)
Pokrovsky, G.I., Nasedkin, N.A., Sinelshchikov, S.I.: Investigation of soil compression at different deformation rates. Soil Sci. **1** (1938) (in Russian)
Pyke, R.: Nonlinear soil model for irregular cyclic loadings. J. Geotech. Eng. Div. **105**, 715–726 (1979)
Rakhmatullin, H.A., Sagomonyan, A.Y., Alekseev, N.A.: Problems of soil dynamics. M., Mosc. State Univ. Publishing House, Moscow (1964) (in Russian)
Rakhmatullin, H.A., Demyanov, Yu.A.: Strength under intense short-term loads. Fizmatgiz, Moscow (1961)
Ramberg, W., Osgood, W.R.: Description of stress-strain curves by three parameters. Techn. Note 902. National Advisory Committee For Aeronautics, Washington DC (1943)
Redpath, B.B.: Seismic refraction exploration for engineering site investigations. Techn. Rep. TR E-73–4, U.S. Army Engineer Waterways Experiment Station, Explosive Excavation Research Laboratory, Livermore, CA. NTIS AD-768710 (1973)
Richart, F.E.: Dynamic stress-strain relationships for soils. Proc. 9th ICSMFE, Tokyo **3**, 605–612 (1978)
Rodionov, V.N., Adushkin, V.V., Kostyuchenko, V.N.: Mechanical effect of an underground explosion. Nauka, Moscow (1971) (in Russian)
Rykov, G.V., Skobeev, A.M.: Measurement of stresses in soils under short-term loads. Nauka, Moscow (1978)

Sagomonyan, A.Y.: Review of some works on soil dynamics. Bull. Moscow State Univ. **5** (1967) (in Russian)

Saifiddinov, S.: Development of subsidence deformations of wet loesses during seismic motions. Abstract dissert. Ph.D. of Technical Science, Moscow (1987). (in Russian)

Sangrey, D.A., Henkel, D.J., Esrig, M.I.: The effective stress response of a saturated clay soil to repeated loading. Canadian Geotechn. J. **6**, 241–252 (1969)

Seed, H.B.: Landslides during earthquakes due to liquefaction. J. Soil Mech. Found. Div. ASCE **94**(SM5), 1053–1122 (1968)

Seed, H.B.: Soil liquefaction and cyclic mobility evaluation for level ground during earthquakes. Proc. ASCE **105**(GT2), 201–255 (1978)

Seed, H.B., Lee, K.L.: Liquefaction of saturated sands during cyclic loading. J. Soil Mechanics Found. Div. ASCE **92**(SM6), 105–134 (1966)

Seed, H.B., Murarka, R., Lysmer, J., Idriss, I.M.: Relationships of maximum acceleration, maximum velocity, distance from source, and local site conditions for moderately strong earthquakes. Bull. Seism. Soc. Am. **66**, 1323–1342 (1976)

Seed, H.B., Chan, C.K.: Thixotropic characteristics of compacted clays. J. Soil Mech. Found. Div. (1957)

Seed, H.B., Idriss, I.M.: Soil moduli and damping factors for dynamic response analyses. Rep. EERC70–10 EERC, Univ. of California, Berkeley (1970)

Seed, H.B., Idriss, I.M.: Simplified procedure for evaluating soil liquefaction potential. J. Soil Mech. Found. Div. ASCE **97**(SM9), 1249–1273 (1971)

Seed, H.B., Idriss, I.M.: Ground motion and soil liquefaction during earthquakes. Earthq. Eng. Res. Inst. (1982)

Seed, H.B., Idriss, I.M.: Arango I. Evaluation of liquefaction potential using field performance data. J. Geotechn. Eng. Division ASCE **109**(GT3), 458–482 (1983)

Seed, H.B., Wong, R.T., Idriss, I.M., Tokimatsu, K.: Moduli and damping factors for dynamic analyses of cohesionless soils. Rep. UCB/EERC-84/14 (1984)

Seed, H.B., Tokimatsu, K., Harder, L.F., Chung R.M.: Influence of SPT procedures in soil liquefaction resistance evaluations. J. Geotechn. Eng. Div. ASCE **5**(12) (1985)

Seed, H.B., Dickinson, S.E., Mok, C.K.: Recent lessons regarding seismic response analysis of soft and clay sites. In: Proceeding seminar seisman design and retrofit of bridges, California department of transportation, Sacramento (1992)

Seed, H.B.: Considerations in the earthquake-resistant design of earth and rockfill dams 19th Rankine Lecture. Geotechnique **29**(3), 215–263 (1979)

Shemyakin, E.I.: About stress waves in dense rocks. Prikladnaya Mekhanika i Tekhnicheskaya Fizika, Moscow **5**, 83–93 (1963). (in Russian)

Srinivasulu, P., Vaidyanathan, C.V.: Handbook of Machine Foundations. TATA McGraw-Hill Ltd., New Delhi (1976)

Stavnitzer, L.R.: Investigation of dynamic compressibility of soils. Proc. Research Inst. Found. **56**, Stroyizdat, Moscow (1966) (in Russian)

Sun, J.I., Golesorkhi, R., Seed, H.B.: Dynamic moduli and damping ratios for cohesive soils. Rep. UCB/EERC-88/15 (1988)

Tankaeva, L.K.: The nature of structural bonds and quicksand of loesses of the basin of the lower reaches of Vakhsh river. Avtoref. Diss. Ph.D. of Geol. Sci, Moscow (1964). (in Russian)

Thiers, G.R., Seed, H.B.: Strength and stress-strain characteristics of clays subjected to seismic loads. Symp. Vibrat. Effects Earthquakes Soils Found. ASTM STP **450**, 3–56 (1969)

Tokimatsu, K.: Penetration tests for dynamic problems. In: Penetration Testing 1988: Proc. 1st International Symposium on Penetration Testing ISOPT-1, Orlando, pp. 117–136 (1988)

Tokimatsu, K.: Dynamic property of soil from laboratory test, in-situ test and earthquake measurement. In: Proceeding 2nd Symposium On Dynamic Interaction between Soil and Structure, pp. 11–16 (1989)

Towhata, I.: Models for cyclic loading. Mechanics of granular materials. Rep. ISSMFE Techn. Comm. Mech. Gran. Mater. ISSMFE 80–90 (1989)

Tsytovich, N.A.: Soil Mechanics. Higher School, Moscow (1968). (in Russian)
Vasil'ev, Yu.I., Gvozdev, A.A., Ivanova, L.A., Molotov, L.V., Fomichev, V.B., Shcherbo, M.N.: Mechanical properties of soft soil in the stress range up to (5–10) 10^5 Pa or 5–10 kg/cm^2 and the choice of a soil model describing its behavior in strong earthquakes. Seismic Microzonation. Nauka, Moscow 121–129 (1977) (in Russian)
Vasil'ev, Yu.I., Ivanova, L.A., Shcherbo, M.N.: Measurement of stresses and strains in the ground during the propagation of explosive waves. Izvestiya Acad. Sci. USSR, Physics of the Earth **1**, 21–37 (1969) (in Russian)
Voznesensky, E.A.: Dynamic Instability of Soils. Editorial URSS, Moscow (1999). (in Russian)
Vucetic, M.: Normalized behavior of clay under irregular cylic loading. Can. Geotech. J. **27**, 29–46 (1990)
Vucetic, M.: Cyclic threshold shear strains in soils. J. Geotechn. Eng. **120**(12), 2208–2228 (1994)
Vucetic, M., Dobry, R.: Effect of soil plasticity on cyclic response. J. Geotechn. Eng. ASCE **117**(1), 89–107 (1991)
Wilson, J.M.R.: The dynamic properties of soil. M Phil dissertation, Engineering Department, Cambridge University (1985)
Woods, R.D.: Measurement of dynamic soil properties. Proc. ASCE Geotechn. Eng. Div. Spec. Conf. Earthquake Eng. Soil Dynam. Pasadena **1**, 91–178 (1978)
Woods, R.D.: Field and laboratory determination of soil properties at low and high strains. SOA paper. In: Proceeding 2nd International Conference on Recent Advances in Geotechnical Earthquake Engineering and Soil Dynamics, St. Luis 1727–1741 (1991)
Yasuda, S., Yamaguchi, I.: Dynamic shear moduli in the laboratory and the field. In: Proceeding Symposium on Evaluation of Deformation and Strength Characteristics of Sandy Soils and Sand Deposits, JSSMFE, Tokyo, pp. 115–118 (1984)
Yoshida, N., Iai, S.: Nonlinear site response and its evaluation and prediction, pp. 71–90. The effects of Surface Geology on Seismic Motion. Balkema, Rotterdam (1998)
Yoshida, N.: Processing of strain dependent characteristics of soil for nonlinear analysis. In: Proceeding 1st International Conference Earthquake Geotechnological Engineering Tokyo, pp. 473–478 (1995)
Youd, T.L., Harp, E.L., Keefer, D.K., Wilson, R.C.: The Borah Peak, Idaho earthquake of October, 28, 1983—liquefaction. Earthq. Spectra EERI **2**(1), (1985)
Youd, T.L., Perkins, M.: Mapping liquefaction-induced ground failure potential. J. Geotechn. Eng. Div. ASCE **104**(GT4), 433–446 (1978)
Zaruba, Q., Mencl, V.: Engineering Geology. Developments in Geotechnical Engineering. Elsevier, Amsterdam (1976)
Zelikson, A.: Geotechnical models using the hydraulic gradient similarity method. Geotechnique **9**(4), 495–508 (1969)
Zvolinsky, N.V.: Wave processes in inelastic media. Probl. Eng. Seismol. **23**, 4–19 (1982). (in Russian)

Chapter 2
Methods for Calculation of Ground Response During Strong Earthquakes

Since the 1920s, seismologists have been developing quantitative methods to assess the influence of local site conditions on the parameters of ground motion. This chapter analyzes the current state of knowledge in this field, describes physical mechanisms of local site effects and the situations in which they appear.

Linear transformations of seismic waves in subsurface soils, such as, amplification of the waves in the upper softer layers, resonant phenomena and topographic effects have been intensively studied in recent decades and are now quite well understood and taken into account in practice of engineering seismology. Significant progress was also made in studying the nonlinear response of subsurface soils. Seismic observations indicate that during strong earthquakes, significant distortions occur in the waveforms and spectra of seismic waves propagating in soil layers. Seismic waves travel from the earthquake sources tens to hundreds kilometers in dense materials of the Earth's crust and only meters in soils, but often the soils mostly determine parameters of motion on the surface.

In Russia and abroad, nonlinear soil behavior during strong ground motion has been studied in laboratory and field experiments. This chapter provides an overview of these studies, describes numerical methods for calculating soil response and the results of testing various methods on earthquake records in situ. Two- and three-dimensional methods are usually the extensions of the corresponding one-dimensional methods.

2.1 Transformations of Seismic Waves in Subsurface Soils

Figure 2.1 illustrates the propagation of seismic waves from an earthquake source up to the Earth's surface.

Seismic waves radiated by the earthquake source propagate in crystalline rocks, where seismic motion can be considered as a function of the earthquake magnitude

O. V. Pavlenko, *Seismic Waves in Soil Layers*,
https://doi.org/10.1007/978-3-031-52338-0_2

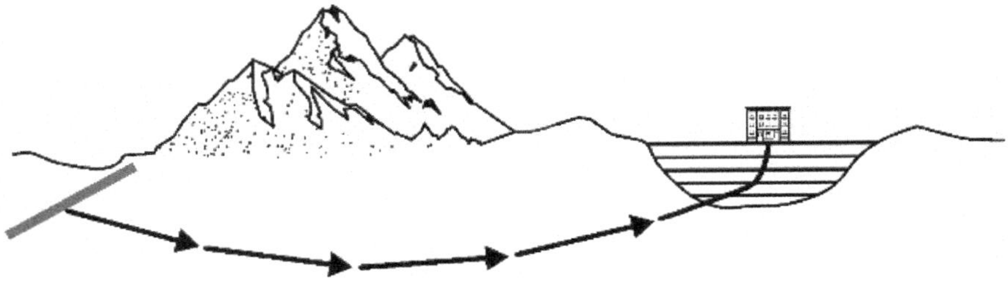

Fig. 2.1 Propagation of seismic waves from an earthquake source to the surface. Refraction leads to almost vertical propagation of the waves in subsurface layers

and distance from the source, then in layers that are treated as basement layers and where seismic waves are not affected by the Earth's surface and subsurface soils. Then the waves enter subsurface soils, in which they usually propagate almost vertically upwards, because their propagation velocities quickly decrease as they approach the surface.

The influence of local geological and ground conditions on the intensity of surface motion and destructions during earthquakes has been known for a long time. MacMurdo (1824) wrote that "buildings based on rocks did not experience such a strong seismic effect as buildings whose foundations were laid in the ground and did not reach bedrock" during the 1819 earthquake in Kutch, India. Mallet (1862) in his report on the 1857 earthquake in Naples described the influence of local geological conditions on the distribution of destructions. Wood and Raid showed that the intensity of ground shaking during the 1906 San Francisco earthquake was associated with local ground and geological conditions. Since these early observations, the influence of local site conditions on motion on the surface during earthquakes has been noted everywhere.

The influence of soft low-impedance surface layers on seismic motion is well documented; in Japan, it has been studied since the early 1930s by Ishimoto and Sezawa. Takahashi and Hirano (1941) may have been the first to interpret the observed differences in earthquake records at different sites based on the theory of elasticity. Early seismological evidence of the strong influence of local site conditions on motion on the surface in Japan are summarized by Kanai et al. (1956).

In the USA, studies of local amplification of seismic motion by soil layers were carried out by Gutenberg (1957) in Pasadena area. He calculated coefficients of local site amplification based on records of microseisms at sites with different subsurface soils. Hudson (1971) compared the amplification estimated by Gutenberg with the amplification observed during the 1971 San Fernando earthquake and concluded that Gutenberg's predictions were not applicable to strong motion.

In the USSR, the influence of subsurface soils on the intensity of seismic motion was studied by Gzelishvili (1946), Tsshoher (1949), Safaryan (1957), Popov (1959), Gorshkov (1960), Byus and Tskhakaya (1960) and others. To estimate the effect of soils on the earthquake intensity in balls,[1] field investigations of the aftereffects of strong earthquakes in the USSR and abroad were carried out. The observations refer to various geotechnical and soil conditions.

According to Suyehiro (1932), the 1923 earthquake in Kuanto (Japan) was 1.5–2 times more intense in the lowland parts of the city on loose silty alluvial soils than in the elevated areas on deluvial dense soils. During the 1944 earthquake in Tatami area, most of the destructions occurred in areas located in river floodplains, on muddy and swampy clay deposits. Less damage was observed on sand dunes, and even weaker damage was found on high terraces composed of gravelly deposits, on the deluvial plateau, and on bedrock.

Macroseismic data from many earthquakes in New Zealand indicate that on sandy wet soils the destructions were usually 1–2 ball units higher than on dense rocks at the same epicentral distances. More destructions occurred on soft clays than on dense clays.

According to the empirical relation by S. V. Medvedev, a decrease in *P*-wave velocities in surface layers by 4 times increases the earthquake intensity by 1 ball unit (Medvedev 1962). Destructions at sandy and clayey sites increase with increasing the thickness of soil layers.

The intensity of the Chatkal earthquake (western Kyrgyzstan, M_w ~ 7.5) on November 2, 1946 at sites located on young alluvial sandy-loamy deposits was one ball unit higher than at elevated sites composed of dense clays and marls. At rock outcrops, the intensity of motion was one unit less. In Arkit place, at a site composed of sands and gravels (the groundwater level ~ 10–15 m), the intensity of the earthquake was one unit higher than at a site composed of dense limestones. The intensity of the earthquake in Namangan (Uzbekistan) varied by a unit at sites of gravel-pebble and loess-like soils, whereas, at homogeneous gravel-pebble sites, the intensity changed only slightly with increasing the groundwater level (from 10 to 2 m) (Medvedev and Shebalin 1967).

During the 1948 Ashkhabad earthquake, in Bezmein area (Turkmenistan) at a site composed of homogeneous loams, the intensity of motion depended on the groundwater level (5–15 m). The areas composed of pebbles had seismic intensity one unit lower than the areas composed of loams (the groundwater level was 4–5 m at both cases) (Medvedev and Shebalin 1967).

Variations of the earthquake intensity on the territory of Ashkhabad city were generally consistent with soil and hydrogeological conditions (Medvedev and Shebalin 1967). In deluvial-proluvial loams, an increase in the groundwater level increased seismic intensity, but at the groundwater level of ~ 10 m and more, it did

[1] Ball in Russia is a measure of earthquake intensity, estimated at a site based on a non–instrumental 12-unit scale MSK-64 (Medvedev–Sponheuer–Karnik). The earthquake characterization is as follows: 1–4—weak, do not cause destructions; 5–7—strong, destroy old buildings; 8—destructive, buildings partially collapse; 9—devastating, most buildings collapse, significant cracks appear on the surface; 10—destructive, destroying bridges and pipelines, landslides occur; 11—disasters, all structures are destroyed, landscape changes; 12—severe disasters, large changes in the landscape over a large area. It is assumed that an increase in seismic intensity by 1 ball unit corresponds to doubling of the acceleration amplitudes on the surface.

not influence the intensity. Increasing the groundwater level from 10 to 4 m increased the earthquake intensity by half a ball unit.

On thin (2–4 m) loose surface sandy and loamy layers covering dense bedrock, the intensity of the 1949 Khait earthquake (Tajikistan, $M_w \sim 7.5$) was 1 ball unit higher than on similar soils of larger (~10 m) thickness. Heterogeneity of horizontally layered sediments (overlapping conglomerates, pebbles, loess) did not cause an increase of intensity in ball units compared to sites with homogeneous deposits (Medvedev and Shebalin 1967).

In San Francisco area, local variations of amplification of seismic motion on unconsolidated sediments reached two orders of magnitude during the 1906 San Francisco and the 1989 Loma Prieta earthquakes. In Mexico City, very soft surface clays caused catastrophic destructions with human losses and huge economic damage during the 1985 Guerro-Michoacan earthquake, the epicenter of which was located ~ 350 km from the city (Bard 1995). Almost all devastating earthquakes of last decades (Armenia 1988; Iran 1990; Philippines 1990; Northridge 1994; Kobe 1995; Chi-Chi 1999; Tokachi-oki 2003; Tohoku 2011, etc.) provided additional evidence of the significance of local site effects.

Numerous macroseismic observations of earthquake damage carried out over the past decades indicate local rises in intensity on soft sediments by 2–3 units on MM (Modified Mercalli) or MSK (Medvedev–Sponheuer–Karnik) scales. These types of site effects have always attracted attention of earthquake engineers and seismologists, because most of the urbanized areas in the world are located in river valleys, on young soft sedimentary deposits. Large cities, such as, Los Angeles, San Francisco, San Salvador, Caracas, Lima, Bogota, Tokyo, Osaka, Kobe, Kathmandu, Manila, Thessaloniki, Mexico City are located in river valleys in earthquake-prone areas (Bard 1995).

The prevalence and significance of these effects stimulated numerous instrumental studies, in which records obtained on sediments and crystalline basements or nearby rock outcrops were compared; theoretical and numerical investigations of seismic wave propagation in complex environments were carried out, aimed at better understanding and quantification of these effects. Their importance is evident from a large number of review papers on the topic (Aki and Irikura 1991; Aki 1993; Bard 1995; Faccioli 1995; Yoshida and Iai 1998; Finn 2000, etc.).

Aki and Irikura (1991) discuss characterization of surface shaking in future earthquakes, accounting for ground conditions. This is not an easy problem because of a variety of geological environments and irregular shapes of geological structures formed by weathering, erosion, sedimentation and other processes. In Japan and the USA, the influence of local site conditions on peak accelerations, peak velocities, response spectra and other parameters was studied (Hayashi et al. 1971; Kuribayashi et al. 1972; Katayama et al. 1978; Kawashima et al. 1986; Seed et al. 1976; Mohraz 1976; Trifunac 1976; Boore et al. 1980; Joyner and Boore 1981, etc.). To establish the relationship between the amplification of seismic waves and ground conditions, subsurface geology was classified into 5 geological epochs: (1) Quaternary sediments, (2) Pliocene, (3) Miocene and Cretaceous sediments, (4) Franciscan formations and Mesozoic granites, and (5) Pre-Cretaceous metamorphic rocks. Rather smooth power-law dependences were obtained of the average amplification on the average age of geological formations: younger sediments provide higher amplification than older rocks at frequencies up to 12 Hz (at lower frequencies the dependence is clearer). However, the relationship between the age and the amplification is significant only in average; for individual measurements standard deviations are large.

Thus, seismic observations show us that the intensity of surface motion during earthquakes depends on the composition and thickness of soil layers, their saturation with water and the groundwater level, as well as on the magnitude and frequency content of the earthquake. The proximity of the groundwater enhances the seismic effect, but the effect is weaker on water saturated than on wet soils.

As a result, the following mechanisms of seismic wave transformations in subsurface soils have been established:

- transition of seismic waves to near-surface layers with lower seismic velocities and densities leads to increasing amplitudes of seismic motion on the surface, according to the law of conservation of the density of seismic energy flow;
- resonance phenomena, i.e., capturing of seismic waves by soil layers with lower velocities and densities also leads to increasing amplitudes of motion on the surface;
- nonlinearity of stress–strain relations in soil layers usually leads to a decrease of the amplitudes of motion on the surface (however, in water-saturated subsurface non-cohesive soils, there may be an increase of the amplitudes; examples are given in Chapter 3).

The action of the three mechanisms usually leads to the fact that weak seismic motion is amplified, whereas stronger motion is damped in soil layers (Yoshida and Iai 1998).

All the three mechanisms are related to each other and usually considered in the algorithms for calculation of motion on the ground surface.

Amplification of seismic motion due to the first two mechanisms depends on the ratios of seismic wave lengths and the thicknesses of soil layers. When the thickness of a soft layer is small compared to the wavelength, the amplitude of motion in the layer is determined only by its seismic rigidity; with increasing the thickness of the layer, resonant effects may appear, and the additional increase in the amplitude can be three- or four-fold.

In cases of horizontally layered sedimentary structures, only body waves propagating up and down in the soil columns can be captured by softer layers. If the structure is two- or three-dimensional, i.e., includes lateral inhomogeneities (e.g., a sedimentary basin), capturing of surface waves by such inhomogeneities can also be observed (Bard 1995).

Let's consider a one-dimensional problem and a number of situations that illustrate the influence of soil layers on characteristics of motion on the surface. In linear approximation, to calculate the response of a system of soil layers with one degree of freedom, transfer functions are used, which determine amplification or attenuation of spectral components of motion incident to soil layers at the boundary of the soil thickness with the bedrock. Nonlinear soil behavior can be approximated by introducing "equivalent linear" parameters of soil layers and iterative calculations.

A uniform absolutely elastic soil layer on a rigid bedrock (Fig. 2.2a). As known (Aki and Richards 2002), horizontal displacements in a layer can be expressed by the formula:

$$u(z,t) = Ae^{i(\omega t + kz)} + Be^{i(\omega t - kz)}, \tag{2.1.1}$$

where ω is the circular frequency of ground shaking, k is the wave number, A and B are the amplitudes of seismic waves travelling up and down, respectively. The boundary conditions at the free surface are zero shear stresses and strains:

$$\tau(0,t) = G\gamma(0,t) = G\frac{\partial u(0,t)}{\partial z} = 0, \tag{2.1.2}$$

From (2.1.1) to (2.1.2), we find the solution A = B and $u(z,t) = 2A\cos kz\, e^{i\omega t}$, i.e., a standing wave with an amplitude $2A$ cos kz, or a superposition of waves propagating up and down. The transfer function has the form (Fig. 2.2b):

$$F(\omega) = \frac{u_{\max}(0,t)}{u_{\max}(H,t)} = \frac{2Ae^{i\omega t}}{2A\cos kHe^{i\omega t}} = \frac{1}{\cos kH} = \frac{1}{\cos(\omega H / v_s)}. \tag{2.1.3}$$

Thus, displacements on the surface of the layer are amplified infinitely at certain frequencies depending on the thickness and S-wave velocity in the layer.

A uniform damped elastic soil layer on a rigid bedrock. Let the behavior of the layer be described by the Kelvin–Voigt model, i.e., a viscoelastic system with

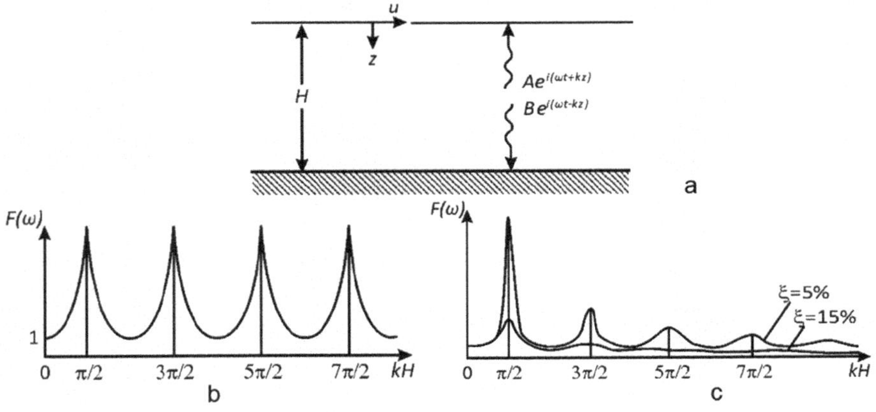

Fig. 2.2 **a**—A uniform absolutely elastic soil layer of thickness H on a rigid bedrock; **b**—the transfer function of the uniform elastic layer on a rigid bedrock; **c**—the transfer function of a uniform damped elastic soil layer on a rigid bedrock

parallel viscous and elastic elements. The horizontal displacements are expressed by the formula (Kramer 1996):

$$u(z,t) = Ae^{i(\omega t + k^* z)} + Be^{i(\omega t - k^* z)}, \tag{2.1.4}$$

where $k^* = k_1 + ik_2$ is a complex wave number. Similarly to the previous case, we get the solution:

$$F(\omega) = \frac{1}{\cos k(1-ih)H} = \frac{1}{\cos[\omega H / v_s(1+ih)]}, \tag{2.1.5}$$

if the attenuation parameter h is small. The wave propagation velocity is complex, and as in the previous case, the wave amplification depends on the thickness of the layer and S-wave velocity, i.e., on the frequency; it is maximal at frequencies corresponding to the fundamental frequencies of the layer (Fig. 2.2b).

Attenuation influences the response of the layer at higher frequencies more than at lower frequencies.

A uniform damped elastic soil layer on an elastic half-space (Fig. 2.3a). In this case, the energy of standing waves partially returns from the layer down to the half-space. The transfer function has the form (Kramer 1996):

$$F(\omega) = \frac{1}{\cos k_s^* H + i\alpha_z^* \sin k_s^* H} = \frac{1}{\cos(\omega H / v_{ss}^* + i\alpha_z^* \sin(\omega H / v_{ss}^*))}, \tag{2.1.6}$$

where $\alpha_z^* = \rho_s v_{ss}^* / \rho_r v_{sr}^*$ is the complex ratio of impedances, ρ_s and v_{ss}^* are the density and S-wave velocity in the layer, ρ_r and v_{sr}^* are the density and S-wave velocity in the half-space.

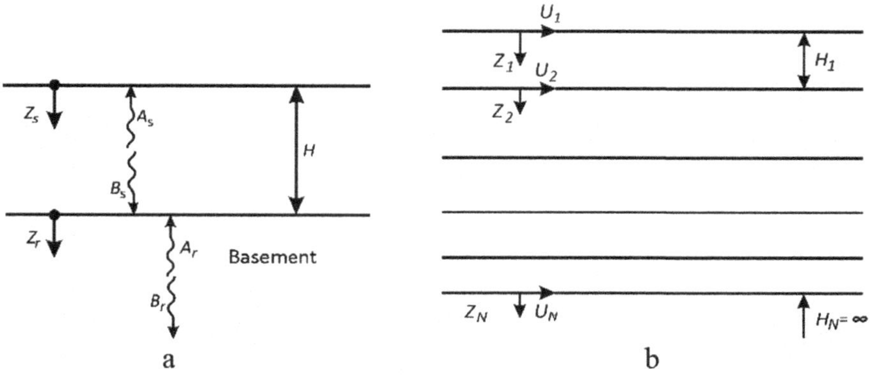

Fig. 2.3 **a**—A uniform damped elastic soil layer on an elastic half-space; **b**—a system of uniform damped elastic layers on an elastic half–space

In practice, the fundamental frequencies of the layer shaking usually vary from 0.2 Hz for thick layers to ~ 10 Hz or more for thin layers.

From the above formulas we can see that peak amplitudes are determined mainly by the attenuation and the contrast of the impedances of subsurface soils and the bedrock; to a lesser extent, they depend on the parameters of input motion, such as, types of seismic waves, incidence angles, etc.; for 2-D and 3-D structures, by their geometry (in cases of small attenuation, the dependence on the geometry becomes very important).

The results of experimental and theoretical studies indicate that maximum values of amplification usually lie in the range 6–10; in some cases (Mexico City, San Francisco), they reach ~ 20 (Bard 1995).

A system of uniform damped elastic layers on elastic half-space (Fig. 2.3b). Let's consider a system of N horizontal layers, which behavior is described by Kelvin–Voigt model. The general solution of the wave equation is described by the expression

$$u(z,t) = Ae^{i(\omega t + k^* z)} + Be^{i(\omega t - k^* z)}, \tag{2.1.7}$$

where the amplitudes A and B of seismic waves propagating up and down are interrelated by the recurrent relations:

$$A_{m+1} = \frac{1}{2} A_m (1 + \alpha_m^*) e^{ik_m^* x_m} + \frac{1}{2} B_m (1 - \alpha_m^*) e^{-ik_m^* x_m}$$

$$B_{m+1} = \frac{1}{2} A_m (1 - \alpha_m^*) e^{ik_m^* x_m} + \frac{1}{2} B_m (1 + \alpha_m^*) e^{-ik_m^* x_m}, \tag{2.1.8}$$

α_m^* is the complex ratio of the impedances at the layer boundaries m and $m + 1$. From the condition that shear stresses are zero on the surface, we obtain $A_1 = B_1$. If we apply the recurrent formulas sequentially to all the layers from the 1st to the m-th, we can express the wave amplitudes in all the layers through A_1 and B_1. The transfer function connecting the amplitudes in the i-th and j-th layers has the form:

$$F_{ij}(\omega) = \frac{|u_i|}{|u_j|} = \frac{a_i(\omega) + b_i(\omega)}{a_j(\omega) + b_j(\omega)}. \tag{2.1.9}$$

In this case the transfer function is much more complicated than in the case of a single layer on a half-space, because of multiple reflections in the layers. To calculate it, computer programs are used; one of the most famous is SHAKE (Schnabel et al. 1972) with an equivalent linear model of soil behavior under dynamic loading.

Application of the equivalent linear model to calculate soil response in strong ground motion is based on the assumption that the model provides a satisfactory approximation, if the dynamic properties of the soil are selected in accordance with the average strains occurring in the soil during strong motion. Since the calculated strain level in each layer depends on the equivalent linear parameters of the layer, iterative calculations are required to ensure that the parameters used in the calculations correspond to the strain levels in all the layers.

In conditions of sinusoidal dynamic loading, the hysteretic curves of such a viscoelastic material are ellipses, whose axes relate to each other as imaginary and real parts of the complex shear modulus. The equivalent linear shear modulus G is defined as the average shear modulus, and the equivalent damping h is defined as the damping providing the same energy losses in one loading cycle as the real hysteretic curve.

Deconvolution. Equation (2.1.9) shows that motion in any layer can be determined from known motion in any other layer. This allows deconvolution, i.e., calculation of motion at the boundary of the soil thickness with the bedrock based on known motion at the free surface; this can be very useful in many cases. Though deconvolution theoretically should provide a single solution for a linear elastic medium, some practical difficulties often arise. Some of them are related to the calculation process, especially in cases when strains are high and multiple iterations are required; the others are due to inaccuracies of the assumptions that all vibrations in the layers are caused by shear waves propagating vertically.

Silva (1988, 1989) showed that approximately 75% of the energy of motion on the surface (which corresponds to ~ 87% level of amplitudes) can be attributed to vertically propagating shear waves at frequencies up to 15 Hz, while the rest belong to scattered and surface waves. He proposed to perform deconvolution using pre-filtered (by low-pass filter with a boundary frequency of 15 Hz, to avoid unrealistically large accelerations at depth) surface records, apply iteration procedures and use 87%-, instead of 100%-, level of amplitudes of input motion. Silva also noted that deconvolution should be carried out with great caution, and the results should be carefully evaluated.

Since equivalent linear analysis allows deconvolution, which is practically important, approaches are being developed to increase the accuracy of the equivalent linear method (Yoshida and Iai 1998). Sugito et al. (1994) proposed application of frequency-dependent soil parameters in the calculations by introducing empirical amplification coefficients for each spectral component; thus, they achieved a better agreement between the calculated and observed high-frequency components. Though physical justifications of the proposed frequency dependence are debatable, sometimes the proposed procedures significantly improve the estimates. A computer program using this method is called FDEL. Suetomi and Yoshida (1996) clarify that only shear moduli are frequency-dependent, not damping.

Peak values of velocities and accelerations are closely related to the amplification of seismic waves in soil layers.

Not long ago it was believed that peak accelerations virtually do not experience the influence of sediments, but peak velocities noticeably increase on sediments (Aki 1988; Boore et al. 1989, etc.). However, other evidence has also appeared. For example, records obtained in Mexico City during the 1985 Michoacán earthquake showed that surface accelerations are approximately 4 times higher in sedimentary basin zones than in the nearby foothills.

As shown by Campbell (1989), in statistical analysis carried out earlier, the obtained estimates are probably biased, since strong motion records obtained on thick sedimentary layers were mainly used, and only a few records were made on thin-layered ($H < 15$ m) sediments, for which fundamental frequencies exceed 2–3 Hz. In these cases, peak accelerations may increase. Such thin-layered structures are widespread in mountain areas with high seismicity.

During some recent strong earthquakes (including the 1995 Kobe earthquake), a noticeable decrease of peak accelerations was observed on liquefied sandy soils (Chap. 3, Sect. 3.1).

Sugito et al. (1991) calculated amplification on sediments by comparing motion on sediments and on rock during the 1989 Loma Prieta earthquake and concluded that the amplification decreased to ~ 1, if peak acceleration increased to ~ 300 cm/s^2 and peak velocity to ~ 30 cm/s. Suetomi and Yoshida (1998) conducted similar analysis based on records of the 1995 Kobe earthquake. They clearly saw the tendency to decrease amplification on soils with increasing accelerations, and noted that on soft soils it happens faster.

The dependence of the amplification of seismic motion on soil conditions was studied by many research teams, and conclusions were made that nonlinear soil behavior significantly influences the amplitudes of motion on the surface, especially on soft soils.

At the same time, the amplification factor has a very large spread, which is obviously due to the nonlinear soil behavior; therefore, any dependence of the acceleration on soft soils on the acceleration on rock cannot be recommended for use in engineering practice, though such dependence may reflect some general trends.

Thus, peak ground accelerations on sediments cannot be predicted by simple recalculations of the corresponding peak accelerations on bedrock.

Comparison of records obtained on rock surface and on the boundary of the crystalline basement and sediments (Fig. 2.4) indicates that seismic records on rock are also amplified, i.e., they possess their own frequency-dependent amplification. Therefore, some seismologists believe that local site effects on rock should also be investigated by measuring profiling parameters, as was done for sediments (Archuleta and Steidl 2001).

Many researchers recognize *the intensity of seismic motion* as a more useful characteristic than peak accelerations or peak velocities. Several papers, starting from (Borcherdt 1970), have shown a rather high correlation between amplification of weak motion and local intensity of motion during earthquakes, so the intensity is probably a more universal characteristic of seismic hazard than peak acceleration.

Seismic observations show that in some cases the combined effects of a seismic source, propagation path and ground conditions can lead to abnormally strong motion on the surface. For example, Bard et al. (1988), Kawase and Aki (1989) conclude that the effects of thick sedimentary columns overlaid by thin clayey surface layers with horizontal variations in their thickness can explain the observed spatial variations of the amplitudes and duration of strong motion in Mexico City during the 1985 Michoacán earthquake.

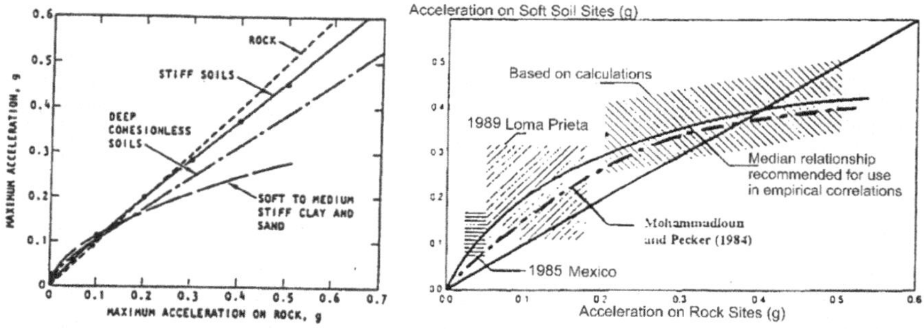

a b

Fig. 2.4 Relations between peak accelerations on rock and on soft soil: **a**—relations published in 1983 show a quasi-systematic decrease in peak accelerations on soft soils during strong earthquakes ($PGA > 0.1$ g) (from Seed and Idriss 1983); **b**—relations obtained in 1990 after the earthquakes in Mexico (1985) and Loma Prieta (1989) are compared with the relations obtained by Mohammadioun and Pecker (1984) in numerical simulations. *PGA* on soft soils increase to ~ 0.4 g (from Idriss 1990)

Since abnormally intense and/or prolonged seismic motions are of great interest for strong-motion seismology, regular studies of such cases are necessary. In Chap. 4, mechanisms of generation of abnormally high accelerations during the 2003 Tokachi-oki and 2011 Tohoku earthquakes are discussed.

When such effects are detected, a possible way to account for them in seismic zonation is the probabilistic approach, i.e., assessing the probability of such effects at certain sites.

In time domain, the influence of subsurface soils on seismic motion is seen from changes of the waveforms, peak accelerations, velocities and *duration of seismic motion,* which is most noticeable in cases of two-dimensional sediment structures. The interference of seismic waves captured in the waveguides, like lower-velocity layers or sedimentary basins leads to resonant effects, the parameters of which are related to the geometric and mechanical characteristics of the structures. Resonances are rather simple in one-dimensional cases (vertical resonance of body waves) and more complicated in cases of 2- or 3-dimensional structures. In frequency domain, the resonances produce spectral peaks, as shown in Fig. 2.5 for 1- 2- and 3-dimensional cases. Peak frequencies are related to the thicknesses of the layers and seismic wave velocities; in 2- and 3-dimensional cases they are also related to the horizontal dimensions of the structures.

To date, some number of strong motion records are accumulated that allow statistical conclusions about the relation between the duration of surface motion and ground conditions. Theodolidis and Bard (1995), Trifunac et al. (1999) and further observations report a significant increase in the duration of motion on sediments, especially at long periods. This issue is closely related to the effects of horizontal heterogeneities and diffraction of surface waves (Bard 1995). It was shown theoretically and experimentally that horizontal heterogeneities in sediments cause significant variations of the intensity of seismic motion at distances comparable to the

Fig. 2.5 Comparison of the response spectra of one-dimensional, two-dimensional and three-dimensional models for the case of an elliptical basin (from Aki and Irikura 1991). For two- and three-dimensional models, the vibrations were calculated in the centre of the basin

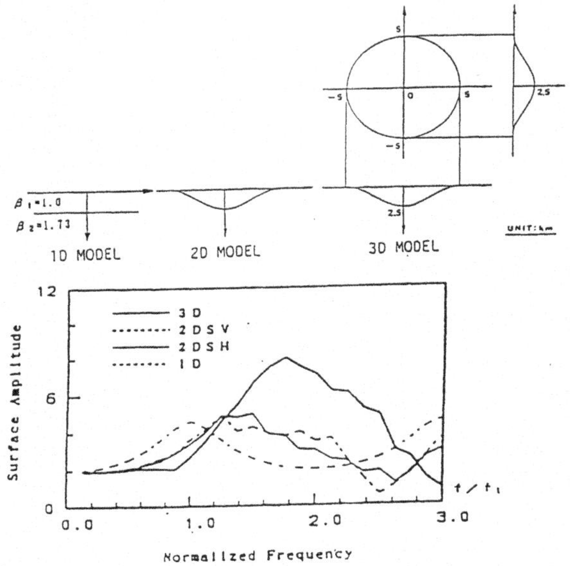

wavelengths (sometimes, even at tens of meters); therefore, they are important for engineering seismology.

The diffraction of seismic waves at internal boundaries and the influence of vertical boundaries are considered in (Bard 1995). In the 1980s, in a number of studies using numerical simulation methods, a rather strong influence of the topography of internal boundaries (like alluvial sedimentary basins) on seismic wave fields was established. At internal boundaries, local surface waves are generated, which are then captured by soft soil layers; as a result, amplification of seismic waves increases compared to simple one-dimensional case. However, there are still few experimental facts proving the existence of these surface waves; there are also few cases when one-dimensional models cannot explain the observed amplifications (Bard 1995).

Locally generated surface waves were observed in Santa Clara Valley in California (Frankel et al. 1991), where they caused long-duration motions during the aftershocks of the Loma Prieta earthquake. In Japan, Osaka area, the appearance of such waves during the 1995 Kobe earthquake depended on the azimuth of incoming waves (Pitarka et al. 1998); in Kanto basin near Tokyo, these phases were generated along the boundaries of the sedimentary basin and often were stronger than *S*-waves (Phillips et al. 1993; Hisada 1994).

The common feature of these cases is that the waves were recorded in large sedimentary basins, with the periods of shaking $T > 1$ s and reaching 6 s, the fundamental modes of surface waves are long-period, and their travel times from the boundaries to the center of the basin can reach tens of seconds. It is relatively simple to identify such phases by using data recorded at several stations.

For smaller structures with sedimentary deposits of less than several hundred meters in depth and less than ~ 10 km in width, the situation is different: since resonant frequencies are higher and travel times smaller, locally generated surface waves are superimposed on *S*-waves and become much more difficult to separate: this requires several dense seismic groups.

Diffracted waves were quite clearly identified at the Euroseistest site near the city of Thessaloniki (Greece) in a 5 km-wide graben with a sedimentary layer ~ 250 m thick (Bard 1995). Other cases when a one-dimensional model could not explain the observed amplification of seismic waves indirectly confirm the existence of locally generated surface waves, such as, in Ubaye Valley in French Alps (Jongmans and Campillo 1993) and in Marino district of San Francisco (Liu et al. 1992; Graves 1993). On the other hand, no clear evidence of 2- and 3-dimensional effects has been found at seismic test sites on the Turkish Plateau and in the Ashigara Valley, where one-dimensional models provide similar satisfactory results as the 2- and 3-dimensional models, which can be due to relatively high absorption in the medium, preventing the development and propagation of surface waves.

The results of the works by Caillot (1992) and Dimitriu et al. (1998) indirectly support the existence of diffraction effects: statistical analysis of strong motion records obtained in Italy and Greece showed that the duration of seismic motion is significantly longer at alluvial sites than on rock at frequencies $f < 2$–3 Hz. This is apparently due to the diffraction of low-frequency waves in thick sedimentary deposits (Bard 1995).

Similar to diffraction effects are the effects of amplification of motion in narrow (tens of meters) zones localized along strong lateral heterogeneities, i.e., in areas where softer material is adjacent to harder one (ancient faults, debris zones, etc.). Macroseismic observations describe significantly higher damage in such areas. The examples can be found in (Moczo and Bard 1993); it is shown that such heterogeneities also generate local surface waves in the softer medium. However, only few detailed studies have been conducted and few numerical models have been used to calculate such configurations.

The influence of surface topography on seismic wave field, 2- and 3-dimensional analysis of ground response. Instrumental data indicate that surface topography significantly affects the amplitudes and frequency content of ground motion; a review of such observations is given for example in (Geli et al. 1988). After earthquakes caused destructions in mountain areas it was often reported that buildings on hilltops suffered much greater damage than buildings located in lowlands. The examples are the 1909 earthquakes in Lambesc (France), 1976 Friuli and 1980 Irpinia (Italy) earthquakes, 1985 Chilean earthquake (Celebi 1987), and 1995 Kozani earthquake (northern Greece).

Observations of unusual surface effects in mountain areas, such as, fallen trees or dislodged large boulders that indicate vertical accelerations > 1 g, can be explained by the effects of surface topography and critical exit angles of *SV* waves (Nava et al. 1989; Umeda 1990).

Theoretical and numerical models predict amplification of seismic motion at the peaks, i.e. at "convex" areas, and deamplification at "concave" areas, such as valleys and foothills (Bard 1995). The size of these effects depends on the types, incidence and azimuth angles of incident seismic waves, as shown in (Pedersen et al. 1994). On hillsides, we may expect a complex distribution of areas of amplification and deamplification of seismic motion.

As noted in the review paper by Bard (1995), the topographic effects in earthquakes are stipulated by three main physical mechanisms:

(1) the dependence of the intensity of surface motion on the incidence angle (especially for *SV*-waves at incidence angles close to critical ones) (Fig. 2.6). Kawase and Aki describe this as a possible explanation for the damage distribution during the 1987 Whittier Narrows earthquake in California (Kawase and Aki 1990).

(2) focusing and defocusing of seismic waves reflected from the topographic surfaces. Sanchez-Sesma et al. (1995) theoretically studied the effect during the propagation of *SH* waves in a medium bounded by two intersecting planes. Considering the simplest situations defined by the angles and the numbers of repeatedly reflected waves (n–1), they found that the amplitude of motion at the vertex is 2n times higher than the amplitude of the incident wave, because there is no phase shift between the incident and reflected waves (Fig. 2.7).

(3) the diffraction of the body and surface waves reflected from the elements of topography and their interference with the incident waves. The effect was observed in Greece in field experiments with dense seismic array (Pedersen et al. 1994).

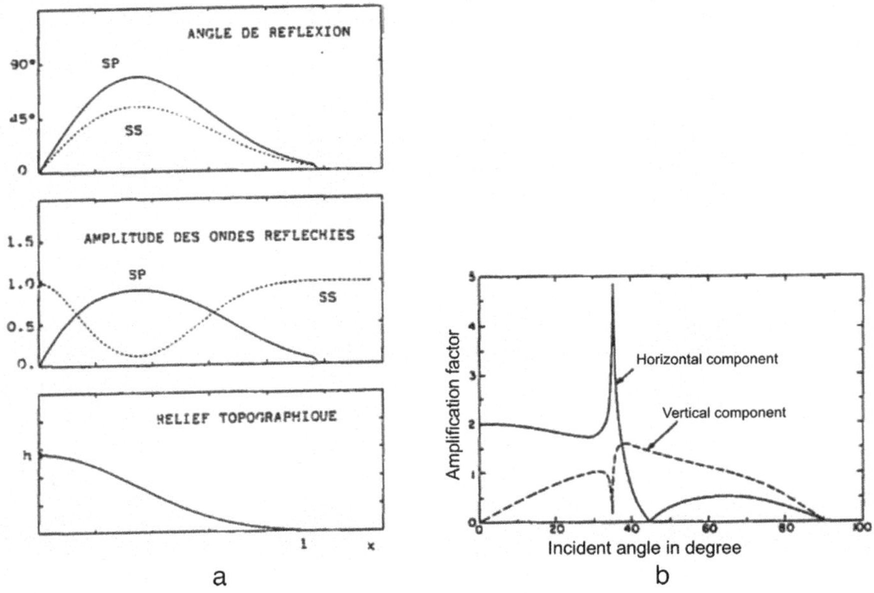

Fig. 2.6 **a**—The dependence of the reflection angle and the amplitudes of the reflected waves on the location of the receiver for the terrain shown below, and on the incidence angle for obliquely incident *SV* waves (from Bard 1995); **b**—the dependence of the amplification of motion on the incidence angle

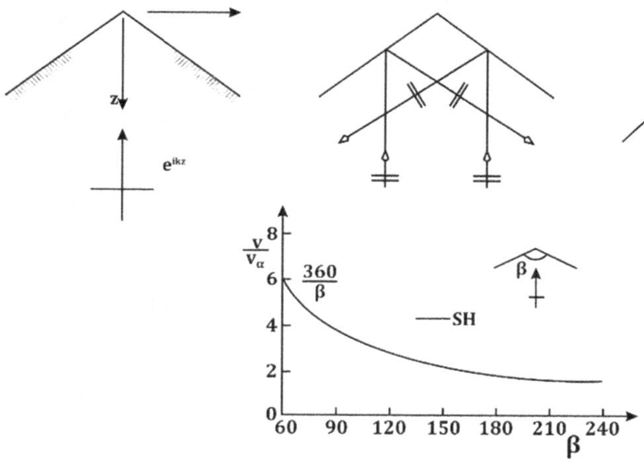

Fig. 2.7 Propagation of vertically incident *SH*-waves in a medium bounded by two intersecting surfaces (from Sanchez-Sesma 1990). If the wedge angle is $2\pi/n$, n different waves exist propagating in each point of the medium: an incident wave, a reflected wave and repeatedly reflected ($n \geq 4$) waves. All the waves interfere positively at the vertex, and the resulting amplitude is n times larger than the amplitude of the incident wave. The cases corresponding to $n = 3$ and $n = 4$ are shown, the last figure shows the resulting motion at the vertex as a function of the wedge angle

Geli et al. (1988) in their review work conclude that the mechanisms can be rather complicated, and as a result, it is difficult to describe topographic effects quantitatively. The observed manifestations of topographic effects qualitatively agree with the theory in the existence of amplification at the tops of hills and deamplification at valley bottoms. Amplification is larger on horizontal than on vertical components, and within two horizontal components, it is larger for the motion perpendicular to the ridge axis. Amplification is very roughly related to the "sharpness" of the topography: the steeper the slope, the higher the top amplification.

Amplification and deamplification of seismic motion due to surface topography are frequency-dependent; i.e., they are observed in certain frequency bands. The maximum effect corresponds to the wavelengths comparable to the horizontal dimensions of the topographic features (Bard 1995).

In terms of quantitative estimates, the situation is ambiguous. There are cases when small amplification was observed at the peaks, in the agreement with the results of numerical simulation (Rogers et al. 1974), but there are also many other cases when the observed amplifications are significantly higher (~ 10 times and more) than those predicted by the theory for complex two- and three-dimensional models. At the same time, significant variations of the amplitudes of motion along the slopes were observed: at distances of less than 200 m, with altitude differences of a few tens of meters, the amplitudes of motion can differ by an order of magnitude, which agrees with the previous observations of the earthquake damage.

P.-Y. Bard notes that detailed field studies with dense seismic arrays and geotechnical surveys should be performed to allow significant advances in the understanding of surface topography effects. These problems are important, because earthquake prone areas are often mountain areas, and villages were often built on elevations for safety; at the same time, topographic effects are closely related to landslides and rockslides often accompanying earthquakes and entailing great economic damage and human casualties (Bard 1995).

Definite manifestations of topographic effects are observed in Kamchatka region in Russia (Pavlenko 2013), where many strong-motion stations are installed on the hills. Figure 2.8 shows acceleration time histories of the subduction-type earthquake on June 8, 1993, recorded by stations of the strong motion network of Kamchatka: NKS (Nikol'skaya hill), PET (Petropavlovsk), MSN (Mishennaya hill), and their Fourier amplitude spectra.

Station PET is a reference base station of the Kamchatka strong-motion network. Based on soil conditions and seismic exploration results, it was classified as a rock site. Peak velocities and peak accelerations recorded at PET during earthquakes are usually 1.5–2.5 times smaller than those recorded at the neighboring stations.

The records of PET station were simulated as records on rock with local amplification estimated based on the results of shallow geophysical investigations and deep crustal seismic explorations in Petropavlovsk-Kamchatsky area (Balesta et al. 1988); it is quite similar to the amplification at "generic rock site" of the USA (Boore and Joyner 1997).

The acceleration time histories at stations NKS and MSN were simulated as at the rock sites with the local site amplification similar to station PET. Comparison of the waveforms and spectral compositions of the recorded and simulated acceleration time histories shows that at these sites the local site effects are large (Fig. 2.8).

We can assume that amplification of horizontal components at 0.5–10 Hz at stations NKS and MSN was related to the topographic effects, because both stations were located near the tops of the hills at elevations of 95 and 372 m above the sea level, respectively.

At NKS station we observed 5–7 fold amplification of horizontal oscillations at ~ 2–7 Hz with central frequency f_c ~ 4 Hz. These frequencies may correspond to oscillations of a structure with horizontal dimensions of ~ 200–400 m (approximate transverse size of the Nikol'skaya hill) composed of rocks with V_S ~ 800–1600 m/s (such V_S values were obtained in the shallow geophysical investigations at the PET site located at the neighboring hill).

The Mishennaya hill represents an amphibole-andesite extrusion with average V_S values higher than that of the Nikol'skaya hill. At MSN, we observed 5-fold to 9-fold amplification of horizontal oscillations at 1–3 Hz with central frequency f_c ~ 1.5 Hz, which corresponds to the oscillations of a structure with horizontal dimensions of ~ 1000–2000 m (size of the hill) composed of rocks with average V_S ~ 1500–3000 m/s.

Thus, at NKS and MSN sites oscillations with wavelengths comparable to horizontal dimensions of the hills were amplified, and at both sites the amplification was larger at horizontal components than at vertical ones.

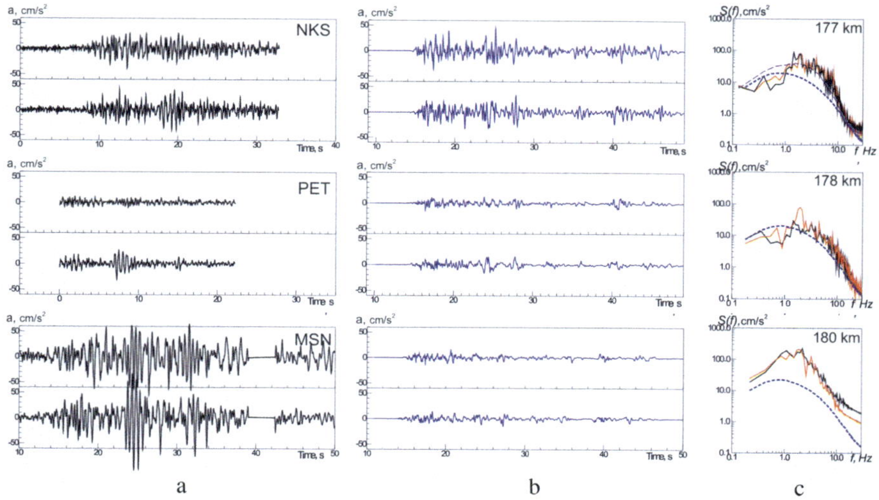

Fig. 2.8 Acceleration time histories of the earthquake on June 8, 1993: **a**—recorded by stations NKS, PET, MSN and **b**—simulated; **c**—the corresponding Fourier amplitude spectra: solid lines show spectra of the recorded NS and EW components, dash smooth lines indicate average spectra for the series of 250 simulated acceleration time histories. Epicentral distances are shown in the upper right corners of the spectra plots. For NKS station, thin dash line indicates the average spectrum simulated with the amplification factor estimated empirically based on records of the weaker earthquake on November 13, 1993

2.2 Observations and Studies of Nonlinear Soil Behavior in Strong Ground Motion

Nonlinear soil behavior in strong ground motion has been intensively studied during last decades, but still remains an urgent problem in engineering seismology. Long ago it was noticed that on soft sediments, earthquake records of different intensities qualitatively differ by their spectral composition and the waveforms. During strong earthquakes, nonlinear phenomena at soft soil sites were observed, such as, soil compaction and subsidence, or loosening and lifting, residual deformations and liquefaction. However, high intensity of motion does not always lead to noticeable manifestations of soil nonlinearity; the effect depends on the composition and thickness of soil layers, their saturation with water, and the groundwater level.

The development of seismologists' ideas about the behavior of soils during strong motion is described by Bard (1995). After the pioneer works of Hardin and Drnevich (1972), engineering seismologists have long been convinced of strong nonlinearities in the behavior of soft soils during earthquakes, characterized by reduction of shear moduli and increase of damping with increasing shear strains. From numerous laboratory tests with soil samples of different types, a sufficiently low strain threshold of ~ 10^{-4} was established, above which it was believed that changes in the elastic properties of soils began. The nonlinear properties of soils should have caused a significant decrease in the resonant frequencies of soil layers, in the amplification and peak accelerations, especially in the high-frequency range, due to increased damping.

At the same time, many seismologists believed that strong motion records can be interpreted with linear viscoelastic models up to 0.3–0.4 g; in some cases, up to 1.0 g (Murphy et al. 1971). They explained the differences in the waveforms of acceleration time histories during strong and weak motions by source and path effects, such as, abnormally strong reflections from the Moho boundary (Sommerville and Yoshimura 1990) and the directivity effects (Joyner and Boore 1988). Thus, till the late 80s–early 90s, there was a significant gap between seismologists and geotechnical engineers regarding the nonlinear soil behavior.

With accumulation of seismic observations, this gap gradually disappeared. During the earthquakes of 1985 in Mexico and 1989 in Loma Prieta (California), seismologists observed manifestations of nonlinearity (Aki and Irikura 1991), while geotechnical engineers came to the conclusion that effects of nonlinearity are important at higher strains and accelerations than they previously thought (Bard 1995).

In the 1970–1980s, it was difficult to find evidence of the nonlinear soil response in records of strong earthquakes, but in recent decades the databases on strong motion has been growing rapidly, and the nonlinearity in soil response can be identified for example, by characteristic wave forms in acceleration time histories—spikes (Archuleta 1998).

Difficulties in establishing nonlinear soil behavior were related to the fact that identification of nonlinearity requires understanding and taking into account the effects of the source, propagation path, and local site conditions. A good example is the 1994 Northridge earthquake (USA). Numerous recordings obtained on loose alluvial sediments indicate a decrease in the amplification during the mainshock compared to the aftershocks, evidently caused by significant nonlinearity of soil response (Field et al. 1997). However, another explanation, rather complicated has been proposed, based on a linear model of seismic wave propagation (O'Connel 1999).

Chin and Aki (1991), taking into account possible source and path effects, found nonlinearity of soil behavior in the epicentral zones of the 1989 Loma Prieta earthquake. According to Chin and Aki (1991), the nonlinearity of soil response was within the range predicted for sediments by the results of geotechnical engineering studies. A decrease in the differences in amplifications between rocks and sediments with an increase of accelerations was noted by various researchers (Sadigh 1983; Idriss 1985; Abrahamson 1988; Campbell 1981).

Instrumental observations and comparison of weak and strong motion records provided evidence of nonlinear soil behavior. In papers (Caillot and Bard 1990; Chang et al. 1991; Darragh and Shakal 1991; Beresnev et al. 1995), the authors report a decrease in the amplification of motion and in some cases, a decrease in resonant frequencies at soil sites when peak ground accelerations exceeded ~ 0.2 g. Vivid evidence of the nonlinearity of soil behavior is provided by recordings of the 1995 Kobe earthquake at Port Island vertical array site, where acceleration time histories show almost complete disappearance of high-frequency components of motion on the surface and quick attenuation of horizontal accelerations in the liquefied surface layers (Aguirre and Irikura 1997). In field studies of nonlinear soil response using real records of strong motion, the authors usually showed the dependence of the shear moduli and damping in soft soils on the level of strains.

At the same time, comparison of local site effects in weak and strong motion indicates their similarity in some regions: in San Francisco area, during weak motion seismic wave amplification in the frequency range of 0.5–2.5 Hz correlates well with the variations of the intensity of motion during the 1906 and 1957 San Francisco earthquakes and 1989 Loma Prieta earthquake (Borcherdt 1970, 1990; Borcherdt and Wentworth 1995). Similar results were obtained for sedimentary basins in Los Alamos in the range 0.1–5 Hz (Rogers et al. 1984, 1985) and in Garm in the range 0–50 Hz (Tucker and King 1984): amplification during weak motion correlates with that during strong motion (Rogers et al. 1984, 1985).

During the 1985 Michoacan earthquake in Mexico, the intensity of motion on soft soils above the lake bed was 8–50 times higher than at rock sites on the hills (Singh et al. 1988). The linearity of the ground response was confirmed by the coincidence of the predominant frequencies during weak and strong motions (Kobayashi and Midorikawa 1986).

A significant increase in seismic motion in San Francisco Bay area during the 1989 earthquake and records of the 1985 Michoacan earthquake (Singh et al. 1988) forced Idriss (1991) to revise the well-known Seed-Idriss ratio (Seed et al. 1976; Seed and Idriss 1983), linking peak accelerations on rocks and on soft soils (Fig. 2.4). According to the dependence proposed in 1983, the acceleration curves on soft soils and rocks intersected at ~ 0.13 g (Fig. 2.4a); in the new proposed dependence, this level was raised to ~ 0.4 g (Fig. 2.4b), which is consistent with the estimates of Mohammadioun and Pecker (1984).

Darragh and Shakal (1991) observed quasi–linear behavior of dense soils in the acceleration range 0.006–0.43 g. Bard concludes that, accounting for these observations, we can expect noticeable manifestations of nonlinearity of soft soils in cases, when peak accelerations on rock exceed ~ 0.1–0.2 g (Bard 1995). In the regions of moderate seismicity, peak accelerations achieve 0.1–0.3 g, and manifestations of nonlinearity are weak, i.e., seismic records on sediments usually show weakened high-frequency components (Bard 1995).

However, these conclusions about manifestations of soil nonlinearity should be revised in detail, as suggested by Borcherdt and Wentworth (1995), since they are based on few and probably unrepresentative observations.

According to many seismologists, saturation of soils with water is one of the most important factors determining the nonlinear soil response and soil liquefaction. The pore pressure built-up is the most likely mechanism for explaining the observed phenomena.

Aki and Irikura conclude (1991) that the effects of nonlinearity of soft soil response are quite widespread, which is confirmed by the following seismological observations: (1) systematic discrepancies in frequency-dependent amplification during weak and strong motion; (2) decrease of peak accelerations with increasing the intensity of motion; (3) results of processing records of strong earthquakes showing that peak accelerations are independent of local site conditions near the epicenter and become dependent on them at epicentral distances larger than 50 km (Aki and Irikura 1991).

Seismic observations deployed in Japan after the 1995 catastrophic Kobe earthquake using the network of seismic vertical arrays KiK-net provided seismologists with valuable research material and allowed a significant progress in studying soil response. Chapters 3 and 4 describe the behavior of soft soils in the near-fault zones of strong earthquakes. The approach proposed in Chap. 3, i.e., the construction of numerical models of soil behavior, allows us to clarify many disputable questions concerning the thresholds of strains, at which soil nonlinearity appears, reduction of resonant frequencies of soil layers at various intensities of motion, the influence of underground water, etc.

In Russia, field studies of nonlinear wave effects accompanying the propagation of intense seismic waves were initiated at the end of the 1970s. Field experiments with seismic vibrators were carried out at test sites to study nonlinear interactions of seismic waves (Chap. 5), generation of higher harmonics of the radiated frequencies and the constant components of seismic wave fields, seismic solitons, recurrence phenomena and other nonlinear wave effects (Alyoshin et al. 1981; Gushchin and Shalashov 1981; Lund 1983; Shalashov 1984; Groshkov and Shalashov 1986; Dimitriu 1988; Groshkov et al. 1991; Nikolaev et al. 1994). The aims of the experiments was the application of methods developed in nonlinear acoustics for studying nonlinear phenomena in the range of seismic frequencies, studying nonlinear seismic effects for their possible practical use for the construction of new types of receiving and transmitting antennas (Gushchin and Shalashov 1981), and development of new methods for seismic exploration, monitoring of the stress-state of the medium and stress accumulation in seismically active regions (Nikolaev 1987).

In the first experiments, the generation of combinational-frequency harmonics of frequencies radiated by powerful vibrators and accumulation of nonlinear distortions during the propagation of intense seismic waves were studied. The experiments were carried out at the test site Uznozh in Belarus (on soft sediments), and seismic vibrators SV-10/100 with peak force of 100 kN were used.

In one of the experiments, two vibrators separated by 0.5, 10, 25 and 100 m from each other worked synchronously at a distance of ~ 50 m (both) from a deep borehole, where the recording devices were installed at depths of ~ 500, ~ 1000, ~ 1500 and ~ 2000 m—to test the accumulation of nonlinear distortions during the propagation of intense body waves in the sediments. One of the vibrators radiated seismic waves at a frequency of 64 Hz, and the other radiated a sweep signal at frequencies 54–34 Hz, of 8 s duration. The processing of the records was aimed at detecting signals of frequencies 10–30 Hz in the recordings; to better separate these frequencies, a low-pass filter with a cutoff at 31 Hz was included in the recording circuit. The researchers could not detect the combinational-frequency waves and accumulation effects at the vertical component; however, at the horizontal components, a combinational shear wave was detected, possessing an apparent frequency slightly higher than 10–30 Hz. This wave was formed in the area adjacent to the vibrators, because it disappeared when the vibrators were spaced over distances of more than 10 m.

The absence of longitudinal combinational waves was explained by the fact that only a small part of the vibrator's power was spent on the formation of longitudinal waves, and a slight increase in the apparent frequency of the shear combinational waves was due to the fact that frequencies 10–30 Hz were not optimal for combinational waves (Program: Development ... 1981). As shown in Chap. 5, probably the main reason was the predominance of cubic, rather than quadratic nonlinearity in the ground response; combinational harmonics were probably generated during three-frequency interactions and precisely at horizontal components, not at vertical ones.

In the experiments investigating interactions of surface waves, seismic receivers were located at distances up to 100 m from the vibrator, and combinational-frequency waves, as well as the second and third harmonics of the frequencies radiated by the vibrator were recorded. Nonlinear distortions of the propagating waves reached tens of percent (Alyoshin et al. 1981; Solovyov 1987; Beresnev et al. 1987). Gushchin and Shalashov (1981) recorded nonlinear distortions of seismic body waves radiated by a rather small industrial vibrator. The second, third and fourth harmonics of the radiated frequency were recorded, and the parameter of quadratic nonlinearity of the medium (sandy soils) was estimated based on the amplitude of the second harmonic: $n = 2000$.

Effective parameters of the nonlinearity of ground response were determined as the ratios of the parameters of elasticity of the 3rd and 2nd, or 4th and 2nd orders (for quadratic and cubic nonlinearities, respectively). Complicated structure of the earth's materials, the presence of pores and inclusions of various compressibility and viscosity stipulate substantial difference between their nonlinear properties and the nonlinear properties of compact materials studied in nonlinear acoustics (Gushchin and Shalashov 1981). Parameters of quadratic nonlinearity can be estimated based on the dependence of seismic wave velocity on pressure (Zarembo and Krasilnikov 1966), and the conclusion was made that the parameter of nonlinearity increases with decreasing pressure (or the depth of the soil layer). At depths of more than ~ 10 km, the parameter of quadratic nonlinearity is $n \approx 10$; such values are typical for compact materials with low porosity (metals, liquids). At depths of several kilometers, n increases to ~ 100, at depths of less than 1 km, $n \approx 10^3$. High values of n in the upper layers are due to an increase of the porosity; abnormally high values of $n \approx 10^3$–10^4 are typical for subsurface weathered layers (Gushchin and Shalashov 1981; Shalashov 1984; Groshkov et al. 1991).

Abnormally high n values were explained by a model of nonlinear deformation of a heterogeneous elastic-viscous medium: the model phenomenologically accounts for a large number of inelastic viscous bonds with various relaxation times, connecting micro- and macroblocks of the medium. It was shown theoretically that in conditions of loading or unloading, softer bonds experience larger strains; as a consequence, the effective nonlinearity parameters significantly increase (Groshkov and Shalashov 1986). Thus, nonlinear corrections in state equations of geomaterials become comparable with the linear terms at strains of ~ 10^{-7}–10^{-4}, occurring in most cases of seismic wave excitations by the vibrators or strong earthquakes (Gushchin and Shalashov 1981). According to the estimates by Zimenkov and Nazarov (1993), nonlinear corrections should be taken into account at strains of ~ 10^{-8}; according to Van Den Abele and Johnson (1996)—at strains of ~ 10^{-7}.

Direct measurements and studies of the constant component of seismic wave field near the vibrator plate and at ~ 0.5–17 m from it were carried out by the researchers of the Schmidt Institute of Physics of the Earth (Moscow). A vibrator SV-10/100 was used as a source, and strain gauges (tensometers), i.e., cylindrical membrane sensors, as receivers. In an intense seismic field around the vibrator, the static (constant) component of the seismic wave field was found and measured; it is so called seismic-radiation stress σ_p indicating the ground extension around the vibrator. The constant components of intense seismic wave fields were also recorded by a sensitive strainmeter in field experiments at distances up to 500 m from the vibrator; their values being proportional to the squares of the vibrational velocity, as predicted by theory (Zarembo and Krasilnikov 1966). Alternating surface strains "extension—compression" that were recorded within ~ 500 m around the vibrator apparently related to the movement of the underground waters (Nikolaev et al. 1994).

In field experiments with small explosions, stress–strain relations were obtained in soft dry (humidity of ~ 10–15%) sandy loams, loams, loesses above the groundwater level (Vasil'ev 1987). The measurements were carried out at depths of 1.5–2 m, and it was found that the stress–strain relations did not reach the saturation, i.e., the horizontal asymptote of $\tau_{\max}$ (Fig. 1.2) at least, up to stresses of ~ 0.4–0.6 MPa. The weakest soils, such as deluvial loams and wet loams, at depths down to ~ 1 m behaved linearly up to ~ 0.05–0.1 MPa.

According to the estimates by Vasil'ev (1987), in soft dry soils at depths of ~ 2 m, shear stresses usually do not exceed 0.3–0.4 MPa during short–term explosions (which corresponds to ~ 9 ball units of seismic intensity). Such factors as saturation of soils with water, increase of the duration of motion during earthquakes compared to explosions, and superposition of the reflections from the boundaries of the layers can increase the stress level (Vasil'ev 1987; Zaitsev 1982, 1985).

The results presented in Chap. 3 allow the estimation of the linearity threshold of ~ 0.01 MPa for loose subsurface soils. In the near-fault zones of the 1995 Kobe and 2000 Tottori earthquakes stresses achieved 0.02–0.04 MPa in the upper meters and ~ 0.1 MPa at depths of more than ~ 30 m. Theoretical estimations of the response of elastic–plastic soil models (confirmed by experimental data) indicate that the increase of the intensity of input motion to soil layers leads to the decrease of the amplification of motion, shifting the spectral maxima to the low-frequency range, and the appearance of additional spectral maxima (Gvozdev et al. 1982).

In the field experiments, the difficulties in studying nonlinear wave effects were revealed. Some of them were related to the source, seismic vibrator. In the operation mode "monochromatic radiation", the vibrator radiated the main frequency signal along with its higher harmonics; at the same time, movements of the vibrator plate influenced nearby soils and changed their properties and the waveforms of the radiated signals. Also, seismic waves propagated in horizontally layered media, and the generated wave fields were rather complicated. In the experiments with prolonged intense vibrations, massive movements of the underground water occurred. Local soil conditions below the receivers varied from point to point. Thus, it was difficult to interpret the obtained results and draw unambiguous conclusions about the quantitative and qualitative characteristics of soil nonlinearity.

2.3 Methods for Calculation of Ground Response and Their Testing on Earthquake Records

Calculation of ground response in earthquakes is a classical problem in engineering seismology. We calculate the propagation of a vertically incident seismic shear wave in a system of horizontal soil layers bounded above by the free surface and below by a semi-infinite elastic medium representing the bedrock. The problem was solved by Kanai for the case of sedimentary layers with Kelvin-Voigt viscoelasticity. He has shown that the motion of a point on the surface of a sedimentary thickness is mainly determined by shear waves propagating vertically upwards from the underlying half-space (Kanai 1952). Based on this solution and the theory of multiple reflections, various procedures have been developed to calculate motion on the surface of soil thickness from motion on the underlying rock or nearby rock outcrops.

Since the 1970–80s, an "equivalent linear method" is used to calculate the soil response, but for input motion that is sufficiently intense to cause damage to structures, we cannot assume linear behavior over the entire range of strains: in this case, we would imply stresses many times higher than the strength of soil materials. To circumvent this difficulty, we approximate the soil response by the response of a linear model, whose properties are chosen in accord with the average strains that occur at each depth in the model during its seismic excitation.

The algorithms of the equivalent linear analysis were developed by Idriss and Seed (1968) and Schnabel et al. (1972). The calculations are performed in frequency domain iteratively. At the first step, trial values for average strains are chosen, then the soil properties (shear moduli and damping) are determined in accordance with the trial values of strains, and the response of the model is calculated. If the calculated strains substantially differ from the trial values, the cycle is repeated, while the properties of the medium are changed in such a way as to bring the calculated strains closer to the observed ones. This approach is implemented in SHAKE computer program—the first program for calculating soil response that has received wide practical application. After conducting test calculations, Idriss and Seed confirmed higher efficiency of the equivalent linear model compared to bilinear model in describing soil behavior (Idriss and Seed 1968).

Currently, equivalent linear models are used in many computer programs: SHAKE, QUAD-4, FEADAM, LUSH, FLUSH, FDEL, ERA, and others; all of them are based on the assumption that nonlinear soil response to dynamic loading can be approximated with satisfactory accuracy by the response of a linear elastic model with attenuation. Though the equivalent linear approach is convenient for calculations, allows deconvolution and at small strains provides acceptable solutions to many practical problems, it still remains an approximation to the real soil response, which is inherently nonlinear.

The observations show that equivalent linear models are hardly applicable for calculating soil response in strong earthquakes. Iterations allow the approximation of the nonlinear soil behavior to some extent; however, the properties and parameters of the soil are assumed constant throughout the earthquake, and changes in the rheological properties of the soil that actually occur during an earthquake are not taken into account.

An alternative approach is to evaluate the soil response by direct numerical integration methods in time domain. When integrating the equations of motion with a small timestep, we can use any linear or nonlinear stress–strain relation, or an advanced constitutive model. At each step of calculations, to determine the soil response at a given time moment, the selected stress–strain relation is applied.

In the calculations with nonlinear models of soil behavior, various researchers used various nonlinear stress–strain relations. The Ramberg–Osgood relation was used by Streeter et al. (1974), Constantopoulos (1973) and Faccioli et al. (1972). Papastamatiou used elastic–plastic ratios, etc.

Joyner and Chen proposed a method based on the representation of a sedimentary thickness as a system of horizontal soil layers, where the mass of each layer is lumped equally at the top and bottom (Joyner and Chen 1975). The essential part of the method is a rheological model suggested by Iwan (1967), which accounts for the nonlinear hysteretic soil behavior and has considerable flexibility for incorporating laboratory results on the dynamic behavior of soils.

In computer programs of nonlinear analysis of soil behavior, the stress-strain relations are represented by the following cyclic nonlinear models:

- the hyperbolic model—in programs DESRA, (Lee and Finn 1978), TARA (Finn et al. 1986; Finn and Yogendrakumar 1989),
- the Ramberg–Osgood model—in program CHARSOIL (Streeter et al. 1974),
- the Hardin-Drnevich-Kundall-Pike model (HDCP)—in program TESS1,
- Martin-Davidenkov's model—in program MASH (Martin and Seed 1978),
- Iwan's model—in program NONLI3 (Joyner and Chen 1975).

Some of the programs are based on the effective stress analysis (Finn et al. 1977; Ishihara and Towhata 1980) and allow the analysis of total stresses. The difference in the results of total and effective stress analysis is due to the difference in the mechanisms of dilatancy. If the dilatancy is absent or can be neglected, effective stresses experience only small fluctuations during an earthquake, while their average values remain approximately constant, and in the calculations we can use soil properties estimated from the initial stresses; this approximation is used in total stress analysis. In the calculation methods based on effective stress analysis, dilatancy is allowed; in this case, average effective stresses change monotonously, and the soil properties should be evaluated at each time moment depending on the current effective stress values.

The basic equation used in effective stress analysis was proposed by Biot (1941). The equation known as the "Biot equation" relates movements of the soil skeleton u, movements of water U (or water relative to the soil skeleton $w = n(U - u)$, where n is porosity), and the pore pressure p: u–U–p, or u–w–p. To reduce the number of unknowns and simplify practical applications, some assumptions are accepted. For applications in dynamic and consolidation analysis, the Biot equation was improved and modified by many researchers, and it is the base of computer programs that include various nonlinear and constitutive models.

Programs DESRA (Lee and Finn 1978), TARA and their modifications (Finn et al. 1986; Finn and Yogendrakumar 1989) are based on the assumption that soil behavior is elastic at small loading increments, but shear moduli change with each increment. The programs allow estimation of the possible pore pressure level and the probability of soil liquefaction, and they can be used in calculation of soil behavior after liquefaction. Pore pressure changes are associated with volumetric strains in the soil in drained conditions, when diffusion is taken into account.

Programs DYSAC2, DYNAFLOW and SWANDYNE, based on advanced constitutive models, account for sliding surfaces and are considered as the most accurate (Arulanandan et al. 1995). The equations of motion of liquid and solid phases are related to the equation of matter conservation; generation and dissipation of the pore pressure are defined by deformations of the solid skeleton according to the Biot formula (Biot 1956).

Various methods can be used to integrate the equations of motion: the so-called implicit and explicit techniques, integration along characteristics, and others. The implicit technique is described, for example, in (Chen and Joyner 1974). It is based on the approach used by Idriss and Seed (1967) to calculate the response of a system of sedimentary layers possessing bilinear stress–strain relations: the soil profile is divided into N layers, and the mass of each layer is lumped equally at the top and bottom. The equations of motion are written as a system of $N + 1$ equations, which are solved at each time step. For large N, calculations demand significant time, and for this reason most existing computer programs currently use an explicit integration method, which is also called the finite difference method.

Many researchers compared the results of strong motion simulations using equivalent linear and nonlinear analysis (Joyner and Chen 1975; Martin and Seed 1978; Iwasaki et al. 1980; Kokusho 1982; Dikmen and Ghaboussi 1984, etc.). For the same ground conditions and motions at the basement, both approaches, i.e., calculation of ground motion for a system with lumped masses (nonlinear analysis) and with equivalent linear models provide close estimates of motion on the surface. Nevertheless, the authors conclude that the results of nonlinear analysis better agree with the observations than the results of equivalent linear analysis.

Linearity as the internal feature of the equivalent linear model leads to the appearance of resonances. The soil rigidity is changing during an earthquake, and high amplifications predicted by equivalent linear models are not observed. The application of effective strains in the equivalent linear analysis leads to "under–damping" or "over–damping", i.e., generally, to wrong assessment of the strain levels (Yoshida and Iai 1998). At the same time, for the equivalent linear analysis we should know only the dependences of elastic moduli and damping on strain $G(\gamma)$ and $h(\gamma)$, while the nonlinear analysis requires knowledge of stress–strain relations and parameters of soil layers obtained from field experiments or laboratory tests.

Differences in the calculations with equivalent linear and nonlinear analysis depend on the degree of nonlinearity of the soil response. If the strains are rather low (in cases of soils with high elastic moduli and/or weak input motion), both methods provide close estimates. If the strains are large enough and approach the strength limit, then nonlinear methods are preferable.

Both equivalent linear and nonlinear methods have been successfully used in calculations of soil response. Interpretation of the results of calculations requires knowledge of the assumptions of each method, understanding of its operation and limitations. None of the methods can be considered as mathematically rigorous, because the accuracy of calculations is limited by the varieties of engineering and geological conditions, approximate knowledge of soil properties and uncertainties in experimental results defining input parameters (Yoshida and Iai 1998).

Methods for two- and three-dimensional analysis. One-dimensional calculations are successfully applied at flat or nearly flat areas with horizontal crystalline basements. Such conditions can be found everywhere, and in practice, one-dimensional analysis is usually used. However, as shown in Sect. 2.1, in some situations one-dimensional approximations are unacceptable, and two- or three-dimensional analysis of soil response is necessary. At the same time, S-wave approximation is also insufficient, and we should consider the propagation and conversions of $P - SV$ waves. Two–dimensional analysis is also used to calculate soil-structure interactions (Kramer 1996).

Two- and three-dimensional analyses use calculation algorithms in frequency domain (equivalent linear approach) and in time domain (nonlinear approach). Finite element method is often used, when a medium is represented as an ensemble of discrete elements whose dimensions are determined by a number of nodal points; in this case, the response of the medium is described by a set of responses at the nodal points (Kramer 1996).

For both equivalent linear and nonlinear models, estimation of the soil response is influenced by the sampling method. Using coarse mesh with large cells in the finite element method leads to filtering out high-frequency components. We should remember that the maximum size of the element is limited to 1/8–1/5 of the smallest wavelength that needs to be taken into account in the analysis. To reduce the counting time, the smallest allowable number of nodal points is taken. In practice, minimization of the number of the elements often means minimization of the size of the studied area, but decreasing the size of the area means increasing the influence of boundary conditions.

The boundary conditions are specified in such a way as to correctly simulate the wave attenuation at the boundaries of the studied region. For example, in cases when the boundaries of soils with harder rocks or the basement are far from the studied area, the energy of propagating seismic waves is removed from the studied area.

Boundary conditions of three types are usually used: (1) stresses and strains at the boundaries are zero; (2) part of the energy is absorbed at the boundaries, or (3) all the energy of incident waves is absorbed at the boundaries (Kramer 1996).

Two-dimensional equivalent linear analysis is carried out in the same way as one-dimensional analysis. A studied system of sedimentary layers is described by a two-dimensional model, finite element method is used, the input motion is represented by Fourier series, and the equations of motion are solved for each spectral component. For individual elements, iterations are carried out in order to find soil parameters corresponding to the strains.

The most famous computer program of that type is FLUSH developed by Lysmer et al. (1975). An example of calculation of the response of an earth fill dam using the equivalent linear model is given in (Kramer 1996).

Two-dimensional nonlinear analysis is used to calculate the behavior of hill-sides, building's foundations and structural elements. Integration of the equations of motion in time domain and the finite element method are used. In the well-known program TARA-3 (Finn and Yogendrakumar 1989), a hyperbolic model with the extended Meising rules and pore pressure generation mechanism are adopted to the two-dimensional case. The simplicity of the hyperbolic stress–strain relation allows a higher computation speed compared to more complicated models; at the same time, the relation reflects all the features of soil response.

The program DYNAFLOW (Prevost 1981) uses an advanced constitutive model, and an agreement with the observations was obtained in calculating the response of earth dams. In the calculations of strains and pore pressure, numerous sliding surfaces are taken into account. A consortium of Japanese Construction Firms has developed the program DIANA, which allows static and dynamic two-dimensional analysis with various advanced constitutive models, and other similar programs have also been developed (Kramer 1996).

The disadvantages of two-dimensional equivalent linear analysis are the same as in the one-dimensional case: an excessive amplification and difficulties in estimations of the effective parameters of the medium. Various modes of motions due to additional degrees of freedom in the two-dimensional case complicate the calculations of maximum strains and require the use of some other parameters (Poisson's ratio, for example) in addition to shear moduli, which results in more complicated loading paths.

As in the one-dimensional case, the equivalent linear approach is limited to the analysis of total stresses and cannot account for effective stresses. Generally, two-dimensional nonlinear models possess more possibilities for calculating pore pressure and consequently, effective stresses and strains. However, the accuracy of the estimation of these parameters depends on the relevance of the applied models of soil behavior.

Three-dimensional analysis of soil response is carried out in cases when two-dimensional analysis does not provide a good enough approximation, i.e., when soil conditions change in three directions, or we calculate the response of a three-dimensional structure. Three-dimensional calculations are performed similarly to two-dimensional ones. The finite element method, equivalent linear and nonlinear approaches are used.

The procedures for three-dimensional calculations using equivalent linear models are almost identical to two-dimensional cases, except that three-dimensional models have a larger number of nodal points with more degrees of freedom. For such calculations, computer program TLUSH is often used, which is a version of FLUSH program expanded to three-dimensional case.

Three-dimensional calculations of nonlinear models are also carried out similarly to two-dimensional calculations. Program DYNAFLOW has the ability to calculate three-dimensional models. It has been successfully used for three-dimensional (as well as two-dimensional) calculations of the response of earth dams. The possibilities and limitations of various approaches in three-dimensional analysis are similar to those of two-dimensional analysis (Kramer 1996).

Aki and Irikura (1991) describe some cases of 2-D and 3-D modeling of the response of sedimentary structures. Following the classification by Shinozaki (1988) and Horike et al. (1990), numerical methods for calculating seismic response of irregular structures have been developing since the 1970s in seven directions, such as, wave function expansion methods (Trifunac 1971; Sanchez-Sesma 1985; Lee 1988), finite element methods (Lysmer and Drake 1971; Zama 1981), finite difference methods (Boore 1972; Virieux 1984), discrete wave number methods (Aki and Larner 1970; Bouchon and Aki 1977; Bard and Bouchon 1980a, b, 1985), boundary integration methods (Wong and Jennings 1975; Sanchez-Sesma and Esquival 1979; Dravinski 1983), discrete wave number and boundary element methods (Bouchon 1985; Kawase 1988), and ray and beam methods (Hong and Helmberger 1978; Lee and Langston 1983; Nowack and Aki 1984; Ihnen and Hadley 1986). Most of the methods are still under active development, as each has its own advantages and can be effectively applied for solving certain classes of problems. The results obtained by the finite difference method by Vidale and Helmberger (1988) and Yamanaka et al. (1989) show excellent agreement between the simulations and observations. Combination of a 2-D structure and a 3-D source was calculated by Khair et al. (1989) using a hybrid method of boundary integration and finite elements.

The Aki-Larner method for calculation the response of two-dimensional structures was used by Horike (1988) to simulate seismic motion at frequencies of ~ 1 Hz in Kyoto sedimentary basin during the 1984 Nagano earthquake. The calculated motions were not intense enough to explain the observed ones, and Horike et al. (1990) extended the Aki-Larner method to the 3-D case; it was shown that the maximum amplification in the 3-D case can be almost 2 times higher than in the 2-D case.

2.4 Models of In Situ Soil Behavior Based on Vertical Array Records

To date, many computer programs have been created to calculate the ground response in various engineering and geological conditions; however, in every case, certain simplifications and assumptions are applied concerning the soil properties and behavior. The accuracy of the calculations is verified in the experiments with shaking tables or by analyzing in situ earthquake records.

Records of seismic vertical (borehole) arrays are the best tool for testing various methods for calculating soil response. Acceleration time histories of strong motion recorded at various depths represent real vibrations in soil layers, including incident and reflected waves. Observations of this kind have been carried out in Japan, the USA, and Taiwan.

The first studies, in which strong motion records made by surface and borehole arrays are compared with the calculations of soil response are described in (Saada and Bianchini 1987; Midorikawa 1992; Ishihara et al. 1998; Iai et al. 1993; Arulanandan and Scott 1993) and others. The results of simulations of soil response during strong earthquakes at first vertical array sites at the experimental test sites Chiba, Shin-Fuji (Japan), Lotung (Taiwan), Wildlife and Treasure Island (USA), vertical arrays of Kansai Energy Company (Japan), Port Kushiro and Port Island (Japan) were analyzed. Below, the results of these studies are briefly described.

IIS Chiba Experiment Station and Shin-Fuji Electric Substation site have been operating since the late 1970s. The vertical array Chiba of IIS (Institute of Industrial Science) consists of 4 three-component accelerometers installed at depths of 1, 5, 10 and 40 m; the vertical array Shin-Fuji consists of 2 accelerometers installed on the surface and at a depth of 28 m. Subsurface soils are sands and clays with S-wave velocities $V_S \sim 125$–140 m/s; below velocities increase to ~ 420 m/s at 40 m in Chiba and to ~ 780 m/s at 28 m in Shin Fuji. The arrays did not record strong earthquakes (maximum recorded strains are ~ 0.1–0.4%); the records were simulated using programs SHAKE, FDEL and programs of nonlinear analysis with Ramberg–Osgood and Yoshida-Ishihara models (Fig. 2.9). A rather good agreement was obtained between the recorded and calculated acceleration time histories (nonlinearity of soil response was not high) (Yoshida and Iai 1998).

Technical Research Center, Kansai Electric Power Co. Seismic devices are located in Amagasaki City, east of Hyogo Prefecture, Japan. Figure 2.10 shows the locations of three vertical array sites, such as, SGK, TKS and KNK, in the vicinity of Osaka Bay. The accelerometers are installed at depths of 0, 24.9 and 97 m (SGK), 0, 25 and 100 m (TKS) and 0, 25 and 100 m (KNK). Peak accelerations recorded by SGK array during the Kobe earthquake in January 1995 exceed 600 cm/s^2. Before the earthquake, the soil profiles were studied by seismic logging (Fig. 2.11).

Records of the Kobe earthquake were simulated by DYNEQ program of equivalent linear analysis, but no agreement of the calculations and observations was obtained (Fig. 2.12) (Yoshida and Iai 1998).

The calculations were repeated by correcting (1) the groundwater level at SGK (assuming it closer to the surface); (2) increasing S-wave velocity in gravel layers at depths of 3–7 m. However, the accelerations calculated on the surface were significantly lower than those recorded (Fig. 2.12b); this was explained by the nonlinear behavior of the gravel layer.

After the earthquake, measurements of S-wave velocities were repeated down to depths of 100 m (Soeda et al. 1999) and the soil response was calculated again using SHAKE program (Figs. 2.13 and 2.14).

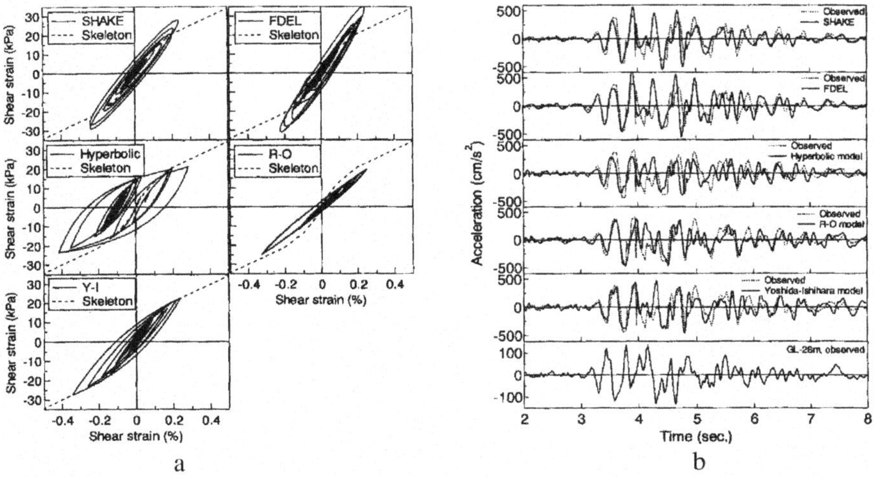

Fig. 2.9 **a**—Stress–strain relations used in calculations of soil response (R–O—Ramberg–Osgood model, Y–I Yoshida-Ishihara model); **b**—the corresponding calculated and recorded acceleration time histories (from Yoshida and Iai 1998)

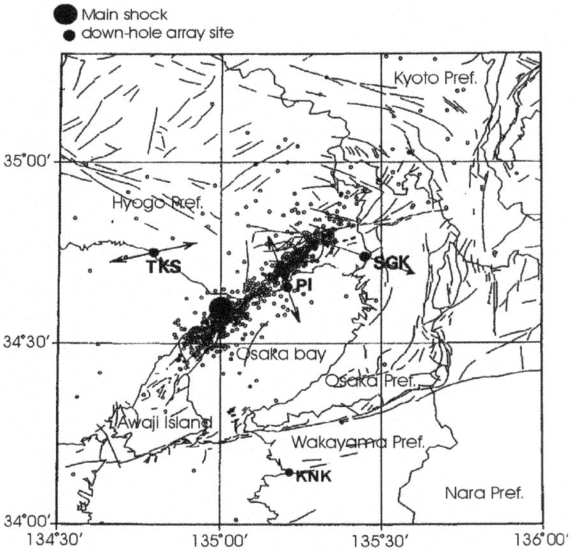

Fig. 2.10 The locations of seismic vertical arrays Port Island (PI), SGK, TKS, and KNK in the vicinities of the Osaka bay and the epicenters of the mainshock and aftershocks of the Kobe earthquake of January 17, 1995 (from Sato et al. 1996)

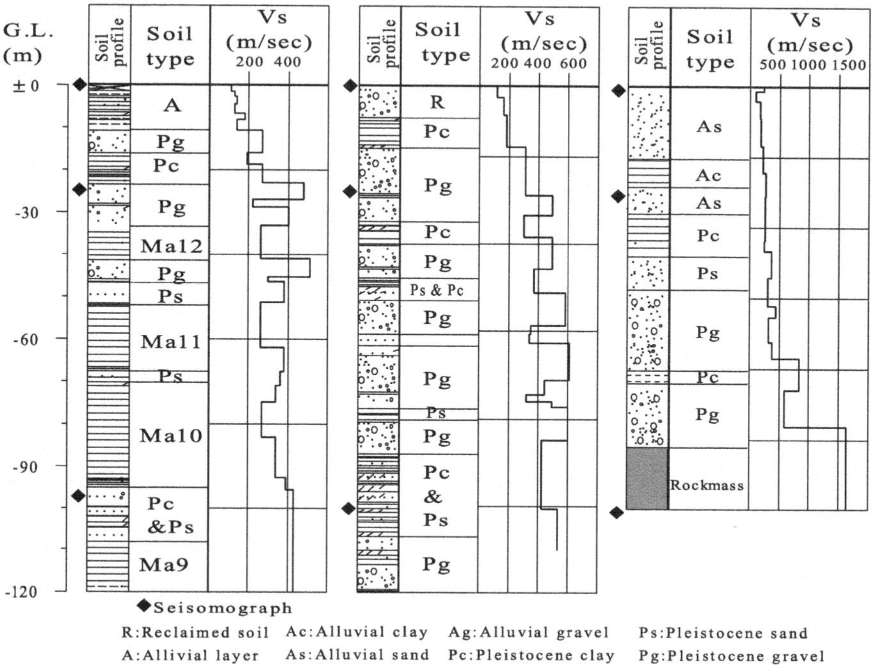

Fig. 2.11 Soil profiles and *S*-wave velocities at SGK, TKS and KNK vertical array sites (from Soeda et al. 1999)

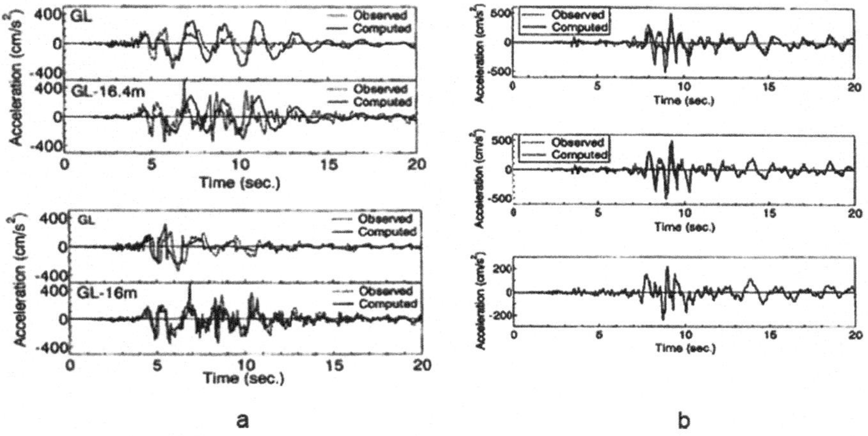

Fig. 2.12 Acceleration time histories of the Kobe earthquake recorded at Port Island (**a**) and SGK (**b**) sites and calculated with: **a**—equivalent linear analysis (top), effective stress analysis (bottom); **b**—calculations using parameters of the soil profile measured before (top) and after (middle) the earthquake and observed record at GL-24.9 m depth (bottom) (from Yoshida and Iai 1998)

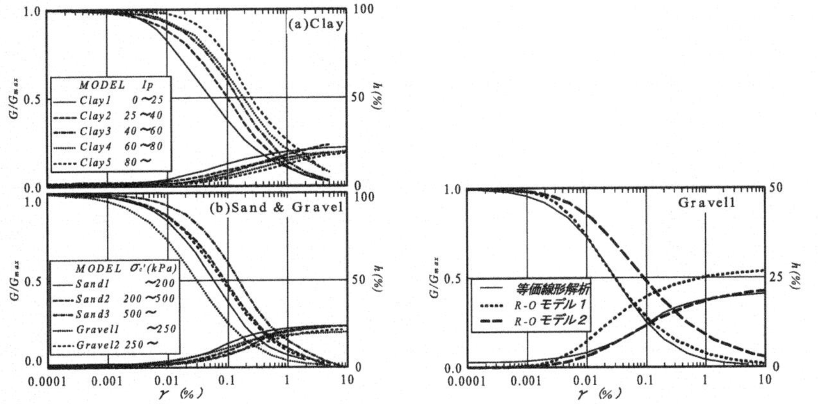

Fig. 2.13 The relations $G(\gamma)$ and $h(\gamma)$, used for the equivalent linear analysis of records of the 1995 Kobe earthquake at SGK, TKS, and KNK (from Soeda et al. 1999)

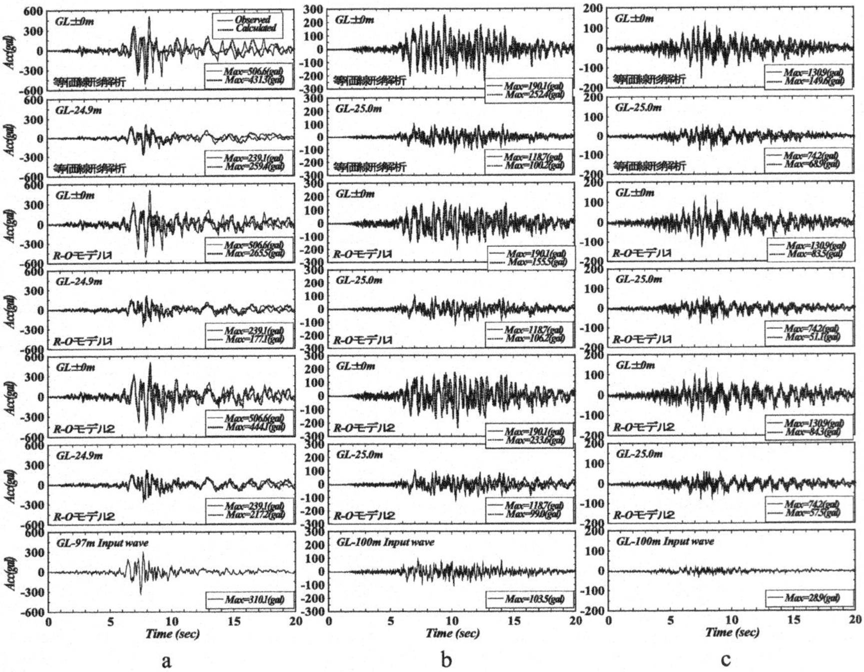

Fig. 2.14 Comparison of the recorded and calculated accelerograms of the Kobe earthquake at SGK (**a**), TKS (**b**) and KNK (**c**) vertical array sites. Calculations were performed by equivalent linear analysis (above) and nonlinear analysis with two versions of the Ramberg–Osgood model (below) (from Soeda et al. 1999)

The authors found better coincidence of the recorded and calculated accelerograms and noted the importance of knowing the exact values of *S*-wave velocities in the soil profile (Soeda et al. 1999).

Port Island. Acceleration time histories of the 1995 Kobe earthquake obtained by Port Island (PI) vertical array at the artificial land of Port Island represent the most thoroughly studied records that characterize the nonlinear soil response. Figure 2.10 shows the location of Port Island site in the near-fault zone; the profiling data and the accelerograms are described below, in Chap. 3.

PI vertical array consists of 4 accelerometers installed on the surface and at depths of 16, 32, and 83 m. Acceleration time histories recorded at different depths significantly differ from the surface records; the most noticeable differences are seen in subsurface layers at 0–16 m; whereas, velosigrams and seismograms look rather similar at all depths, with a slight amplification on the surface. Below 32 m, the motions are close to linear, while the nonlinearity is seen in the upper layers, especially in the upper 16 m.

Yoshida and Iai note that no amplification is observed at horizontal components, obviously due to the nonlinear behavior and liquefaction of the upper layers; whereas, vertical components are amplified. The results of simulations of the soil response with equivalent linear analysis and effective stress analysis (Yoshida and Iai 1998) are presented in Fig. 2.12a. The authors draw the following conclusions:

- significantly nonlinear behavior occurred in the Holocene clay layer at 17–28 m (strains ~ 3–5%); as a result, motions in the overlying artificial layer smoothed;
- equivalent linear analysis provides an agreement of the calculations and observations on the surface, but looking at records at 16 m, we can conclude that the shear modulus used for the calculations in the clay layer is underestimated, and the shear modulus in the overlying layer of artificial soil is overestimated;
- effective stress analysis provides significantly better agreement of the recorded and calculated accelerograms than the equivalent linear analysis;
- the occurrence of cyclic mobility of soils made waveforms complicated.

Kazama et al. (1998) calculated stress–strain relations at depths of 4, 20 and 54 m (Fig. 2.15) at PI site and showed that elastic moduli estimated from these relations are consistent with the elastic moduli obtained during geotechnical studies at this site (Kokusho and Matsumoto 1998). According to their estimates, the soil liquefaction began 5–7 s after the beginning of strong motion, and a significant decrease of shear moduli occurred in the upper layers. Studies of the aftershock records showed that reduced values of shear moduli still persisted a few minutes after the mainshock, which is obviously due to the gradual dissipation of excess pore pressure. Aftershocks recorded a month after the mainshock indicated the restoration of the elastic moduli in the soil layers, however, it is not clear when this happened.

Aguirre and Irikura performed spectral analysis of the mainshock and weak seismic events that occurred before and after it (Aguirre and Irikura 1997). Significant variations were found in the ratios of spectra of motion on the surface and at different depths at the beginning of strong motion and during the liquefaction of the upper

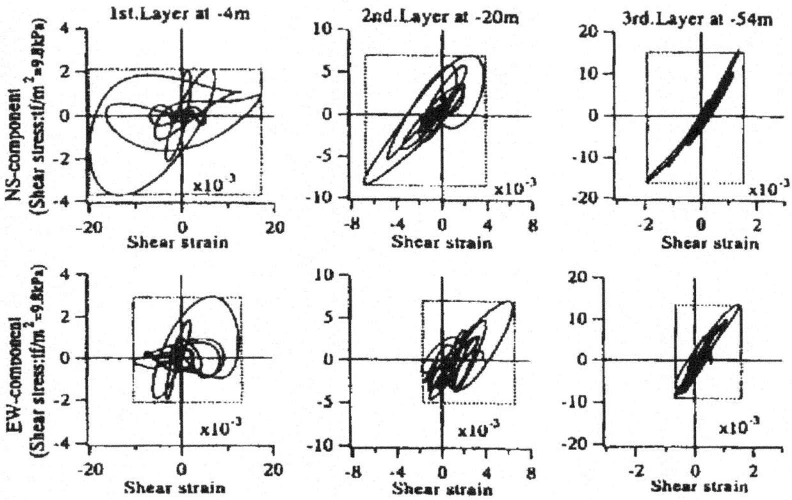

Fig. 2.15 Stress–strain relations in soil layers during the 1995 Kobe earthquake obtained from records of Port Island vertical array site (from Kazama et al. 1998)

layers. Spectral ratios calculated before and after the earthquake markedly differ from each other; after the earthquake, the peak frequencies of the ratios decreased. *S*-wave velocity profiles were inverted to determine the velocity structure before, during, and after the mainshock; it was found that at depths of 5–16 m, the velocities decreased by ~ 20% as a result of the earthquake; the liquefied state of the upper layer continued for at least 3 h, but not more than 24 h. The rigidity of the soil decreased close to zero when liquefaction happened and then gradually increased following a trend that resembles a consolidation curve. Significant nonlinearity of the soil response during the mainshock yielded a large reduction of the horizontal surface ground motions, so that the observed horizontal peak accelerations were only about 25% of the peak accelerations expected from the linear theory; on vertical components, the nonlinearity was not significant (Aguirre and Irikura 1997). In Chap. 3 we will return to the analysis of these records.

Lotung Experimental site. Surface and vertical seismic arrays were deployed on the east coast of Taiwan in Lotung area, and records of earthquakes with magnitudes 4.5–7.0 and with peak accelerations of 30–210 cm/s^2 were obtained (maximum strains did not exceed ~ 0.06%). Based on these records, shear moduli depending on strains and nonlinear stress–strain relations in soil layers were estimated. Calculations of the acceleration time histories using the equivalent linear analysis, programs SHAKE and FDEL, gave good results.

Wildlife test-site (USA). Seismic vertical array Wildlife recorded accelerations during the earthquake that occurred on November 27, 1987 ($M_w = 6.8$); at the same time, pore pressure was recorded. The soil profile down to ~ 10 m consists of alternating layers of silt, silty sand and silty clays with a gradual increase in *S*-wave velocities with depth. Approximately 10 s after the beginning of strong motion, pore pressure began to increase; at the same time, low-frequency components noticeably increased in the acceleration time histories on the surface and at a depth of 7.5 m, and spikes appeared. The stress–strain relations in the layer 0–7.5 m were estimated by calculating separately strains and stresses (Zeghal and Elgamal 1994); they resembled stress–strain relations that describe soil behavior under cyclic mobility.

The obtained pore pressure records allowed effective stress analysis, and the calculations showed an increase in the excess pore pressure, which was close to the observations. Equivalent linear analysis was used to calculate the soil response, but it did not show such good agreement with the observations as effective stress analysis (Keane and Prevost 1989).

Kushiro Port. The vertical array includes two accelerometers, one is installed on the surface, and the other at a depth of 77 m; the soils are dense water–saturated sandy deposits. The array recorded the 1993 Kushiro-oki earthquake of $M_w = 7.8$; peak accelerations at the horizontal components reached ~ 470 cm/s^2 on the surface and ~ 210 cm/s^2 at 77 m. Oscillations with a period of ~ 1.5 s and sharp peaks—spikes are clearly seen in the surface records.

The soil response was modeled using effective stress analysis taking into account dilatancy (Iai et al. 1995); soil parameters were determined in laboratory experiments with frozen in situ samples. A good agreement between the calculations and observations was obtained, and it was concluded that the maximum shear strains caused cyclic mobility of sandy soils that occurred at depths of less than 10 m.

Treasure Island is an artificial island in San Francisco Bay, the upper part of which is composed of man-made fill. Seismic instruments installed on the island recorded the 1989 Loma Prieta earthquake, and a significant amplification of seismic motion was observed on artificial soils compared to nearby rock outcrops.

Calculations were carried out using SHAKE program, where motion recorded at rock outcrops was used as input motion to fill layers. The results were evaluated by comparing spectra of the calculated and recorded accelerograms. A sufficiently good agreement was obtained for EW component, and unsatisfactory agreement for NS component. Nonlinear analysis gave a better agreement between the calculations and observations (Seed et al. 1992), and effective stress analysis showed a better agreement than total stress analysis (Finn 1993).

Analyzing the problem of predicting nonlinear soil response, Yoshida and Iai came to the following conclusions (Yoshida and Iai 1998):

- nonlinear characteristics of soils, such as, the dependences of shear moduli and damping on strains $G(\gamma)$ and $h(\gamma)$ estimated from earthquake records, are consistent with similar characteristics obtained in laboratory;
- motion on the surface can be estimated with sufficient accuracy, if an appropriate calculation method is applied and soil characteristics are known;
- effective stress analysis provides more accurate estimates of soil response than total stress analysis, and nonlinear analysis provides more accurate estimates than equivalent linear analysis;

- equivalent linear analysis may be applicable at strains < 0.5%, but it is not applicable at larger strains and in cases of soil liquefaction;
- amplification of accelerations on sediments is observed for weak input motion; as input motion increases, de-amplification occurs.

Yoshida and Iai (1998) propose to conduct further research in the following directions:

- the index or standard test method is required that characterizes soil behavior at strains > 1%, when soils show degradation behavior (there are few data from laboratory or field experiments in this range);
- to study in detail transformations of seismic waves in soil layers, we should collect records of strong earthquakes, especially records with simultaneous recordings of pore pressure, and dense vertical array data;
- relevant index of ground motion is necessary to evaluate the relationships between ground motion and structural damage due to earthquakes. Peak acceleration is widely used as such a parameter, but obviously, this parameter is not perfect. According to some studies (for example, Iwata et al. 1991), the correlation of the earthquake damage with peak velocity or spectral intensity is higher than with peak acceleration or displacement; perhaps it depends on the structure type.

Calculations of the soil response using various computer programs are also analyzed in (Saada and Bianchini 1987; Midorikawa 1992; Ishihara et al. 1989; Iai et al. 1993; Arulanandan and Scott 1993, etc.); the authors conclude that there is no program that would work well in all situations. In cases of strong soil nonlinearity, there are often large discrepancies between the calculations and observations.

******

Thus, the research results indicate that soil behavior during strong earthquakes is still insufficiently studied, is not always well described by the existing models and often remains misunderstood.

In the following chapters, a method is proposed aimed at more detailed studying of in situ soil behavior in strong ground motion based on vertical array records.

References

Abrahamson, N.A.: Statistical properties of peak ground accelerations recorded by the SMART 1 array. Bull. Seism. Soc. Am. **78**, 26–41 (1988)

Aguirre, J., Irikura, K.: Nonlinearity, liquefaction, and velocity variation of soft soil layers in PI, Kobe, during the Hyogo-ken Nanbu earthquake. Bull. Seism. Soc. Am. **87**, 1244–1258 (1997)

Aki, K.: Local site effects on weak and strong ground motion. Tectonophysics **218**, 93–111 (1993)

Aki, K., Larner, K.L.: Surface motion of a layered medium having an irregular interface due to incident plane SH waves. J. Geophys. Res. **75**, 933–954 (1970)

Aki, K., Irikura, K.: Characterization and mapping of earthquake shaking for seismic zonation. Proc. 4th Int. Conf. Seismic Zonation **1**, 61–110, Aug 25–29, Stanford, California (1991)

Aki, K., Richards P.: Quantitative seismology. University Science Books (2002)

Aki, K.: Local site effect of ground motion. Earthquake Eng. Soil Dyn. II **20**, 103–155 (1988). Recent advances in ground-motion evaluation. In: Lawrence Von Thun, J. (ed.) Am. Soc. Civil Eng. Geotechnical Special Publications

Alyoshin, A.S., Gushchin V.V., Krekov M.M., et al.: Experimental studies of nonlinear interactions of surface waves. Dokl. USSR Acad. Sci. **260**(3), 574–575 (1981) (in Russian)

Archuleta, R.J.: Direct observations of nonlinearity in accelerograms, pp. 787–792. The Effects of Surface Geology on Seismic Motion. Balkema, Rotterdam (1998)

Archuleta, R.J., Steidl, J.H.: Borehole array data: open windows into site response. In: 10th International Conference on Soil Dynamics and Earthquake Engineering, vol. Ext. Abst. vi. Philadelphia, USA (2001)

Arulanandan, K., Scott, R.F. (eds.) In: Proceedings of International Conference on the Verification of Numerical Procedures for the Analysis of Soil Liquefaction Problems. Davis, California (1993)

Arulanandan, K., Manzari, M., Zeng, X., Fagan, M., Scott, R.F., Tan, T.S.: Significance of the VELACS project to the solution of boundary value problems in geotechnical engineering. In: 3d Int. Conf. Rec. Adv. Geotechn. Earthq. Eng. Soil Dyn. St. Louis, MO. **2**, 825–832 (1995)

Balesta, S.T., Gontovaya, L.I., Kargopol'tsev, A.A., Pushkarev V.G., Senyukov, S.L.: Seismic model of Avachinsky volcano (based on KMPV-GSZ data). Volcanol. Seismolog. **2**, 43–55 (1988) (in Russian)

Bard, P.-Y., Bouchon, M.: The seismic response of sediment-filled valleys. Part I. The case of incident SV waves. Bull. Seism. Soc. Am. **70**, 1263–1286 (1980)

Bard, P.-Y., Bouchon, M.: The seismic response of sediment-filled valleys, Part II. The case of incident SH waves. Bull. Seism. Soc. Am. **70**, 1921–1941 (1980)

Bard, P.-Y., Bouchon, M.: The two-dimensional resonance of sediment-filled valleys. Bull. Seism. Soc. Am. **75**, 519–541 (1985)

Bard, P.-Y., Campillo, M., Chàvez-Garcia F.J., Sanchez-Sesma F.J.: Strong ground motion in Mexico City during the great Michoacan earthquake. Part B: A theoretical investigation of large- and small-scale amplification effects. Earthq. Spectra **4**(3), 609–633 (1988)

Bard, P.-Y.: Effects of surface geology on ground motion: Recent results and remaining issues. In: Proceedings of 10th ECEE, Duma, pp. 305–324. Balkema, Rotterdam (1995)

Beresnev, I.A., Wen, K.-L., Yeh, Y.T.: Nonlinear soil amplification: its corroboration in Taiwan. Bull. Seism. Soc. Am. **85**, 496–515 (1995)

Beresnev, I.A., Solovyov, V.S., Shalashov G.M.: Nonlinear and parametric phenomena in the seismic fields of harmonic vibration signals. In: Problems of Nonlinear Seismics, pp. 180–186. Nauka, Moscow (1987) (in Russian)

Biot, M.A.: General theory of three-dimensional consolidation. J. Appl. Phys. **12**, 155–164 (1941)

Biot, M.A.: Theory of propagation of elastic waves in a fluid-saturated porous solid. J. Acoust. Soc. Am. **28**(2), 168–191 (1956)

Boore, D.M.: A note on the effect of simple topography on seismic SH waves. Bull. Seism. Soc. Am. **62**, 275–284 (1972)

Boore, D.M., Joyner, W.B.: Site amplifications for generic rock sites. Bull. Seism. Soc. Am. **87**, 327–341 (1997)

Boore, D.M., Joyner, W.B., Oliver, A.A., Page, R.A.: Peak acceleration, velocity, and displacement from strong motion records. Bull. Seism. Soc. Am. **70**, 305–321 (1980)

Boore, D.M., Seekins, L., Joyner, W.B.: Peak acceleration from the 17 October 1989 Loma Prieta earthquake. Seism. Res. Lett. **60**, 151–166 (1989)

Borcherdt, R.D., Wentworth C.M.: Strong ground motion generated by the Northridge earthquake of January 17/1994: implications for seismic design coefficients and seismic zonation. In: Proceedings of International Conference Seismic Zonation, pp. 964–972. Nice, France (1995)
Borcherdt, R.D.: Effects of local geology on ground motionnear San-Francisco Bay. Bull. Seism. Soc. Am. **60**, 29–61–2213 (1970)
Borcherdt, R.D.: Influence of local geology in the San Francisco Bay region, California, on ground motion generated by the Loma Prieta earthquake of October 17, 1989. In: Proceedings of International Symposium Safety and Urban Life Facilities Tokyo Institute Technology. Tokyo, Japan (1990)
Bouchon, M.: A simple, complete numerical solution to the problem of diffraction of SH waves by an irregular surface. J. Acoust. Soc. Am. **77**, 1–5 (1985)
Bouchon, M., Aki, K.: Discrete wavenumber representation of seismic source wavefield. Bull. Seism. Soc. Am. **67**, 259–277 (1977)
Byus, E.I., Tskhakaya, A.D.: Seismological foundations of seismic zonation in the Caucasus. Bull. Counc. Seism. USSR Acad. Sci. **8** (1960) (in Russian)
Caillot, V., Bard, P.-Y.: Characterizing site effects for earthquake regulations in the French seismicity context: a statistical analysis, earthquake soil movement; bases and foundations; structure-soil interaction; models of seismic forces influence. In: 9th European Conference on Earthquake Engineering Moscow, vol. 4-B, pp. 27–36 (1990)
Caillot, V.: Quantification statistique et étude expérimentale de mouvements sismiques: application àl'évaluation du risqué. Ph.D. Thesis, Univ. J. Fourier. Grenoble I (1992) (in French)
Campbell, K.W.: Near-source attenuation of peak horizontal acceleration. Bull. Seism. Soc. Am. **71**, 2039–2070 (1981)
Campbell, K.W.: The dependence of peak horizontal acceleration on magnitude, distance, and site effects for small-magnitude earthquakes in California and eastern North America. Bull. Seism. Soc. Am. **79**, 1311–1346 (1989)
Celebi, M.: Topographical and geological amplifications determined from strong-motion and aftershock records of 3 March 1985 Chile earthquake. Bull. Seism. Soc. Am. **77**, 1147–1167 (1987)
Chang, C.-Y., Mok, C.M., Power, M.S., Tang, Y.K., Tang, H.T., Stepp, J.C.: Development of shear modulus reduction curves based on Lotung downhole ground motion data. In: Proceedings of 2nd International Conference Recent Advances Geotechnical Earthquake Engineering Soil Dynamics, pp. 111–118 (1991)
Chen, A.T.F., Joyner W.B.: Multi-linear analysis for ground motion studies of layered systems. Rep. USGS-GD-74-020, NTIS PB232-704/AS. Clearinghouse, Springfield, VA 22151 (1974)
Chin, B.-H., Aki, K.: Simultaneous study of the source, path, and site effects on strong ground motion during the 1989 Loma Prieta earthquake: a preliminary result on pervasive nonlinear site effects. Bull. Seism. Soc. Am. **81**(5), 1859–1884 (1991)
Constantopoulos, I.V.: Amplification studies for a nonlinear hysteretic soil model. In: Res. rep. R73-46. Department of Civil Engineering Massachusetts Institute Technology (1973)
Darragh, R.B., Shakal, A.F.: The site response of two rock and soil station pairs to strong and weak ground motion. Bull. Seism. Soc. Am. **81**(5), 1885–1899 (1991)
Dikmen, S.U., Ghaboussi, J.: Effective stress analysis of seismic response and liquefaction theory. J. Geotech. Eng. **110**, 628–644 (1984)
Dimitriu, P.P.: Self-modulation and recurrence phenomena in vibrator-induced steady-state sinusoidal ground vibrations. Phys. Earth Planet. Inter. **50**, 74–82 (1988)
Dimitriu, P.P., Papaioannou Ch.A., Theodulidis N.P.: EURO-SEISTEST strong-motion array near thessaloniki, Northern Greece: A study of site effects. Bull. Seism. Soc. Am. **88**(3) 862–873 (1998)
Dravinski, M.: Amplification of P, SV and Rayleigh waves by two alluvial valleys. Soil Dyn. Earthq. Eng. **2**, 66–77 (1983)
Faccioli, E.E., Santoyo, V., Leon, T.: Microzonation criteria and seismic response studies for the city of Managua. Proc. Earthq. Eng. Res. Inst. Conf. **1**, 271–291. Managua, Nicaragua (1972)

Faccioli, E.: Induced hazards: earthquake triggered landslides. Proc. 5th Int. Conf. Seism. Zonation **3**, 1908–1933, Nice, France (1995)
Field, E.H., Johnson, P.A., Beresnev, I.A., Zeng, Y.: Nonlinear ground-motion amplification by sediments during the 1994 Northridge earthquake. Nature **390**(6660), 599–602 (1997)
Finn, W.D.L.: State-of-the-art of geotechnical earthquake engineering practice. Soil Dyn. Earthq. Eng. **20**(1–4), 1–15 (2000)
Finn, W.D.L., Lee, K.W., Martin, G.R.: An effective stress model for liquefaction. GED, ASCE **103**(GT6), 517–533 (1977)
Finn, W.D.L., Yogendrakumar, M.: TARA-3FL-program for analysis of liquefaction induced flow deformations. Department of Civil Engineering University British Columbia, Vancouver (1989)
Finn, W.D.L., Yogendrakumar, M., Yoshida, N., Yoshida, M.: TARA-3: a program for nonlinear static and dynamic effective stress analysis. Soil Dyn. Group. University of British Columbia, Vancouver (1986)
Finn, W.D.L.: Evaluation of liquefaction potential. In: Soil Dynamics Lectures University British Columbia (1993)
Frankel, A., Hough, S.E., Friberg, P., Busby, R.: Observations of Loma Prieta aftershocks from a dense array in Sunnyvale California. Bull. Seism. Soc. Am. **81**, 1900–1922 (1991)
Geli, L., Bard, P.-Y., Jullien, B.: The effect of topography on earthquake ground motion: a review and new results. Bull. Seism. Soc. Am. **78**, 42–63 (1988)
Gorshkov, G.P.: Problems of seismotectonics and seismic zonation of the territory of the Republic of China. Bull. Counc. Seism. USSR Acad. Sci. **7** (1960) (in Russian)
Graves, R.W.: Modeling three-dimensional site response effects in the Marina District Basin, San Francisco California. Bull. Seism. Soc. Am. **83**, 1042–1063 (1993)
Groshkov, A.L., Kalimullin, R.R., Shalashov, G.M.: Cubic nonlinear seismic effects, Doklady Acad. Sci. USSR **315**(1), 65–68 (1991) (in Russian)
Groshkov, A.L., Shalashov, G.M.: Equations of nonlinear dynamics of elastic-relaxation medium, Doklady Acad. Sci. USSR **290**(4), 825–827 (1986) (in Russian)
Gushchin, V.V., Shalashov, G.M.: On the possibility of using nonlinear seismic effects in the problems of vibrational transmission of the Earth. Study of the Earth by Non-explosive Seismic Sources, pp. 144–155. Nauka, Moscow, (1981) (in Russian)
Gutenberg, B.: Effects of ground on earthquake motion. Bull. Seism. Soc. Am. **47**, 221–250 (1957)
Gvozdev, A.A., Zvolinsky, N.V., Kovshov, A.N.: Seismic vibrations of the Earth's surface depending on local conditions. Izvestiya Acad. Sci. USSR, Phys. Earth **9**, 10–29 (1982) (in Russian)
Gzelishvili, I.A.: Seismic microzonation of Tbilisi. Doklady Acad. Sci. Georgian SSR **VII**(4) (1946) (in Russian)
Hardin, B.O., Drnevich, V.P.: Shear modulus and damping in soils: measurement and parameter effects. Proc. Am. Soc. Civil Eng., J. Soil Mech. Found. Div. **98**, 603–624 (1972)
Hayashi, S., Tsuchida, H., Kurata, E.: Average response spectra for various subsoil conditions. In: Proceedings of 3rd Joint Meeting U.S.-Japan Panel on Wind and Seismic Effects. UJNR, Tokyo (1971)
Hisada, Y.: An efficient method for computing green's functions for a layered half-space with sources and receivers at close depths. Bull. Seism. Soc. Am. **84**, 1456–1472 (1994)
Hong, T.L., Helmberger, D.V.: Glorified optics and wave propagation in non planar structures. Bull. Seism. Soc. Am. **68**, 1313–1330 (1978)
Horike, M.: Analysis and simulation of seismic ground motion observed by an array in a sedimentary basin. J. Phys. Earth **36**, 135–154 (1988)
Horike, M., Uebayashi, H., Takeuchi, Y.: Seismic response in three-dimensional sedimentary basin due to plane S-wave incidence. J. Phys. Earth **38**, 261–284 (1990)
Hudson, D.E.: Strong-Motion Instrumental data on the San Fernando Earthquake of Feb. 9, 1971. Techn. Rep. Caltech EERL.1971.001. California Inst. of Technology (1971)
Iai, S., Morita, T., Kameoka, T., Matsunaga, Y., Abiko, K.: Response of a dense sand deposit during 1993 Kushiro-Oki earthquake. Soils Found. **35**(1), 115–131 (1995)

Iai, S., et al.: Simultaneous analysis on liquefaction. In: Proceedings of Symposium Remedial Measures against Liquefaction. JSSMFE, pp. 170–190 (1993)
Idriss, I.M., Seed, H.B.: Response of Horizontal Soil Layers During Earthquakes. Department of Civil Engineering University, Berkeley, California (1967)
Idriss, I.M., Seed, H.B.: Seismic response of horizontal soil layers. Proc. Am. Soc. Civil Eng. J. Soil Mech. Found. Div. **94**, 1003–1031 (1968)
Idriss, I.M.: Evaluating seismic risk in engineering practice. In: Proceedings of 11th International Conference Soil Mechanical Foundation Engineering San Francisco, pp. 265–320. Balkema, Rotterdam, The Netherlands (1985)
Idriss, I.M.: Response of soft soil sites during earthquakes. In: Proceedings of H.B. Seed Memorial Symposium, Berkeley, California (1990)
Idriss, I. M.: Earthquake ground motions at soft soil sites. In: International Conferences on Recent Advances in Geotechnical Earthquake Engineering Soil Dynamics, vol. 3 (1991)
Ihnen, S.M., Hadley, D.M.: Prediction of strong ground motion in the Puget Sound Region: the 1965 Seattle earthquake. Bull. Seism. Soc. Am. **76**, 905–922 (1986)
Ishihara, K., Towhata, I.: One-dimensional soil response analysis during earthquake based on effective stress method. J. Fac. Eng. Univ. Tokyo **V**(4), 656–700 (1980)
Ishihara, K., Cubrinovski, M., Takeshi, N.: Characterization of undrained behavior of soils in the reclaimed area of Kobe. Soils Found. Spec. Issue on Geotechnical Aspects of the January 17 1995 Hyogoken-Nambu Earthquake **2**, 33–46 (1998)
Ishihara, K., et al.: Effective stress analysis of ground and soil structures. In: Proceedings of Symposium on Behavior of Ground and Soil Structures during Earthquake. JSSMFE, pp. 50–136 (1989)
Iwan, W.D.: On a class of models for the yielding behavior of continuous and composite systems. J. Appl. Mech. **34**, 612–617 (1967)
Iwasaki, T., Kawashima, K., Tatsuoka, F.: Effect of soil nonlinearity on earthquake response of ground. Civil Eng. J. PWRI **22**(12), 27–32 (1980) (in Japanese)
Iwata, T., et al.: Practical evaluation of seismic intensity for earthquake control sensor. In: Proceedings of 21st JSCE Earthquake Engineering Symposium, pp. 613–616 (1991)
Jongmans, D., Campillo, M.: The response of the Ubaye valley (France) for incident SH and SV waves: comparison between measurements and modeling. Bull. Seism. Soc. Am. **83**, 907–924 (1993)
Joyner, W.B., Boore, D.M.: (1988). Measurement, characterization, and prediction of strong ground motion. In: Proceedings of America Social Civil Engineering Geotechnical Engineering Division Specialist Conference Park City, Utah 43–102 (1988)
Joyner, W.B., Boore, D.M.: Peak horizontal acceleration and velocity from strong motion recording including records from the 1979 Imperial Valley, California. Earthquake. Bull. Seism. Soc. Am. **71**, 2011–2038 (1981)
Joyner, W.B., Chen, T.F.: Calculation of nonlinear ground response in earthquakes. Bull. Seism. Soc. Am. **65**(5), 1315–1336 (1975)
Kanai, K.: Relation between the nature of surface layer and the amplitudes of earthquake motions. Bull. Earthq. Res. Inst. Tokyo Univ. **30**, 31–37 (1952)
Kanai, K., Takahasi, R., Kawasumi, H.: Seismic characteristics of ground. In: Proceedings of the World Conference on Earthquake Engineering California, pp. 31-1–31-16 (1956)
Katayama, T., Iwasaki, T., Saeki, M.: Statistical analysis of earthquake acceleration response spectra. Collected Papers, 275. Japan. Soc. Civil Eng. 29–40 (1978)
Kawase, H.: Time-domain response of a semicircular canyon for incident SV, P and Rayleigh waves calculated by the discrete wave number boundary element method. Bull. Seism. Soc. Am. **78**, 1415–1437 (1988)
Kawase, H., Aki, K.: A study of the response of a soft basin for incident S, P and Rayleigh waves with special reference to the long duration observed in Mexico City. Bull. Seism. Soc. Am. **79**, 1361–1382 (1989)

Kawase, H., Aki, K.: Topography effect at the critical SV-wave incidence: possible explanation of damage pattern by the Whittier Narrows, California, earthquake of 1 October 1987. Bull. Seism. Soc. Am. **80**, 1–22 (1990)
Kawashima, K., Aizawa, K., Takahashi, K.: Attenuation of peak ground acceleration, velocity and displacement based on multiple regression analysis of Japanese strong motion records. Earthq. Eng. Struct. Dynamics **14**, 199–215 (1986)
Kazama, M., Yamaguchi, A., Yanagisawa, E.: Seismic behavior of an underlying alluvial clay on man-made islands during the Hyogoken-Nanbu earthquake. Soils Found. Spec. Issue on Geotechnical Aspects of the January 17 1995 Hyogoken-Nambu Earthquake **2**, 23–32 (1998)
Keane, C.M., Prevost, J.: An analysis of earthquake data observed in the Wildlife liquefaction array site, Imperial County, California. Proc. 2nd U.S.-Japan Workshop on Liquefaction, Large Ground Deformation and Their Effects of Lifelines. New York, Techn. Rep. NCEER-89-0032, 176–192 (1989)
Khair, K.R., Datta, S.K., Shah, A.H.: Amplification of obliquely incident seismic waves by cylindrical alluvial valleys of arbitrary P ans SV waves. Bull. Seism. Soc. Am. **79**, 610–630 (1989)
Kobayashi, H., Midorikawa, S.: Study of site effects in Mexico City using microtremors. Proc. Japan. Symp. Earthq. Eng. **7**, 355–360 (1986) (in Japanese)
Kokusho, T., Matsumoto, M.: Nonlinearity in site amplification and soil properties during the 1995 Hyogoken-Nambu earthquake. Soils Found. Spec. Issue on Geotechnical Aspects of the January 17 1995 Hyogoken-Nambu Earthquake **2**, 1–9 (1998)
Kokusho, T.: Dynamic deformation characteristics of soil and nonlinear response of ground. Rep. 301. Central Electric Research Institute, pp. 207–240 (1982) (in Japanese)
Kramer, S.L.: Geotechnical Earthquake Engineering, pp. 230–238. Prentice Hall, Upper Saddle River, N.J. (1996)
Kuribayashi, E., Iwasaki, T., Iida, Y., Tuji, K.: Effects of seismic and subsoil conditions on earthquake response spectra. In: Proceedings of International Conference on Microzonation, pp. 499–512 (1972)
Lee, V.W.: Three-dimensional diffraction of elastic waves by a spherical cavity in an elastic half-space, I: closed form solutions. Soil Dyn. Earthq. Eng. **7**, 149–161 (1988)
Lee, J.-J., Langston, C.A.: Three-dimensional ray tracing and the method of principal curvature for geometric spreading. Bull. Seism. Soc. Am. **73**, 765–780 (1983)
Lee, M.K.W., Finn, W.D.L.: DESRA-2, dynamic effective stress response analysis of soil deposits with energy transmitting boundary including assessment of liquefaction potential. Soil Mech. Ser. **38**, Department of Civil Engineering University of British Columbia, Vancouver (1978)
Liu, H.P., Warrick, R.E., Westerlund, R.E., Sembera, E.D., Wennerberg, L.: Observation of local site effects at a downhole-and-surface station in the Marina district of San Francisco. Bull. Seism. Soc. Am. **82**, 1563–1591 (1992)
Lund, F.: Interpretation of the precursor to the 1960 Great Chilean earthquake as a seismic solitary wave. Pure Appl. Geoph. **121**(1), 17–26 (1983)
Lysmer, J., Drake, L.A.: The propagation of Love waves across nonhorizontally layered structures. Bull. Seism. Soc. Am. **61**, 1233–1252 (1971)
Lysmer, J., Udaka, T., Tsai, C.-F., Seed, H.B.: FLUSH—a computer program for approximate 3-D analysis of soil-structure interaction problems Rep. EERC 75-30. Earthq. Eng. Res. Center, University of California, Berkeley (1975)
MacMurdo, J.: Paper relating to the earthquake which occurred in India in 1819. Phil. Mag. **63**, 105–177 (1824)
Mallet, R.: Great Neapolitan Earthquake of 1857: The First Principles of Observational Seismology. Royal Society (1862)
Martin, P.P., Seed, H.B.: A computer program for the nonlinear analysis of vertically propagating shear waves in horizontally layered soil deposits. In: Rep. EERC 78-23, University of California. Berkeley, California (1978)
Medvedev, S.V., Shebalin, N.V.: An earthquake can be resisted. Nauka, Moscow (1967)

Medvedev, S.V.: Engineering seismology. In: State Publishing House of papers on Construction, Architecture and Building Materials, Moscow (1962)
Midorikawa, S.: A statistical analysis of submitted predictions for the Ashigara Valley blind prediction test. In: Proceedings of the International Symposium on the Effects of Surface Geology on Seismic Motion, vol. 2, pp. 65–77. Odawara, Japan (1992)
Moczo, P., Bard, P.-Y.: Wave diffraction, amplification and differential motion near strong lateral discontinuities. Bull. Seism. Soc. Am. **83**, 85–106 (1993)
Mohammadioun, B., Pecker, A.: Low-frequency transfer of seismic energy by superficial soil deposits and soft rocks. Earthq. Eng. Struct. Dyn. **12**, 537–564 (1984)
Mohraz, B.: A study of earthquake response spectra for different geological conditions. Bull. Seism. Soc. Am. **66**, 915–935 (1976)
Murphy, J.R., Davis, A.H., Weaver, N.L.: Amplification of seismic body waves by low-velocity surface layers. Bull. Seism. Soc. Am. **61**(1), 109–145 (1971)
Nava, S.J., Munsey, J.W., Johnson, A.C.: First fault plane identification in the southern Appalachians: the mbLg Vonore, Tennessee, earthquake of March 27, 1987. Seis. Res. Lett. **60**(3), 119–129 (1989)
Nikolaev, A.V., Pavlenko, O.V., Yakovlev, A.P.: Quasi-static deformations of the Earth's surface caused by vibration and nonlinear properties of terrestrial materials. Izvestiya Acad. Sci. USSR Phys. Earth **12**, 3–11 (1994) (in Russian)
Nikolaev, A.V.: Problems of nonlinear seismology. In: Problems of Nonlinear Seismics, pp. 5–20. Nauka, Moscow (1987) (in Russian)
Nowack, R., Aki, K.: The two-dimensional Gaussian beam synthetic method; testing and applications. J. Geoph. Res. **89**, 7797–7819 (1984)
O'Connel, D.R.H.: Replication of apparent nonlinear seismic response with linear wave propagation models. Science **283**(5410), 2045–2050 (1999)
Pavlenko, O.V.: Simulation of ground motion from strong earthquakes of Kamchatka Region (1992–1993) at rock and soil sites. Pure Appl. Geoph. **170**(4), 571–595 (2013)
Pedersen, H., Le Brun, B., Hatzfeld, D., Campillo, M., Bard, P.-Y.: Ground-motion amplitude across ridges. Bull. Seism. Soc. Am. **84**, 1786–1800 (1994)
Phillips, W.S., Kinoshita, S., Fujiwara, H.: Basin-induced Love waves observed using the strong-motion array at Fuchu Japan. Bull. Seism. Soc. Am. **83**, 65–84 (1993)
Pitarka, A., Irikura, K., Iwata, T., Sekiguchi, H.: Three-dimensional simulation of the near-fault ground motion for the 1995 Hyogo-ken Nanbu (Kobe), Japan, Earthquake. Bull. Seism. Soc. Am. **38** (1998)
Popov, V.V.: Engineering and geological criteria for detailed seismic zonation. Proc. Inst. Phys. Earth **5** (1959) (in Russian)
Prévost, J.H.: DYNAFLOW: a nonlinear transient finite element analysis program. Princeton University, Princeton, New Jersey (1981) last revision (2004)
Program: Development of Physical Basement for Seismic Methods. Moscow, Schmidt Institute of Physics of the Earth (1981) (in Russian)
Rogers, A.M., Katz, L.J., Bennett, T.J.: Topographic effects on ground motion for incident P waves: a model study. Bull. Seism. Soc. Am. **64**, 437–456 (1974)
Rogers, A.M., Borcherdt, R.D., Covington, P.A., Perkins, D.M.: A comparative ground response study near Los Angeles using recordings of Nevada nuclear tests and the 1971 San Fernando earthquake. Bull. Seism. Soc. Am. **74**, 1925–1949 (1984)
Rogers, A.M., Tinsley, J.C., Borcherdt, R.D.: Predicting relative ground response. Evaluating earthquake hazards in the Los Angeles region—an earth-science perspective. U.S. Geological Survey Profess. Paper **1360**, 221–247 (1985)
Saada, A., Bianchini, G.S. (eds.): In: Proceedings of International Workshop on Constitutive Equations for Granular Non-cohesive soils. Case Western Reserve University, Cleveland (1987)

Sadigh, K.: Considerations in the development of site-specific spectra. In: Proceedings of Conference XXII, A Workshop on Site Effects of Soil and Rock on Ground Motion and their Implications for Earthquake Resistant Design. Santa Fe, New Mexico, U.S. Geol. Survey Open-File Rep. 83-845, pp. 423–458 (1983)

Safaryan, A.N.: On the methodology of seismic zonation and micro-zonation. Proc. Inst. Construction. Acad. Sci. Georgian SSR **VI** (1957) (in Russian)

Sanchez-Sesma, F.J.: Diffraction of elastic SH waves by wedges. Bull. Seism. Soc. Am. **75**, 1435–1446 (1985)

Sanchez-Sesma, F.J., Esquivel, J.A.: Cround motion on alluvial valleys under incident plane SH waves. Bull. Seism. Soc. Am. **69**, 1107–1120 (1979)

Sanchez-Sesma, F.J., Rodriguez-Zuniga, J.L., Perez-Rocha, L.E.: Seismic response of shallow alluvial valleys: the use of simplified models. Bull. Seism. Soc. Am. **85**, 890–899 (1995)

Sánchez-Sesma, F.J.: Elementary solutions for the response of a wedge-shaped medium to incident SH and SV waves. Bull. Seism. Soc. Am. **80**, 737–742 (1990)

Sato, K., Kokusho, T., Matsumoto, M., Yamada, E.: Nonlinear seismic response and soil property during 1995 Hyogoken Nanbu earthquake. Soils Found. Spec. Issue on Geotechnical Aspects of the January 17, 1995 Hyogoken Nanbu earthquake 41–52 (1996)

Schnabel, P.B., Lysmer, J., Seed, H.B.: SHAKE—A computer program for earthquake analysis of horizontally layered sites. In: Earthquake Engineering Research Center, University of California, Berkeley. Report EERC 72-12 (1972)

Seed, H.B., Murarka, R., Lysmer, J., Idriss, I.M.: Relationships of maximum acceleration, maximum velocity, distance from source, and local site conditions for moderately strong earthquakes. Bull. Seism. Soc. Am. **66**, 1323–1342 (1976)

Seed, H.B., Idriss, I.M.: Arango I. Evaluation of liquefaction potential using field performance data. J. Geotechn. Eng. Division ASCE **109**(GT3), 458–482 (1983)

Seed, H.B., Dickinson, S.E., Mok, C.K.: Recent lessons regarding seismic response analysis of soft and clay sites. In: Proceedings of Seminar Seismic Design and Retrofit of Bridges, California Department of Transportation, Sacramento (1992)

Shalashov, G.M.: Cross-modulation of acoustic waves on cubic nonlinearity of solids. Acoust. J. (3), 386–390 (1984) (in Russian)

Shinozaki, Y.: High frequency response of a sediment-filled valley for incident plane SH waves. In: Proceedings of 2nd Workshop IASPEI/IAEE Joint Working Group on Effects of Surface Geology on Seismic Motion v1-v8 (1988)

Silva, W.J.: Soil response to earthquake ground motion. EPRI Report NP-5747. Electric Power Research Institute. Palo Alto, California (1988)

Silva, W.J.: Site dependent specification of strong ground motion. Dynamic Soil Properties and Site Characterization. In: Proceedings Workshop sponsored by the Natural Science Foundation and the Electric Power Research Institute Palo Alto, CA, pp. 6-1–6-80 (1989)

Singh, S.K., Mena, E., Castro, R.: Some aspects of source characteristics of the 19 September 1985 Michoacan earthquake and ground motion amplification in and near Mexico City from strong motion. Bull. Seism. Soc. Am. **78**, 451–477 (1988)

Soeda, Y., Tamai, H., Nakatsu, N., Takezawa, K., Maegawa, F.: Nonlinearity and irregularity on strong seismic motions by vertical arrays during the 1995 Kobe Earthquake. Earthq. Geotechn. Eng. 225–230. Balkema, Rotterdam (1999)

Solovyov, V.S.: Experimental study of nonlinear seismic phenomena. Problems of nonlinear Seismics. Nauka, Moscow, 164–179 (1987) (in Russian)

Somerville, P.G., Yoshimura, J.: The influences of critical Moho reflections on strong ground motions recorded in San Francisco and Oakland during the 1989 Loma Prieta Earthquake. Geoph. Res. Lett. **17**, 1203–1206 (1990)

Streeter, V.L., Wylie, E.B., Richart, F.E.: Soil Motion computations by characteristics method. Proc. Am. Soc. Civil Eng. J. Geotech. Eng. Division **100**, 247–263 (1974)

Suetomi, I., Yoshida, N.: Effect of frequency characteristics on earthquake motion to the nonlinear response of ground. In: Proceedings of 51st Annual Conference JSCE (I-B), pp. 352–353 (1996)

Suetomi, I., Yoshida, N.: Nonlinear behavior of surface deposit during the 1995 Hyogoken-Nambu earthquake. Soils Found. **2**(Special), 11–22 (1998)

Sugito, M., Kiremidjian, A.S., Shah, H.C.: Nonlinear ground motion amplification factors based on local site parameters. Proc. 4th Int. Conf. Seismic Zonation, Stanford **II**, 221–228 (1991)

Sugito, M., Aida, N., Masuda, T.: Frequency dependent equilinearized technique for seismic response analysis of multi-layered ground. J. Geotech. Eng. Proc. JSCE **493**, 49–58 (1994) (in Japanese)

Suyehiro, K.: Engineering seismology. Notes on American lectures. Proc. Am. Soc. Civil Eng. **58**(4) (1932)

Takahashi, R., Hirano, K.: Changes in the vertical intensity of geomagnetism that accompanied the eruption. Bull. Earthq. Res. Inst. Tokyo Univ. **19**, 83–103 (1941)

Theodolidis, N.P., Bard, P.-Y.: Horizontal to vertical spectral ratio and geological conditions: an analysis of strong motion data from Greece and Taiwan (SMART-1). Soil Dynam. Earthq. Eng. **14**(3), 177–197 (1995)

Trifunac, M.D.: Surface motion of a semi-cylindrical alluvial valley for incident plane SH waves. Bull. Seism. Soc. Am. **61**, 1755–1770 (1971)

Trifunac, M.D.: Preliminary analysis of the peaks of strong earthquake ground motion—dependence of peaks on earthquake magnitude, epicentral distance and recording site conditions. Bull. Seism. Soc. Am. **66**, 189–219 (1976)

Trifunac, M.D., Hao, T.Y., Todorovska, M.I.: On the reoccurrence of site specific response. Soil Dynam. Earthq. Eng. **18**(8), 569–592 (1999)

Tsshoher, V.O.: Seismics in the problems of urban planning. In: Proceedings of Physico-technical Institute Turkmen Branch of the USSR Academy Science, Ashgabat (1949) (in Russian)

Tucker, B.E., King, J.L.: Dependence of sediment-filled valley response on the input amplitude and the valley properties. Bull. Seism. Soc. Am. **74**, 153–165 (1984)

Umeda, Y.: The bright spot of an earthquake. In: Proceedings of International Symposium Earthquake Source Physics and Earthquake Precursors, Tokyo (1990)

Van Den Abeele, K.E.-A., Johnson, P.A.: Elastic pulsed wave propagation in media with second- or higher-order nonlinearity. Part II. Simulation of experimental measurements on Berea sandstone. J. Acoust. Soc. Am. **99**, 3346–3352 (1996)

Vasil'ev, Yu.I.: A new problem of engineering seismology—calculation of fluidity zones in a soil layer during a strong earthquake. In: Problems of Nonlinear Seismics, pp. 145–149. Nauka, Moscow (1987) (in Russian)

Vidale, J.E., Helmberger, D.V.: Elastic finite difference modeling of the 1971 San Fernando, California, earthquake. Bull. Seism. Soc. Am. **78**, 122–141 (1988)

Virieux, J.: Wave propagation in heterogeneous media: velocity-stress finite-difference method. Geophysics **49**, 1933–1957 (1984)

Wong, H.L., Jennings, P.C.: Effect of canyon topography on strong ground motion. Bull. Seism. Soc. Am. **65**, 1239–1257 (1975)

Yamanaka, H.K., Sea, K., Samano, T.: Effects of sedimentary layers on surface-wave propagation. Bull. Seismol. Soc. Am. **79**, 631–644 (1989)

Yoshida, N., Iai, S.: Nonlinear site response and its evaluation and prediction. Effects Surf. Geol. Seismic Motion 71–90. Balkema, Rotterdam (1998)

Zaitsev, L.P.: On vibrations arising by a shear wave in a plastic layer covering an elastic half-space. Izvestiya Acad. Sci. USSR, Phys. Earth **11**, 13–24 (1982) (in Russian)

Zaitsev, L.P.: On the motion of a medium with an arbitrary depth distribution of the plasticity limit induced by a shear wave. Izvestiya Acad. Sci. USSR, Phys. Earth **9**, 3–17 (1985) (in Russian)

Zama, S.: Behavior of the elastic waves propagating through the irregular structure. Bull. Earthq. Res. Inst. Tokyo Univ. **56**, 741–777 (1981)

Zarembo, L.K., Krasilnikov, V.A.: Introduction to nonlinear acoustics. Nauka, Moscow (1966) (in Russian)

Zeghal, M., Elgamal, A.-W.: Analysis of site liquefaction using earthquake records. J. GT, ASCE **120**(6), 996–1017, (1994)

Zimenkov, S.V., Nazarov, V.E.: Nonlinear acoustic effects in rock samples. Izvestiya Acad. Sci. Phys. Earth **1**, 13–18 (1993) (in Russian)

Chapter 3
Nonlinear Soil Behavior During Recent Strong Earthquakes

As shown above, for a detailed study of soil behavior in strong ground motion, records from in situ observations are needed. There still remain significant differences between the conditions of in situ dynamic loading of soils and laboratory and field conditions. Any information on the in situ soil behavior is extremely important; especially, observations with seismic vertical arrays.

Since Seed and Idriss (1970) directly observed nonlinear soil response in a borehole during the 1965 Seattle earthquake, the borehole observations have continually evolved to the point where we see entire networks of borehole instruments, such as, KiK-net in Japan, DART in Taipei basin, SCEC in Los Angeles basin, Bay Bridges in San Francisco and others. A seismic vertical array usually consists of 2–4 three-component accelerometers installed in a borehole on the surface and at various depths, down to ~ 100–200 m.

As noted by Archuleta and Steidl (2001), earthquake records obtained by seismic vertical arrays provide us answers to many important questions: Where in the soil column did the response become nonlinear? At what level of strain did the nonlinear response occur? How is it manifested in the observed ground motion? What soil properties that can be measured in situ are most important in affecting nonlinearity? Does the soil return to a state with the same linear or nearly linear response? In what time period? etc. (Archuleta and Steidl 2001).

Strong earthquakes of last decades, such as the 1985 Michoacan earthquake, 1989 Loma Prieta earthquake, 1994 Northridge earthquake, 1995 Kobe earthquake, 2000 Tottori earthquake, 1999 Chi-Chi earthquake and others provided new experimental data on soil behavior in strong ground motion and on soil liquefaction. At the same time, they caused new discussions about manifestations of nonlinearity of soil response, ranges of applicability of linear models, features of plastic behavior of clays, etc. (Lomnitz et al. 1995; Aguirre and Irikura 1997; Field et al. 1997; O'Connel 1999, etc.). The strong-motion database is growing quickly, but (as shown in Chap. 2) the interpretation of the obtained records is sometimes difficult, and new approaches

O. V. Pavlenko, *Seismic Waves in Soil Layers*,
https://doi.org/10.1007/978-3-031-52338-0_3

for data processing are needed to extract from the records more information about soil behavior.

This chapter describes a method developed by the author in collaboration with Professor Kojiro Irikura (DPRI, Kyoto University, Japan) for estimating stresses and strains induced in soil layers by strong motion, based on vertical array records. The method was used to construct models of soil behavior during recent strong earthquakes: the 1995 Kobe earthquake, 2000 Tottori earthquake, 1999 Chi-Chi (Taiwan) earthquake, 2003 Tokachi-oki earthquake, and 2011 Tohoku earthquake. Models of soil behavior, i.e., vertical distributions of stresses and strains induced in soil layers by the earthquake, allow us to trace changes in the behavior of separate layers during strong motion and evaluate changes of the shear moduli. The estimates are based on real records, so they are free from assumptions about the features of soil behavior in conditions of strong motion.

3.1 Models of Nonlinear Soil Behavior in the Near-Fault Zones During the 1995 Kobe (Japan) Earthquake

The Great Hanshin Earthquake, or Kobe earthquake, with a moment magnitude $M_w = 6.9$ and a source depth of 18 km occurred on January 17, 1995, at 05:46 local time in the southern part of Hyogo Prefecture, Japan. The epicenter was located at the northern end of the Awaji Island, 20 km away from the center of Kobe. It was recorded by seismic stations throughout Japan, including vertical arrays PI (Port Island), SGK and TKS, located 2, 6 and 16 km from the fault plane. Figure 2.10 (Chap. 2) shows the locations of the sites, the major principal axes, and the epicenters of the mainshock and the aftershocks summarized by Disaster Prevention Research Institute of Kyoto University. The vertical array at Port Island consists of four three-component accelerometers installed at GL-0 m, GL-16 m, GL-32 m, and GL-83 m; the arrays at SGK and TKS sites consist of three three-component devices at GL-0 m, GL-24.9 m, and GL-97 m, and GL-0 m, GL-25 m, and GL-100 m, respectively. The soil profiles and the accelerometers in the boreholes are shown in Fig. 3.1, and Fig. 3.2 shows the acceleration time histories of the mainshock recorded at the three sites.

As noted by many authors (Sato et al. 1996; Yasuda et al. 1996; Aguirre and Irikura 1997; Ishihara et al. 1998; Kazama et al. 1998; Kokusho and Matsumoto 1998; Soeda et al. 1999; Pavlenko and Irikura 2002a, b, c, 2003, etc.), vertical array records of the 1995 Kobe earthquake undoubtedly testified to the soil nonlinearity at the recording sites. Partially analysis of records of the Kobe earthquake is described in Chap. 2, Sect. 2.4. Here is a brief overview of some other studies.

Methods of back calculation, multireflection analysis, and inversion of records of the 1995 Kobe earthquake at PI, SGK, TKS, and KNK vertical array sites were used to estimate the optimum values of *S*-wave velocity (*Vs*) and damping ratio in sedimentary layers during the mainshock and the aftershocks (Sato et al. 1996). The

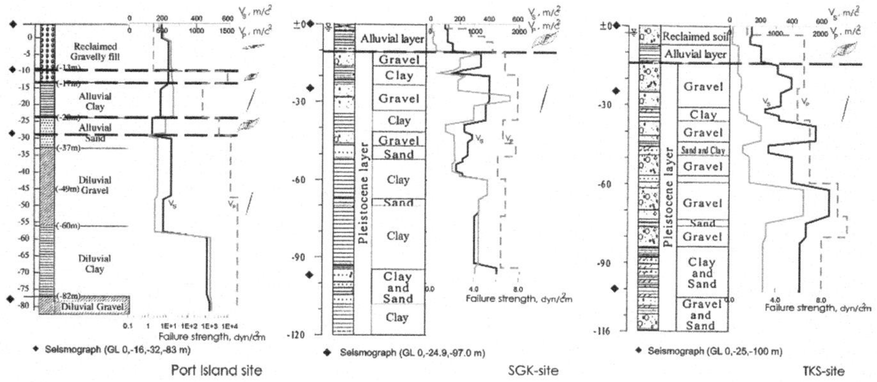

Fig. 3.1 Soil profiles and types of stress–strain relations in soil layers at horizontal and vertical components (left and right, respectively) at Port Island, SGK and TKS sites. Rhombs on the left show the locations of the accelerometers

optimized values of V_S during the mainshock were found to be smaller than their initial values obtained by the *PS*-wave logging, especially in the upper liquefied layers at PI. To trace the development of liquefaction, the records were divided into two parts, and it was found that liquefaction was still developing after the first period, 4.1–9.22 s. Analysis of the aftershock with a maximum surface acceleration of ~ 10 Gal at PI, which occurred 2 min after the mainshock, showed that V_S for Pleistocene and Holocene layers underlying the reclaimed fill had returned to their initial values, while V_S in the fill layer had almost the same reduced values as in the mainshock, indicating that the fill was still under liquefied conditions (Sato et al. 1996). However, the computed horizontal accelerations were slightly different from the observed values and quite different at the level GL-25 m at TKS site. Hence, it was difficult to simulate nonlinear seismic amplification by using multireflection analysis with optimized V_S values for a long-duration time history, including completely different types of waves at TKS (Sato et al. 1996).

Kokusho and Matsumoto (1998) estimated maximum accelerations and maximum shear strains as functions of depth. A clear strain dependency was found in the back calculated shear moduli and damping, which is consistent with the laboratory test results for each kind of considered soil. However, they could not obtain a good agreement between the observed and simulated acceleration time histories, and concluded that further study was needed for a better estimation of the nonlinear soil response.

The seismic behavior of an alluvial clay layer underlying reclaimed soil at PI was studied by Kazama et al. (1998). The shear stress–strain relation of the clay layer was estimated from the mainshock by calculating separately the shear-strain and shear-stress evolution from filtered ($f \leq 1.2$ Hz) acceleration records above GL-16 m and below GL-32 m clay layer. Stress–strain relations corresponding to the low-frequency components were estimated for six successive time intervals within ~ 10 s of strong motion, and it was concluded that a significant decrease of the shear moduli occurred in the upper layers.

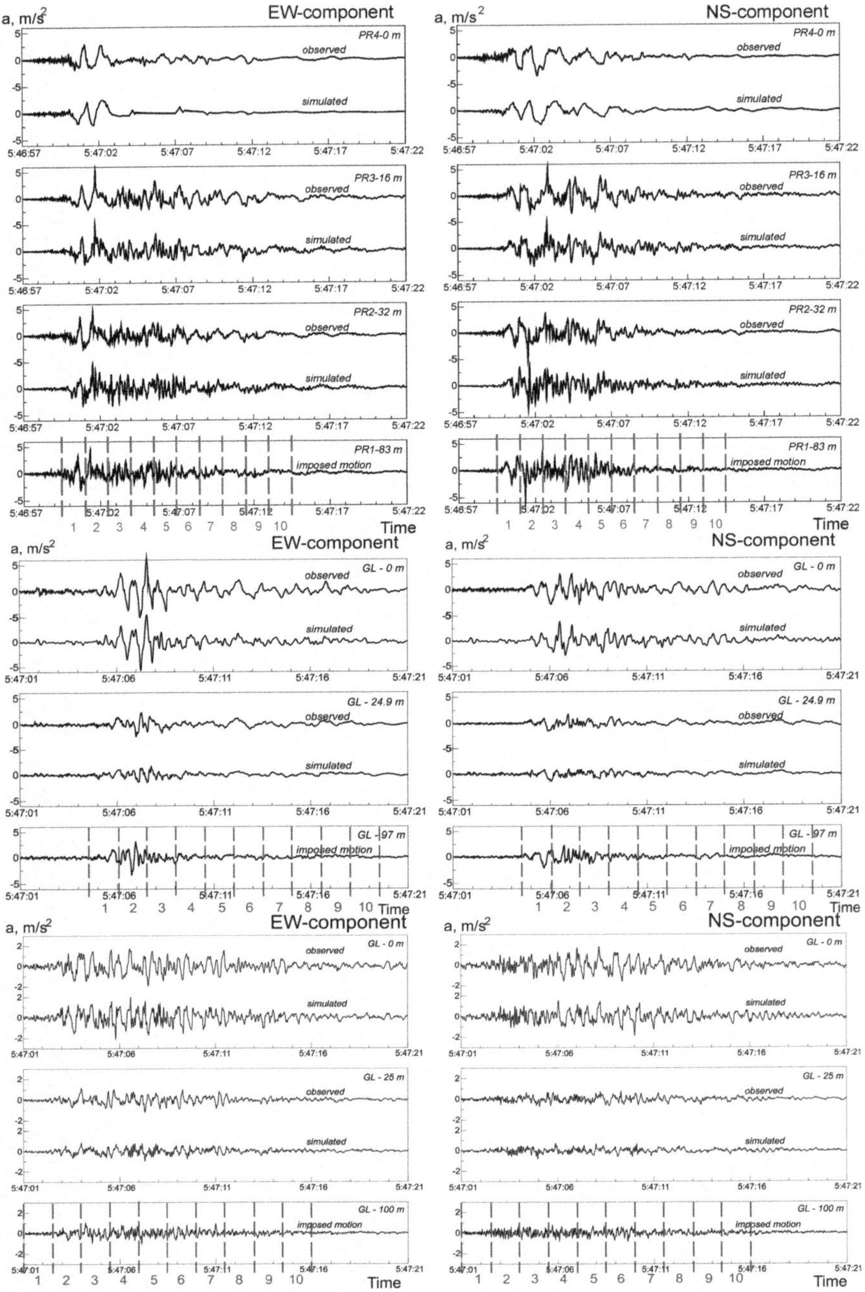

Fig. 3.2 Acceleration time histories of the mainshock at Port Island (upper), SGK (middle), and TKS (bottom), observed and simulated. The dotted lines show time intervals used in the calculations

Thus, the application of conventional programs like SHAKE or NONLI3 to calculate the soil response in the near-fault zones could not provide an agreement with the observations. Thus, we suggested a method to estimate stresses and strains induced by strong motion in soil layers at various depths, based on the records of seismic vertical arrays. We assumed that different soil layers may behave differently in conditions of their dynamic loading, depending on their composition, depth, water-saturation, and that soil behavior (rheological properties of soil layers) may change during strong motion.

The directional drifts of the accelerometers were checked by calculating the horizontal orbits of the long-period particle motions for the mainshock and the aftershocks at different depths at the three sites. At Port Island, N19° W rotation at GL-83 m was detected and corrected; at SGK site, a reverse of NS component and N6° W rotation at GL-24.9 m and a reverse of NS component and N34° E rotation at GL-97 m were detected and corrected. At TKS site, N23° W rotation at GL-25 m and N9° E rotation at GL-100 m were detected and corrected. All these corrections agree in general with the conclusions of other authors.

The soil profiles are similar at the three sites (Fig. 3.1). The deepest layers are Pleistocene gravely soils, and the upper layers consist of alternating Pleistocene gravel/clay layers, Holocene sand/clay layers and fill (Sato et al. 1996).

The profiling data at Port Island used for the simulations were taken from (Agguirre and Irikura 1997). Shear wave velocity, shear moduli degradation, and maximum shear strains at SGK and TKS sites were taken from (Soeda et al. 1999), These data were used to calculate the shear stress in failure $\sigma_{\max}$ and the low-strain shear modulus $G_{\max}$ in different layers, following Seed et al. (1984) and Sun et al. (1988).

Parameters of soil layers used in the simulations are shown in Table 3.1. The coefficient of linear damping α is taken from (Aguirre and Irikura 1997). *P*-wave velocities, when exceeding ~ 1000 m/s, indicate saturation of soils with water. The underground water level is about 13 m at Port Island and ~ 3–5 m at SGK and TKS.

To estimate stress–strain relations in soil layers, we applied the modified algorithm of calculation of the propagation of vertically incident shear waves in soil layers up to the surface, based on the lumped mass method of Joyner and Chen (1975), and combined it with the method of "trial and error" to find stress–strain relations showing the best fit to the observations. Records of the deepest devices were used as inputs to soil layers, and the behavior of the layers was described by normalized, as proposed by Hardin and Drnevich (1972), stress–strain relations: stress was normalized by multiplying by $1/\tau_{\max}$, and strain was normalized by multiplying by $G_{\max}/\tau_{\max}$.

For the calculations, the studied strata from the surface down to the location of the deepest device was divided into groups of layers, for which certain types of stress–strain relations were assumed (Fig. 3.1). Three main types of stress–strain curves were considered:

(1) Similar to those obtained in laboratory experiments by Hardin and Drnevich (1972), to describe the behavior of dense soils at depths (18–26 m and below 34 m at Port Island, below 11 m at SGK, below 14 m at TKS);

Table 3.1 Parameters of soil layers used in the simulations at Port Island, SGK and TKS sites

Layer thickness *h*, m	τ_{max}, bar	V_S, m/s	ρ, g/cm^3	α	V_P, m/s
			Port Island		
2	3.25	194	1.8	0.07	260
3	5.15	194	1.8	0.07	350
7.6	12.1	237	1.8	0.08	780
3.4	13.7	237	1.8	0.08	1480
3	14.9	237	1.8	0.02	1480
8	16.4	191	1.5	0.5	1180
5	6.62	138	1.85	0.5	1350
1	0.673	138	1.85	0.5	1530
17	2.84	254	1.85	0.5	1530
11	2.11	208	1.85	0.5	1610
18	1187	469	1.8	0.5	1610
4	1386	486	1.9	0.5	2000
			SGK		
2	0.114	98	1.4	0.07	420
1	0.157	117	1.4	0.07	840
2	0.29	117	1.7	0.07	840
2	0.22	117	1.7	0.07	1180
1	0.266	149	1.7	0.07	1390
2	0.503	149	1.6	0.1	1390
1	0.62	149	1.6	0.1	1390
1	4.2	342	2	0.1	1390
5	2.67	342	2	0.1	1610
1	2.14	222	2	0.1	1610
2	0.76	154	2	0.5	1610
1	3.09	400	2	0.5	1610
6	2.51	400	2	0.5	2000
3	7.18	400	2	0.5	1900
3	3.98	375	2	0.5	1900
6	3.91	375	1.7	0.5	1900
3	1.416	231	1.7	0.5	1540
3	2.166	286	2	0.5	1540
7	2.204	255	2	0.5	1870
2	2.25	222	2	0.5	1390
4	1.99	177	2	0.5	1390

(continued)

Table 3.1 (continued)

Layer thickness h, m	τ_{max}, bar	V_S, m/s	ρ, g/cm^3	α	V_P, m/s
2	2.68	222	2	0.5	1390
7	5.19	389	2	0.5	1600
7	4.4	333	2	0.5	1600
20	4.4	303	2	0.5	1520
3	4.11	455	2	0.5	2170
			TKS		
3.4	0.191	140	1.7	0.07	400
3.75	0.163	130	1.7	0.07	1600
6.25	0.702	200	1.6	0.07	1600
5.6	0.648	310	1.9	0.1	1700
6.0	0.973	400	1.9	0.5	1700
5.05	0.8	330	1.8	0.5	1420
2.95	0.84	230	1.7	0.5	1420
4	1.493	320	1.7	0.5	1420
6.7	2.709	560	1.9	0.5	1700
4.8	1.125	250	1.8	0.5	1700
11.8	1.675	405	1.9	0.5	1700
13.2	6.414	650	1.9	0.5	2300
7.8	3.167	500	1.9	0.5	2500
18.7	2.92	460	1.8	0.5	1800

(2) Those of "soft" type, similar to type (1), but with larger slope, being close to horizontal for large strains, for liquefied soils (the upper 13 m at Port Island);
(3) Those of "hard" type, declining to the stress axis at large strains, for water-saturated soils (13–18 m and 27–32.5 m at Port Island, the upper 11 m at SGK, the upper 14 m at TKS). The terms "soft" and "hard" type stress–strain curves were introduced by Zvolinsky (1982).

Relations of type (1) were defined by three power functions, as described by Joyner and Chen (1975); relations (2) and (3) were defined parametrically, by 4th degree polynomials. Sets of such curves were generated, and item-by-item examination was applied to find groups of curves showing the best-fit approximation to the observed data. To account for temporal changes in the soil behavior, the records were divided into intervals of 1.5 s duration. Within each interval, the stress–strain relations were assumed to be stationary, and vary for different intervals. Calculations were performed (i.e., the "best-fit" stress–strain relations in the layers were determined) successively, interval by interval. No discontinuities occur at the boundaries of the intervals, because in the next interval the whole cycle of loading (or unloading) was recalculated for the new curve from its beginning. Horizontal components of records were processed.

For successive 1.5 s time intervals, the groups of stress–strain relations were found, showing the best-fit approximation to the observed records. Figures 3.2, 3.3, 3.4 and 3.5 show the observed and simulated records and the obtained stress–strain relations at different depths for ten successive time intervals at the three sites.

The results of simulations with the "best-fit" stress–strain relations show a good agreement between the observed and simulated records (Fig. 3.2).

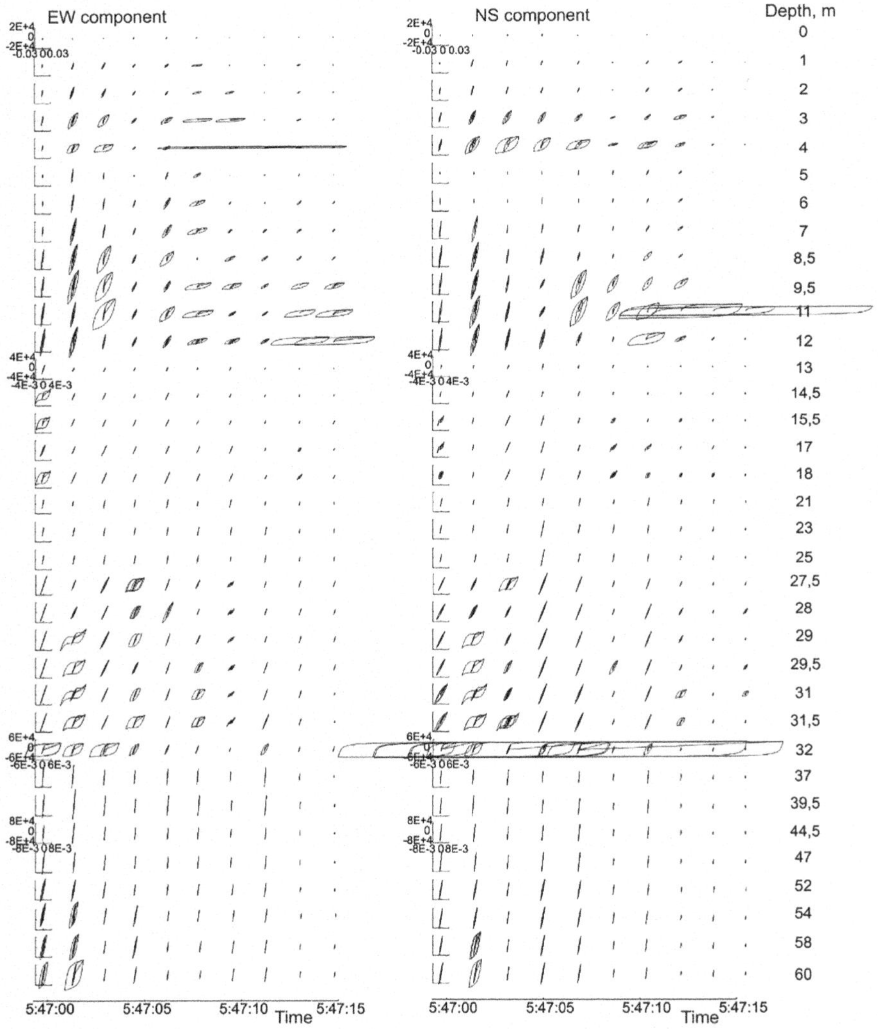

Fig. 3.3 The obtained stress–strain relations changing with time at Port Island site. The axes scales of the stress–strain relations are the same for all time intervals at given depths and below. Shear stresses are in Pa, strains in strain

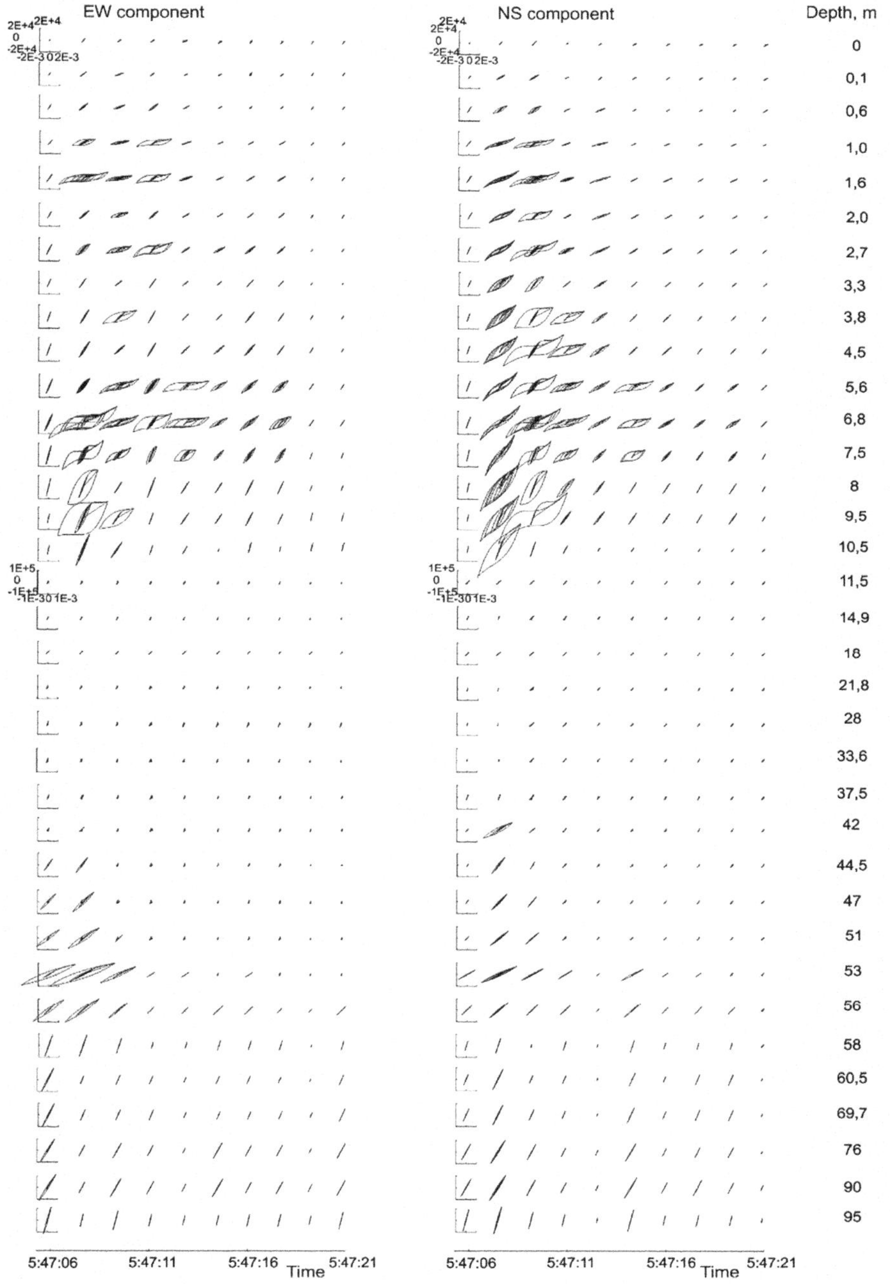

Fig. 3.4 The obtained stress–strain relations changing with time at SGK site. The axes scales of the stress–strain relations are the same for all time intervals at given depths and below. Shear stresses are in Pa, strains in strain

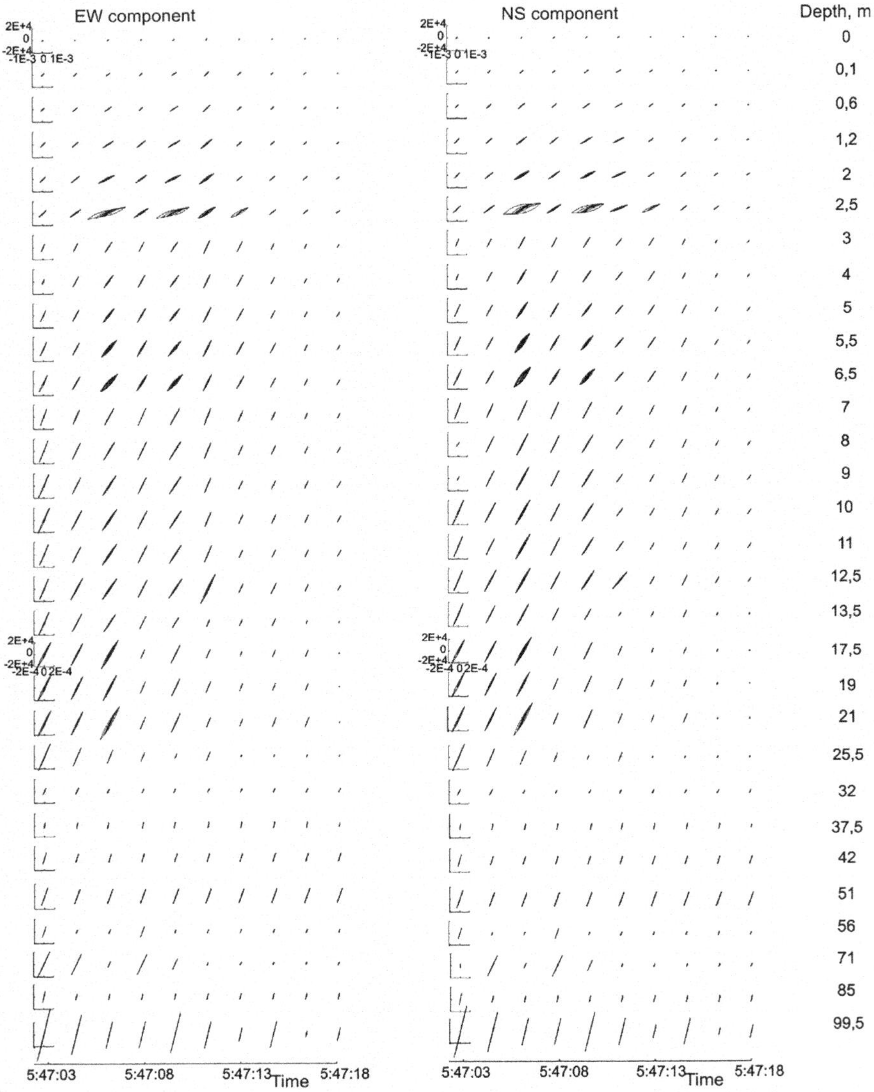

Fig. 3.5 The obtained stress–strain relations changing with time at TKS site. The axes scales of the stress–strain relations are the same for all time intervals at given depths and below. Shear stresses are in Pa, strains in strain

The most intense and complicated motion and liquefaction were observed at Port Island. The strongest horizontal accelerations were measured at the deepest point of 83 m, such as, ~ 500 Gal at EW component and ~ 600 Gal at NS component. According to the profiling data (soil composition, depth, saturation with water) and the results obtained in the previous analysis (Kawase et al. 1995), the soil profile down to 83 m was divided into five groups of layers, for which certain types of stress–strain relations were assumed (Fig. 3.1).

Since liquefaction occurred in the upper layers, curves of type (2) were applied for surface soils above the level of the underground water (0–13 m). Soil layers at depths 13–18 m and 27–32.5 m are reclaimed gravely fill and alluvial sand saturated with water; therefore, curves of type (3) were used to describe the soil behavior in these layers (Fig. 3.1). Peaks, or sharp increases in the amplitudes of the observed records at depths of 16 and 32 m confirm that curves of type (3) are relevant in this case. Stress–strain curves of type (1), which are considered to be typical for dense soils at depth, were assumed for alluvial clay layers at 18–27 m and for deep diluvial layers below 32.5 m.

The most noticeable changes in the soil behavior were observed in surface layers, in the upper 13 m: the stress–strain relations became more and more sloping with time, showing a substantial progressive reduction of the shear moduli and liquefaction (Fig. 3.2a). At depths 13–18 m and 27–32.5 m, the obtained stress–strain curves show a slight reduction and the following recovery of the shear moduli. Below 32.5 m, no changes in the soil behavior are observed (Figs. 3.2a and 3.3).

At SGK and TKS sites no liquefaction occurred, though sand boils observed after the quake around the TKS site indicate that liquefaction took place in the vicinities of this site (Figs. 3.2b, c, 3.4 and 3.5).

Maximum accelerations recorded at SGK site were also high (Fig. 3.2b), up to 650 Gal on the surface at EW component. At this site, the soil behavior can be described by a relatively simple model: the soil profile was divided into two groups of layers, such as, the near-surface alluvial layer (0–11 m) and deeper layers below 11 m, which are mostly gravels and clays. High values of *P*-wave velocities in the upper layers indicate the presence of water (Fig. 3.1); therefore, curves of type (3) are appropriate for the upper layers. Curves of type (1) were used for the deeper layers. The behavior of the layers below 11 m is stationary, whereas the stress–strain relations describing the behavior of the upper layers change with time: slopes of the curves decrease, then increase again, indicating reduction and recovery of the shear moduli in the upper layers (Fig. 3.6).

At TKS site, maximum recorded accelerations were about 200 Gal, and the soil profile is represented by a water-saturated reclaimed fill, an alluvial layer in the upper part (0–14 m), and by gravels, clays, and sands in the deeper parts. Since the soil conditions are similar to those at SGK site, the choice of curves was also similar: curves of type (3) were selected for the upper 14 m, and curves of type (1) were chosen for the deeper parts. The observed and simulated accelerograms and the obtained stress–strain relations are shown in Figs. 3.2c and 3.5. The most intense motions took place in the layers below the level of the underground water, at depths 4–7 m.

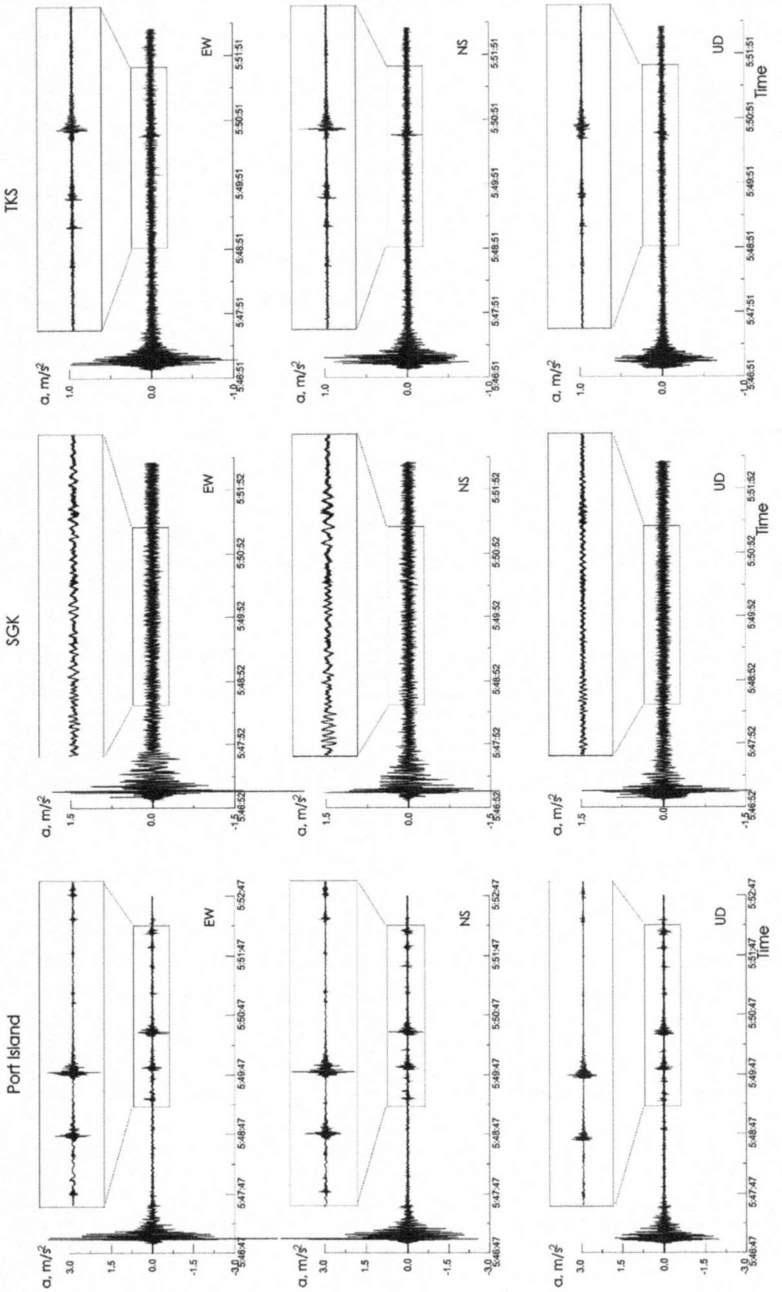

Fig. 3.6 Acceleration records of the 1995 Kobe earthquake and its aftershocks at sites PI, SGK, and TKS at depths of 32 m, 24.7 m, and 25 m, respectively. Reverberated *S* waves and basin-induced surface waves are seen in horizontal components at PI and SGK sites

At TKS site, soil behavior in the upper layers changed with time similar to SGK site: we observe reduction and the following recovery of the shear moduli, whereas, the behavior of layers below 14 m is stationary (Fig. 3.5).

The obtained stress–strain relations seem to represent a fairly good approximation to reality, because they show a good agreement between the observed and simulated data, they give a description of the process which is physically correct, and similarity of the stress–strain relations for two horizontal components.

An important factor is the accuracy of the profiling data, allowing a better agreement of the calculations and observations. Parameters of the soil profiles at SGK and TKS sites have been verified by repeated measurements and determined with higher accuracy than the parameters at Port Island site, estimated by inversion methods by Aguirre and Irikura (1997).

The number of the recording devices is another important factor, determining the accuracy of the description. Four three-component accelerometers at Port Island provide more detailed information than three devices at SGK and TKS, and allow a more detailed description of the soil behavior.

The 1995 Kobe earthquake was accompanied by several aftershocks that occurred just after the mainshock. Records of the aftershocks were used to study recovering of shear moduli in the upper layers after strong motion and liquefaction at PI (Fig. 3.6).

Weak signals like aftershocks were found more difficult to process than strong ones, because in weak motion, the working intervals of the stress– strain curves are located close to zero. As a result, many curves look similar, and the difference between the records simulated with different curves is small. Therefore, we used larger data lengths than in strong motion (10–40 s instead of 1.5 s), and obtained the best-fit stress–strain curves averaged over these time intervals. However, the parameters of the neighboring curves were close to each other, and the inaccuracies in their estimation related to inadequate choice of the curves were smaller than in the case of the mainshock.

Long-period waves recorded at horizontal components at PI and SGK sites (Fig. 3.6) may be standing waves due to *S*-wave reverberation in the basin sediments, probably including basin induced surface waves (Hatayama et al. 1995; Pitarka et al. 1998). Both sites PI and SGK belong to the Osaka basin, and the corresponding depths to the basement are 1000–1200 m and 1200–1400 m, respectively (Hatayama et al. 1995). The amplitudes of these waves achieved ~ 5 cm/s^2 at SGK site and ~ 3 cm/s^2 at PI, and their time period was about 4–6 s. Basin-induced waves were not recorded at TKS site because of its location outside the limits of the Osaka basin, on the other side of the fault plane (Hatayama et al. 1995). The aftershocks were superimposed on the background of basin-induced surface waves, which occurred after the mainshock.

The first aftershocks at Port Island were recorded $\sim$ 5 min after the main-shock, when the subsurface soils were still in a liquefied state (Figs. 3.7 and 3.8).

Progressive liquefaction within the upper 0–13 m of the reclaimed soil during the mainshock can be traced by the evolution of the stress–strain relations: their slope decreases with time, indicating a substantial reduction of shear moduli to approximately near-zero values. Stress–strain relations obtained for aftershocks at

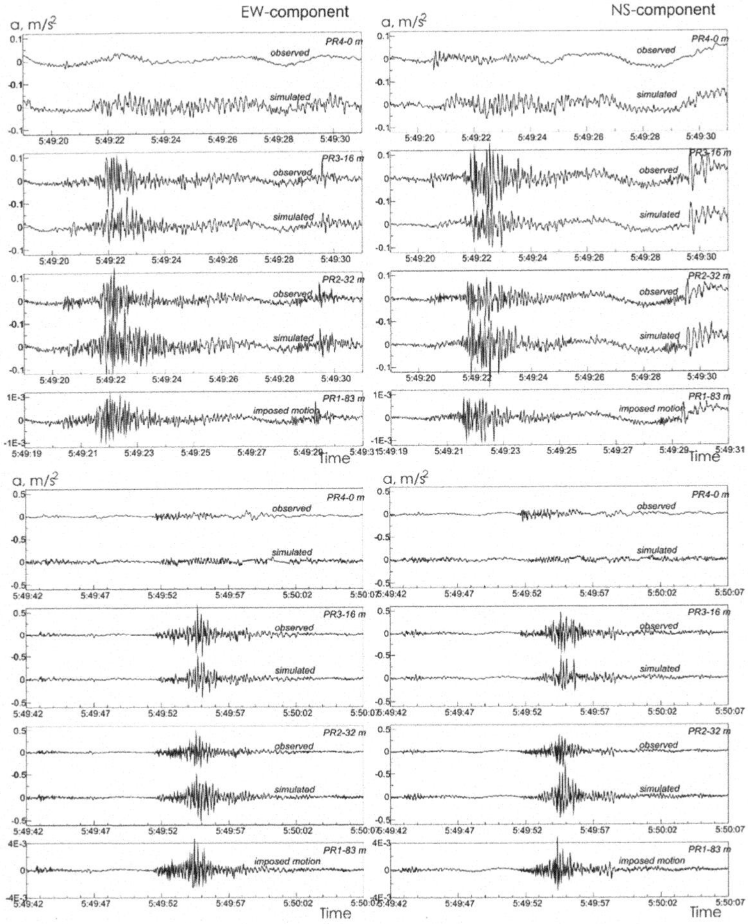

Fig. 3.7 Acceleration time histories of the first five aftershocks, recorded and calculated, at PI site

PI site also show large slopes for the upper 0–13 m; however, these slopes are slightly steeper than those obtained for the final parts of the mainshock. This can indicate the beginning of shear-moduli recovery a few minutes after the mainshock (Fig. 3.8). At SGK and TKS sites the imposed accelerations were not as high as at PI (300 Gal and 100 Gal, respectively). Changes in soil behavior occurred in the upper meters. Stress–strain relations characterizing soil behavior during the aftershocks indicate a complete recovery of the shear moduli in the upper layers just after the end of strong motion (Figs. 3.9 and 3.10).

The difference between the aftershock records in liquefied (PI) and nonliquefied (SGK and TKS sites) soils is clearly seen in Figs. 3.7, 3.9 and 3.10. At SGK and TKS sites, the acceleration amplitudes on the surface were higher than at PI site, and amplification of seismic waves in the upper layers was observed; whereas at PI,

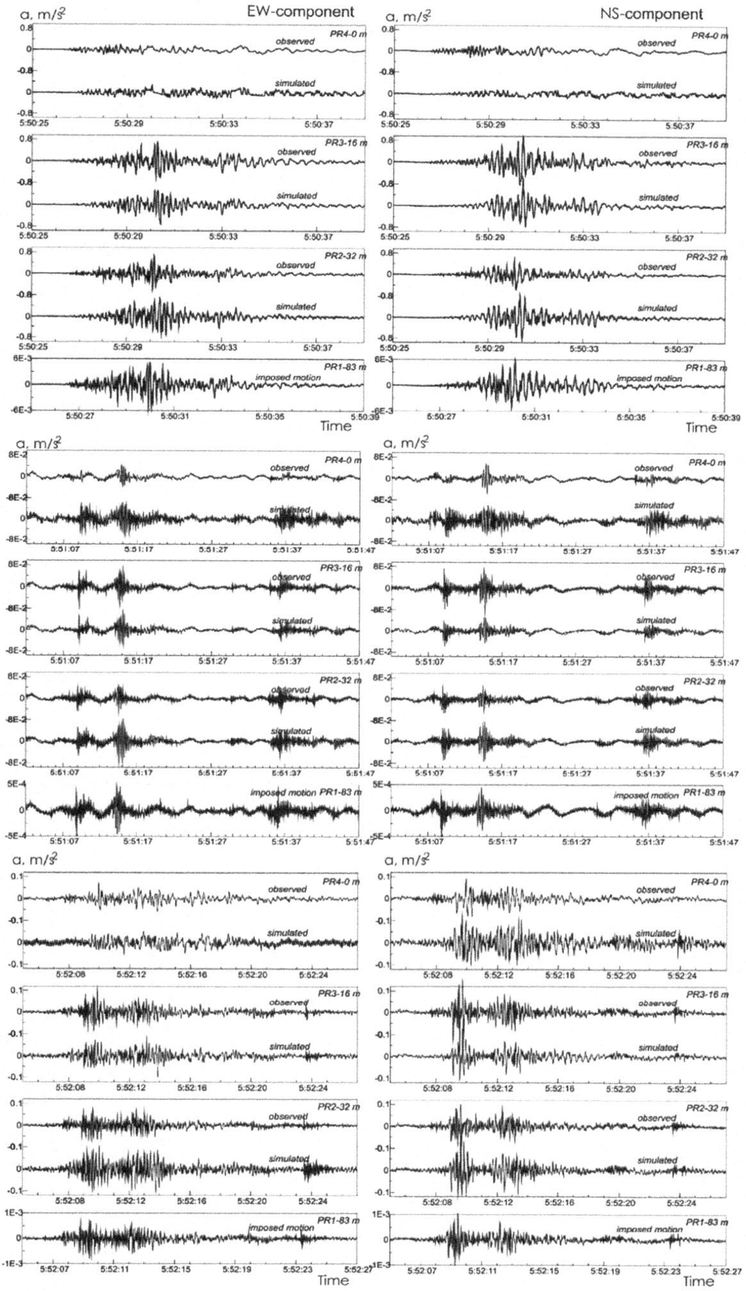

Fig. 3.7 (continued)

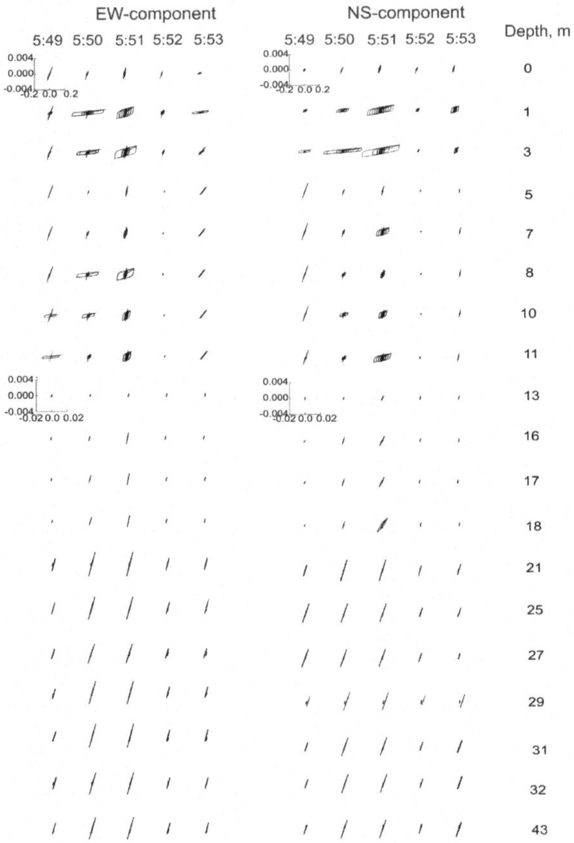

Fig. 3.8 The obtained stress–strain relations for the mainshock and five aftershocks at PI at EW and NS components. The stress–strain curves are scaled in relative units, which are the same for all time intervals at a given depth. Stresses and strains are scaled as proposed by Hardin and Drnevich (1972): stress was normalized by multiplying by $1/\tau_{max}$, and strain was normalized by multiplying by G_{max}/τ_{max}

where soils were still liquefied, the acceleration amplitudes were very small on the surface for all aftershocks except the last one, which showed higher amplitudes. This confirms that ~ 5 min after the termination of the strong motion, recovery of shear moduli in the liquefied layers was already noticeable.

The obtained stress–strain relations during the mainshock and the aftershocks were used to estimate shear moduli at different depths (Fig. 3.11). We calculated G_{sec}, which approximate average values of G_{tan} over the entire stress–strain loops (Kramer 1996). As explained above, one stress–strain curve was used to describe the behavior of soils within a certain group of layers at a given time interval. However, the working areas of the stress–strain curves differ from layer to layer within a group, and the estimates of G_{sec} vary within each group from layer to layer. Weighted-mean values of G_{sec} normalized by G_{max} (G/G_{max}) were determined, which characterize average values of the shear moduli within given groups of layers in given time intervals.

In the calculations, there are two main sources of possible errors: (1) inexact determination of stress–strain relations in the layers and (2) scattering of shear moduli

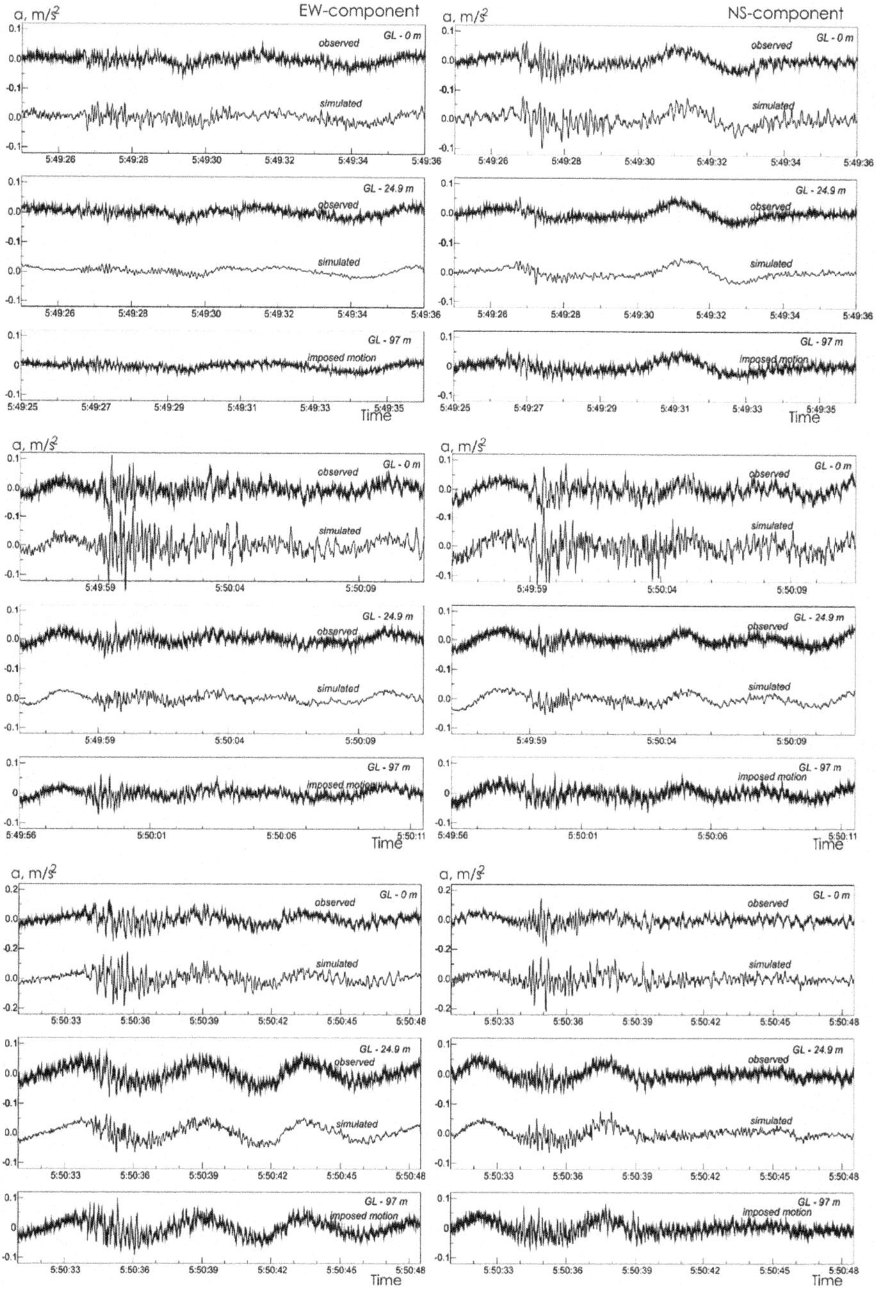

Fig. 3.9 Acceleration time histories of the first three aftershocks, recorded and calculated, at SGK site

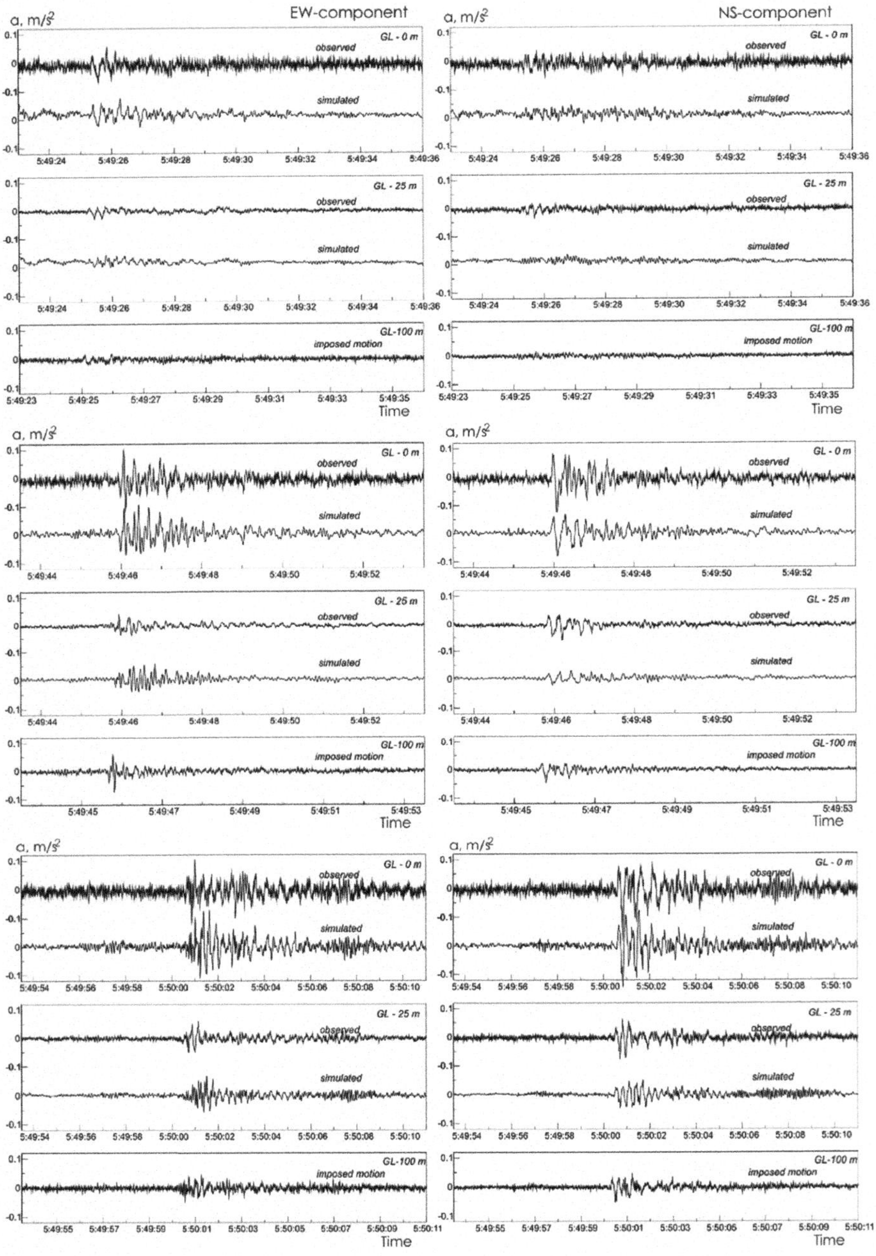

Fig. 3.10 Acceleration time histories of the first three aftershocks, recorded and calculated, at TKS site

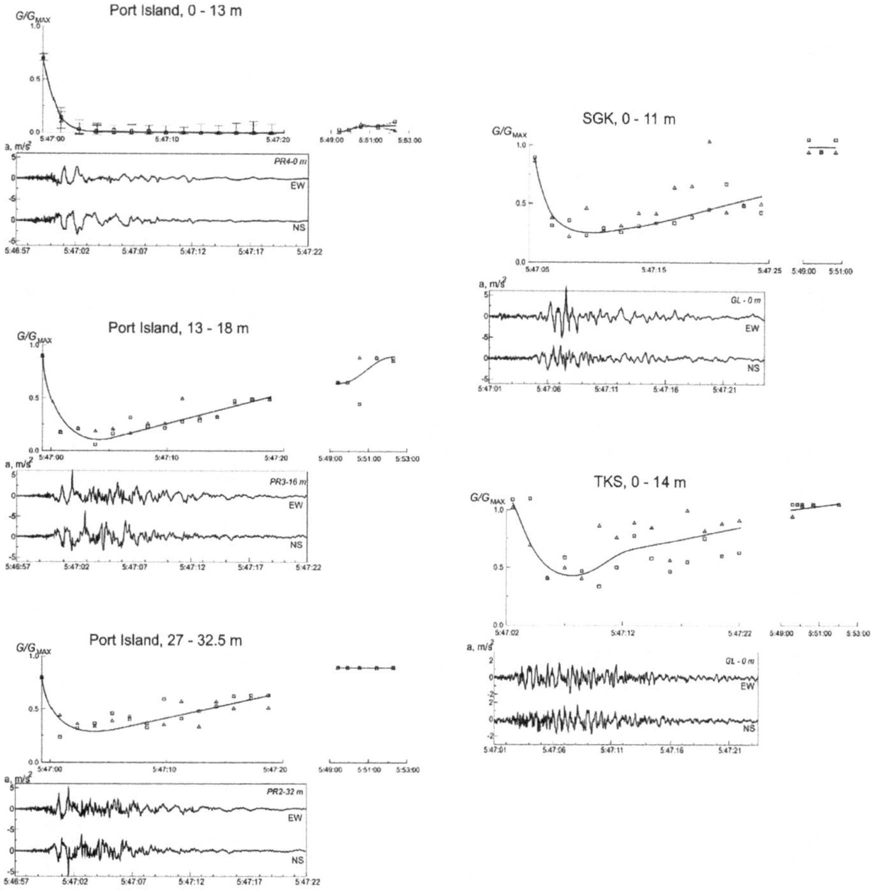

Fig. 3.11 Evolution of shear moduli in subsurface soils at PI, SGK, and TKS sites during the 1995 Kobe earthquake and its aftershocks. The observed and calculated accelerograms for the corresponding groups of layers are also shown

estimates within groups of layers. Figure 3.11 shows confidence limits characterizing the spread of the obtained estimates of shear moduli in individual layers within groups of layers.

In the upper 0–13 m at PI, shear moduli decreased gradually during the mainshock (Fig. 3.11). The decrease was most noticeable in the first 3–5 s of strong motion, when it achieved ~ 80–90% of the initial value. Shear moduli reached their minima during the following 5–7 s and showed a tendency to increase a few minutes after a decrease in the intensity of strong motion.

Scattering of shear moduli estimates corresponding to aftershocks indicate that the recovery proceeded nonuniformly, which can be attributed to the movement of the ground waters. The upper subsoil layer is a decomposed granite soil called "Masado," which is a sandy soil with much gravel, silt, and clay, and the contents of gravels and fines are ~ 0%–65% and 5%–35%, respectively (Yasuda et al. 1996).

According to many researchers, the presence in soil of even a small quantity of clay particles (1.5–2% according to Gumensky 1965) imparts thixotropic properties to the soil (Boswell 1949; Akkerman 1958; Gumensky 1965). Soils like "Masado" containing small quantities of fines, possess intermediate properties between noncohesive and cohesive (quasi-thixotropic) soils.

For such weakly cohesive soils, a quick decrease in rigidity even at rather low levels of dynamic loading is typical, due to their mixed composition. In such soils, the Coulomb friction is less than in a pure sand, and the coagulation lattice is weak and usually discontinuous. The decreased permeability hampers dissipation of the pore pressure and facilitates liquefaction. In such soils, gradual and slow hardening after termination of the dynamic loading includes two processes, such as, weak compression of the soil accompanied by the loss of water and thixotropic strengthening of the coagulation lattice. The compression proceeds in accord with the pore pressure dissipation, and its rate is determined by the filtering properties of a soil, whereas the thixotropic strengthening takes several hours (maximum, 1 day) and proceeds rather nonuniformly during the first few minutes owing to continued water movement (Voznesensky 1999).

This recovery mechanism consisting of the two processes, pore pressure dissipation and thixotropic strengthening, seems to be probable at PI, considering the soil composition, nonuniformity of the recovery during the first few minutes and the total time of the recovery estimated by Aguirre and Irikura (1997). They showed that the aftereffects of liquefaction remained at least 3 h after the mainshock but no more than 24 h.

At depths of 13–18 m changes in the shear moduli during the first few seconds of strong motion were as large as in the surface layer (Fig. 3.11). The soil composition was similar to that in the upper layers, but the soils were saturated with water. A decrease of the rigidity in these layers started with the beginning of strong motion and continued for 3–4 s, that is, until the intensity of the imposed motion decreased. After that, the compression and probably thixotropic strengthening took place. The maximum decrease of shear modulus was about 80%, and the obtained stress–strain relations show its recovery during the following few minutes. The same tendency can be revealed from the stress–strain curves presented by Kazama et al. (1998), that is, its decrease during the first 1–4 s of strong motion and the following recovery during the next 6–9 s. Our results also agree with the results obtained by Sato et al. (1996), who identified an 80% and 50% reduction of V_S for saturated and unsaturated fill layers, respectively. At depths of 27–32.5 m at PI, shear moduli decreased by 50–60% of the initial values during the first few seconds of strong motion (Fig. 3.11), which also agrees on the whole with the estimates by Sato et al. (1996).

At SGK and TKS sites shear moduli decrease of ~ 60–70% and ~ 50% of the initial values was observed in the upper layers, and the recovery was completed during the following minutes. These estimates are generally consistent with the estimates by Sato et al. (1996).

Thus, subsurface weakly-cohesive soils at Port Island, SGK and TKS behaved as dynamically unstable systems. The sensitivity of the soils to dynamic loadings decreases with depth, and no significant changes in the soil behavior during strong motion were observed in denser underlying layers.

Vertical components. Similar analysis was performed for the vertical components (*P*-waves, approximately). To estimate stress–strain relations in soil layers, the propagation of vertically incident seismic *P*-waves in horizontally layered sediments was calculated; a modified algorithm by Joyner and Chen (1975), implicit method, was used. The vertical components were processed by the same way as horizontal ones, with some modifications in the technique. The compression stress in failure E_{max} and the low-strain compression moduli were estimated approximately from the *P*-wave velocity according to formulas of the theory of elasticity. Assumptions based on the profiling data were made about the types of the stress–strain relations in different soil layers (Fig. 3.1). Sets of stress–strain curves were generated, and item-by item examination was applied to find groups of curves, showing the best-fit approximation to the observed records; the intervals of 1.5 s duration were analyzed; the results are presented in Figs. 3.12 and 3.13.

As seen from the figures, on vertical components nonlinear distortions are not as large as on horizontal ones, because the acceleration amplitudes of *P*-waves are smaller than of *S*-waves; at the same time, *P*-wave velocities in soil layers are noticeably higher (especially, in water-saturated upper layers), and shorter time intervals (when nonlinear distortions are accumulated) are required for the wave to reach the surface.

The most intense motion and liquefaction were observed at PI. We can try to improve the accuracy of simulations, taking into account changes in the groundwater level caused by strong motion: a better agreement between the observations and simulations is achieved if we increase the groundwater level, as shown in Fig. 3.12.

The obtained results are generally consistent with the conclusions of Yang Jun and Sato (2000) about an incomplete water saturation of subsurface layers, but cannot confirm their conclusion that the intensity of motion on vertical components is greatly influenced by small changes in the degree of water saturation of the soil. The stress–strain relations in subsurface soils show a decrease in the compression moduli due to soil liquefaction. At SGK and TKS sites, soil behavior estimated by the records of the vertical components is rather stationary and close to linear.

Variations of the rigidity of soil layers at the three sites during strong motion are shown in Fig. 3.14. E_{sec} moduli were calculated, which approximate average values of E_{tan} over the working intervals of the stress–strain loops (Kramer 1996). The mean values were estimated characterizing the average values of the compression moduli within groups of layers in each time interval.

The most noticeable changes of the moduli were observed in the upper 13 m at PI: the compression moduli reduced by ~ 80% of their initial values. In the layers at 13–18 m and 27–32.5 m, a reduction and recovery of the rigidity is observed, proceeding synchronously at the horizontal and vertical components; the recovery started just after the end of strong motion. Below 32.5 m, no substantial changes in the soil behavior were observed.

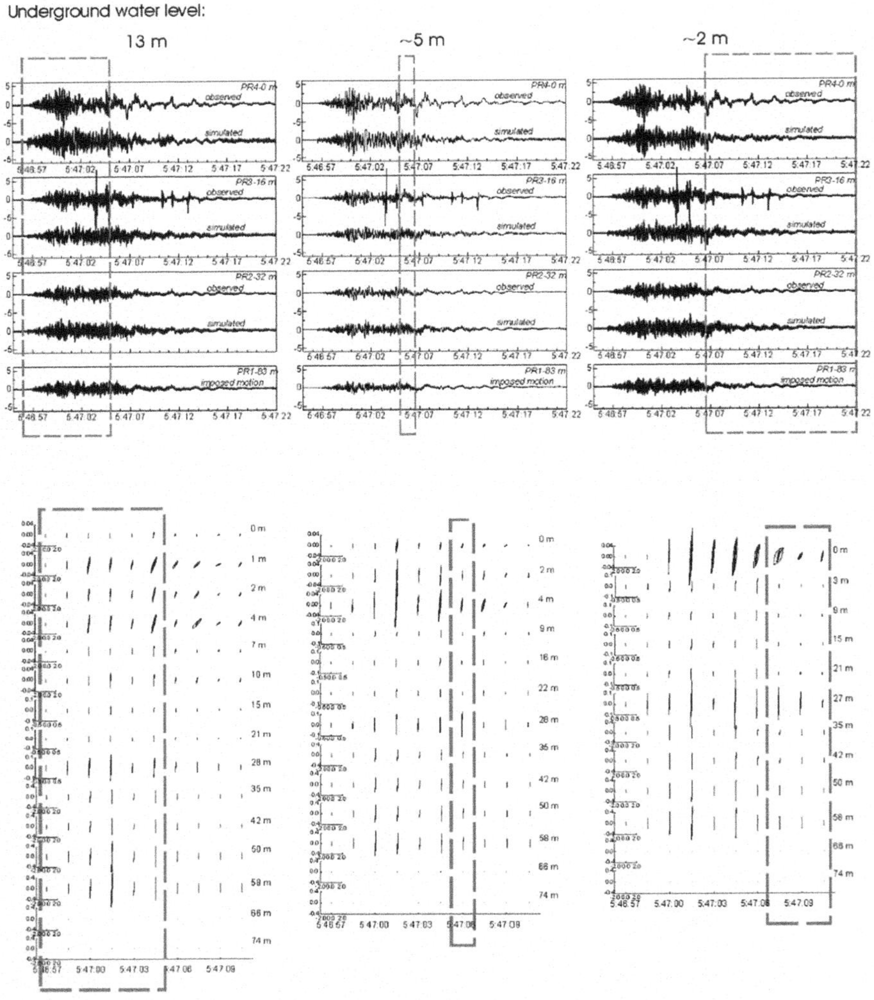

Fig. 3.12 Acceleration time histories of the mainshock at PI, observed and simulated, and the obtained time-dependent stress–strain relations for vertical components. Three levels of the underground water are considered, such as, 13 m, ~ 5 m, and ~ 2 m. The axes scales of the stress–strain relations are in relative units, the same for all time intervals at given depths

At SGK and TKS sites, changes in the rigidity occurred simultaneously at the horizontal and vertical components in the upper water-saturated alluvial and reclaimed soil layers (Fig. 3.14); the recovery started with the decrease of the intensity of motion and finished in a few minutes. In general, changes in the compression moduli of the soil layers during strong motion were noticeably lower than changes in the shear moduli.

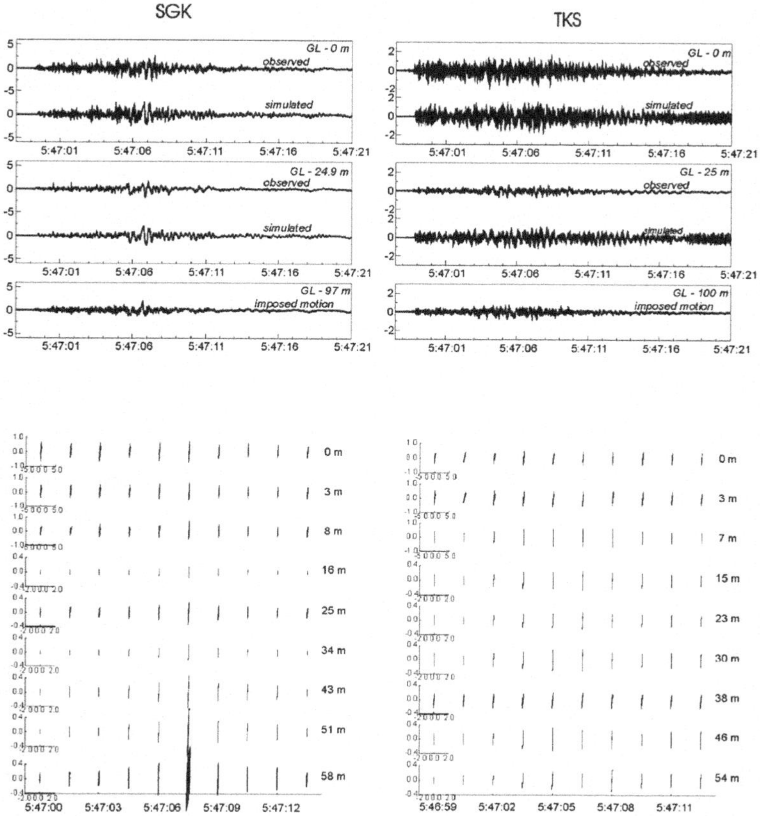

Fig. 3.13 Acceleration time histories of the mainshock at SGK and TKS sites, observed and simulated, and the obtained time-dependent stress–strain relations for vertical components. The axes scales of the stress–strain relations are in relative units, the same for all time intervals at given depths

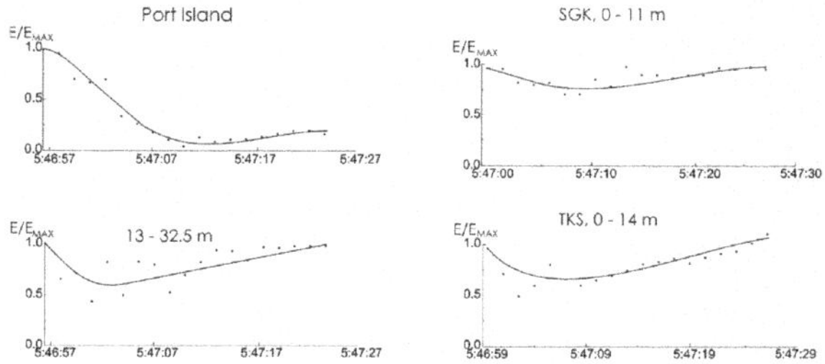

Fig. 3.14 Evolution of compression moduli in subsurface soils at PI, SGK, and TKS sites during the 1995 Kobe earthquake and its aftershocks

3.2 Models of Nonlinear Soil Behavior During the 2000 Tottori (Japan) Earthquake

After the 1995 Kobe earthquake, the next strong earthquake that caused strong motion in Japan occurred in the Tottori Prefecture at approximately 1:30 p.m. on October 6, 2000. The earthquake had an estimated moment magnitude $M_w = 6.7$ and a source depth of ~ 11 km. The fault rupture was ~ 30 km wide by 10 km deep, nearly vertical. The upper boundary of the rupture plane passed below the surface at a depth of ~ 1 km.

The earthquake was recorded by vertical arrays of the Kiban-Kyoshin Digital Strong- Motion Seismograph Network KiK-net at 220 sites located at epicentral distances of 7–626 km. Figure 3.15 shows the locations of the mainshock and the aftershocks tracing the fault plane of the Tottori earthquake, and the nearby KiK-net sites. The vertical arrays contain two accelerometers, installed on the surface and at a depth of ~ 100–200 m or more.

The records were used to study soil response in the near-fault zones; we chose stations located within ~ 100 km from the epicenter, where maximum recorded accelerations exceeded 100 Gal, assuming that at larger distances or smaller accelerations the soil response was virtually linear. The list of the studied KiK-net sites with the epicentral distances, maximum recorded accelerations, and soil conditions is given in Table 3.2.

The profiling data at the KiK-net sites include information on the composition of the soil layers and *P*- and *S*-wave velocities (Fig. 3.16), whereas, shear stress in failure

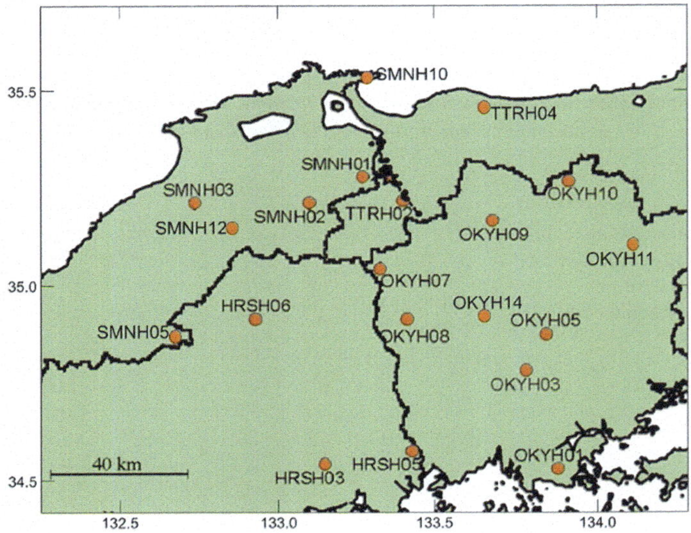

Fig. 3.15 Locations of the mainshock (star) and the aftershocks (black points) of the 2000 Tottori earthquake and nearby KiK-net sites (circles)

Table 3.2 Maximum recorded accelerations and the epicentral distances of the studied KiK–net sites located in the near-fault zones of the Tottori earthquake

Site code	PGA, Gal	Epicentral distance, km	V_{S30}, m/s	d/V_S
TTRH02	927.2	7	374.8	11/210
SMNH01	720.4	8	453.0	11/290
SMNH02	564.0	24	502.7	12/330
OKYH07	179.7	26	929.2	6/510
SMNH10	226.4	31	967.3	4/500
OKYH09	283.8	32	511.0	14/320
TTRH04	218.1	33	254.3	30/254.3
OKYH08	238.5	41	694.2	6/400
OKYH14	443.0	45	693.2	3/120
SMNH12	258.6	46	590.2	4/120
OKYH10	280.7	53	553.7	8/250
HRSH06	240.3	57	276.9	19.5/234
SMNH03	154.8	57	434.6	7/240
OKYH05	149.0	65	607.9	4/240
OKYH03	129.5	69	307.3	8/124
OKYH11	139.1	74	532.0	10/280
SMNH05	121.5	79	710.5	4/300
HRSH05	131.0	80	365.2	9/280

(τ_{max}) and attenuation parameters used in the calculations were estimated based on the records of twenty aftershocks occurred within one month after the mainshock. Inversions were performed using a genetic algorithm; the ranges for shear stress in failure τ_{max} in the layers were selected accounting for the composition of the layers, their depth and *S*-wave velocities: τ_{max} ~ 0.4–1.3 bar were taken for the upper layers, while in the deeper layers, τ_{max} was estimated based on the lithostatic pressure increasing with depth.

According to the profiling data (soil composition, *P* and *S*-wave velocities), at sites with soft subsurface soils (TTRH02, SMNH01, TTRH04, OKYH10, HRSH06, SMNH03, OKYH03, OKYH11, HRSH05) the soil profiles were divided into groups of layers, such as, the upper softer layers of ~ 5 m thickness and more, where *S*-wave velocities $V_S < 300$ m/s and the underlying denser layers with $V_S > 300$ m/s. The behavior of the upper layers was described by hard-type stress–strain relations, which are typical for water-saturated soils ($V_P > 600$ m/s), while the behavior of the underlying denser layers was described by stress–strain relations obtained in laboratory experiments by Hardin and Drnevich (1972). At other sites with denser soils, the behavior of all the layers was described by one normalized stress–strain relation.

The propagation of vertically incident seismic waves in soil layers was calculated; records of the accelerometers located at depth were used as input motion, and calculations were performed for time intervals of 1.5 s duration.

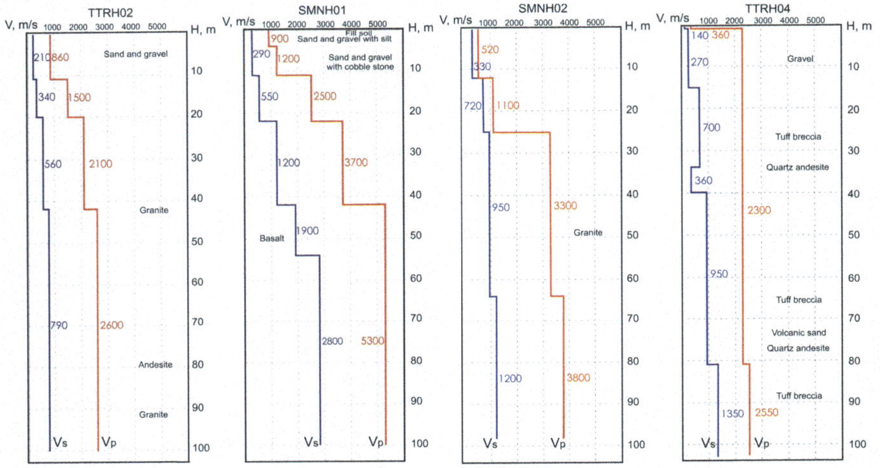

Fig. 3.16 The profiling data at some of the studied KiK-net sites closest to the fault

The results of the simulations, i.e., acceleration time histories and models of soil behavior at soft-soil sites closest to the epicenter are presented in Figs. 3.17, 3.18, 3.19, 3.20, 3.21, 3.22, 3.23, 3.24, 3.25 and 3.26.

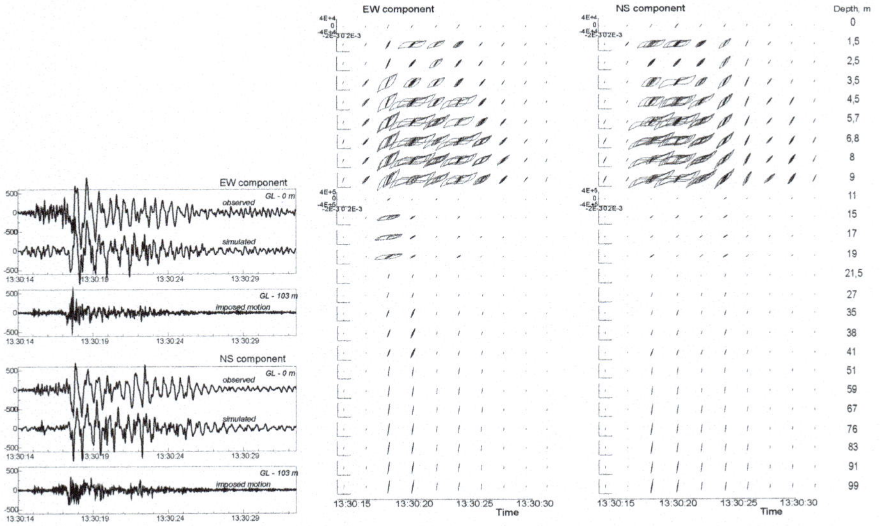

Fig. 3.17 Recorded and simulated acceleration time histories of the 2000 Tottori earthquake at TTRH02 site and the corresponding stress–strain relations, changing with time during strong motion. The axes for the stress–strain relations are the same for all time intervals at a given depth and below. Shear stresses are in kPa, shear strains in strain

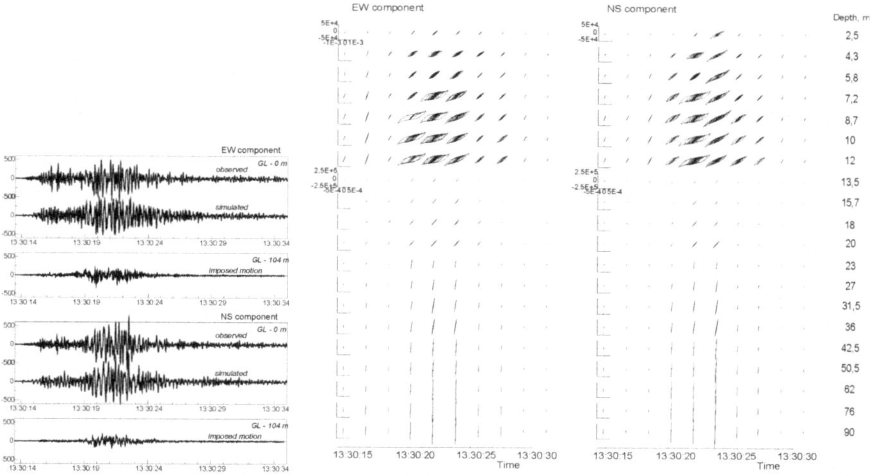

Fig. 3.18 Recorded and simulated accelerograms and stress–strain relations at SMNH01 site

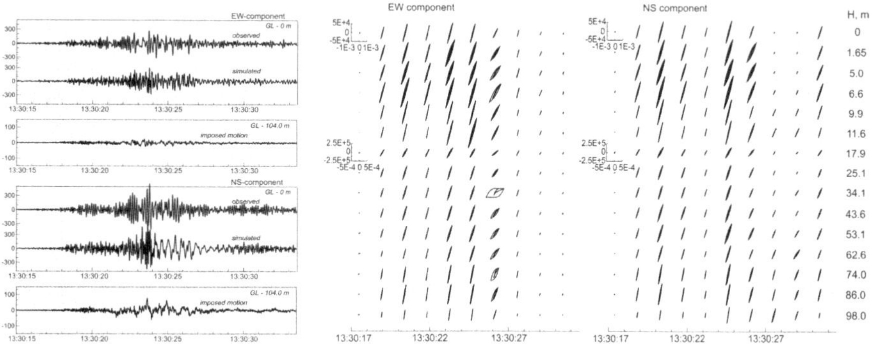

Fig. 3.19 Recorded and simulated accelerograms and stress–strain relations at SMNH02 site

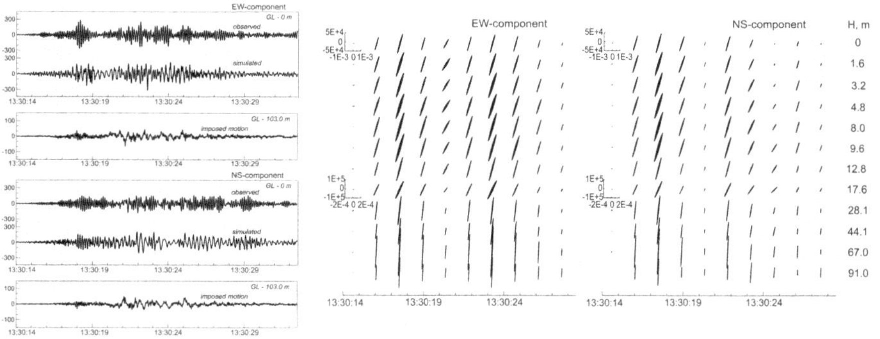

Fig. 3.20 Recorded and simulated accelerograms and stress–strain relations at OKYH09 site

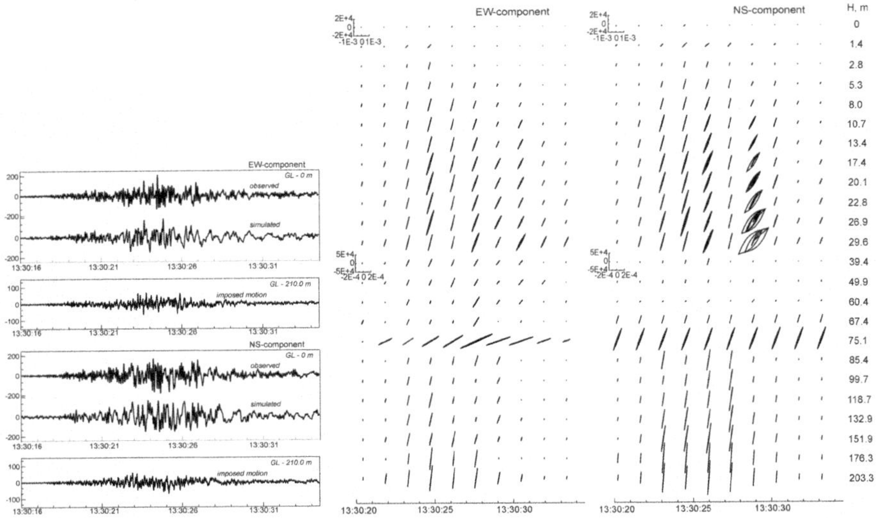

Fig. 3.21 Recorded and simulated accelerograms and stress–strain relations at TTRH04 site

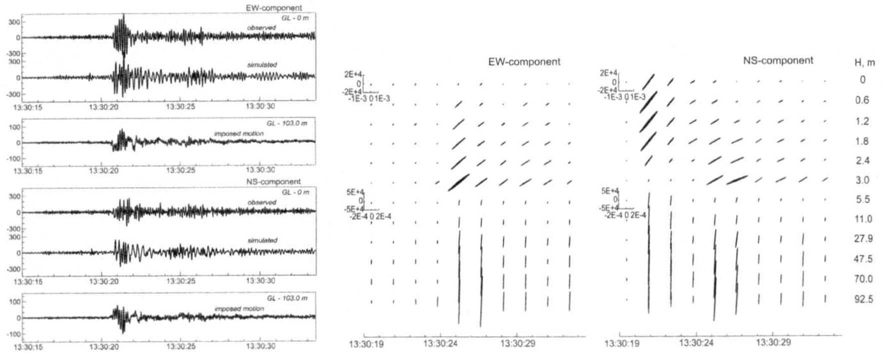

Fig. 3.22 Recorded and simulated accelerograms and stress–strain relations at OKYH14 site

Acceleration records at TTRH02 and SMNH01 sites (7 and 8 km from the epicenter) show clear evidence of substantial nonlinearity of soil response, such as, differences in the shapes and spectra of the records on the surface and at depth, emphasizing of low-frequency components of motion on the surface (Figs. 3.17 and 3.18). Other sites located at larger distances from the fault, show weaker manifestations of soil nonlinearity (Figs. 3.19, 3.20, 3.21, 3.22, 3.23, 3.24 and 3.25), and at HRSH05 site the soil response looks as virtually linear (Fig. 3.26).

As seen from the figures, a fairly good agreement was obtained between the recorded and calculated acceleration time histories at the near-fault and remote sites.

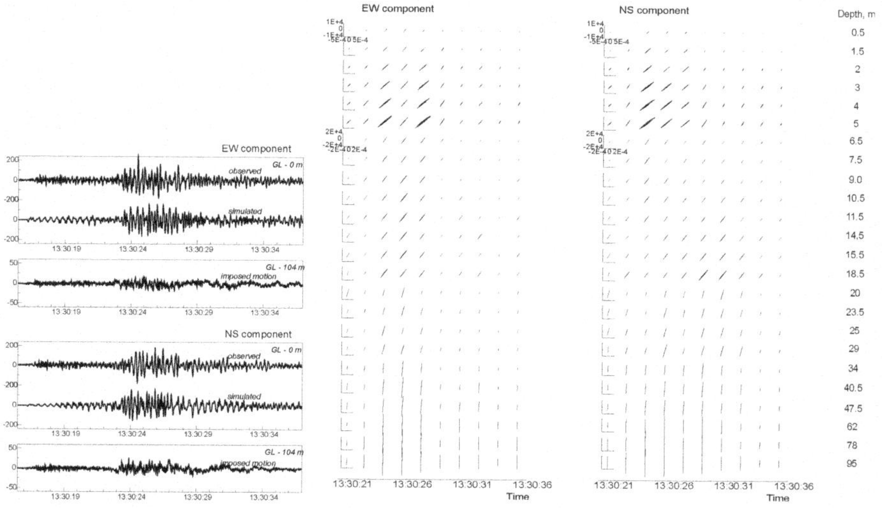

Fig. 3.23 Recorded and simulated accelerograms and stress–strain relations at HRSH06 site

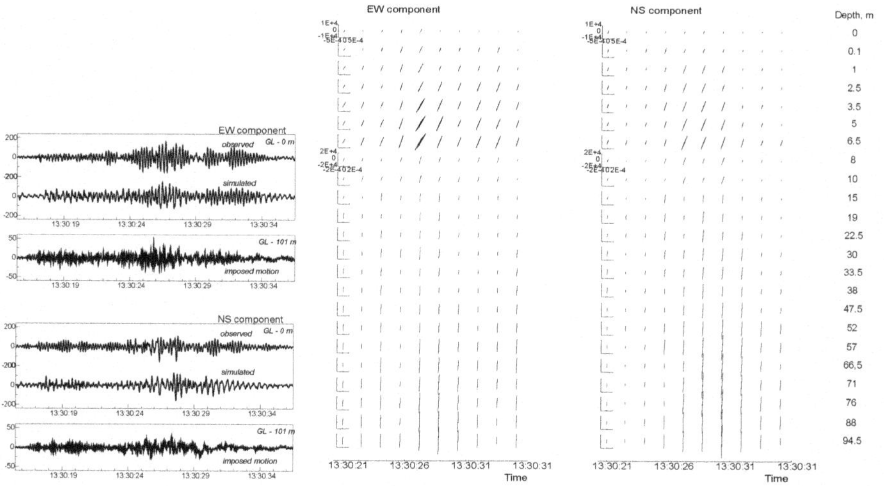

Fig. 3.24 Recorded and simulated accelerograms and stress–strain relations at SMNH03 site

At the near-fault sites TTRH02 and SMNH01, the rheological properties of the upper layers changed during strong motion (Figs. 3.17 and 3.18). At other sites, the changes are less pronounced. As seen from the figures, the changes are observed mainly in the upper subsurface layers, whereas soil behavior at depth is more stable.

Figure 3.27 shows the estimates of shear moduli changes in the upper subsurface layers during the earthquake. The shear moduli were estimated as the slopes of the stress–strain relations averaged over the groups of the upper layers and motions within the studied time intervals.

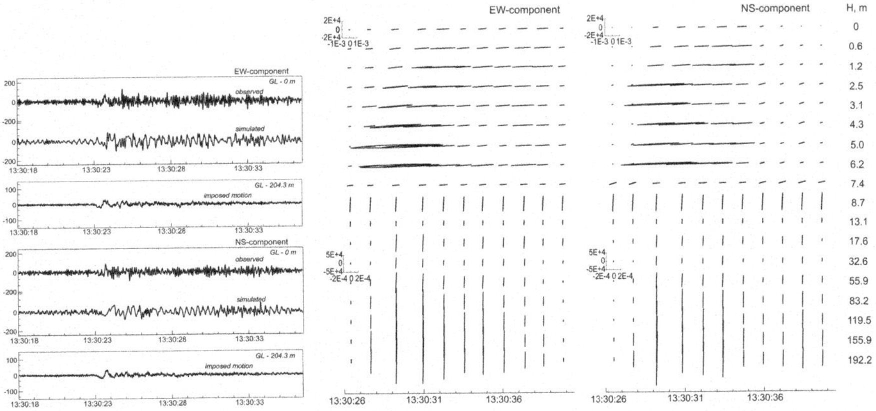

Fig. 3.25 Recorded and simulated accelerograms and stress–strain relations at OKYH03 site

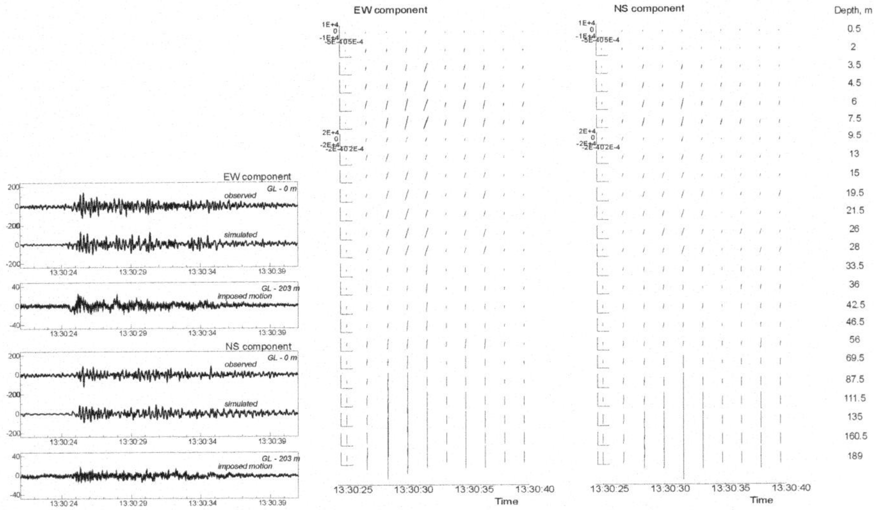

Fig. 3.26 Recorded and simulated accelerograms and stress–strain relations at HRSH05 site

Maximum decrease of the shear moduli reaching ~ 60% of the initial values was observed at TTRH02 and SMNH01 sites. Shear moduli reduction was also observed at TTRH04, OKYH14 and HRSH06 sites, where acceleration time histories contain groups of intense waves; shear moduli reduced at the arrivals of the intense waves. At other sites, shear moduli slightly vary, reducing and recovering during strong motion, which probably related to the waveforms of the acceleration time histories: they represent long-duration motion of approximately constant intensities.

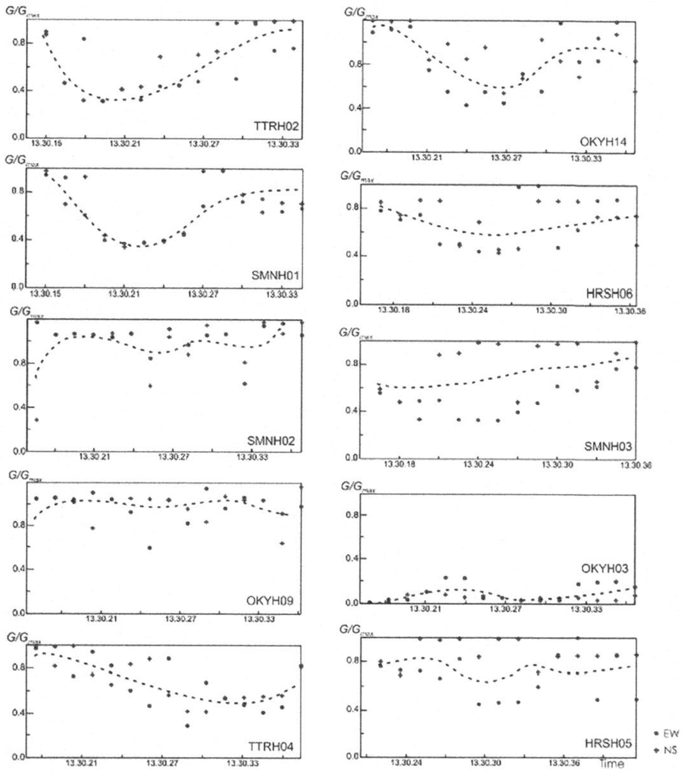

Fig. 3.27 Temporal changes of shear moduli at the studied KiK-net sites during the 2000 Tottori earthquake

3.3 Models of Nonlinear Soil Behavior During the 1999 Chi-Chi (Taiwan) Earthquake

A destructive earthquake of a moment magnitude $M_w = 7.6$ and depth of ~ 7.5 km occurred in central Taiwan with its epicenter near the town of Chi-Chi on September 21, 1999. The earthquake had a predominantly thrust focal mechanism, and the surface rupture was mapped over a distance of ~ 100 km in NS direction. Field observations show the greatest surface offset to be more than 8 m vertically, occurring near the northern end of the fault.

The earthquake triggered more than 400 strong motion devices across Taiwan (Fig. 3.28). The strong-motion wavefield was captured by a dense network of surface stations, and the obtained records represent one of the most complete strong-motion databases to date, which allows us to study the radiation and propagation of seismic waves from the earthquake source, as well as local effects at rock and soil sites. The availability of such a complete near-source strong-motion dataset being generated by the Chi-Chi earthquake, combined with broadband teleseismic displacement waveforms, well-distributed Global Positioning System (GPS) data, and the considerable volume of geological and geophysical information collected in Taiwan over the past decades has provided an excellent opportunity to study the source process of this large earthquake, and various versions of slip distribution over the fault plane have been suggested (Ma et al. 2001; Zeng and Chen 2001; Oglesby and Day 2001; Wu et al. 2001; Chi et al. 2001; Wang et al. 2001).

The majority of near-fault records were obtained at soil sites (virtually all stations on the footwall of the fault are soil sites); these records show signs of nonlinearity in the soil response, such as, the predominance of low-frequency components in motion on the surface, smoothed spectral shapes approaching the form $E(f) \sim f^{-k}$, typical for cases of strong nonlinearity (Pavlenko and Irikura 2005), and characteristic spiky waveforms (Archuleta 1998).

Models of nonlinear soil behavior at sites of the Taiwan strong motion network based on stochastic simulations with finite-fault source

The possibilities of studying soil behavior during the 1999 Chi-Chi earthquake are restricted by the fact that only surface records are available. A single vertical array at Dahan site, located on dense soils, did record this earthquake but at a rather large distance (~80 km) from the fault (Chiu 2001; Pavlenko and Loh 2005).

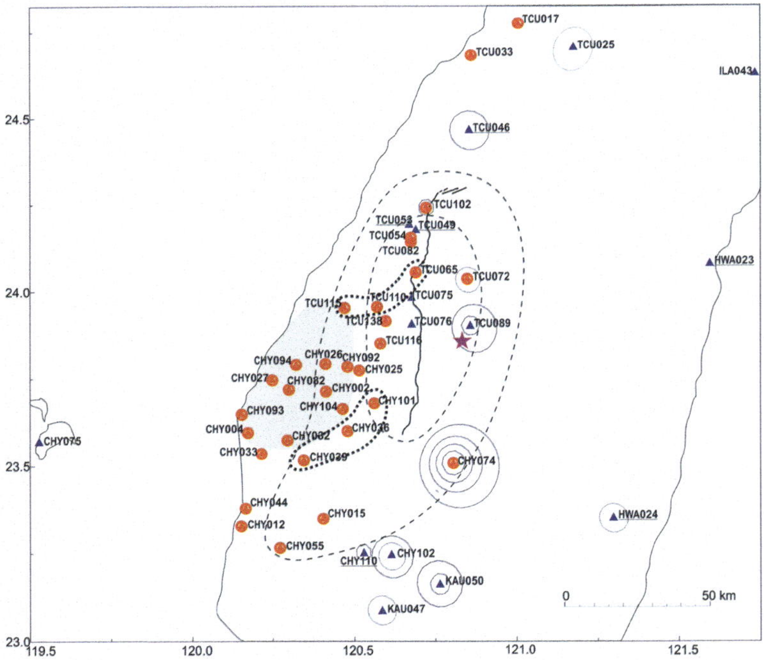

Fig. 3.28 Map showing the epicenter of the 1999 Chi-Chi earthquake (star), the surface rupture (black line) and the locations of eighteen rock (blue triangles) and thirty one soil stations (orange circles) studied in this work. Dashed lines bound the areas where $R_c < 15$ km and $R_c < 50$ km (R_c is described below). Dotted lines mark liquefaction areas, and the area, where basin-induced surface waves were identified, is marked by grey color

Nonlinearity of soil response during the Chi-Chi earthquake has been studied by Roumelioti and Beresnev (2003). In these studies, spectra of acceleration time histories recorded at soil sites are investigated in order to find characteristics of soil nonlinearity. Lower amplification of seismic waves in some frequency bands during strong ground motion compared to weaker motion (aftershocks) is interpreted as an indicator of nonlinearity of soil response, and respective quantitative characteristics are introduced.

The authors simulated records of the Chi-Chi earthquake by the stochastic (Boore 2003) finite fault (Beresnev and Atkinson 1998) method and calibrated them against the data recorded at twenty-four rock sites. Then they simulate soil site records using the linear-response assumption, where the simulated soil site input motions were amplified by weak motion amplification functions, estimated by the spectral ratio technique from the aftershock records. Their comparison with the actual observations allows the authors to reveal an average reduction in strong motion amplification to about 0.5–0.6 of that in weak motions, with an acceleration "threshold" for detectable nonlinearity near 200–300 cm/s^2. The conclusions of this study confirm in the whole earlier obtained average estimates of decreasing amplification of seismic waves during strong motion at soil sites; however, the authors emphasize a large inter-station response variability, which did not allow them to derive a statistically significant difference in weak and strong motion amplifications based on the responses available at the studied sixteen sites.

Accounting for the observed variability of soil response patterns, we concentrate on constructing models of soil behavior at separate stations, instead of studying average regularities, which seems to be more informative in this case. In collaboration with K.-L. Wen, we performed a detailed investigation of the soil response during the 1999 Chi-Chi earthquake. Similar to the approach by Roumelioti and Beresnev (2003), based on the records at rock sites, we estimated input parameters and simulated acceleration time histories at soil sites at the bottoms of soil layers and then calculated their propagation up to the surface, selecting models of soil behavior showing the best fit to the observed acceleration records at the surface.

The epicenter of the 1999 Chi-Chi earthquake, surface rupture and locations of eighteen rock and thirty-one soil stations studied in this work are shown in Fig. 3.28. Strong motion data of the Chi-Chi earthquake were used, disseminated on CD-ROM by Lee et al. (2001a; b), accelerograms provided by the website of the Central Weather Bureau (CWB) of Taiwan and the profiling data provided by the National Center for Research in Earthquake Engineering (NCREE) of Taiwan.

Firstly, accelerograms of the Chi-Chi earthquake at rock sites were calculated. Since the purpose of the study was to simulate soil behavior, we tried to simulate most accurately motion at the "input" to the soil layers. In the calculations, David Boore's stochastic approach (Boore 2003) was used, modified in order to account for the dimensions of the source and slip distribution over the fault plane. In the calculations the slip distribution was used obtained by W.-C. Chi, D. Dreger and A. Kaverina (Chi et al. 2001) who inverted velocity waveforms recorded at 21 near-fault stations and constructed spatial variation in slip on a planar fault model composed of 416 subfaults, each with a size of 3.5 km by 3.5 km. To account for possible temporal source complexity, each subfault was allowed to slip within 10 overlapping 3 s time windows (Chi et al. 2001).

In the simulations, the earthquake source (119 km by 35 km) was represented as a set of subfaults with dimensions, locations, radiation parameters, and the method for summarizing their energy being selected according to recommendations by Gusev and Pavlov (2006), who studied all aspects of earthquake ground motion simulations for many years and achieved success in this area (Bazzurro et al. 2001), i.e., the space–time structure of the slip rate function was represented through a grid of subfaults, where the cell size of the subfaults was defined by site-to-fault distance; onset times and seismic moments of the subfaults were determined by the generalized Haskell source model; time histories of the subfaults were taken uncorrelated; final slip and rupture velocity were defined as random functions and varied along x and y axes; each point subfault was randomly shifted from a node of the perfect regular grid—typically by 0.3–0.6 cell size.

Each subfault was assigned an ω^2 spectrum. Some other parameters used in the simulations, such as, spatial orientation of the fault plane, the nucleation point, the coordinates and depth of the hypocenter, slip velocity and slip distribution over the fault plane, were derived from the paper by Chi et al. (2001).

The subfault dimension of 7 km by 7 km, averaging slips over the adjacent areas of 3.5 km by 3.5 km of the "preferred model" by Chi et al. (2001), was selected; the estimated temporal variations of the rupture process were also averaged, and thus, the final simplified and smoothed representation of the rupture process was obtained (Fig. 3.16).

This representation was used for generating two spatial weight functions, "dip-slip" and "left-lateral strike-slip", indicating the relative contributions of the subfaults in the two directions. Time delays for subfault radiation were calculated based on rupture velocity, which was taken as constant and equal to 2.6 km/s according to (Chi et al. 2001). Subfault radiation was simulated as stochastic, empirically attenuated to the observation site, and summarized with introducing slight random variations in the subfault locations and rupture velocity. Local rise times were taken equal for all subfaults. Thus, acceleration time histories in "dip-slip" and "left-lateral strike-slip" directions were calculated with spatial weight functions (Fig. 3.29) applied to subfault radiation and rotated in NS and EW directions. Simulated acceleration time histories at rock sites were compared to the observed ones, and calibration of the simulation procedure was performed, i.e., parameters characterizing source and path effects were found, which show the best fit to the observations.

Table 3.3 provides information on the studied rock stations: their coordinates, elevations, characteristic distances R_c (discussed below), distances to the closest subfaults r, and weighted (weights being proportional to subfault radiation intensities) average distances to the subfaults r_{av}. Some stations classified in (Lee et al. 2001a; b) as Class B Sites, have shown amplification at low frequencies (the problem is discussed in Lee et al. 2001a; b), which was interpreted as site effects. Such stations (CHY042, CHY052, CHY074, CHY086, CHY109, KAU069, KAU078, ILA063, HWA022, HWA026, HWA046, HWA056, TTN018, TTN026, TTN041, TAP034, and TAP035) were excluded from the list of "rock sites". At the same time, other

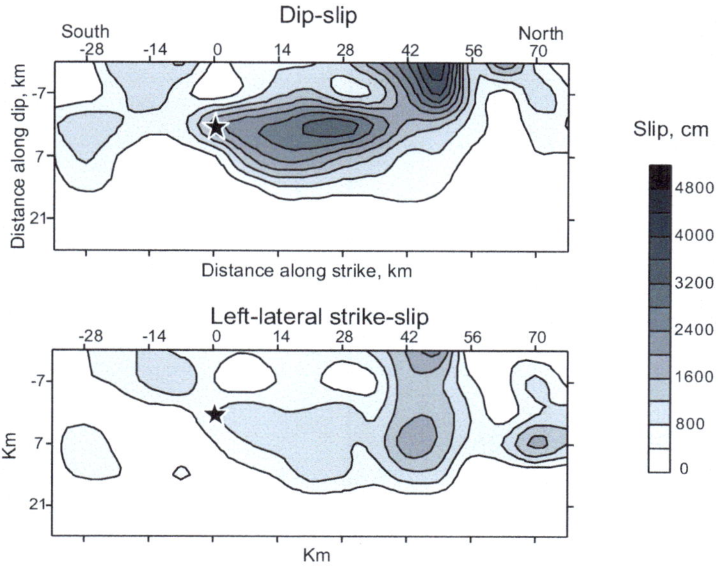

Fig. 3.29 Distribution of slip over the fault plane of the Chi-Chi earthquake used in the simulations: averaged slip distribution of the "preferred model" by Chi et al. (2001). Star indicates the hypocenter

stations were added to this list based on the results of simulations, such as, TCU049, TCH053, TCU075, and TCU076.

Comparison of the observed and simulated accelerograms and their spectra indicates the accuracy of the simulations. As seen from Fig. 3.30, a rather good agreement was achieved between the observed and simulated accelerograms, and some misfit in their spectra, noticeable at stations TCU089, TCU046, CHY102, and island station CHY075, was evidently due to local site effects of wave amplification in some frequency bands.

To trace the effects of directivity, the stations were arranged according to their orientation with respect to the direction of the crack propagation (Chi et al. 2001). Directivity effects are clearly seen as decreased durations of acceleration time histories in forward (northern) direction and increased durations in backward (southern) direction, while eastern and western directions represent intermediate cases (Fig. 3.30).

Parameters used in the simulations are shown in Table 3.4. The spatial orientation and dimensions of the fault plane, coordinates and depth of the hypocenter, slip velocity and slip distribution over the fault plane were derived from the paper by Chi et al. (2001). Frequency-dependent attenuation effects of the propagation path are modeled through the function $Q(f)$ as described in (Boore 2003). The relationship estimated for Taiwan area from coda-waves by Chen et al. (1989) was assumed: $Q(f) = 117f^{0.77}$. Kappa operator was used for additional attenuation of the spectra (Anderson and Hough 1984; Boore 2003): $\kappa = 0.07$ s. Thus, these parameters were taken the same as those used by Roumelioti and Beresnev (2003).

Table 3.3 Locations of the studied Rock Stations

Station code	Latitude (°)	Longitude (°)	Elevation (m)	R_c (km)	Dist. to the closest subfault (km)	Weighted av. dist. to subfaults (km)
North (forward direction)						
TCU049	24.179	120.690	129	10	6.5	28.3
TCU053	24.194	120.669	133	15	5.4	28.9
TCU046	24.468	120.854	231	50	13.3	44.4
TCU025	24.707	121.176	310	50	36.8	80.3
ILA043	24.629	121.735	36	50	88.6	118.9
ILA052	24.609	121.849	20	50	99.9	128.3
East and West (normal to the fault)						
TCU075	23.983	120.678	104	15	6.8	31.3
TCU076	23.908	120.676	99	10	7.5	34.8
TCU089	23.904	120.857	705	12	16.2	38.1
HWA023	24.080	121.596	30	50	79.2	93.5
HWA024	23.352	121.297	260	50	57.7	103.2
CHY075	23.567	119.555	11	50	100.9	135.9
South (backward direction)						
CHY102	23.246	120.614	560	50	33.2	95.0
CHY110	23.252	120.530	215	50	32.7	95.8
KAU050	23.163	120.757	640	50	41.9	103.2
KAU047	23.082	120.583	278	50	50.6	112.6
TTN024	22.973	121.108	277	50	70.1	129.7
KAU077	22.747	120.723	835	50	85.6	147.2

The simulations show that for the near-fault stations, geometrical spreading is one of the most important factors defining the shapes of the acceleration time histories. For example, simulating acceleration time histories at TCU075, TCU076, TCU089, TCU072 and other near-fault stations located within the area marked by a dashed line in Fig. 3.28 with conventional geometrical spreading relationships:

$$1/r \text{ for } r < 50\,\text{km},\ 1/50 \text{ for } 50 \le r < 150\,\text{km} \tag{3.3.1}$$

(Boore, 2003), we obtain too strong attenuation of the simulated accelerograms, which does not satisfy the observations. As seen from Fig. 3.30, the accelerograms at these stations represent successive arrivals of groups of seismic waves of similar intensity, which evidently correspond to radiation from different parts of the fault plane (subfaults). Application of conventional relationships (3.3.1) produces accelerograms with the intensity quickly decreasing with time. However, a good

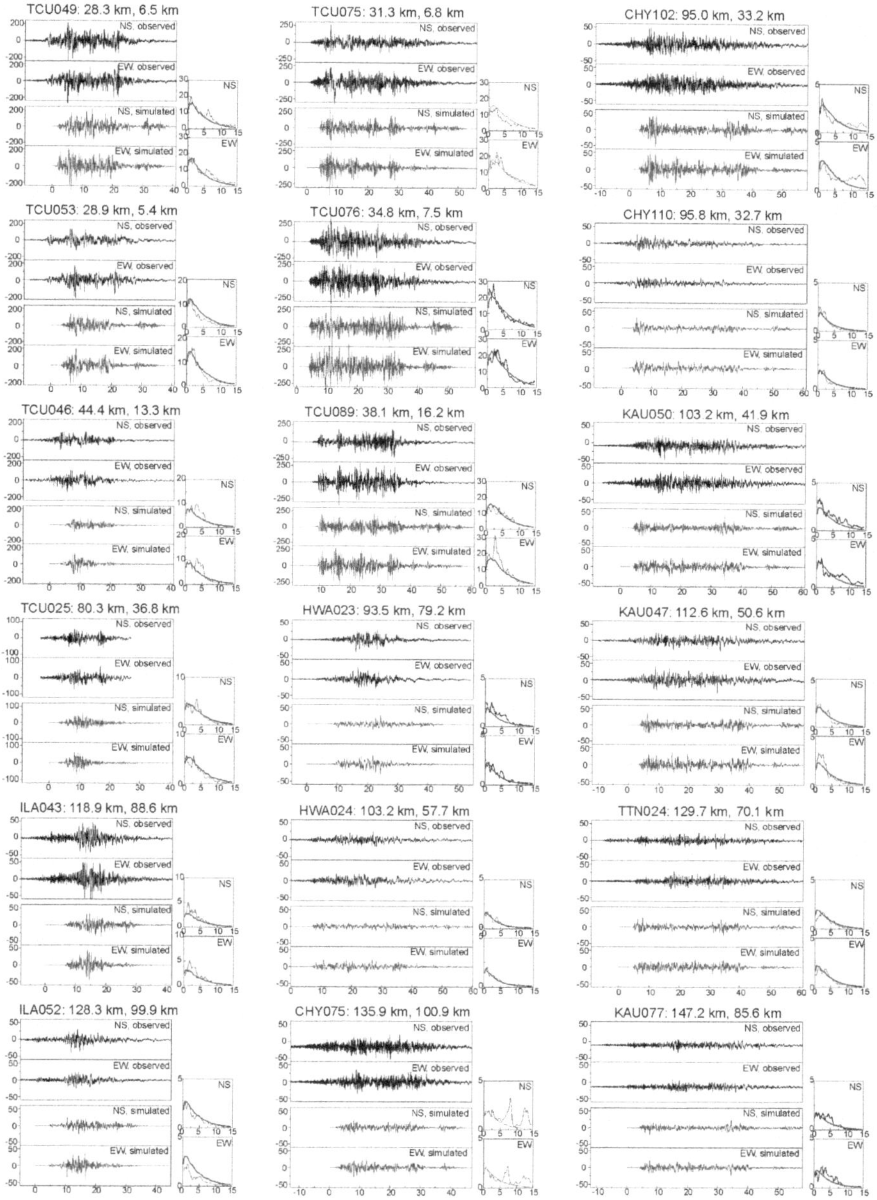

Fig. 3.30 The results of simulations of acceleration time histories of the Chi-Chi earthquake at rock sites: observed and simulated (one of the series) accelerograms and the corresponding spectra (the simulated spectra are averaged over the set of simulated accelerograms). After the station codes weighted average distances to the subfaults and distances to the closest subfaults are shown

Table 3.4 Parameters used in simulations

Fault orientation (strike/dip)	5°/30°
Fault dimensions along strike and dip (km)	119 by 35
Coordinates of the hypocenter (lat., lon.)	23.8689, 120.84
Depth of the hypocenter (km)	8.75
Subfault dimensions (km)	7 × 7
Stress drop $\Delta\sigma$ (bars)	50
Number of subsources summed	85
$Q(f)$	$117{\cdot}f^{0.77}$
Geometrical spreading	$1/r$ for $r < R_c$ km $1/R_c$ for $R_c \le r < 150$ km
Kappa κ, (s)	0.07
Crustal amplification	Boore and Joyner (1997) western North America generic rock site; at underlined stations—hard rock (Boore 2003)
Crustal shear-wave velocity (km/s)	3.2
Rupture velocity (km/s)	2.6
Crustal density (g/cm^3)	2.6

agreement between the observations and simulations can be obtained if describing the geometrical spreading by the relationship:

$$1/r \text{ for } r < R_c \text{ km}, \ 1/R_c \text{ for } R_c \le r < 150 \text{ km}, \tag{3.3.2}$$

where distance R_c varies from 10 km in the vicinity of the fault plane to 50 km beyond the area marked by the dash line in Fig. 3.28. For all the stations, R_c values were found in calculations as showing the best fit to the observations. It is seen from Table 3.3 that $R_c < 50$ km in cases where the average removal of the station from the subfaults is less than 44–50 km. As shown below, the values $R_c < 50$ km were also found for stations located to the south of the fault plane at distances of 50–85 km; the area of decreased R_c values is stretched in southern direction (Fig. 3.28). Evidently, the effect should be due to the geometry of the fault plane and directivity effects.

At the majority of the studied rock sites, the upper-crustal amplification is well described by the parameters established by Boore and Joyner (1997) for the western North America generic rock site. However, at some stations (underlined in Fig. 3.28 and Table 3.3) these parameters provide too higher amplification not corresponding to the observations, so they were changed to "hard rock" amplification parameters (Boore 2003), which better satisfy the observations.

At the second step of the study, acceleration time histories at soil sites were simulated (the coordinates of the soil stations, their elevations, distances R_c, distances to the subfaults, Site Classes and soil composition are described in Table 3.5), and models of soil behavior during the Chi-Chi earthquake were constructed.

Table 3.5 Information on Soil Stations Used in Simulations

Station code	Latitude °	Longitude °	Elevation m	R_c km	Clos. dist. to subft km	Aver. dist. to subft km	V_{S30} m/s	Site class, soil composition	
North									
TCU082	24.15	120.68	84	15	5.7	28.3	476	D	Gravel
TCU054	24.16	120.69	99	20	6.1	28.4	454	D	Silt, gravel
TCU102	24.25	120.72	227	22	7.8	29.6	735	D	Gravel, siltsone
TCU033	24.69	120.86	6	50	16.9	64.2	450	D	Gravel, siltsone
TCU017	24.78	121.01	86	50	29.7	78.1	600	?	Gravel, mudsone
Center									
TCU065	**23.06**	**120.69**	**56**	12	**6.9**	**28.9**	**300**	**D**	Sandy clays, silt, gravel
TCU072	**24.04**	**120.85**	**292**	12	**14.9**	**32.4**	**563**	**D**	Colluvium, breccia, shale
TCU110	23.96	120.57	23	14	2.4	35.5	207	E	Sand, clay, silt
TCU138	**23.92**	**120.60**	**37**	12	**3.8**	**36.2**	**605**	**D**	Gravel
TCU116	23.86	120.58	42	15	3.4	40.3	483	E	Silt, gravel
TCU115	23.96	120.47	15	35	11.4	41.6	190	E	Silt
South									
CHY025	23.78	120.51	42	16	5.9	48.2	277	E	Silt, gravel
CHY092	23.79	120.48	34	25	9.5	49.5	260	E	Silt
CHY101	23.69	120.56	75	10	1.9	53.0	260	D	Silt, sand, gravel
CHY026	**23.80**	**120.41**	**25**	**30**	**16.1**	**53.1**	**226**	**E**	Sand, clay, silt
CHY002	23.72	120.41	26	25	15.0	58.1	225	E	Silt
CHY104	**23.67**	**120.47**	**33**	**30**	**9.5**	**58.7**	**225**	**E**	Sand, clay, silt
CHY094	23.79	120.32	10	50	24.9	59.7	227	E	Silt
CHY036	23.61	120.48	45	20	7.5	63.1	282	D	Silt, clay, sand, gravel
CHY082	23.72	120.30	13	50	26.6	65.4	210	E	Silt
CHY027	23.75	120.25	8	50	32.1	67.6	220	E	Silt

(continued)

Table 3.5 (continued)

Station code	Latitude °	Longitude °	Elevation m	R_c km	Clos. dist. to subft km	Aver. dist. to subft km	V_{S30} m/s	Site class, soil composition	
CHY074	**23.51**	**120.81**	**2413**	**16**	**12.8**	**67.8**	**546**	**C**	Sand, clay, siltstone, breccia
CHY032	23.58	120.29	9	50	25.6	75.6	202	E	Silt, clay, sand
CHY039	23.52	120.34	16	30	20.8	77.3	198	E	Silt, clay, sand
CHY093	23.65	120.15	4	50	41.2	81.0	199	E	Silt, sand
CHY004	23.60	120.17	6	35	38.2	82.5	279	E	Silt, sand
CHY033	23.54	120.22	3	50	33.6	83.7	194	E	Silt, sand
CHY015	**23.36**	**120.41**	**34**	**18**	**25.9**	**89.4**	**234**	**D**	Silt, clay, sand
CHY044	23.38	120.16	2	50	43.0	99.4	193	E	Silt, sand
CHY055	23.27	120.27	8	35	41.5	103.4	240	E	Silt, clay, sand
CHY012	23.33	120.15	1	50	46.6	104.2	202	E	Silt, clay, sand

For the study, soil stations were selected, for which the profiling data were available. At stations TCU065, TCU072, TCU138, CHY026, CHY104, CHY074, and CHY015, soil composition, *P*- and *S*-wave velocities and the results of *SPT-N* testing were available down to ~ 70–140 m (these sites are marked in bold in Table 3.5), while at other twenty four stations, these data were known down to ~ 30–40 m.

Site Classification of Taiwan free-field strong motion stations is described by Lee et al. (2001a; b). Based on the available geologic and geomorphologic data, free-field strong motion station sites were classified using a scheme compatible with the 1997 Uniform Building Code (UBC) provisions. The response spectral shapes (RSS) and the horizontal-to-vertical spectral ratio (HVSR) methods and field inspections were used for checking purposes. However, some problematic sites do exist.

Concentric lines in Fig. 3.28 mark stations with elevations above the sea-level of 200 m and more. The influence of topographic effects on strong motion stations in Taiwan is discussed in (Lee et al. 2001a; b), where stations, obviously influenced by surface topography are listed; none of these stations is studied in this research. Also, as seen from Fig. 3.28 and Table 3.5, all the studied soil sites except TCU072, TCU102 and CHY074, are located in valleys, therefore surface topography does not influence the records; the influence of subsurface topography is discussed below.

According to Lee et al. (2001a; b), ten sites, including TCU082, TCU054, NCU102, TCU065, TCU072, TCU138, CHY015, are categorized as class D sites, station CHY074 as class C site, and the rest twenty stations, such as, TCU110, TCU116, TCU115, CHY025, CHY026, CHY104, etc., as class E sites (Table 3.5).

First, soil behavior was studied at seven stations possessing the most complete information on the profiling data, such as, TCU065, TCU072, TCU138, CHY026, CHY104, CHY074, and CHY015; the results are presented in Figs. 3.31, 3.32, 3.33, 3.34, 3.35, 3.36 and 3.37.

To estimate stresses and strains induced in soil layers during the Chi-Chi earthquake, the propagation of vertically incident "input" seismic waves (simulated accelerograms) up to the surface was calculated; the modified algorithm of nonlinear analysis by Joyner and Chen (1975) was used.

Data on the composition and thickness of the soil layers, *P*- and *S*-wave velocities and *SPT-N* values were provided by the National Center for Research in Earthquake Engineering (NCREE) of Taiwan.

At TCU065, CHY026, CHY104, and CHY015 stations, soil profiles in the upper ~ 50 m, ~ 100 m, ~ 90 m, and ~ 60 m, respectively, represent alternating layers of sandy clays with silt and silty soils with *S*-wave velocities increasing with depth from ~ 160–200 m/s at the surface to 420–600 m/s at 80–100 m. These layers are underlain by denser colluvium soils. At TCU072, TCU138, and CHY074 stations colluvium deposits are closer to the surface. At TCU072 and CHY074, *S*-wave velocities increase with depth from 250 m/s and 360 m/s, respectively, in the upper 8–10 m to ~ 1120 m/s at ~ 70 m. At TCU138 site, soil layers down to ~ 150 m represent mostly dense soils with *S*-wave velocities varying within the interval ~ 400–800 m/s.

Since the level of the underground water was close to the surface at all the stations ($V_P > 1300$ m/s in surface layers), "hard-type" stress–strain relations were applied to describe the soil behavior.

To find the best fit to the observations, the spectra, average intensities (as mean squares of the acceleration time histories), and peak ground accelerations of the observed and simulated accelerograms at two horizontal components on the surface were compared. The observed and simulated accelerograms were analyzed on the whole, without dividing them into time intervals, as was done in the case of vertical array record analysis, because of the evident insufficiency of input information.

Unlike strong-motion sites in Japan (Sect. 3.1 and 3.2), where the upper 10–15 m of soft soils were underlain by denser layers with a noticeable contrast of impedances, at most of Taiwanese stations the soil strata are rather homogeneous with *S*-wave velocities gradually increasing with depth. In such cases, one "hard-type" normalized stress–strain relation is usually used to describe the behavior of the entire soil column. That is, all the soil thickness at all the stations, except TCU065, was treated as one group of layers. With these simple models of soil behavior, a rather good agreement between the observations and simulations was obtained.

At TCU065, a more complicated model was applied to fit the observations: one "hard-type" stress–strain relation was used to describe the behavior of the upper ~ 37 m, and stress–strain relation obtained in laboratory experiments by Hardin and Drnevich (1972) was applied to describe the behavior of the underlying denser layers.

Though "hard-type" stress–strain relations were applied to describe soil behavior at all the sites, however, the stress–strain curves show substantial "hardening" (declinations to the stress axis) mostly at stations closest to the fault plane (TCU065 and TCU072), where the input motion was rather intense. At these stations, vertical distributions of stresses and strains in soil layers and spectra of motion at the bottoms

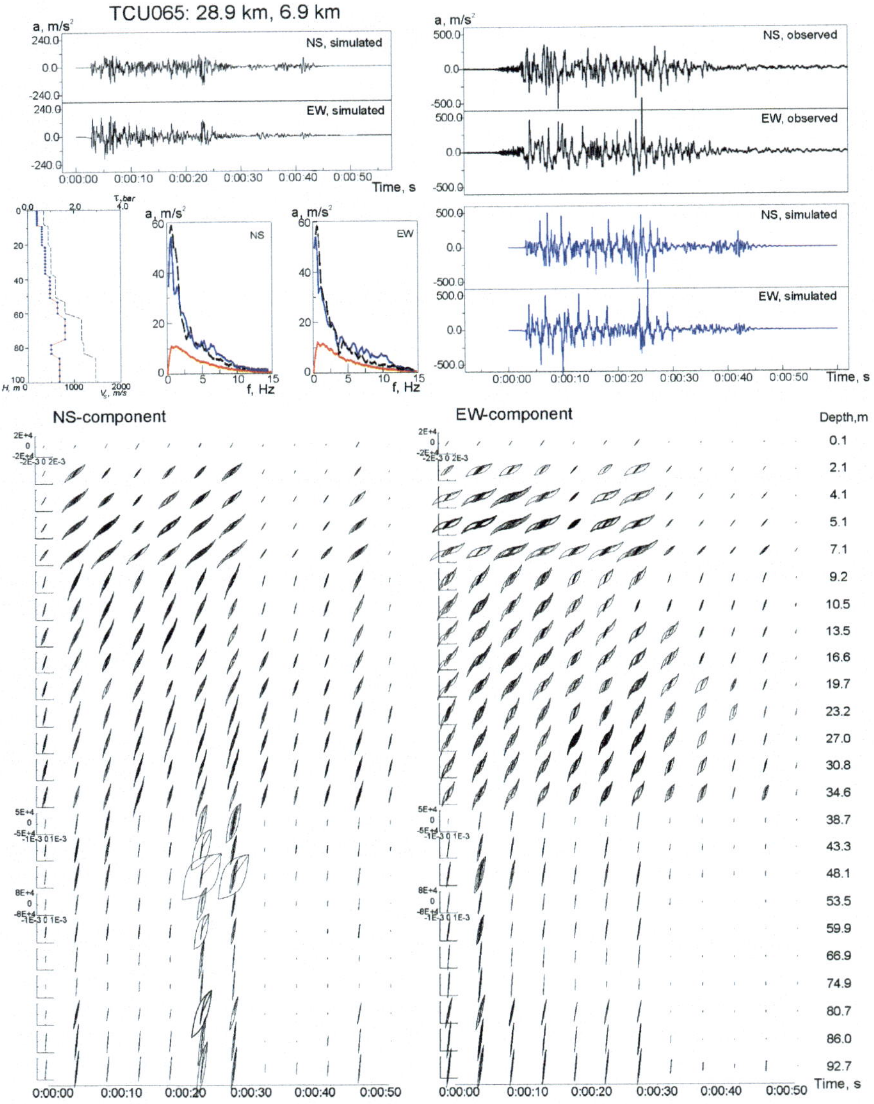

Fig. 3.31 The observed and simulated accelerograms of the Chi-Chi earthquake at TCU065 site: simulated accelerograms at the bottoms of soil layers (upper left corner); the profiling data (V_S—solid, τ_{max}—dashed lines) and spectra of the simulated accelerograms at the bottoms of soil layers (red line) and the observed and simulated accelerograms on the surface (dashed and solid black and blue lines); the observed and simulated accelerograms on the surface (upper right corner), and the corresponding time-dependent vertical distributions of stresses and strains in soil layers

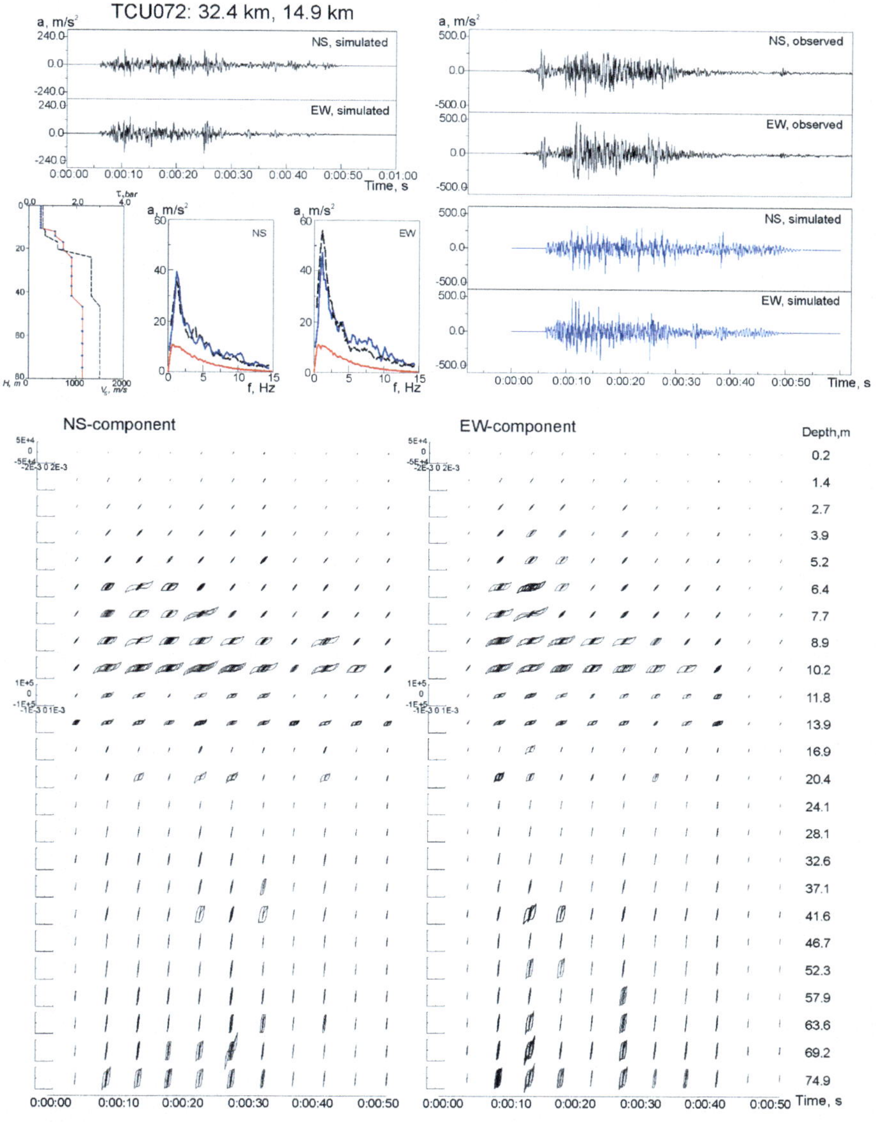

Fig. 3.32 The observed and simulated accelerograms of the Chi-Chi earthquake at TCU072 site

and on the surfaces of the soil profiles indicate resonant phenomena in the upper soil layers induced by the strong motion, which lead to amplification of seismic waves at ~ 1–2 Hz at TCU065 and at ~ 1–3 Hz at TCU072 (Figs. 3.31 and 3.32). The "hard type" of soil behavior at these stations and rather low *S*-wave velocities (~ 200–250 m/s in the upper ~ 10 m) stipulate a substantial amplification of motion

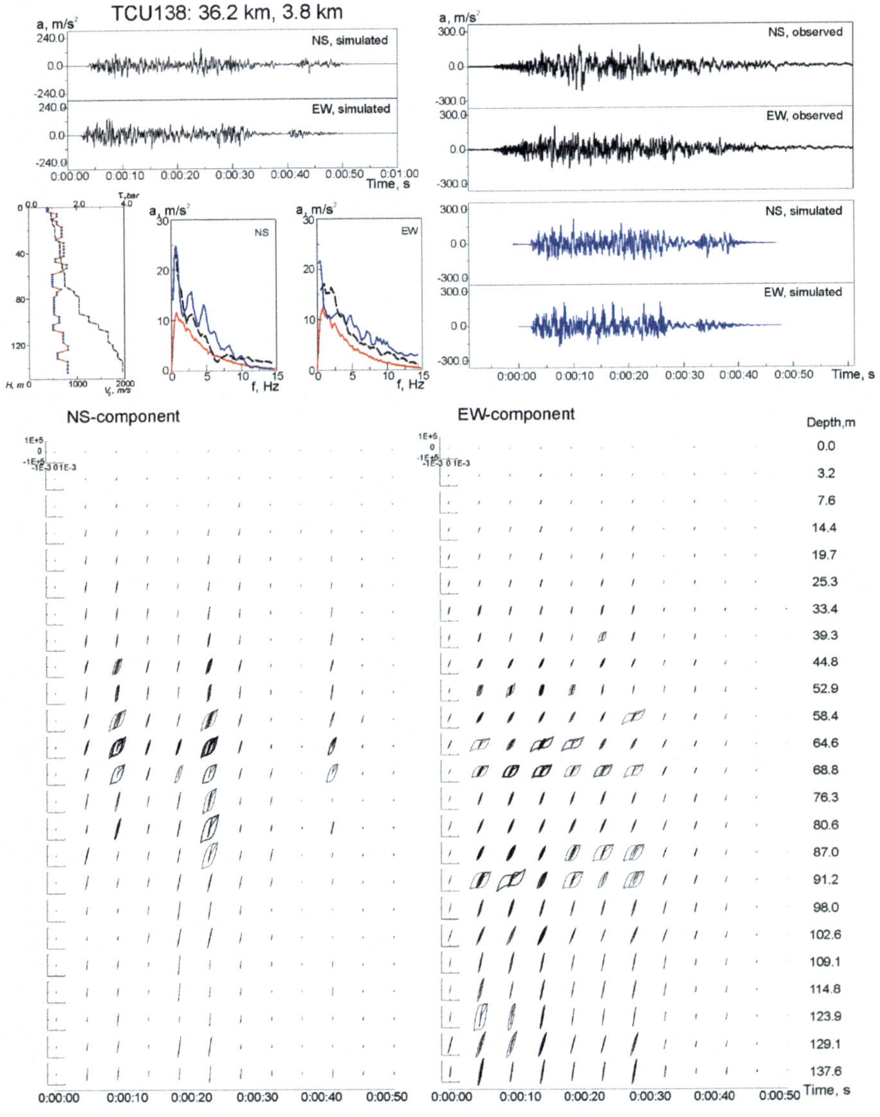

Fig. 3.33 The observed and simulated accelerograms of the Chi-Chi earthquake at TCU138 site

on the surface. At TCU065 station, characteristic spiky waveforms in the acceleration time histories are observed, often related to strong nonlinearity of soil response (Archuleta 1998) and cyclic mobility and liquefaction of sandy surface soils (Lee et al. 2001a; b).

As seen from the figures, resonant vibrations were induced in the upper layers down to 40–60 m at all the soil stations located within ~ 30 km from the fault plane.

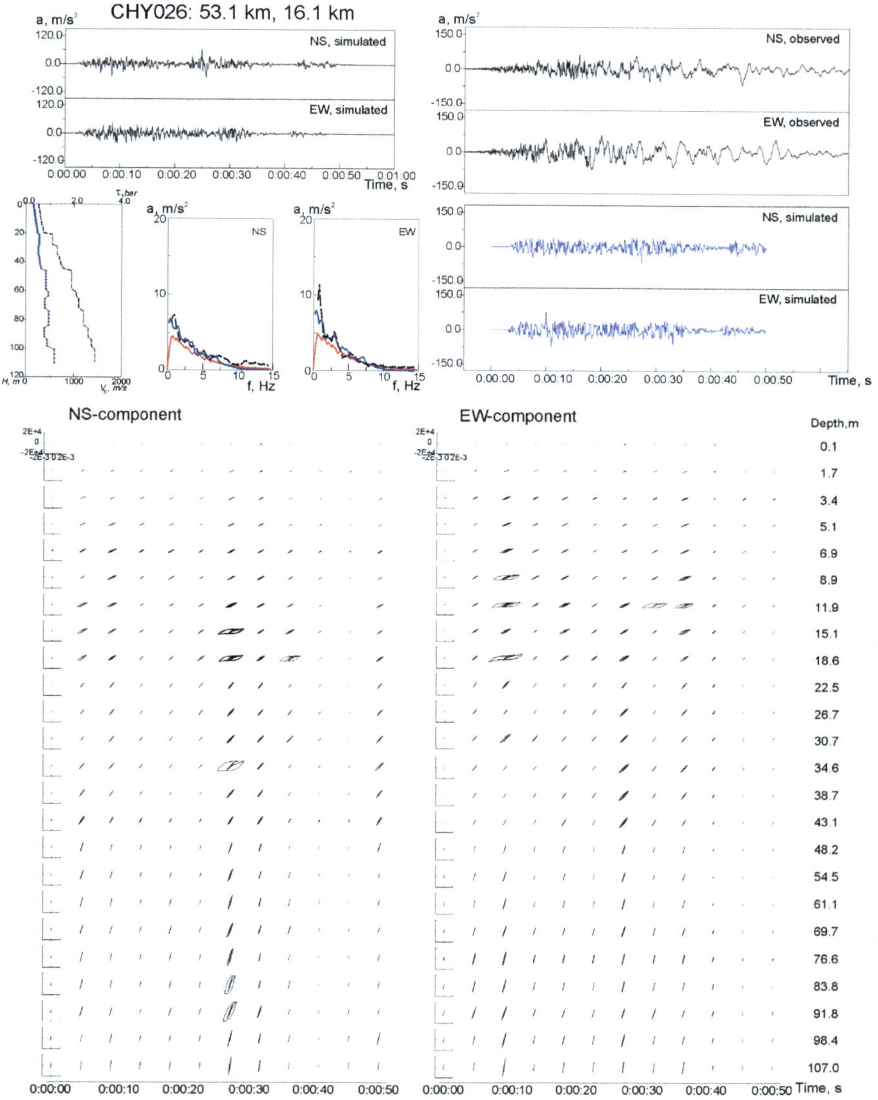

Fig. 3.34 The observed and simulated accelerograms of the Chi-Chi earthquake at CHY026 site

According to our estimates, at TCU065, resonant phenomena occurred in the upper ~ 9 m of sandy clays with silt (V_S ~ 200 m/s) with maximum strains of ~ 0.6%. Lee et al. (2001a; b) report liquefaction in the upper layers at this site. At the same time, resonant vibrations were also induced in the upper ~ 37 m (where V_S increased from ~ 200 m/s to ~ 380 m/s) with maximum strains of ~ 0.2–0.4%.

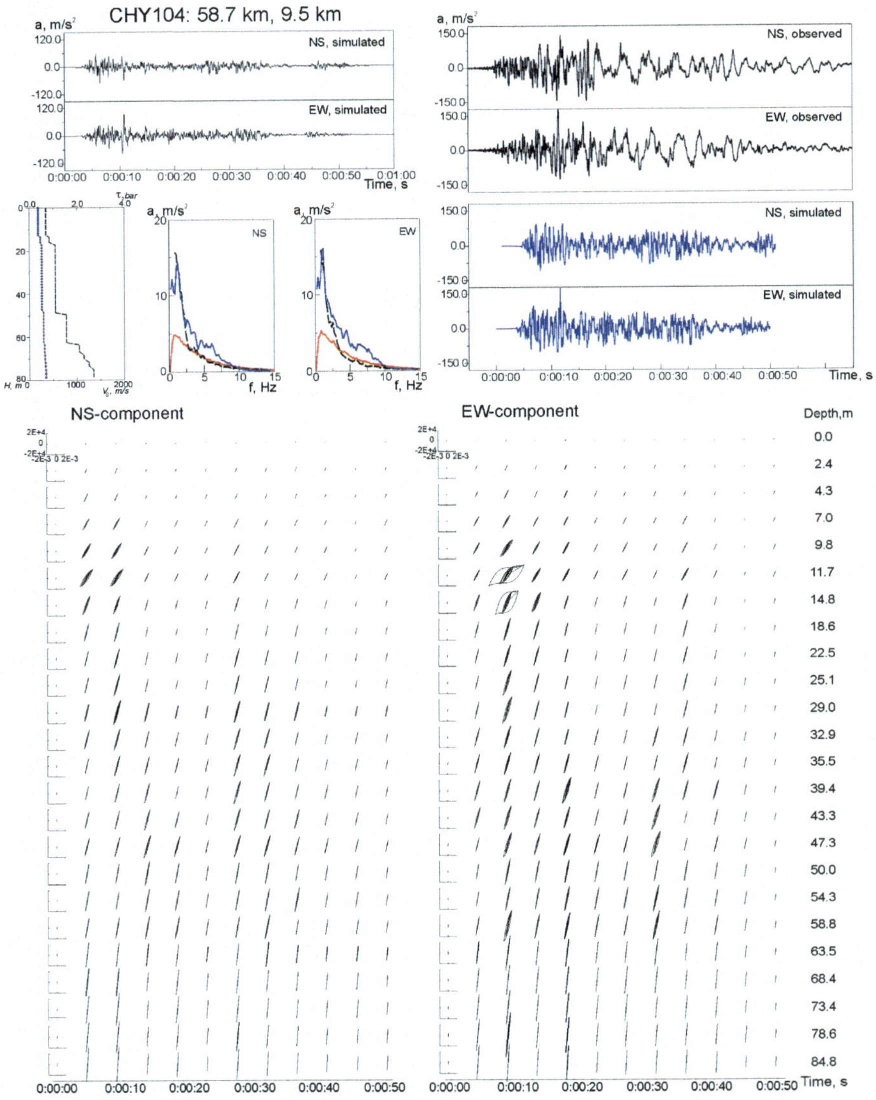

Fig. 3.35 The observed and simulated accelerograms of the Chi-Chi earthquake at CHY104 site

At TCU072 site, resonances occurred in the upper ~ 11 m of the soft colluvium (V_S ~ 250 m/s), below which more dense breccia and shale layers are deposited (V_S ~ 550–700 m/s); maximum strains reached 0.3–0.4% at ~ 8–12 m.

At stations TCU138, CHY074, and CHY015, resonant phenomena were less pronounced and deeper: maximum strains occurred at ~ 40–100 m at TCU138 site (~ 0.1%), at ~ 40 m at CHY074 site (~ 0.2%), and at ~ 30–70 m at CHY015 site

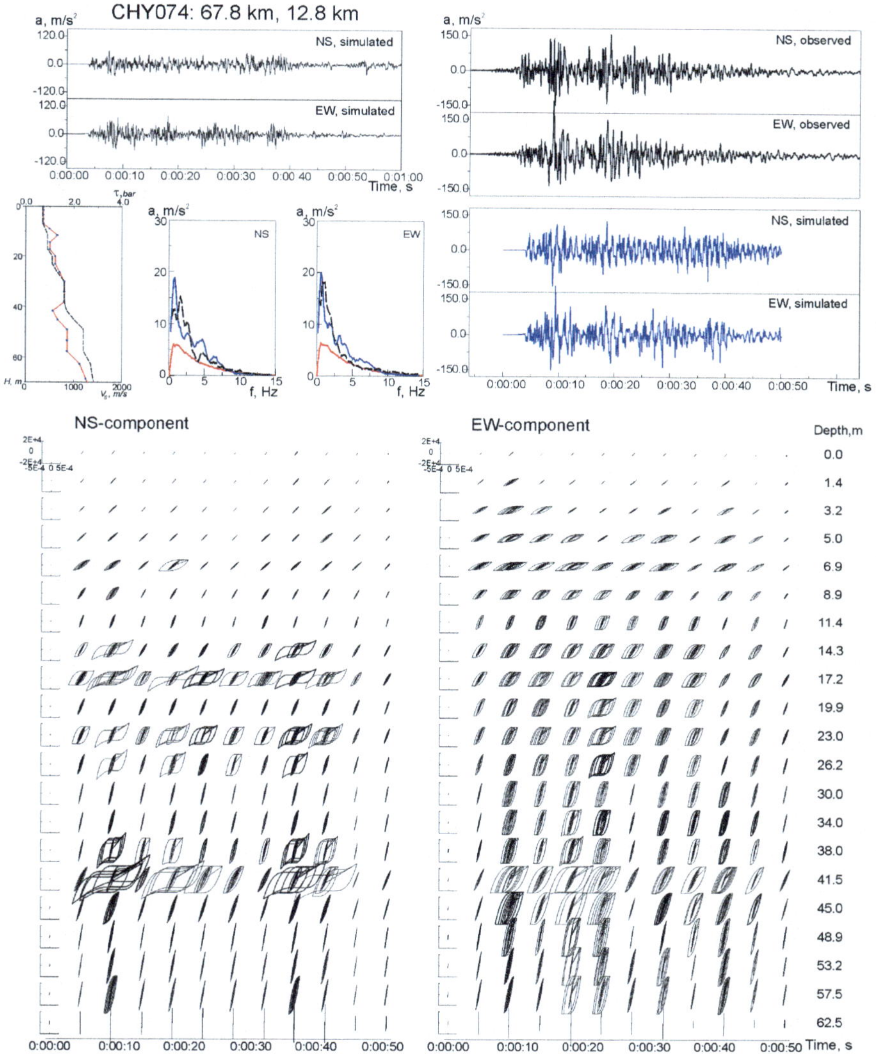

Fig. 3.36 The observed and simulated accelerograms of the Chi-Chi earthquake at CHY074 site

(0.1–0.2%). The corresponding stresses were as high as ~ 0.6–1.2 bars, ~ 0.4 bars, and 0.2–0.4 bars, respectively.

At TCU138 and CHY074 sites, resonating layers correlate with decreased *S*-wave velocities, whereas at CHY015 site the whole upper part of the soil profile down to ~ 60 m (V_S ~ 170–340 m/s), underlain by denser soils (V_S > 600 m/s), was engaged into the resonant excitations.

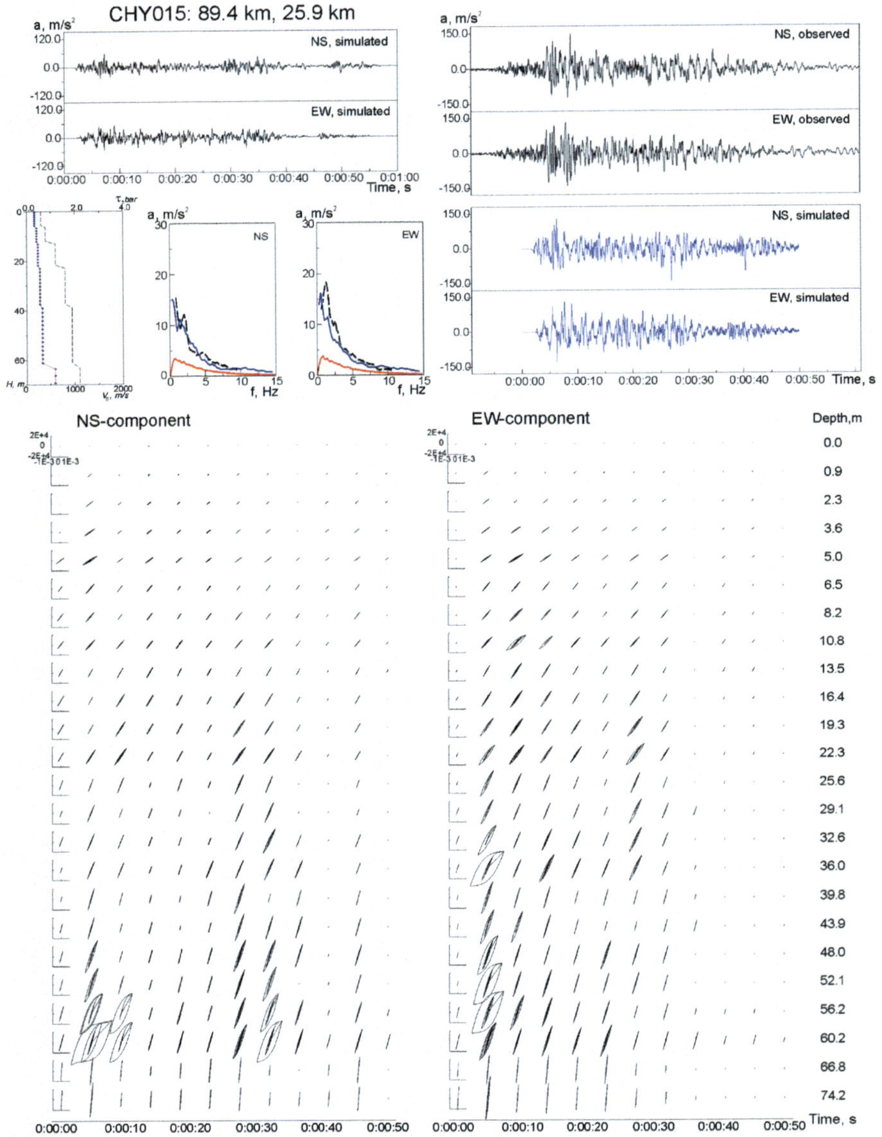

Fig. 3.37 The observed and simulated accelerograms of the Chi-Chi earthquake at CHY015 site

At CHY026 and CHY104 sites, resonant vibrations were induced by strong motion in the upper ~ 20 m (at CHY026 site, where V_S ~ 160–240 m/s) and ~ 12 m (at CHY104 site, where V_S ~ 180–190 m/s) of sandy clays with silt. At these sites, acceleration time histories also show the presence of intense seismic waves with periods of 3–5 s (Figs. 3.34 and 3.35). The stations are located to the southwest of the fault plane, in a valley, and the long-period motions evidently represent standing waves due to *S*-wave reverberation in the basin sediments, i.e., basin-induced surface waves. These waves cannot be modeled in our 1-*D* simulations, and therefore, only the beginning of strong motion (15–20 s) is properly simulated at these sites. Spectra of the recorded accelerograms correspond to the beginning parts of motion.

Thus, resonant phenomena occurred in the upper 40–60 m of soft soils during the Chi-Chi earthquake. Strong manifestations of soil nonlinearity were observed at stations TCU065 (Class D), TCU072 (Class D), and CHY074 (Class C); TCU065 site did show liquefaction (spikes, large strains), and its possible classification as site Class F is discussed in Lee et al. (2001a; b).

Since realistic models of soil behavior at the seven sites possessing most complete information on the profiling data have been obtained, the study was continues with additional twenty-four soil stations, for which the profiling data, *P*- and *S*-wave velocities and *SPT-N* testing results were restricted to ~ 30–40 m. It was shown above that resonant vibrations were induced by the Chi-Chi earthquake in the upper layers (~40–60 m) in the near-fault zones. Therefore, knowledge of soil parameters in the upper ~ 30–40 m was not enough for calculations, and the profiling data should be completed down to at least ~ 70–80 m. The profiling data at depths ~ 30–80 m were reconstructed based on the results by Lin et al. (2006), who conducted microtremor measurements at seven stations spaced at a distance of ~ 120 km along the western coastal plain of Taiwan. Shallow *S*-wave velocity structure under all the studied sites down to ~ 3 km was estimated by the inversion of the dispersion curves obtained by frequency-wavenumber ($f - k$) analysis (Lin et al. 2006).

The most reliable estimates of *S*-wave velocities were obtained for the upper ~ 100 m, and they were verified by *P-S* logging data in the upper 40–70 m at some nearby TSMIP (Taiwan Strong Motion Instrumentation Program) stations. It was also found that the obtained *S*-wave velocity models correctly reflect the depths of major interfaces including the tops of the Pliocene and Miocene (Lin et al. 2006). For the upper ~ 100 m, fairly simple velocity structures and gradual increase of *S*-wave velocities with depth were found (Lin et al. 2006).

The results of simulations of the acceleration time histories at the twenty-four soil stations are presented in Figs. 3.38, 3.39, 3.40, 3.41, 3.42 and 3.43 (EW components).

As seen from the figures, in spite of the shortage of the initial information, in some cases a fairly good agreement was achieved between the observations and simulations. The best agreement was achieved in cases of intense seismic waves and strong nonlinearity of soil response.

At stations TCU115, CHY092, CHY002, CHY094, and CHY093 (Figs. 3.39, 3.40 and 3.42) the final parts of records contain low-frequency intense seismic waves with periods of 3–5 s, which cannot be modelled in our 1-*D* simulations. These oscillations seem to represent the effects of subsurface topography; they resemble basin-induced surface waves, i.e., standing waves due to *S*-wave reverberation in the basin sediments. These waves can also be distinguished at stations CHY082,

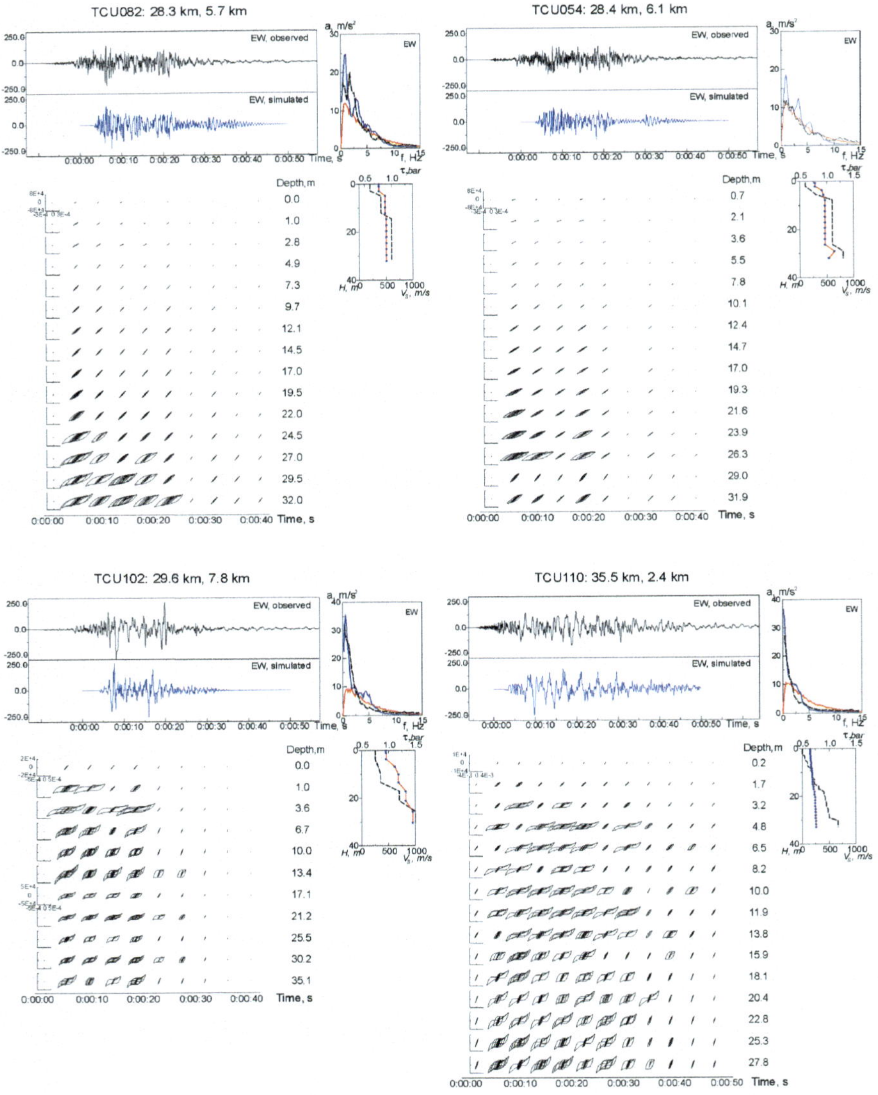

Fig. 3.38 The results of simulations of acceleration time histories of the Chi-Chi earthquake at TCU082, TCU054, TCU102, and TCU110 soil sites (EW components): the observed and simulated accelerograms on the surface (upper left corners); the corresponding vertical distributions of stresses and strains in soil layers changing in time during strong motion (below); spectra of simulated accelerograms at the bottoms of soil layers (red lines) and of the observed (dashed line) and simulated (solid line) surface accelerograms (upper right corner); and the profiling data (below): V_S—solid line and τ_{max}—dashed line

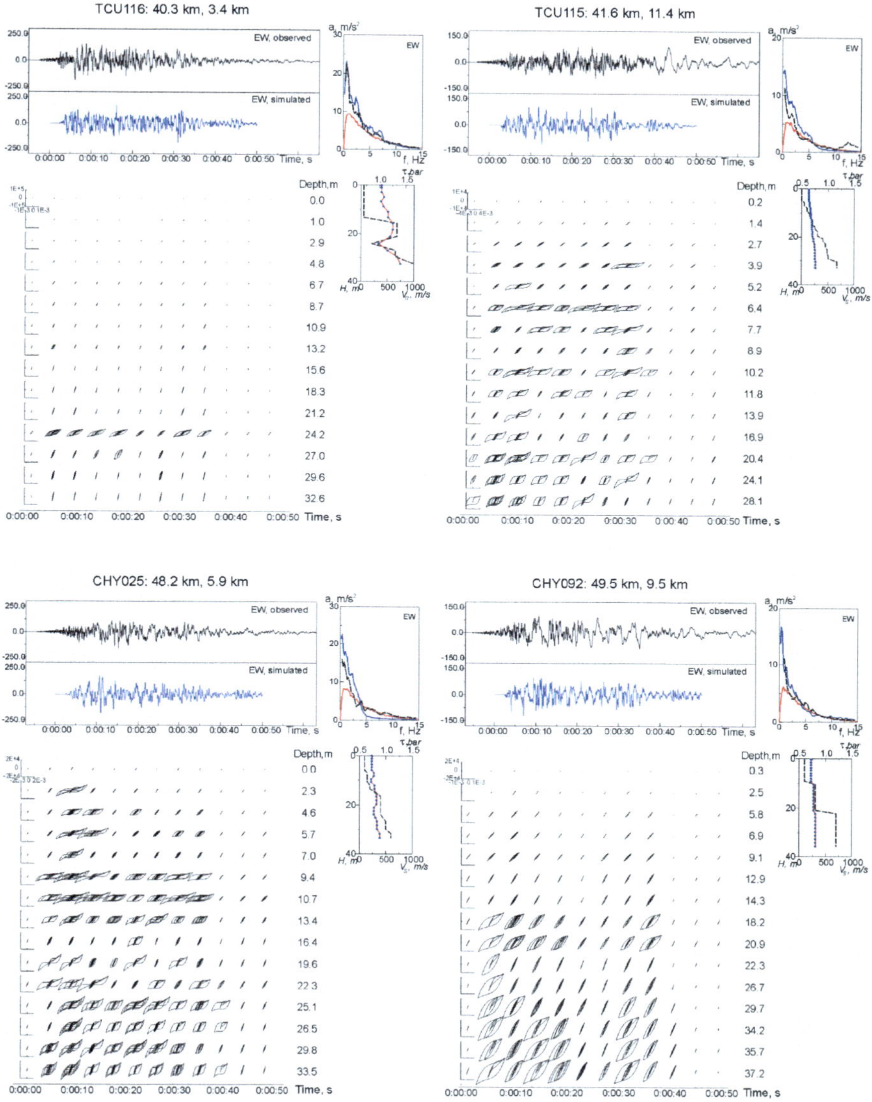

Fig. 3.39 The results of simulations of acceleration time histories of the Chi-Chi earthquake at TCU116, TCU115, CHY025, and CHY092 soil sites (EW component)

CHY027, and CHY032 (Fig. 3.41). All these sites are located in the valley (elevations 4–34 m); this area is marked by grey in Fig. 3.28. At all these sites, the soil profiles in the upper ~ 30–40 m contain predominantly silty soils with sands and clays with average *S*-wave velocities of ~ 190–260 m/s (Table 3.5).

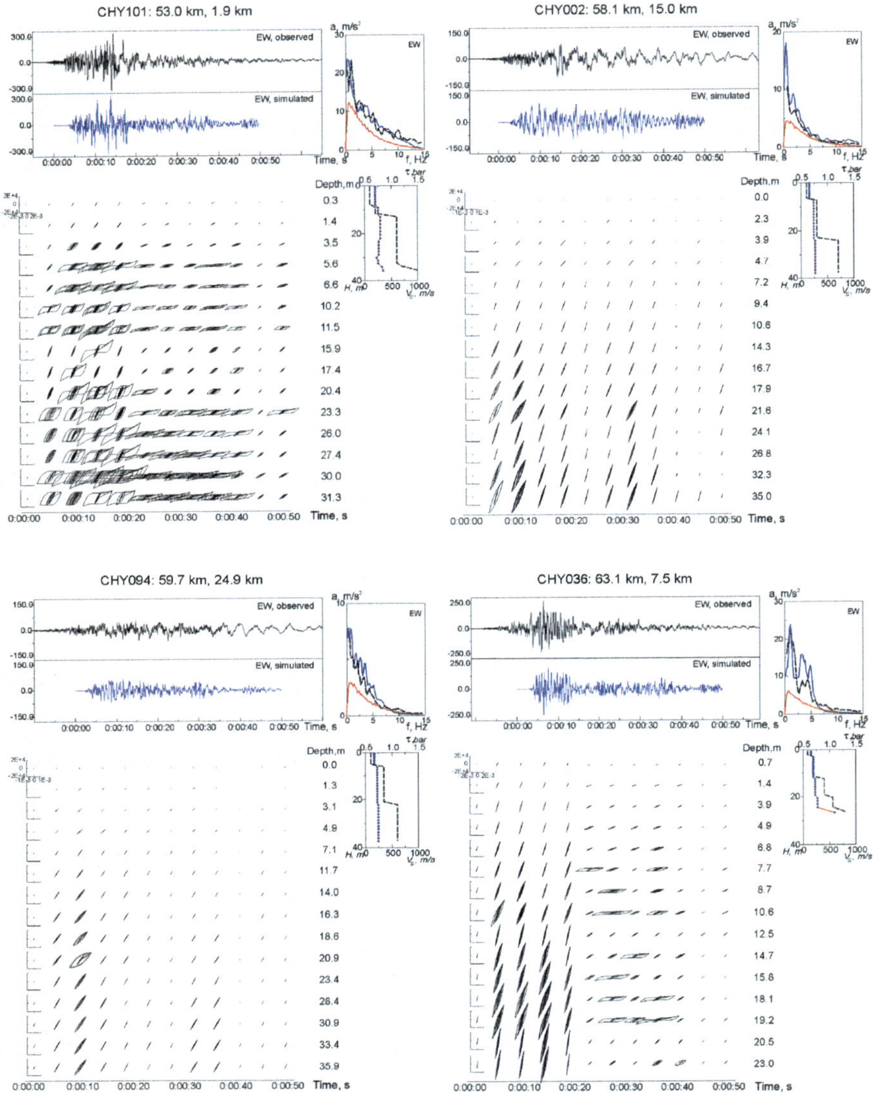

Fig. 3.40 The results of simulations of acceleration time histories of the Chi-Chi earthquake at CHY101, CHY002, CHY094, and CHY036 soil sites (EW component)

Long-period waves are also noticed in the final parts of records at northern sites TCU033 and TCU017 (Figs. 3.41 and 3.42). Lee et al. (2001a; b) describe TCU017 site as a typical tree root problem, where the concrete slab base of the station is affected by tree roots; and it is not clear, if basin waves are affecting records at these stations.

An agreement between the observations and simulations was obtained for the beginning parts of records at both sites, and spectra were estimated based on the

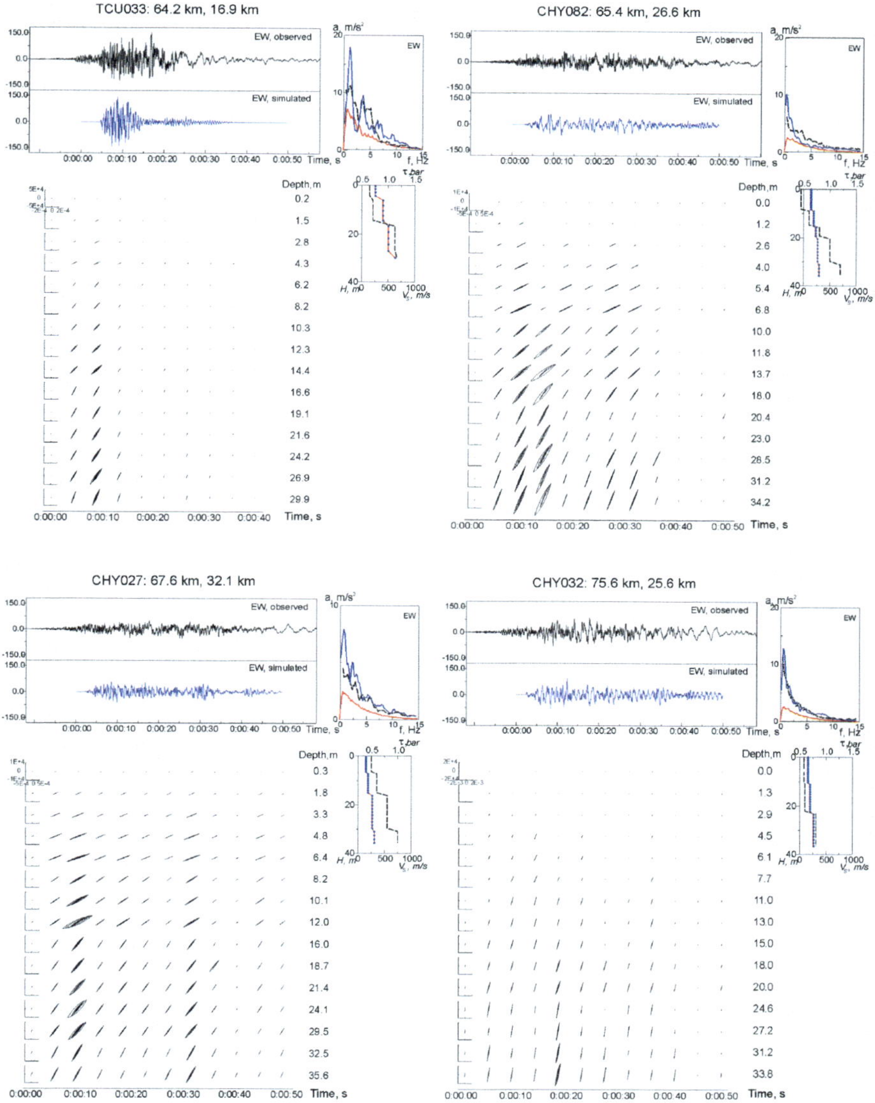

Fig. 3.41 The results of simulations of acceleration time histories of the Chi-Chi earthquake at TCU033, CHY082, CHY027, and CHY032 soil sites (EW component)

beginning parts. At the same time, the duration of the simulated strong motion at both sites is shorter than that of the observed ones. This is a problem at all rock and soil sites of the "northern" group. This can be due to the source mechanism and some inconsistency in the applied models of slip distribution on the fault plane.

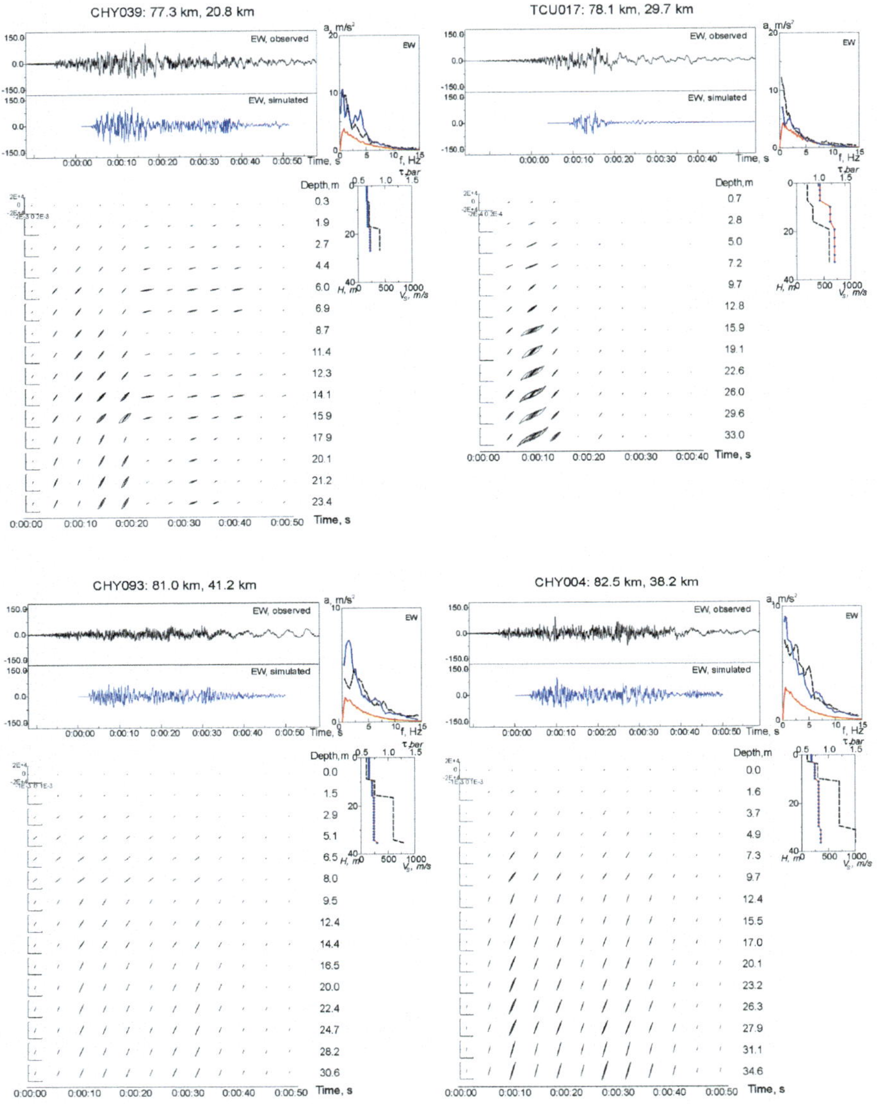

Fig. 3.42 The results of simulations of acceleration time histories of the Chi-Chi earthquake at CHY039, TCU017, CHY093, and CHY004 soil sites (EW component)

At stations CHY101, CHY036, and CHY039 (Figs. 3.40 and 3.42), the recorded accelerograms possess characteristic shapes, such as, intense beginnings (first 15–20 s of strong motion) and final parts of smaller intensity, whereas the intensities of "input" motions at the three sites (simulated accelerograms not shown) remained at almost the same levels during the earthquake. We found that simulation of the

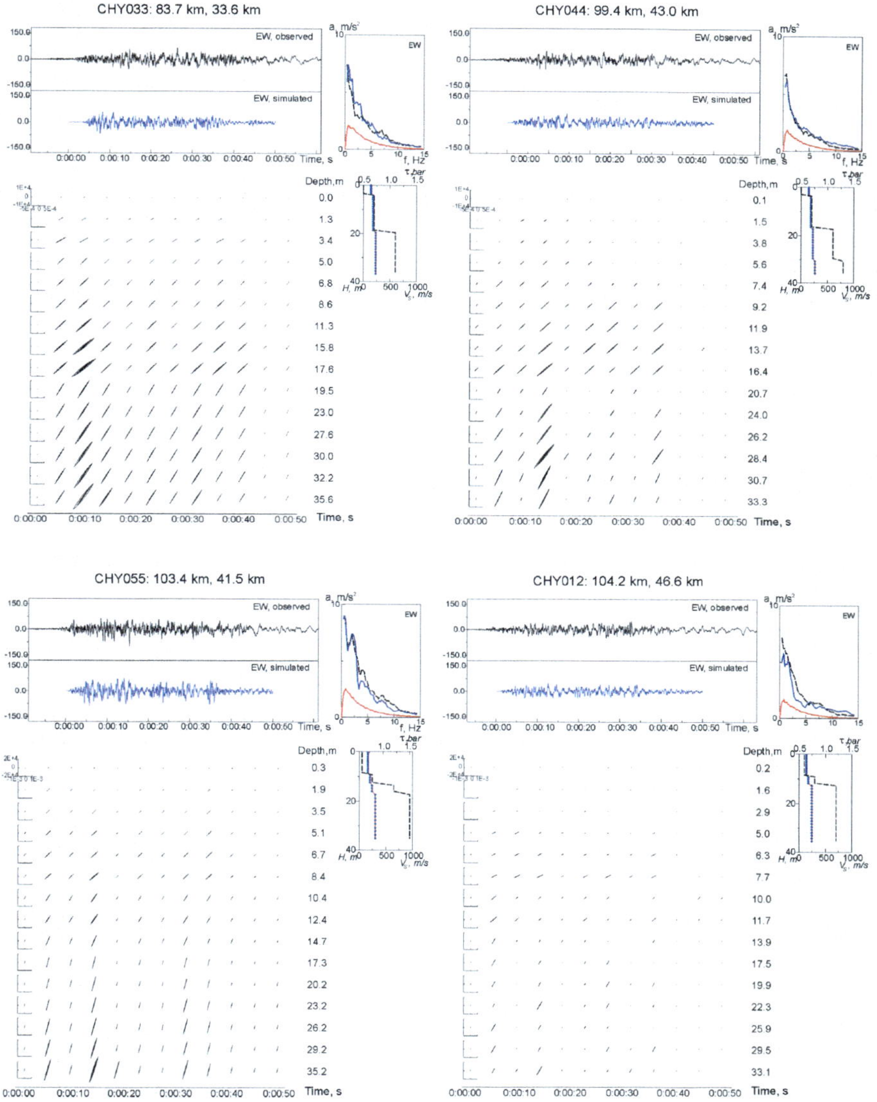

Fig. 3.43 The results of simulations of acceleration time histories of the Chi-Chi earthquake at CHY033, CHY044, CHY055, and CHY012 soil sites (EW component)

recorded accelerograms at the three sites with one stress–strain relation applied at all time intervals during strong motion is not possible; therefore, the beginning and final parts of the records were simulated separately. It was found that for the final parts of records, an agreement between the observations and simulations can only be achieved, if describing soil behavior by stress–strain relations with slopes close to horizontal ones, indicating liquefaction phenomena (Figs. 3.40 and 3.42).

Similarly, soil behavior at Port Island vertical array site during the Kobe earthquake was simulated: at Port Island, liquefaction in the upper soil layers (~ 13 m) occurred, and the behavior of these layers was also described by very sloping stress–strain relations (Sect. 3.1).

Stations CHY101, CHY036, and CHY039 are located close to each other, in a foothill depression stretched at the south-western corner of the fault; the elevations above the sea level are 75 m, 45 m, and 16 m, respectively (Fig. 3.1). The upper parts of the soil profiles represent silty soils with sands and clays with average *S*-wave velocities V_S ~ 200–280 m/s (Table 3.5); below (~ 26–34 m) are gravels and dense clays.

According to our estimates, resonant vibrations were induced in the upper layers during strong motion, and the maximum strains achieved ~ 0.8% at CHY101 site (stresses reached ~ 0.4 bars), ~ 0.6% at CHY036 site (stresses reached ~ 0.4 bars), and ~ 0.3% at CHY039 site (stresses reached ~ 0.1 bars). Evidently, large strains and sloping stress–strain relations may indicate liquefaction in the upper 20–30 m at these sites.

Large strains were also achieved in the upper ~ 30–40 m at TCU065 (~0.6%), TCU110 (~0.6–0.8%), and TCU115 (~0 .6–0.8%) sites (Figs. 3.31, 3.38 and 3.39), possessing similar soil composition in the upper layers: mostly sandy, clayey, and silty soils with V_S ~ 300 m/s at TCU065 site and ~ 190–210 m/s at TCU110 and TCU115 sites (Table 3.5), similar models of soil behavior and signs of liquefaction.

Lee et al. (2001a; b) reported liquefaction occurred at TCU065 and TCU110 sites during the Chi-Chi earthquake, and the constructed models of soil behavior confirm this. Liquefaction effects are not as clearly expressed at TCU115 site as at TCU065 and TCU110 sites, and they involve deeper layers. The constructed models of soil behavior at these sites indicate that resonant vibrations were induced by strong motion in the upper layers and accompanied by large strains and pore pressure build-up. Similar phenomena were also observed at TCU102 (strains in the upper layers ~ 0.2%), CHY025 (strains ~ 0.4%), TCU072 (strains ~ 0.3–0.4%) and CHY026 (strains ~ 0.3–0.4%) sites. At other sites, maximum strains in the upper layers did not exceed ~ 0.2%, according to our estimates.

Thus, resonant phenomena were observed in soft soil layers down to ~ 40–60 m in the near-fault zones during the Chi-Chi earthquake; most clear at stations located within ~ 20–22 km from the fault plane, such as, TCU065, TCU110, TCU115, CHY025, CHY092, CHY101, CHY026, CHY036, and CHY039 sites.

The constructed models of soil behavior were used to estimate reduction of shear moduli in soil layers during strong motion. Normalized shear moduli G/G_{max} were estimated as the ratios of stresses and strains averaged over 4 s time intervals and over layers within the selected depth intervals (Fig. 3.44). Reduction of shear moduli was calculated as a difference between the average value of the shear modulus G for the entire period of strong motion (50 s) and its maximum value G_{max}, expressed in percent of G_{max}; the estimates obtained for NS and EW components were also averaged.

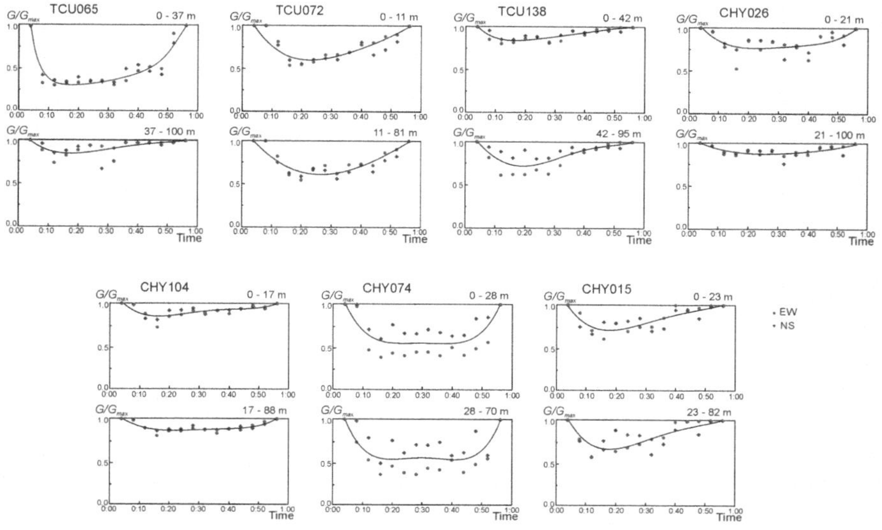

Fig. 3.44 Changes of shear moduli in soil layers at seven sites during the Chi-Chi earthquake

As seen from the figure, shear moduli reduction $\Delta G / G_{\max}$ was as high as ~ 70% in the upper layers at TCU065 site. At TCU072 and CHY074 sites, $\Delta G / G_{\max}$ was also significant, up to ~ 40–50% in the whole soil column, from the surface down to ~ 70–80 m. At other sites, $\Delta G / G_{\max}$ was about ~ 15% (at TCU138 site), 20–25% (at CHY026 site), 10–15% (at CHY104 site), and ~ 30% (at CHY015 site) in the upper layers. It is interesting that $\Delta G / G_{\max}$ did not always decrease with depth: for example, at TCU138 site $\Delta G / G_{\max}$ at depths of ~ 40–95 m is higher than in subsurface layers (in the upper ~ 40 m), which is evidently due to a decrease of *S*-wave velocities at depth. At TCU072, CHY074 and CHY015 sites, $\Delta G / G_{\max}$ also do not decrease with depth in the upper ~ 70 m, which can be related to the inaccuracy of our estimations, or to large values of stresses and strains induced by the strong motion in these layers (Figs. 3.32, 3.36 and 3.37).

Table 3.6 presents $\Delta G / G_{\max}$ values averaged in time (during strong motion) and in depths: in the upper 30 m, the most representative for earthquake engineering, and in the upper 80 m.

Reduction of shear moduli at soil sites during the Chi-Chi earthquake was high: in the upper 30 m, it exceeded ~ 60% just near the fault plane and reached ~ 25–30% at ~ 30 km from the fault; in the upper 80 m, these values were ~ 35% and ~ 20%, respectively. This testifies to a significant influence of strong motion on soft sediments down to ~ 80 m. Thus, during the Chi-Chi earthquake, site effects captured thick soil layers and weakened slowly with increasing distance from the fault. The discussion of these effects is continued in Chap. 5, Sect. 5.3.

Table 3.6 Reduction of shear moduli during the Chi-Chi earthquake at seven stations, for which more complete information on the profiling data was available

Station code	r_{av}, km	r, km	$\Delta G/G_{\max}$ (30 m), %	$\Delta G/G_{\max}$ (80 m), %	V_{S30}, m/s	V_{S80}, m/s	Site class (Lee et al. 2001a, b)
TCU065	28.9	6.9	63	34	300	480	D
TCU072	32.4	14.9	37	35	563	877	D
TCU138	36.2	3.8	10	18	605	630	D
CHY026	53.1	16.1	22	17	226	330	E
CHY104	58.7	9.5	19	15	225	270	E
CHY074	67.8	12.8	44	45	546	735	C
CHY015	89.4	25.9	19	22	234	360	D

Soil response at Dahan vertical array site

Records of the 1999 Chi-Chi earthquake were obtained at Dahan vertical array site, located at about 80 km from the epicenter, in the north-eastern Taiwan, as a part of SMART-2 strong motion array. It was designed mainly for studying earthquake sources and characteristics of near-source ground motions and installed at the campus of the Dahan Institute of Technology in April 1992. The array includes one surface triaxial accelerometer and three downhole triaxial accelerometers at depths of 50, 100, and 200 m.

Figure 3.45 represents the profiling data on *S*- and *P*-wave velocities and the geological cross-section (Chiu 2001) and estimated values of the maximum shear stress $\tau_{\max}$, which were used in our calculations. As seen from the figure, soils in the upper layers are mainly dense gravelly deposits.

Stress–strain relations in soil layers at depths of 0–200 m are reconstructed. Acceleration records of the Chi-Chi earthquake at a depth of 200 m were used as input signals, and the propagation of seismic waves vertically up to the surface was calculated; acceleration time histories on the surface and at depths of 50 and 100 m were simulated and compared to the observed ones.

For the calculations, the soil profile was divided into two groups of layers, for which two different types of stress–strain relations (shown in Fig. 3.45) were assumed. To describe the behavior of the upper 7 m ($V_S \le 200$ m/s), "hard-type" stress–strain relations were used, which are typical for water-saturated soils, because the level of the underground water was close to the surface, as indicated by *P*-wave velocities. In these relations, the stress–strain curves decline to the stress axis at large strains. To describe the behavior of soil layers below 7 m, stress–strain relations similar to those obtained in laboratory experiments by Hardin and Drnevich (1972) were used. Successive time intervals of 3 s duration were analyzed.

Before the simulations, the directional drifts of the accelerometers were checked and corrected by calculating the horizontal orbits of the long-period (0.01–0.2 Hz) particle motions at all depths.

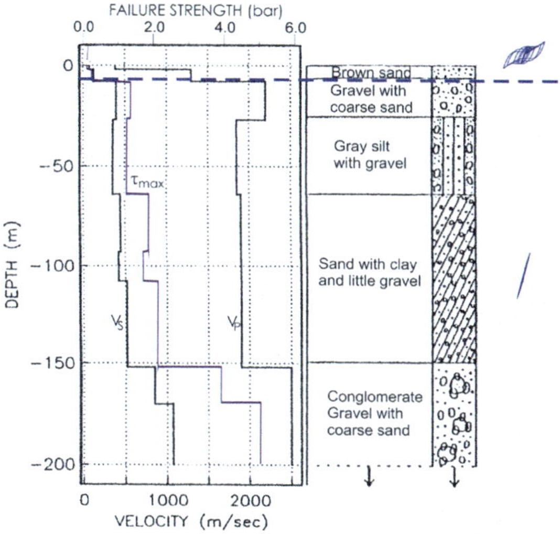

Fig. 3.45 The ground profile (from Chiu 2001) and characteristic types of stress–strain relations at different depths at Dahan downhole array site. Maximum shear stress τ_{max} was estimated based on the profiling data

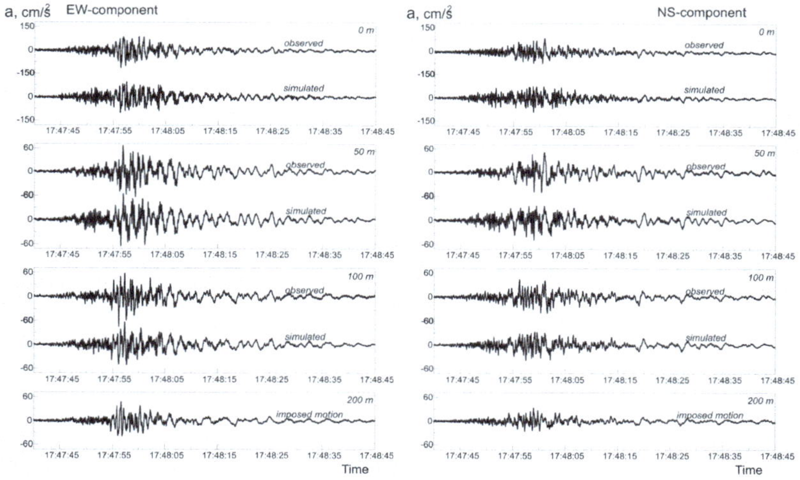

Fig. 3.46 Observed and simulated acceleration time histories of the 1999 Chi-Chi earthquake at Dahan vertical array site

The observed and simulated accelerograms of the mainshock of the Chi-Chi earthquake are presented in Fig. 3.46, and the corresponding stress–strain relations in the soil layers at depths of 0–200 m are shown in Fig. 3.47.

As seen from Figs. 3.46 and 3.47, a rather good agreement was obtained between the observed and simulated accelerograms and a similarity in vertical distributions of stress–strain relations at two horizontal components, testifying to the validity of the constructed models of soil behavior. The shapes of the stress–strain curves in the layers at all depths are virtually linear; therefore, the nonlinear distortions of the propagating seismic waves were weak. According to our estimates, strains did not

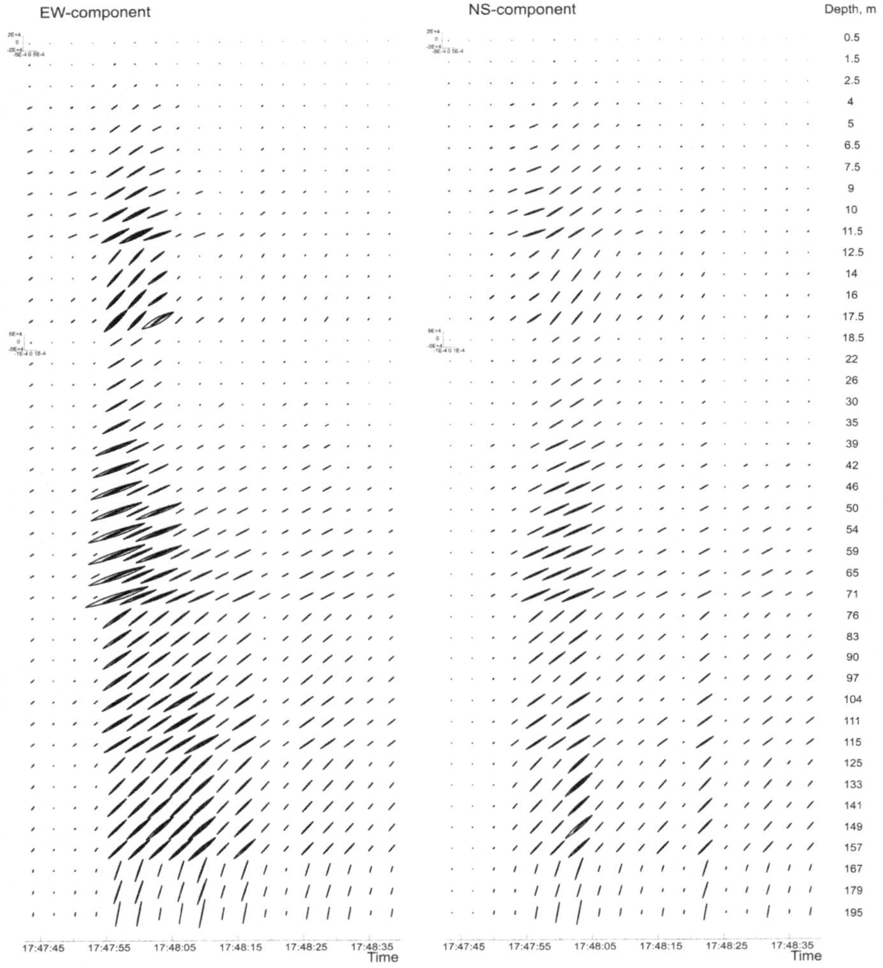

Fig. 3.47 The estimated stress–strain relations for the Chi-Chi earthquake at Dahan vertical array site. Stresses (vertical axes) are given in Pa, strains (horizontal axes) in strain

exceed ~ 0.03% at all depths, and stresses attained values of ~ 10 kPa at 0–7 m and ~ 50–80 kPa below ~ 7 m.

The obtained stress–strain relations were used to estimate changes in the shear moduli of soil layers during the earthquake. Average shear moduli were calculated within 3 s time intervals in the two groups of layers, 0–7 m and 7–200 m, at two horizontal components. In the upper layers, changes of the shear moduli were virtually inconspicuous, and in layers below 7 m, reduction of the shear moduli was less than 3–5%. These small changes in the shear moduli are evidently due to rather large distance of the site from the fault plane and the fact that soils were firm gravelly deposits.

******

Thus, the behavior of subsurface soft soils was significantly nonlinear in the near-fault zones during recent strong earthquakes with magnitudes $M_w \sim 6.7$–7.6, such as, the 1995 Kobe earthquake ($M_w = 6.8$), 2000 Tottori earthquake ($M_w = 6.7$), and 1999 Chi-Chi earthquake ($M_w = 7.6$), and rheological properties of soil layers changed during strong motion. Shear moduli decreased at the beginning and recovered at the end of strong motion; the most noticeable changes were observed in the upper soft layers at sites located close to the fault plane.

Significantly nonlinear soil behavior, different types of nonlinear behavior of different soil layers and the maximum reduction of shear moduli (up to ~ 80–90% of the initial values) were observed within epicentral distances of ~ 1/4L, where L is the length of the fault plane. Reduction of shear moduli decreased to ~ 50–60% at distances of ~ 1/2 L from the fault, and did not exceed several percent at distances of about L.

Nonlinearity of soil response weaken high-frequency components of strong motion and strengthen low-frequency components, so that power spectra of motion on the surface tend to take the form $E(f) \sim f^{-k}$.

During the three earthquakes, almost all sites (except Port Island) located near the fault showed nonlinear soil behavior of the hard type. Strong motion induced resonant phenomena in subsurface softer layers, i.e., large strains and pore pressure build-up, which resulted in the hard-type nonlinearity. In such cases, strong nonlinearity of soil response did not noticeably decrease amplitudes of accelerations on the surface.

Resonant phenomena in the upper softer layers were observed at many sites during the three earthquakes; however, during the stronger Chi-Chi earthquake, they involved deeper layers (~ 40–60 m) than during the Kobe and Tottori earthquakes (~ 10–15 m).

Finally, we see a similarity in the behavior of similar soils during different strong earthquakes, which indicates the possibility to predict soil behavior in future strong earthquakes.

References

Aguirre, J., Irikura, K.: Nonlinearity, liquefaction, and velocity variation of soft soil layers in PI, Kobe, during the Hyogo-ken Nanbu earthquake. Bull. Seism. Soc. Am. **87**, 1244–1258 (1997)

Akkerman, E.: Thixotropy and fluidity of fine-grained soils. Probl. Engin. Geol. **1**, 73–92 (1958)

Anderson, J.G., Hough, S.E.: A model for the shape of the Fourier amplitude spectrum of acceleration at high frequencies. Bull. Seism. Soc. Am. **74**, 1969–1993 (1984)

Archuleta, R.J.: Direct observations of nonlinearity in accelerograms. The Effects of Surface Geology on Seismic Motion, pp. 787–792. Balkema, Rotterdam (1998)

Archuleta, R.J., Steidl, J.H.: Borehole array data: open windows into site response. In: 10th International Conference Soil Dynamics Earthquake Engineering Philadelphia, USA, vol. Ext. Abst. vi (2001)

Bazzurro, P., Sjoberg, B., Luco, N.: Post-elastic response of structures to synthetic ground motions. http://geohazards.cr.usgs.gov/staffweb/nluco/manuscripts/0409--Bazzurro+Sjoberg+Luco.pdf (2001)

Beresnev, I.A., Atkinson, G.M.: FINSIM—a FORTRAN program for simulating stochastic acceleration time histories from finite faults. Seism. Res. Lett. **69**(1), 27–32 (1998)

Boore, D.M.: Simulation of ground motion using the stochastic method. Pure Appl. Geoph. **160**, 635–676 (2003)

Boore, D.M., Joyner, W.B.: Site amplifications for generic rock sites. Bull. Seism. Soc. Am. **87**, 327–341 (1997)

Boswell, P.G.H.: A preliminary examination of the thixotropy of some sedimentary rocks. Qurterly J. Geol. Soc. London **104**(4), 499–526 (1949)

Chen, K.C., Shin, T.C., Wang, J.H.: Estimates of coda Q in Taiwan. Proc. Geol. Soc. China. **32**, 339–353 (1989)

Chi, W.-C., Dreger, D., Kaverina, A.: Finite-source modeling of the 1999 Taiwan (Chi-Chi) earthquake derived from a dense strong-motion network. Bull. Seism. Soc. Am. **91**, 1144–1157 (2001)

Chiu, H.C.: Data files from the SMART-2 strong motion array for the Chi-Chi earthquake. Bull. Seism. Soc. Am. **91**(5), 1391–1392 (2001)

Field, E.H., Johnson, P.A., Beresnev, I.A., Zeng, Y.: Nonlinear ground-motion amplification by sediments during the 1994 Northridge earthquake. Nature **390**(6660), 599–602 (1997)

Gumensky, B.M.: Fundamentals of physics and chemistry of clayey soils and their application in construction. Stroyizdat, Leningrad-Moscow (1965) (in Russian)

Gusev, A.A., Pavlov, V.M.: Wideband simulation of earthquake ground motion by a spectrum-matching, multiple-pulse technique. Preprint (IC/2006/023) Trieste (2006)

Hardin, B.O., Drnevich, V.P.: Shear modulus and damping in soils: measurement and parameter effects. Proc. Am. Soc. Civil Eng., J. Soil Mech. Found. Div. **98**, 603–624 (1972)

Hatayama, K., Matsunami, K., Iwata, T., Irikura, K.: Basin-induced love waves in the eastern part of the Osaka basin. J. Phys. Earth **43**, 131–155 (1995)

Ishihara, K., Cubrinovski, M., Takeshi, N.: Characterization of undrained behavior of soils in the reclaimed area of Kobe. Soils Found. Spec. Issue on Geotechnical Aspects of the January 17 1995 Hyogoken-Nambu Earthquake **2**, 33–46 (1998)

Joyner, W.B., Chen, T.F.: Calculation of nonlinear ground response in earthquakes. Bull. Seism. Soc. Am. **65**(5), 1315–1336 (1975)

Kawase, H., Satoh, T., Fukutake, K., Irikura, K.: Borehole records observed at the Port Island in Kobe during the Hyogo-ken Nanbu earthquake of 1995 and its simulation. J. Constr. Eng., AIJ **475**, 83–92 (1995)

Kazama, M., Yamaguchi, A., Yanagisawa, E.: Seismic behavior of an underlying alluvial clay on man-made islands during the Hyogoken-Nanbu earthquake. Soils Found. Spec. Issue on Geotechnical Aspects of the January 17 1995 Hyogoken-Nambu Earthquake **2**, 23–32 (1998)

Kokusho, T., Matsumoto, M.: Nonlinearity in site amplification and soil properties during the 1995 Hyogoken-Nambu earthquake. Soils Found. Spec. Issue on Geotechnical Aspects of the January 17 1995 Hyogoken-Nambu Earthquake **2**, 1–9 (1998)

Kramer, S.L.: Geotechnical Earthquake Engineering, pp. 230–238. Prentice Hall, Upper Saddle River, N.J. (1996)

Lee, C.-T., Cheng, C.-T., Liao, C.-W., Tsai, Y.-B.: Site classification of Taiwan free-field strong-motion stations. Bull. Seism. Soc. Am. **91**, 1283–1297 (2001a)

Lee, W.H.K., Shin, T.C., Kuo, K.W., Chen, K.C., Fu, C.F.: CWB free-field strong-motion data from the 21 September Chi-Chi, Taiwan, earthquake. Bull. Seism. Soc. Am. **91**, 1370–1376 (2001b)

Lin, C.-M., Wen, K.-L., Chang, T.-M.: Estimation of S-wave velocity model in the western coastal plain of Taiwan. In: 3rd International Conference Urban Earthquake Engineering, Center for Urban Earthquake Engineering, Tokyo, pp. 617–622 (2006)

Lomnitz, C., Chavez-Garcia, F.J., Bard, P.-Y.: Two articles on gravity waves; discussion and reply. Bull. Seism. Soc. Am. **85**(4), 1271–1274 (1995)

Ma, K.-F., Mori, J., Lee, S.-J., Yu, S.B.: Spatial and temporal distribution of slip for the 1999 Chi-Chi, Taiwan, earthquake. Bull. Seism. Soc. Am. **91**, 1069–1087 (2001)

O'Connel, D.R.H.: Replication of apparent nonlinear seismic response with linear wave propagation models. Science **283**(5410), 2045–2050 (1999)

Oglesby, D., Day, S.: Fault geometry and the dynamics of the 1999 Chi-Chi (Taiwan) earthquake. Bull. Seism. Soc. Am. **91**, 1099–1111 (2001)

Pavlenko, O.V., Irikura, K.: Types of elastic nonlinearity of sedimentary soils. Geoph. Res. Lett. **29**(19), 36-1–36-4 (2002)

Pavlenko, O.V., Irikura, K.: Changes in shear moduli of liquefied and nonliquefied soils during the 1995 Kobe earthquake and its aftershocks at PI, SGK, and TKS vertical array sites. Bull. Seism. Soc. Am. **92**(5), 1952–1969 (2002a)

Pavlenko, O.V., Irikura, K.: Nonlinearity in the response of soils in the 1995 Kobe earthquake in vertical components of records. Soil Dyn. Earthq. Eng. **22**, 967–975 (2002c)

Pavlenko, O.V., Irikura, K.: Estimation of nonlinear time-dependent soil behavior in strong ground motion based on vertical array data. Pure Appl. Geoph. **160**, 2365–2379 (2003)

Pavlenko, O.V., Irikura, K.: Identification of the nonlinear behavior of liquefied and non-liquefied soils during the 1995 Kobe earthquake. Geoph. J. Int. **160**, 539–553 (2005)

Pavlenko, O.V., Loh, K.: Nonlinear identification of the soil response at Dahan downhole array site during the 1999 Chi-Chi earthquake. Soil Dyn. Earthq. Eng. **25**(3), 241–250 (2005)

Pitarka, A., Irikura, K., Iwata, T., Sekiguchi, H.: Three-dimensional simulation of the near-fault ground motion for the 1995 Hyogo-ken Nanbu (Kobe), Japan, Earthquake. Bull. Seism. Soc. Am. **38** (1998)

Roumelioti, Z., Beresnev, I.A.: Stochastic finite-fault modeling of ground motions from the 1999 Chi-Chi, Taiwan, earthquake: application to rock and soil sites with implications for nonlinear site response. Bull. Seism. Soc. Am. **93**(4), 1691–1702 (2003)

Sato, K., Kokusho, T., Matsumoto, M., Yamada, E.: Nonlinear seismic response and soil property during 1995 Hyogoken Nanbu earthquake. Soils Found. Spec. Issue on Geotechnical Aspects of the January 17, 1995 Hyogoken Nanbu Earthquake 41–52 (1996)

Seed, H.B., Idriss, I.M.: Soil moduli and damping factors for dynamic response analyses. Rep. EERC70-10 EERC, University of California, Berkeley (1970)

Seed, H.B., Wong, R.T., Idriss, I.M., Tokimatsu, K.: Moduli and damping factors for dynamic analyses of cohesionless soils. Rep. UCB/EERC-84/14 (1984)

Soeda, Y., Tamai, H., Nakatsu, N., Takezawa, K., Maegawa, F.: Nonlinearity and irregularity on strong seismic motions by vertical arrays during the 1995 Kobe Earthquake. Earthq. Geotechn. Eng. 225–230. Balkema, Rotterdam (1999)

Sun, J.I., Golesorkhi, R., Seed, H.B.: Dynamic moduli and damping ratios for cohesive soils. Rep. UCB/EERC-88/15 (1988)

Voznesensky, E.A.: Dynamic Instability of Soils. Editorial URSS, Moscow (1999) (in Russian)

Wang, W.-H., Chang, S.-H., Chen, C.-H.: Fault slip inverted from surface displacements during the 1999 Chi-Chi, Taiwan, earthquake. Bull. Seism. Soc. Am. **91**, 1167–1181 (2001)

Wu, C., Takeo, M., Ide, S.: Source process of the Chi-Chi earthquake: a joint inversion of strong motion data and global positioning system data with a multifault model. Bull. Seism. Soc. Am. **91**, 1128–1143 (2001)

Yang, J., Sato, T.: Interpretation of seismic vertical amplification observed at an array site. Bull. Seism. Soc. Am. **90**(2), 275–285 (2000)

Yasuda, S., Ishihara, K., Harada, K., Shinkawa, N.: Effect of soil improvement on ground subsidence due to liquefaction. Soils Found. Spec. Issue on Geotechnical Aspects of the January 17 1995 Hyogoken-Nambu Earthquake 99–107 (1996)

Zeng, Y., Chen, C.-H.: Fault rupture of the 20 September 1999 Chi-Chi, Taiwan, earthquake. Bull. Seism. Soc. Am. **91**, 1088–1098 (2001)

Zvolinsky, N.V.: Wave processes in inelastic media. Probl. Eng. Seismolog. **23**, 4–19 (1982) (in Russian)

Chapter 4
Soil Hardening and Generation of Abnormally High PGA During Large Earthquakes with Extended Sources

During past strong earthquakes, observations of abnormally high accelerations exceeding 1 g were usually attributed to disturbances in recordings, effects of recording piers, falling objects, impacts, etc., or interferences with nearby weak earthquakes, landslides, or similar possible causes. However, during the 2011 Tohoku earthquake, 19 K-NET and KiK-net stations reliably recorded PGA higher than 1 g, the highest ~ 3 g.

Generation of abnormally high PGA in the near-fault zones of large earthquakes may be associated with the effects of directivity of seismic wave radiation from the sources and their limiting case, shock waves.

Since the 1980s, seismologists have been studying the effects of radiation directivity of extended earthquake sources. Archuleta and Hartzell (1981) discuss "the complications that arise when one is no longer in the far-field but situated close to a finite rupture", when the radiation originates over some area, and the receiver distance, radiation pattern, and arrival times of *P*- and *S*-waves are ill defined. The ground motion cannot be interpreted using the same approach that was appropriate for the far-field. By means of numerical simulation they revealed a strong influence of directivity effects on the accelerations in the near-source zones of the 1979 Imperial Valley earthquake, where PGA ~ 1195 cm/s^2 were recorded.

Directivity effects occur when the propagation of rupture toward a site at a velocity that is almost as large as the shear wave velocity causes most of the seismic energy from the rupture to arrive in a single large pulse of motion, which represents the cumulative effect of almost all of the seismic radiation from the fault (Sommerville et al. 1997).

Recent years with developing strong-motion networks, seismologists more often observe directivity effects, and publications appear where they discuss the generation of shock waves during strong earthquakes due to supershear crack propagation (Bouchon et al. 2001; Bouchon and Vallee 2003; Aagaard and Heaton 2004; Ellsworth et al. 2004; Robinson et al. 2006; Dunham and Bhat 2008; Vallee and Dunham 2012; Zhan et al. 2013, etc.).

O. V. Pavlenko, *Seismic Waves in Soil Layers*,
https://doi.org/10.1007/978-3-031-52338-0_4

In this chapter, we discuss soil behavior and possible mechanisms of generation of high PGA during two large earthquakes with extended sources, such as, the 2011 Tohoku earthquake (M_w ~ 9.0) and 2003 Tokachi-oki earthquake (M_w ~ 8.3). The distributions of PGA in the near-fault zones show mosaic patterns, and the highest accelerations were recorded close to the epicenter, as well as at remote sites.

Models of soil behavior during these earthquakes indicate nonlinearity of soft soil response only at sites located just near the fault plane. Instead, a rather widespread soil hardening was observed, most expressed at remote sites recorded the highest PGA. To explain the observed features of soil response, two possible mechanisms are suggested, that could act together, such as, (1) shock wave fronts generated by rupture propagation along the fault plane induce soil hardening and high PGA; (2) soil compaction and hardening represent soil response to long-lasting dynamic loading during strong motion. Obviously, we may expect similar effects of soil hardening and generation of high PGA during other mega-thrust earthquakes in future.

4.1 Models of Soil Behavior in the Near-Fault Zones of the 2011 Mega-Thrust M9.0 Tohoku Earthquake

The Tohoku mega-thrust earthquake ($M_w = 9$) that occurred on 11 March 2011 off the coast of Japan is the strongest earthquake recorded by a large number of seismic instruments, including stations in the near-fault zones. 19 K-NET and KiK-net stations reliably recorded PGA higher than 1 *g*, the highest ~ 3 *g*, "anomalous strong ground motions that we have never seen previously" (Furumura et al. 2011). Large areas of high PGA are located in central Tohoku rather far from the fault plane (Fig. 4.1). Acceleration time histories possessing the highest PGA resemble impulses with sharp peaks in one or two groups of intense waves at both horizontal and vertical components. Are these high PGA values due to site, source, or path effects? Numerous records of the Tohoku earthquake obtained by strong motion networks throughout Japan may help to clarify this question.

Many researchers constructed models of the source process and studied soil behavior during strong motion. Bonilla et al. (2011) found a wide-spread nonlinearity of soil behavior during the Tohoku earthquake; they concluded that high recorded peak accelerations can be a combination of near-fault and nonlinear site response effects involving fluid mobilization.

Kawase et al. (2011) note that at stations recorded the highest PGA, amplification of seismic waves in soil layers estimated by H/V technique during the mainshock of the Tohoku earthquake exceeded (in some cases $\sim$ 3 times) the similar amplification during weak events. Irikura and Kurahashi (2012) constructed a source model of the Tohoku earthquake that contains five main strong-motion generation areas with high slip velocity; their delays from the origin time were 15.6, 66.4, 68.4, 109.7, and 118.2 s. They used the model to simulate velocity and acceleration time histories of the Tohoku earthquake at coastal sites.

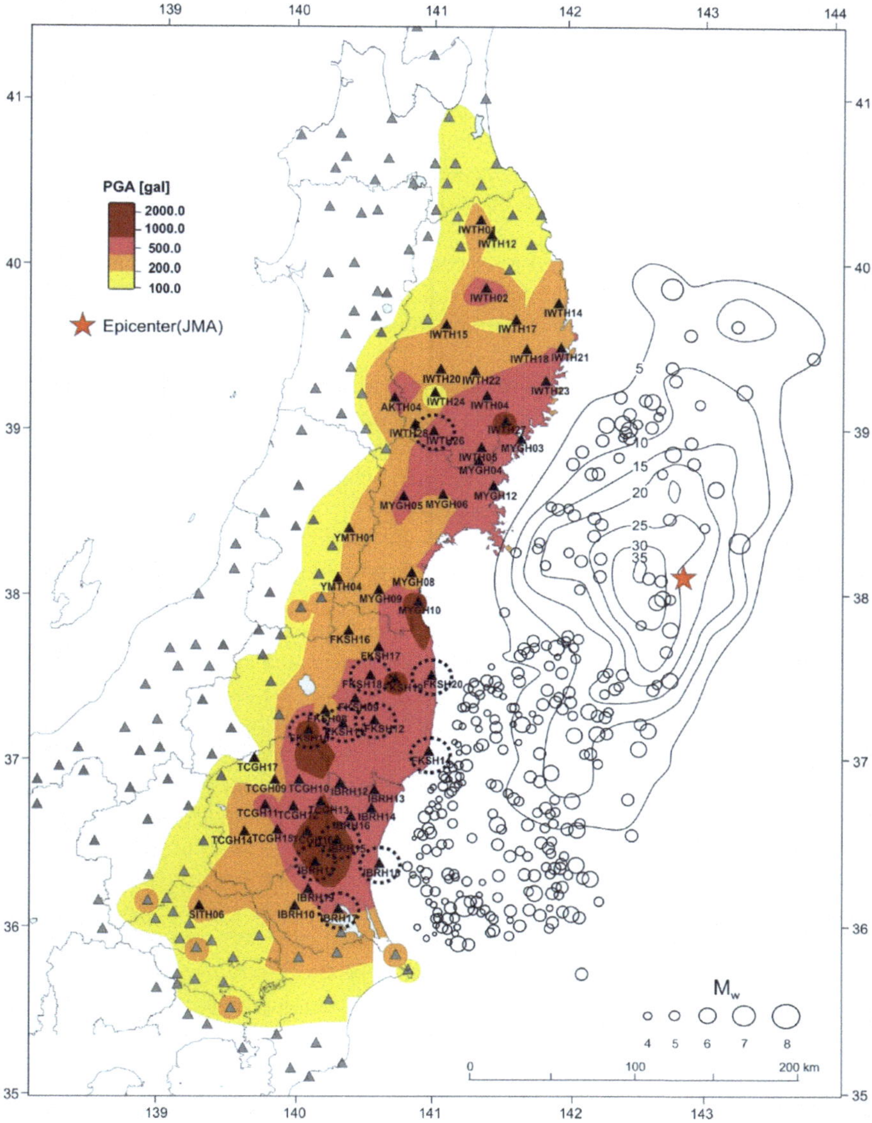

Fig. 4.1 Map showing the distribution of PGA recorded during the 2011 Tohoku earthquake (NIED 2011), locations of the mainshock (star) and aftershocks (circles) (M > 3) occurred in 24 h after the mainshock, and KiK-net stations (triangles). Contour lines depict slip distribution in meters according to the "joint source model" by Koketsu et al. (2011). Dashed circles mark sites with nonlinear soil response

Roten et al. (2013) analyzed ground motions recorded at Onahama port (Iwaki, Fukushima Prefecture) with the peak amplitude of 3-D acceleration vector approaching 2 g and found that, compared to records of weak motions, the response of the site during the Tohoku earthquake was characterized by increased amplification at frequencies above 10 Hz. The occurrence of high-frequency spikes coincided with a rapid increase of pore water pressure in the upper part of sand deposits, and this increase is possibly linked to a burst of high-frequency energy from a large slip patch below the Iwaki region.

Although at some sites located just near the fault zone in coastal areas, we could explain high PGA by bursts of energy from slip patches and pore-pressure effects (Roten et al. 2013), the causes of abnormally high PGA at remote sites cannot be explained by this mechanism.

Based on records of KiK-net vertical arrays, models of soil behavior in the near-fault zones during the Tohoku earthquake were constructed (i.e., vertical distributions of stresses and strains induced in soil layers by the strong motion), shear moduli of the soil layers were estimated and their temporal changes during strong motion were traced at sites with various soil conditions.

Figure 4.1 shows the locations of the mainshock, aftershocks occurred in 24 h after the mainshock, and KiK-net stations, for which models of soil behavior were constructed.

To construct models of soil behavior, we identify groups of layers within the soil profiles that can behave similarly during strong motion and assume "soft" or "hard" types of their behavior; then we choose the duration of the analyzed time intervals. As known from our previous experience in studying soil behavior during past strong earthquakes, groups of soil layers possessing different behavior in strong motion are usually distinguished at sites located within distances of $\sim 1/4\,L$ from the fault plane (where L is the length of the fault plane) in cases, when seismic boundaries exist in the upper meters of the soil profiles; for example, when a $\sim$ 5–20 m soft soil layer is underlain by denser stratum. When the types of stress–strain relations in soil layers are selected, they are checked by preliminary calculations.

At all the studied KiK-net sites except FKSH14, the best agreement with the observations was achieved with using one hard-type stress–strain relation to describe the behavior of all soil layers down to the location of the deep device. The only station, where the groups of surface softer soils and denser soils below were distinguished, was FKSH14 station, located very close to the fault plane (Fig. 4.1); at this site, liquefaction of the upper ~ 8 m was observed.

The profiling data, such as, the composition and thickness of soil layers and *P*- and *S*-wave velocities, were taken from the KiK-net website, whereas other parameters, which are not provided by the website (densities, shear stresses in failure τ_{max}, and damping coefficients), were estimated based on the composition and depth of the layers, accounting for the increase of τ_{max} with depth due to the lithostatic pressure, as described by Joyner and Chen (1975).

Preliminary calculations revealed that the results of simulations (stress and strain estimates, shear moduli, the degree of nonlinearity, etc.) are not too sensitive to some inaccuracies in prescribing τ_{max}, apparently because the simulated accelerograms on the surface are defined by real stresses and strains induced in soil layers by the strong motion, rather than some limiting stress values.

The Tohoku earthquake produced long-duration strong motions (~ 3 min), and to trace changes in the soil behavior, the records were divided into small time intervals. Preliminary calculations with time intervals of 1.5, 3, 5, and 10 s have shown that intervals of ~ 5 s duration were the optimal ones, because soil behavior was rather stable within this interval; at the same time, such duration allowed a reliable estimation of the best-fit nonlinear stress–strain relations.

Sets of parametric nonlinear stress–strain relations (250 curves) were generated and tested item-by-item to identify those which show the best fit to the observed accelerograms on the surface in 32 successive 5 s time intervals (160 s of strong motion). The models were evaluated by the deviations of the simulated records on the surface from the observed ones.

It was found convenient to divide all the records of the earthquake into two parts, such as, records of northern stations, located to the north of FKSH17 site, 37.66 N parallel, containing two groups of intense waves (associated with at least two large ruptures on the fault), and records of southern stations, located to the south of 37.66 N parallel, containing one group of intense waves.

For the analysis we selected records of KiK-net stations with the highest PGA and located close to the fault. Finally, 55 KiK-net sites were selected with epicentral distances of ~ 137–390 km. Tables 4.1 and 4.2 provide information on these sites: the coordinates, soil conditions, epicentral distances, maximum recorded PGA on the surface and at depth, depths of the boreholes, coefficients of amplification of seismic waves in soil layers (the ratios of the intensities of motion on the surface and at depth, averaged over two horizontal components). The last columns contain estimates of the degree of nonlinearity of soil response described below.

The results of the simulations, i.e., constructed models of soil behavior at the KiK-net sites during the Tohoku earthquake are presented in Fig. 4.2a–i (southern stations, close to the source), Fig. 4.3a–i (southern stations, remote), Fig. 4.4a–i (northern stations, close to the source), and Fig. 4.5a–i (northern stations, remote). As seen from the figures, a rather good agreement was obtained between the observations and simulations.

It was shown in Chap. 3 that substantially nonlinear behavior of soft soils during strong earthquakes with magnitudes M_w ~ 6.7–7.6 was observed within the distances of ~ 1/4L from the fault. Within ~ 1/2L, we still observe nonlinearity of soil behavior. Soil behavior gradually becomes linear with further increase of the distance from the fault.

Nonlinearity of soil behavior is usually associated with low-frequency motion on the surface that we can see in acceleration time histories (at FKSH20 site, Fig. 4.2a, for example), as a result of softening of subsurface soils by vertically incident intense seismic waves.

Thus, during the Tohoku earthquake, we could expect nonlinear soil behavior at soft soil sites over a large part of Honshu (within ~ 1/4L from the fault), strong low-frequency components in the acceleration time histories on the surface and reduction of shear moduli during strong motion.

Table 4.1 Information on the KiK-net stations in southern Tohoku and Kanto regions, for which models of soil behavior were constructed

Site code	Latitude (°)	Longitude (°)	d/V_S	V_{S30} (m/s)	Epic. dist. (km)	PGA, surface Gal	PGA, depth Gal	A	nl (%)
FKSH20	37.49	140.99	28/350	357	178	663.0	362.0	2.8	22
FKSH19	37.47	140.72	20/255	338	201	914.0	373.1	5.8	16
FKSH14	37.03	140.97	8/166	237	205	481.4	152.7	3.1	33
IBRH18	36.36	140.62	3/180	559	216	633.6	161.5	6.2	19
FKSH12	37.22	140.57	4/250	449	225	506.6	112.4	5.4	23
FKSH09	37.35	140.43	10/244	585	230	562.6	215.4	3.8	7
FKSH11	37.20	140.34	34/241	240	244	504.7	200.0	2.9	24
IBRH13	36.80	140.57	16/269	335	249	570.8	133.9	6.0	16
FKSH08	37.28	140.21	4/200	562	250	392.7	126.0	3.4	2
IBRH14	36.69	140.55	2/180	829	258	454.2	124.5	3.9	1
IBRH12	36.84	140.32	6/240	486	265	782.4	151.8	5.7	14
FKSH10	37.16	140.09	4/150	487	266	1335.4	194.7	6.2	17
IBRH16	36.64	140.40	5/206	626	272	666.8	131.3	6.2	11
FKSH18	37.49	140.54	12/185	307	277	633.2	127.7	5.4	19
TCGH13	36.73	140.18	4.5/150	563	282	907.5	155.6	6.5	16
IBRH15	36.5	140.3	2/60	450	284	1062.2	144.5	6.3	17
TCGH10	36.86	140.02	4/120	371	286	673.7	205.1	2.9	15
TCGH12	36.70	139.98	4/130	344	299	509.4	195.2	2.5	15
TCGH16	36.55	140.08	18/163	213	301	1304.8	190.3	5.5	9
TCGH09	36.84	139.84	6/180	468	301	164.6	69.4	2.3	1
IBRH11	36.37	140.14	20/197	242	309	1223.9	267.9	5.8	27
TCGH11	36.71	139.77	12/198	329	314	469.6	71.6	8.3	3
TCGH15	36.56	139.86	3/100	423	316	358.1	90.5	4.4	2
IBRH17	36.09	140.31	10/212	301	318	522.8	130.7	5.9	20
TCGH14	36.55	139.62	2/200	849	335	264.3	52.3	5.3	14
IBRH10	36.11	139.99	20/110	144	338	334.4	98.3	4.8	12
SITH06	36.11	139.29	10/183	369	387	275.6	71.8	4.8	4

However, we observe mostly high-frequency acceleration time histories on the surface at soft and dense soil sites (Figs. 4.2, 4.3, 4.4 and 4.5) and soil hardening, instead of softening, at the moments of the highest intensity of motion. This is clearly seen, for example, in the records of IBRH15, TCGH10, TCGH12, and TCGH16 stations (Fig. 4.3e–h): stress–strain relations in the upper layers decline to the stress axes at the moments of the highest intensity of motion, indicating soil hardening. Then, with decreasing the intensity of motion, stress–strain relations decline to the

Table 4.2 Information on the KiK-net stations in northern Honshu, for which models of soil behavior were constructed

Site code	Latitude (°)	Longitude (°)	d/V_S	V_{S30} m/s	Epic dist. km	PGA, surface Gal	PGA, depth Gal	A	nl (%)
MYGH12	38.64	141.44	6/280	748	137	568.3	250.4	3.2	1
MYGH03	38.92	141.64	4/350	934	140	511.4	186.6	3.4	1
MYGH04	38.79	141.33	4/220	850	154	632.0	133.9	4.3	2
IWTH27	39.03	141.53	4/150	486	155	863.6	110.8	6.0	2
IWTH05	38.87	141.35	2/160	429	156	791.3	182.0	4.8	1
IWTH23	39.27	141.82	4/370	923	158	521.0	154.1	3.2	2
MYGH06	38.59	141.07	2/200	593	165	327.8	194.8	1.6	7
IWTH21	39.47	141.93	2/150	521	172	464.3	89.2	5.2	4
MYGH10	37.94	140.89	3/175	348	174	1136.8	221.3	4.4	11
IWTH04	39.18	141.39	5/220	456	175	530.0	109.3	4.8	13
MYGH08	38.11	140.84	1/20	203	177	301.1	213.6	1.8	5
IWTH18	39.46	141.68	2/180	892	183	363.2	82.0	4.7	1
IWTH26	38.97	141.00	4/130	370	188	615.2	125.4	4.2	18
MYGH05	38.58	140.78	2/120	305	189	503.1	199.4	2.7	8
IWTH22	39.33	141.30	10/262	532	192	346.0	83.8	4.9	16
MYGH09	38.01	140.60	2/150	358	198	362.3	167.4	2.5	3
IWTH14	39.74	141.91	2/170	816	200	300.0	52.6	6.4	3
IWTH28	39.01	140.86	4/240	417	201	303.9	83.9	1.9	13
IWTH24	39.20	141.01	2/180	486	202	202.3	115.2	4.3	3
FKSH17	37.66	140.60	6/180	544	205	321.7	93.5	2.2	4
IWTH20	39.34	141.05	2/110	289	209	407.9	198.9	2.9	9
FKSH16	37.76	140.38	4/180	532	221	354.2	107.2	16.3	6
AKTH04	39.17	140.71	2/150	459	222	523.3	52.3	6.7	6
YMTH04	38.08	140.30	6/130	248	225	263.4	53.2	3.9	3
IWTH15	39.61	141.09	4/150	338	227	220.6	78.5	7.2	7
IWTH02	39.83	141.38	5/150	390	230	744.2	43.8	2.9	6
IWTH12	40.15	141.42	8/230	368	259	291.7	53.2	4.8	13
IWTH01	40.24	141.34	10/222	438	271	353.0	54.4	7.5	5

strain axes showing soil softening. The shapes of the stress–strain relations are close to linear ones, i.e. soil behavior is virtually linear.

Figures 4.6 and 4.7 show the profiling data down to ~ 100 m at the KiK-net sites and the estimates of maximum stresses and strains induced by strong motion in the upper ~ 100 m. The estimates were obtained from the constructed models of soil behavior. As seen from the plots, stresses increase with depth, and strains are maximal in the upper layers, which indicates resonant phenomena in the upper layers

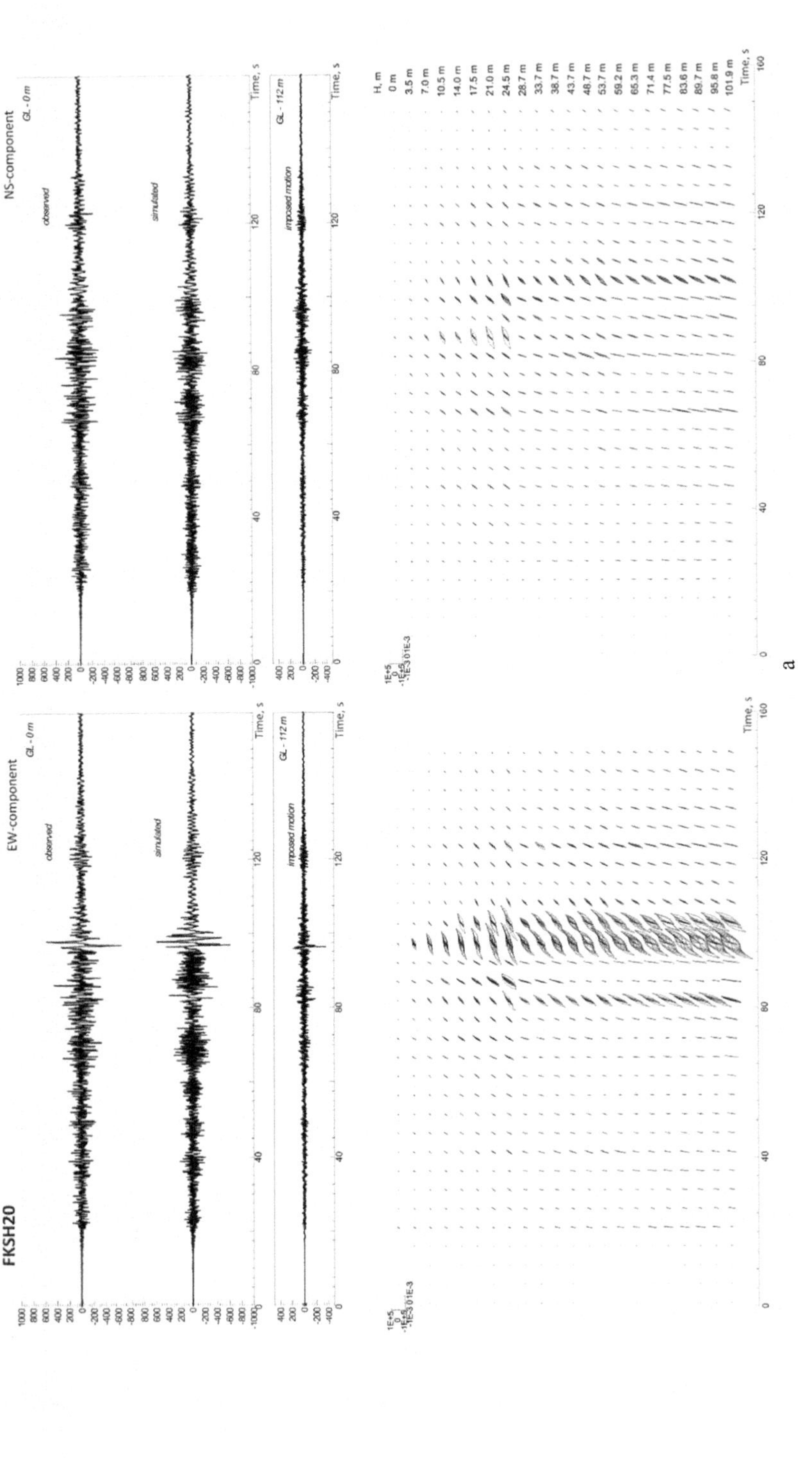

Fig. 4.2 Acceleration time histories of the Tohoku earthquake at southern stations, close to the source, observed and simulated, and estimated stress–strain relations in soil layers, changing with time during strong motion. Accelerations are given in cm/s^2; stresses are given in Pa, and strains in strain

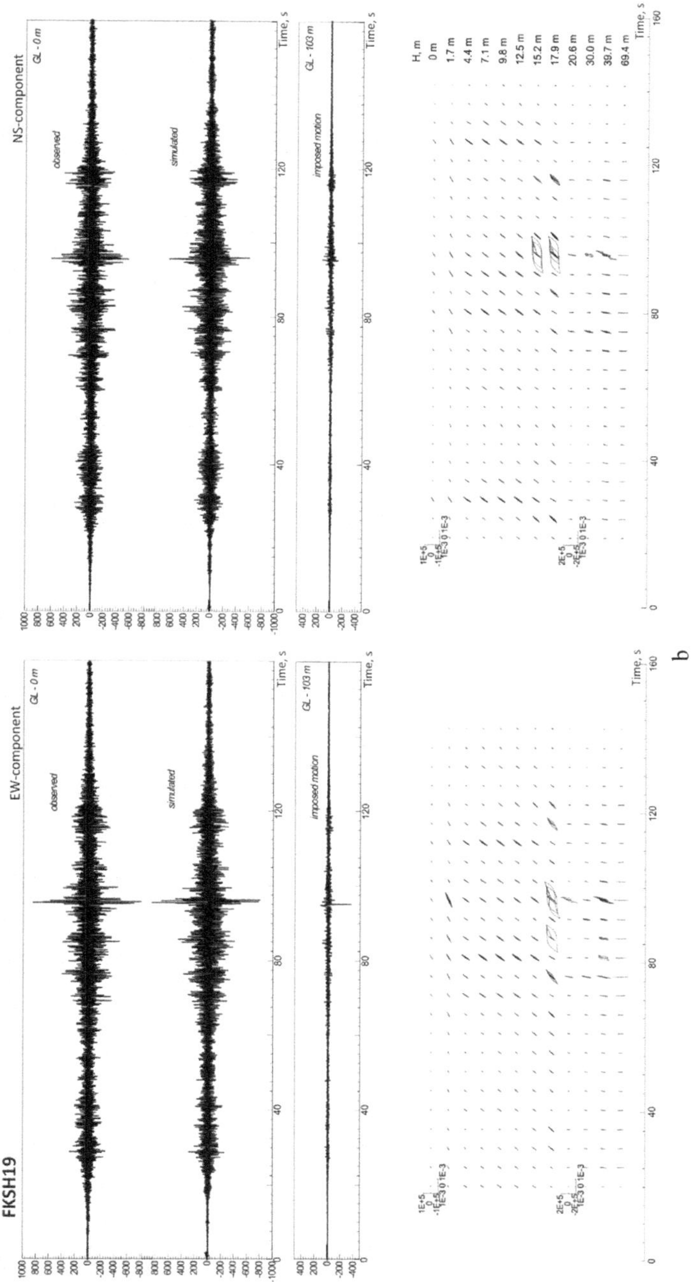

Fig. 4.2 (continued)

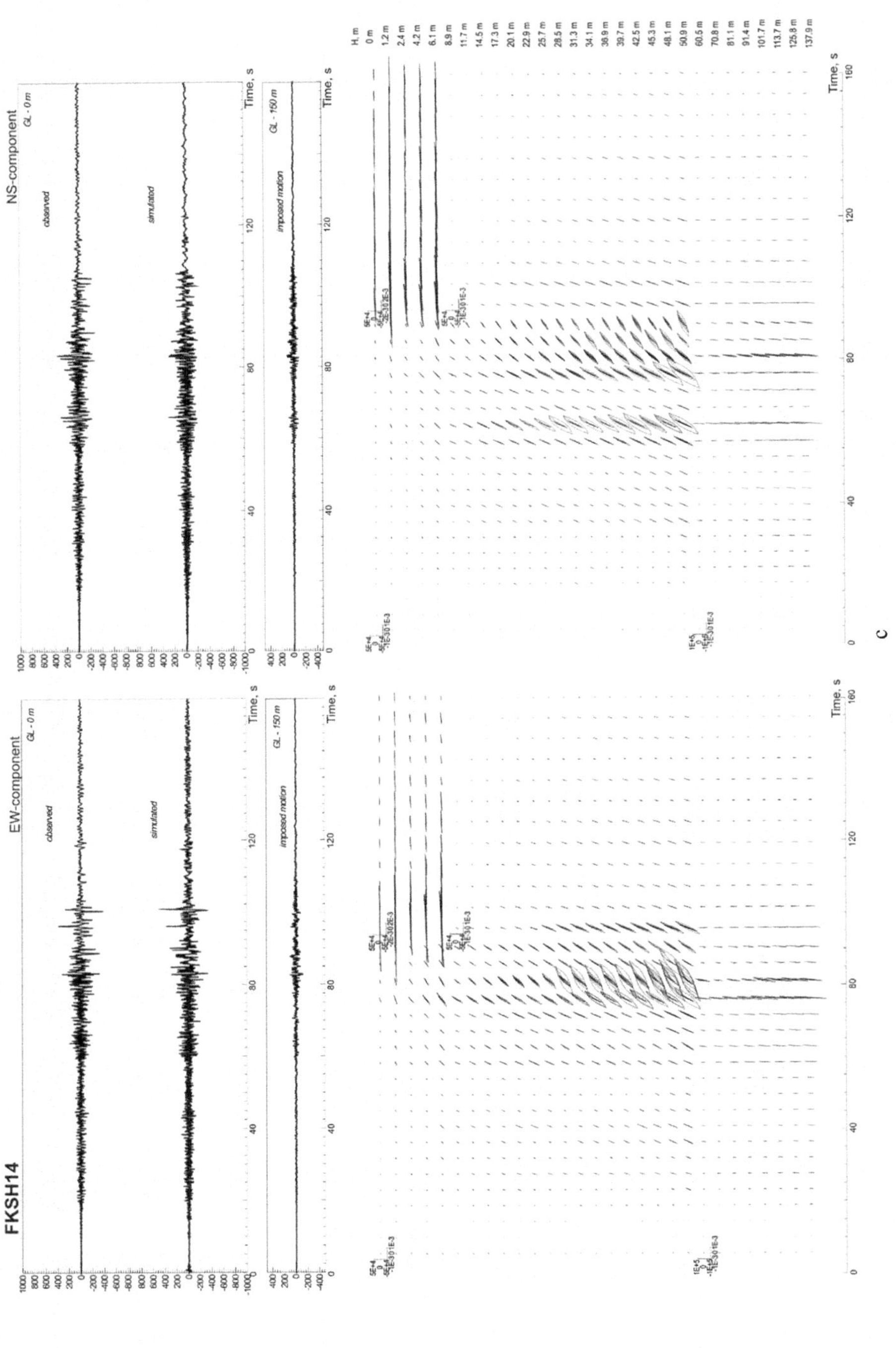

Fig. 4.2 (continued)

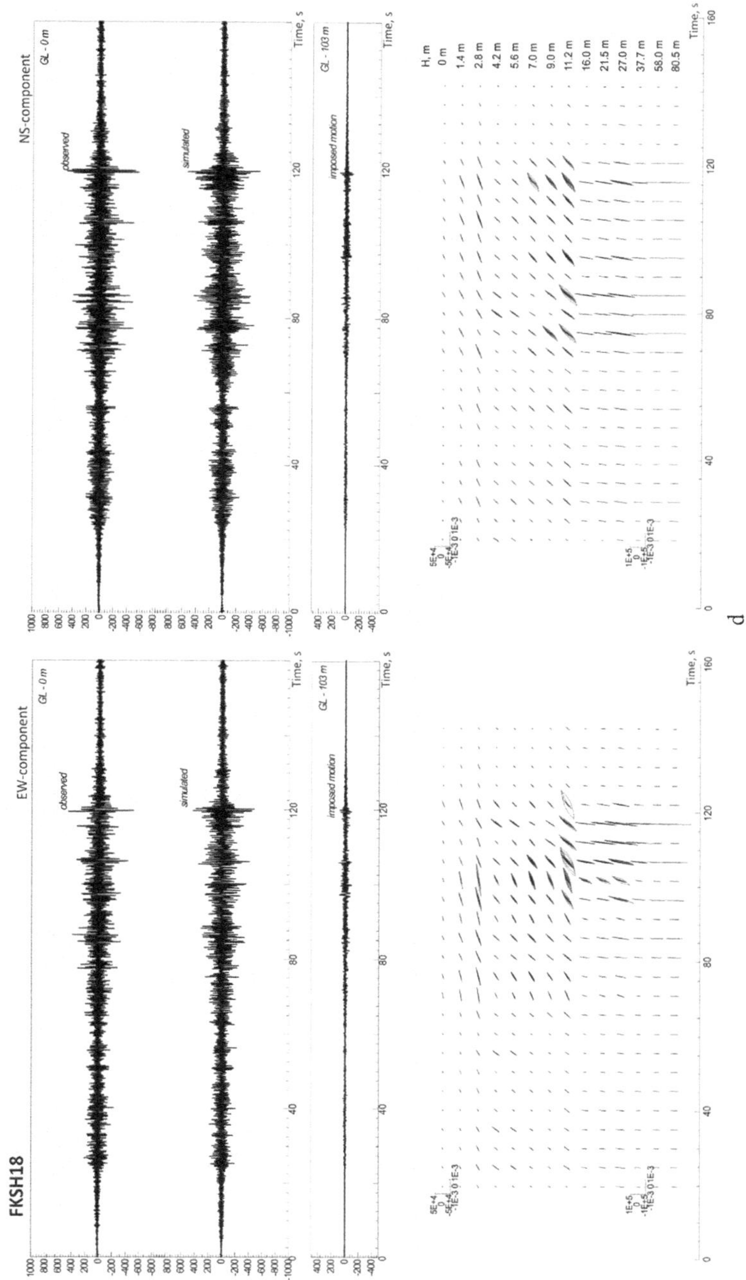

Fig. 4.2 (continued)

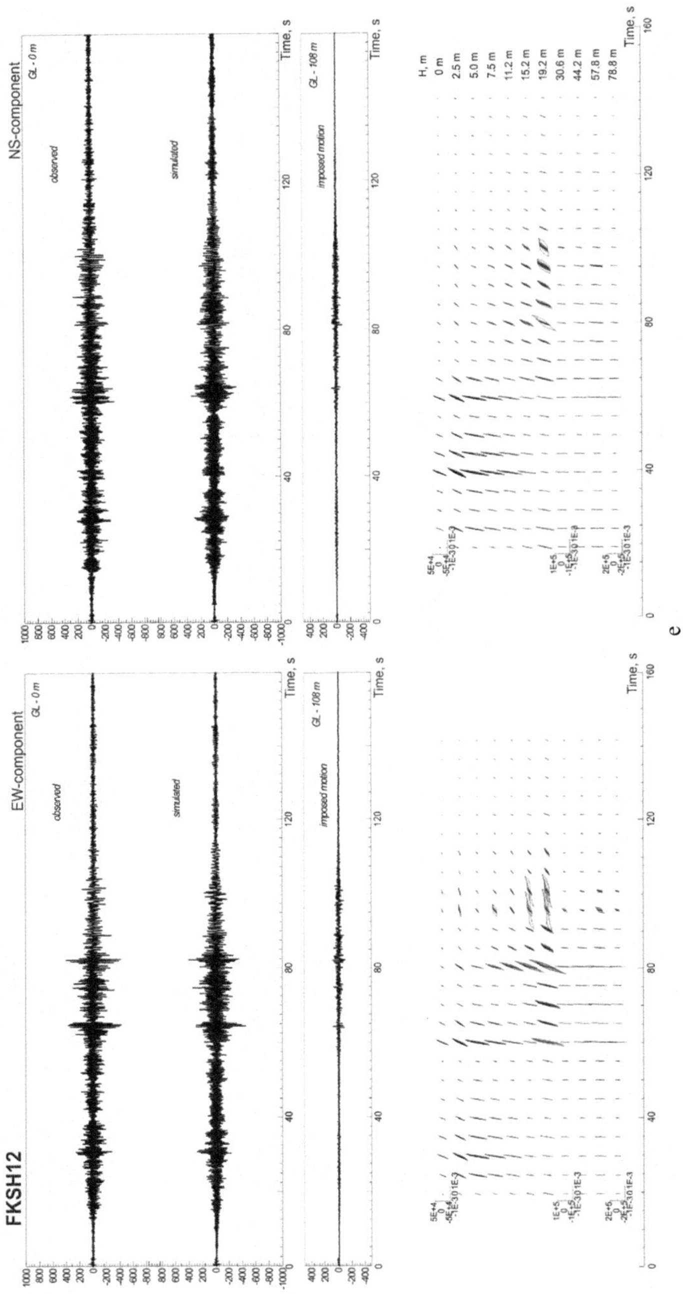

Fig. 4.2 (continued)

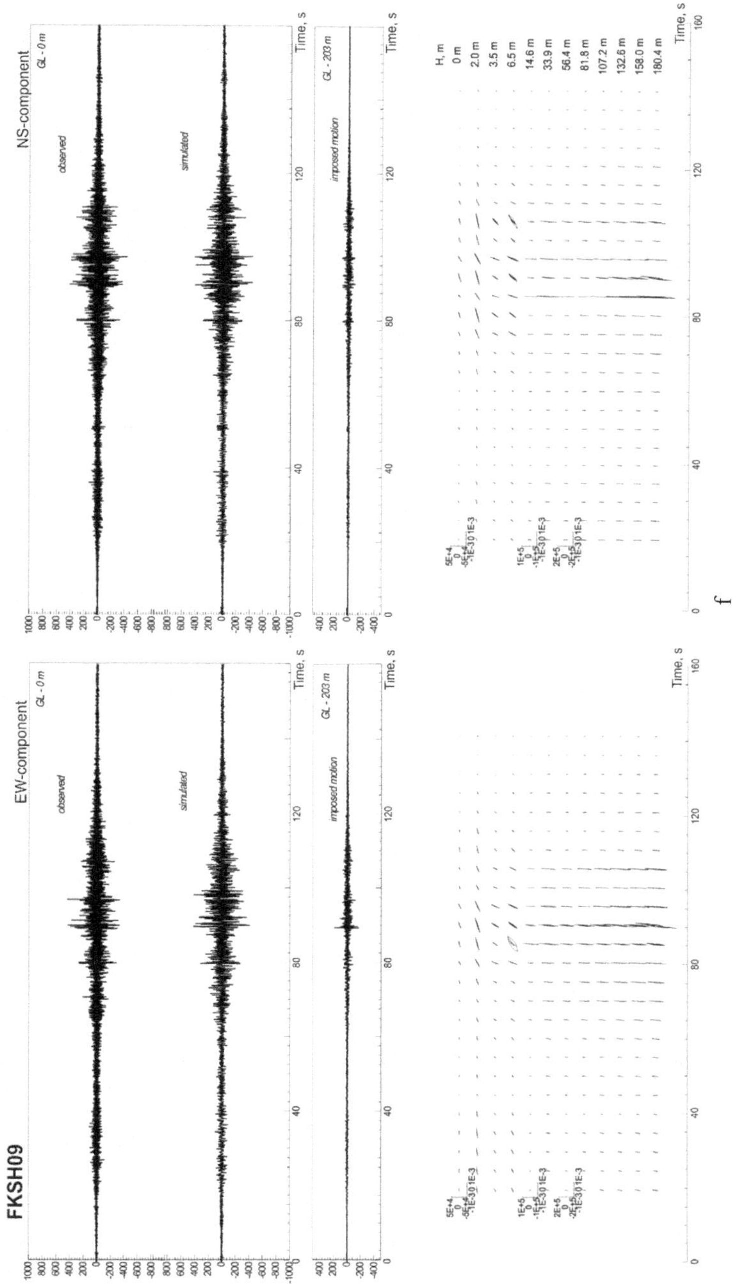

Fig. 4.2 (continued)

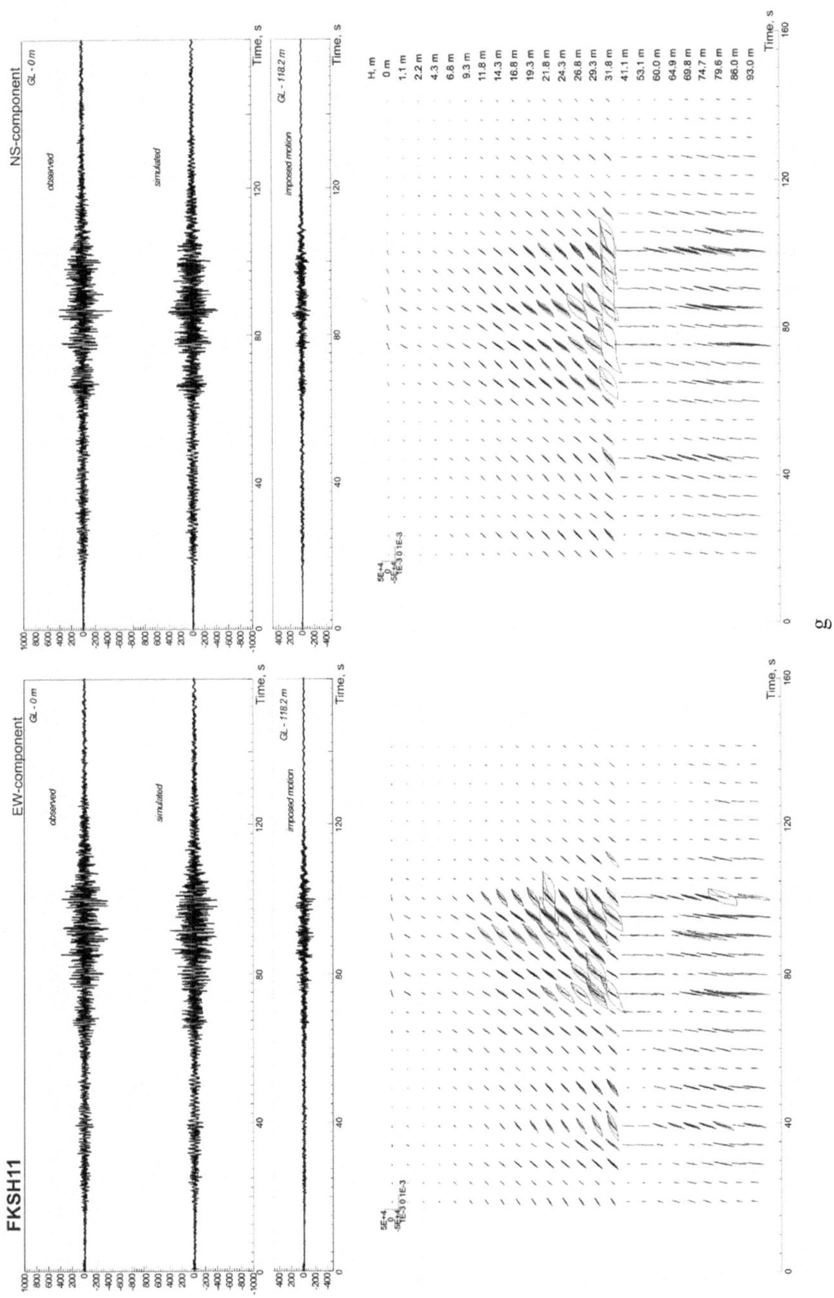

Fig. 4.2 (continued)

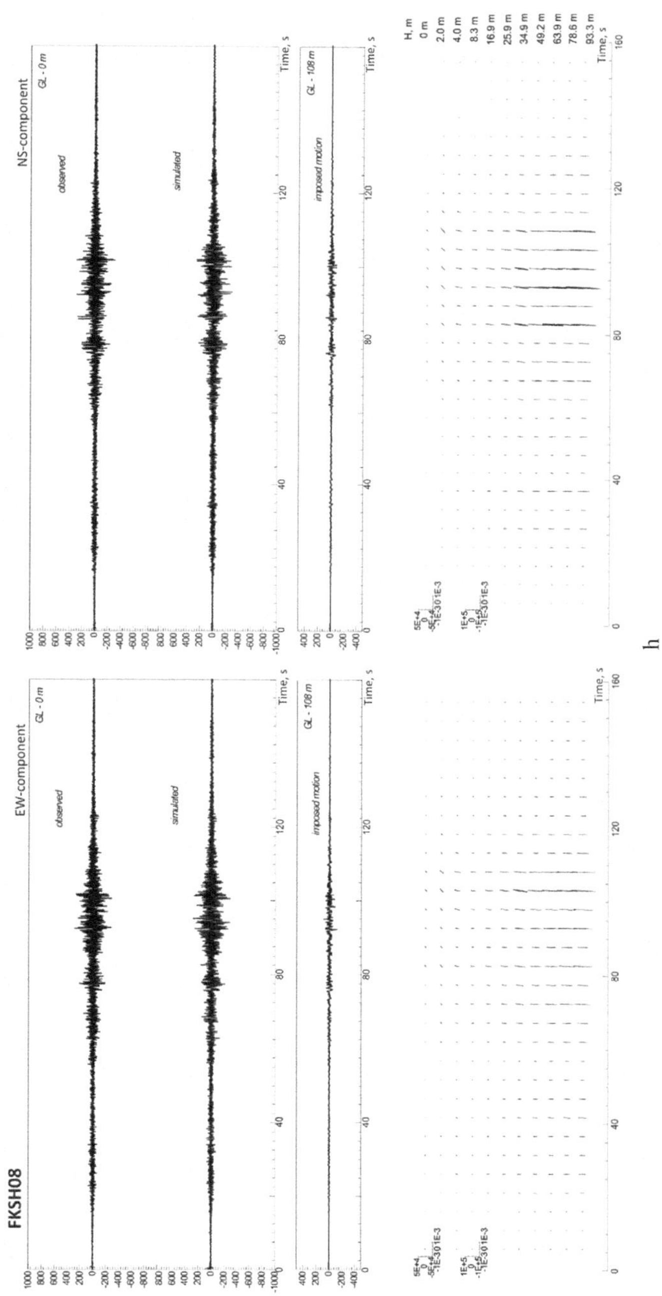

Fig. 4.2 (continued)

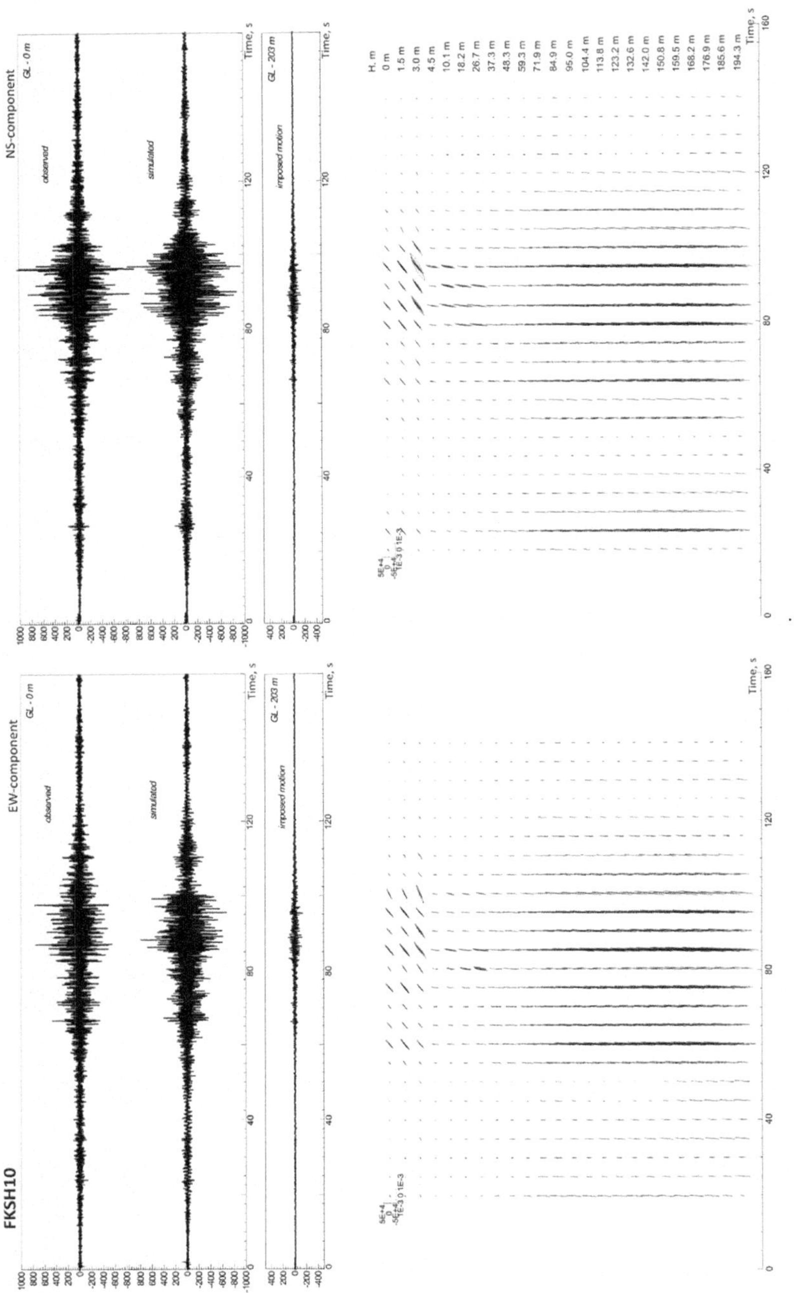

Fig. 4.2 (continued)

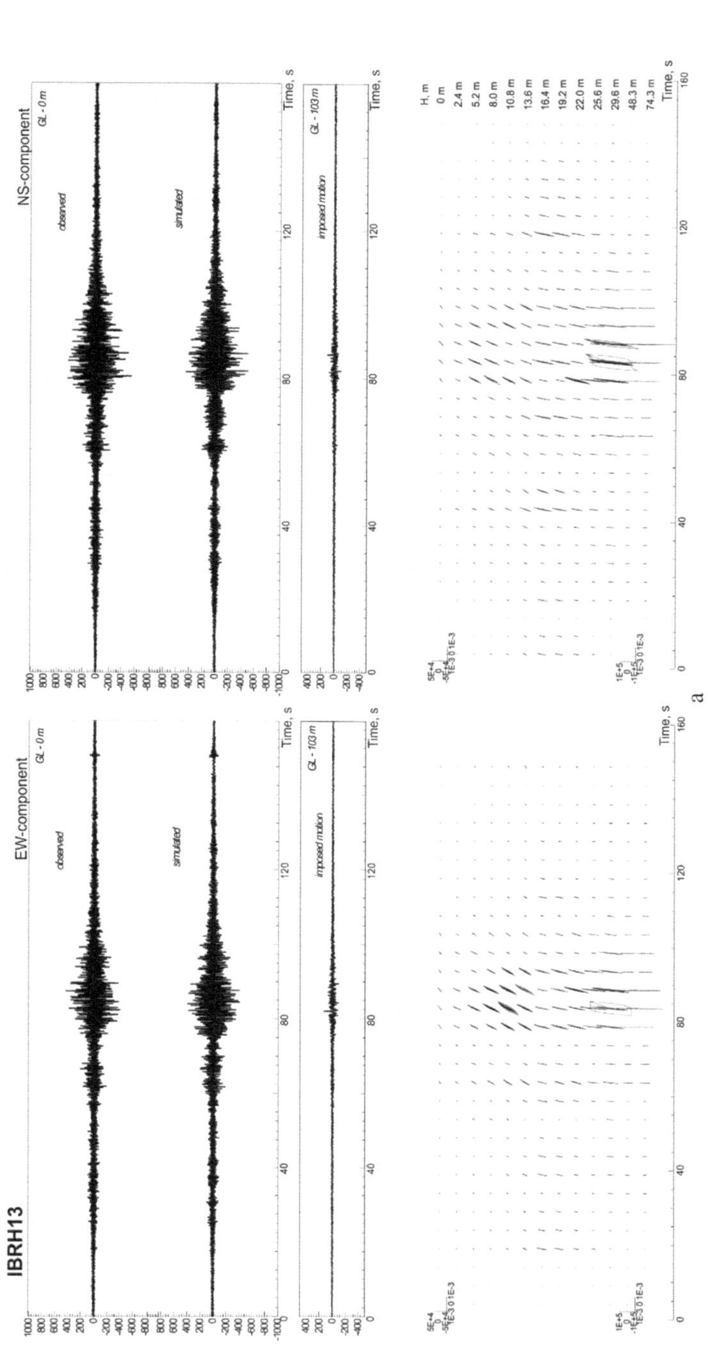

Fig. 4.3 Acceleration time histories of the Tohoku earthquake at southern stations, remote, observed and simulated, and estimated stress–strain relations in soil layers, changing with time during strong motion. Accelerations are given in cm/s^2; stresses are given in Pa, and strains in strain

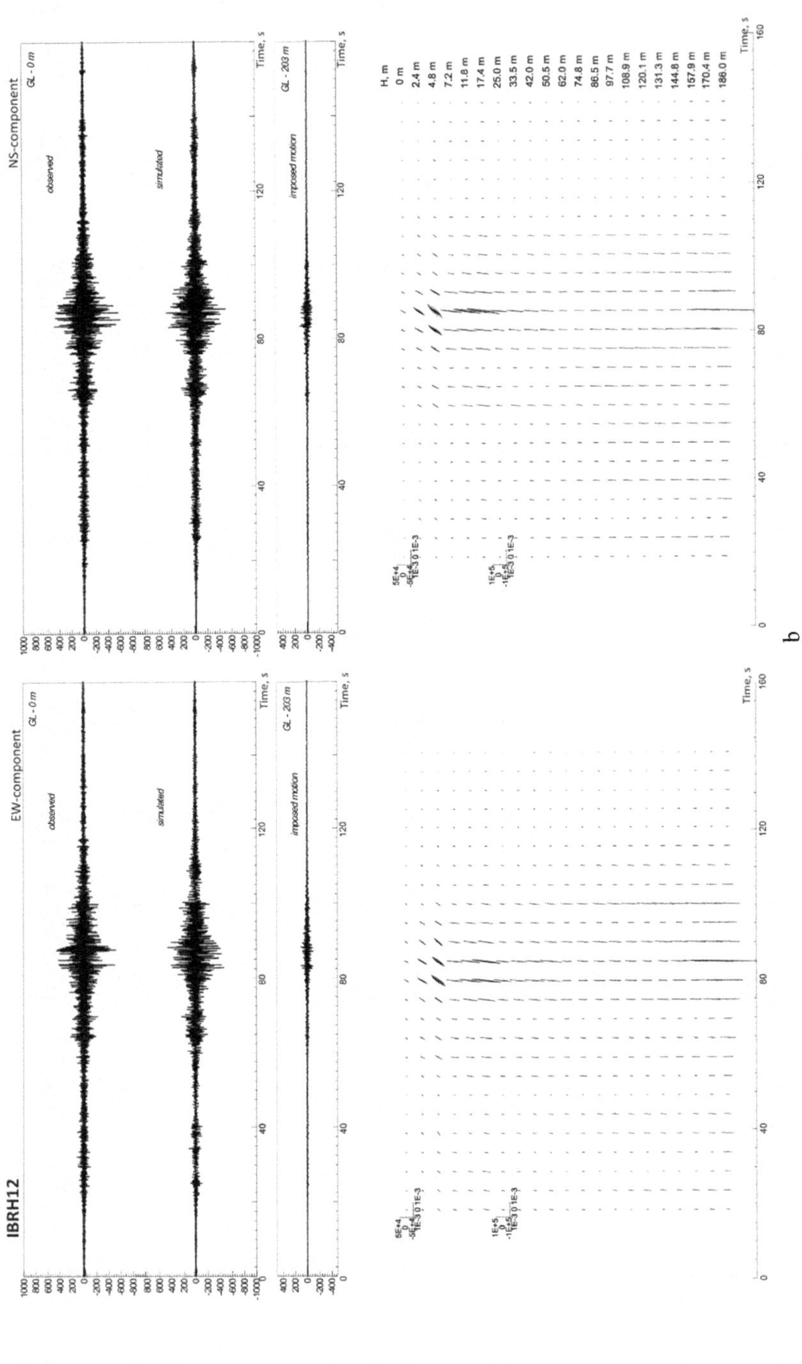

Fig. 4.3 (continued)

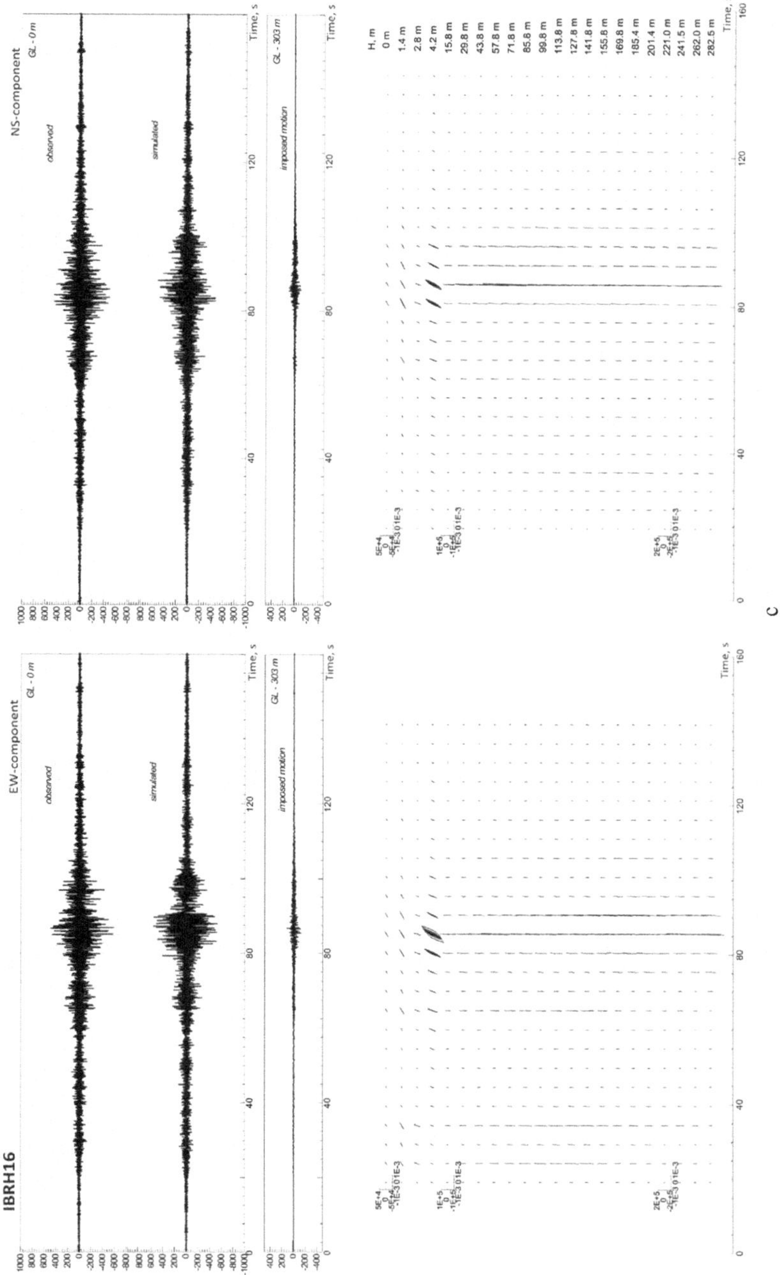

Fig. 4.3 (continued)

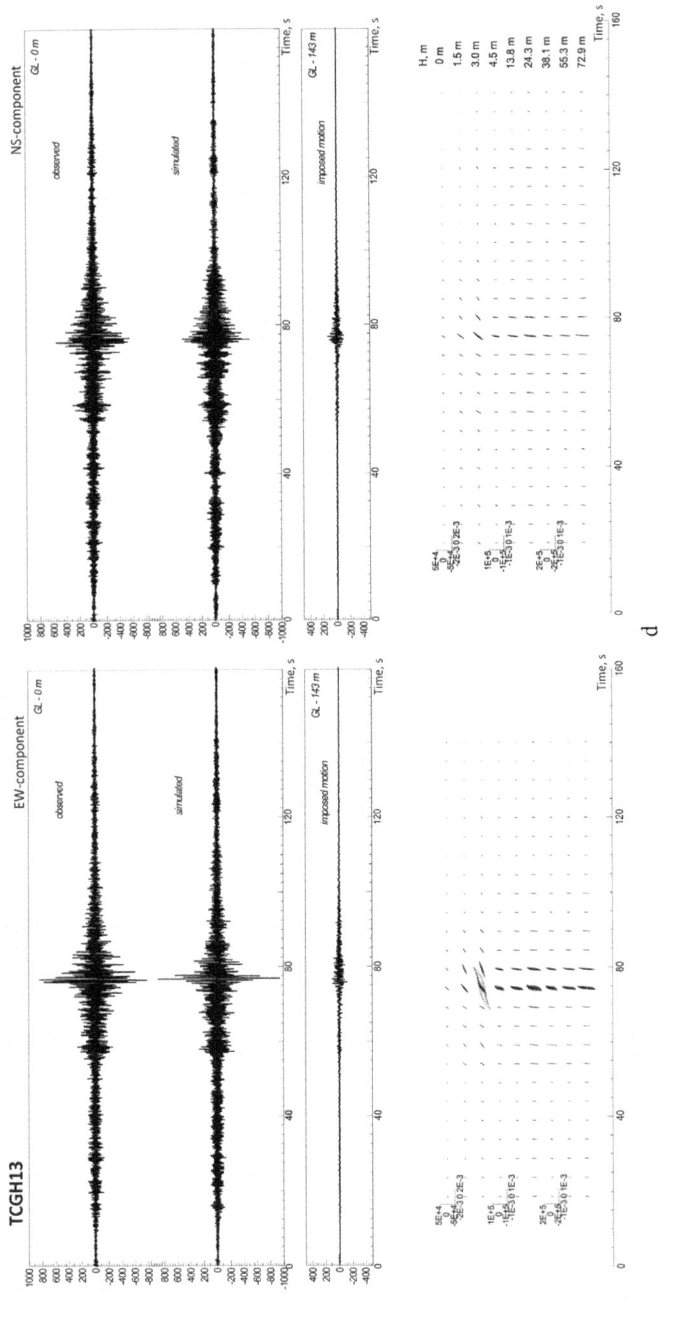

Fig. 4.3 (continued)

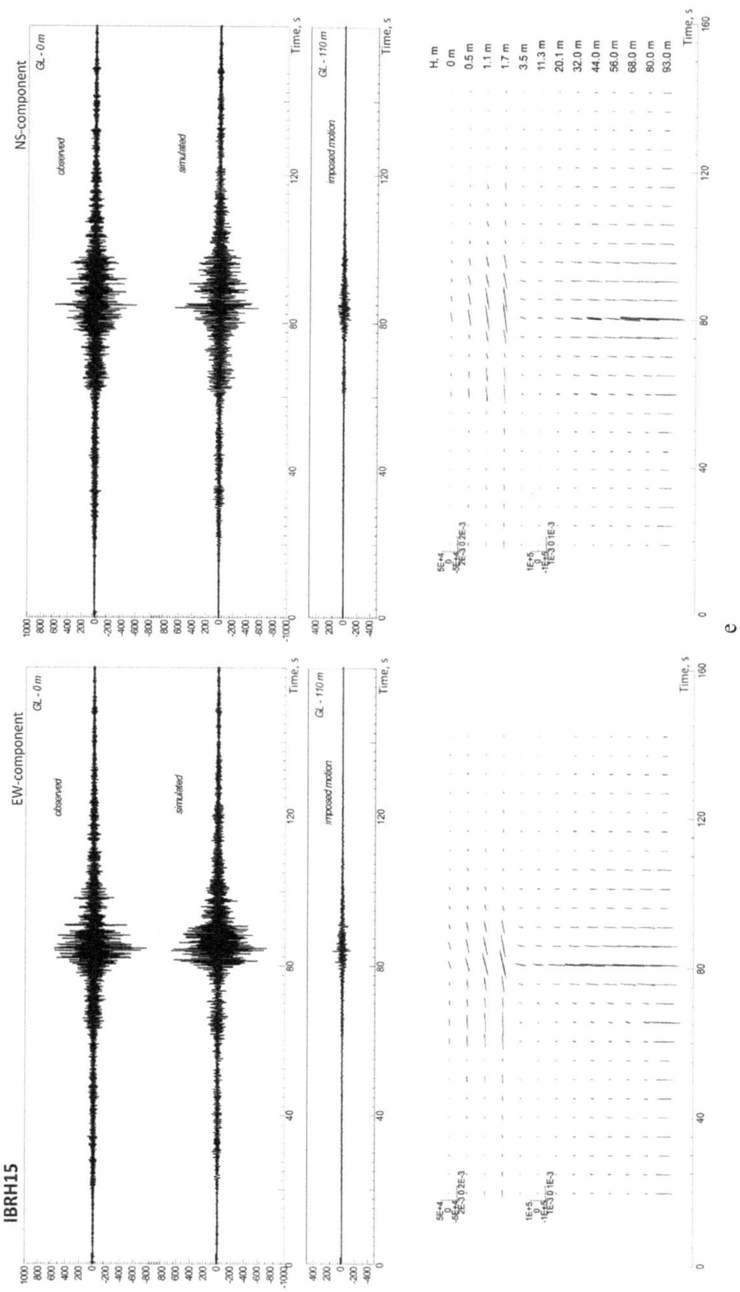

Fig. 4.3 (continued)

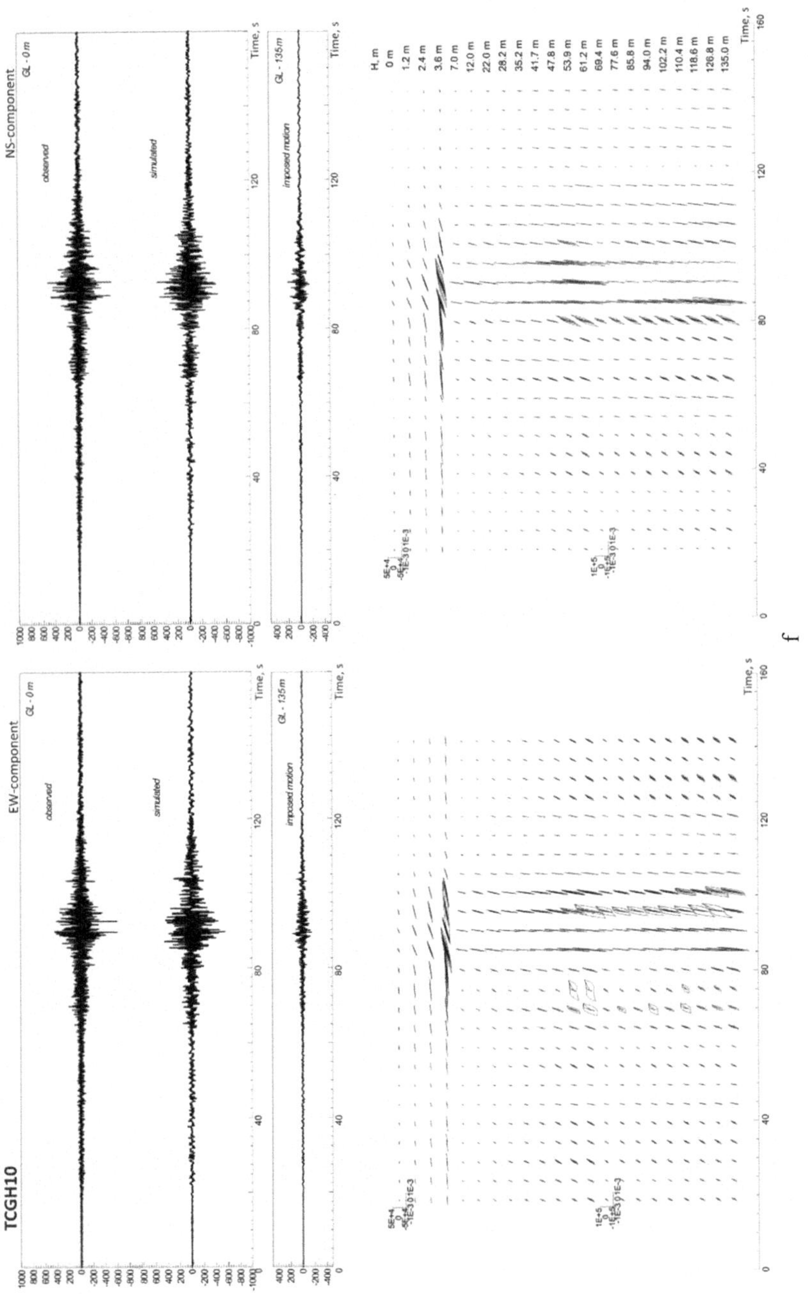

Fig. 4.3 (continued)

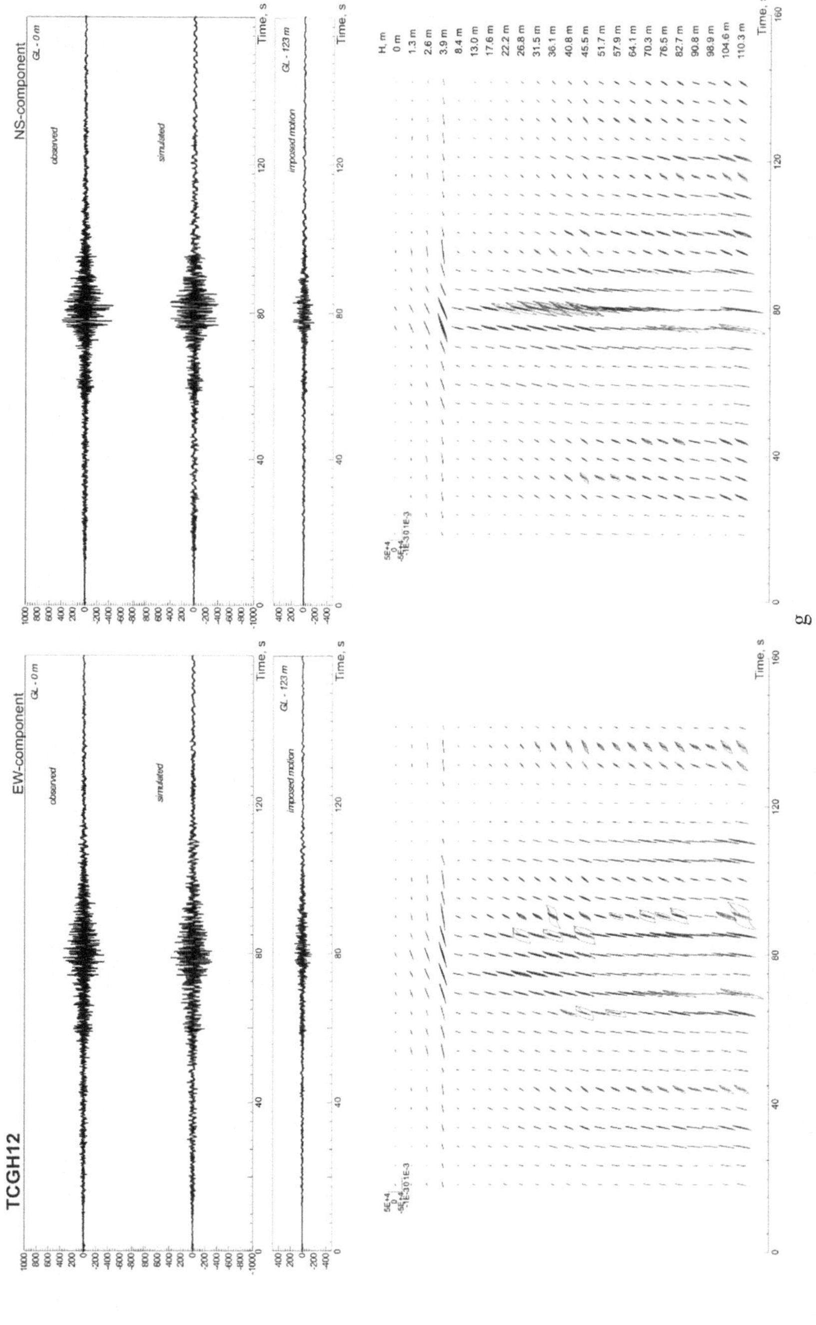

Fig. 4.3 (continued)

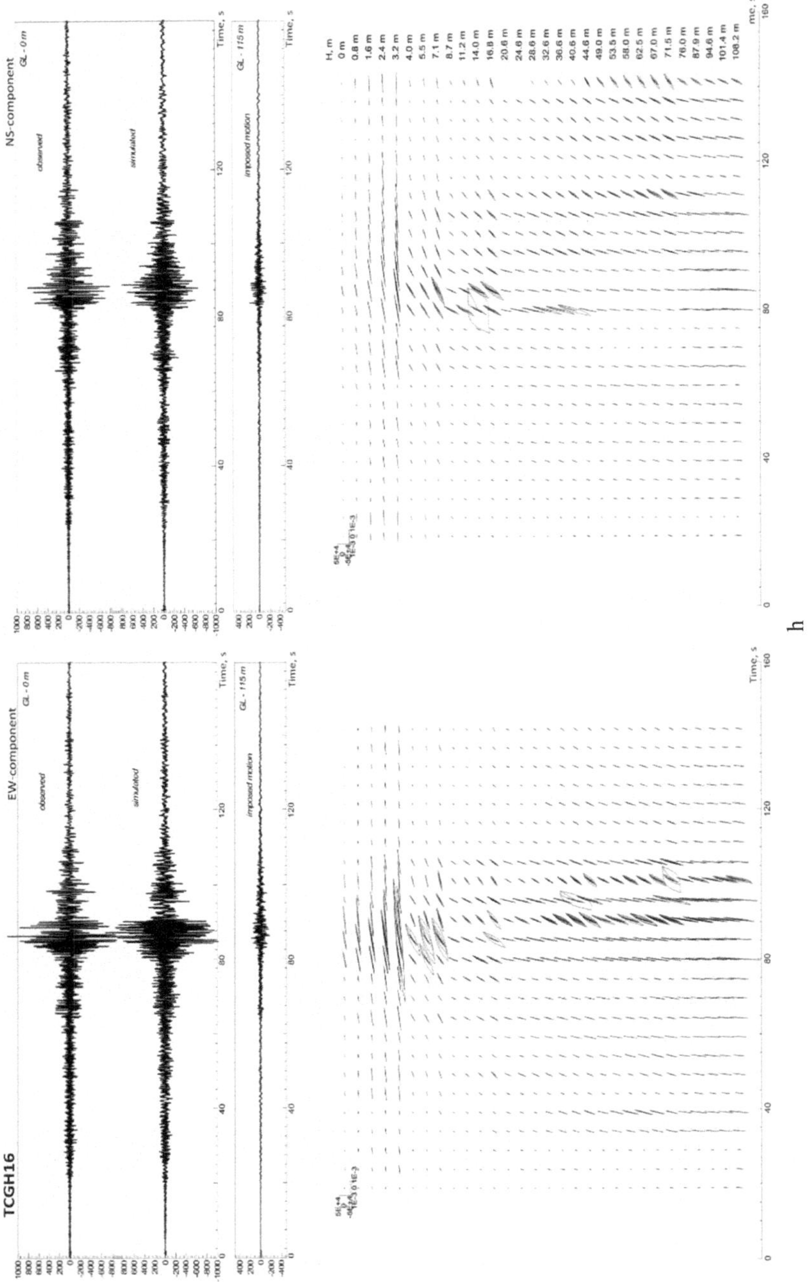

Fig. 4.3 (continued)

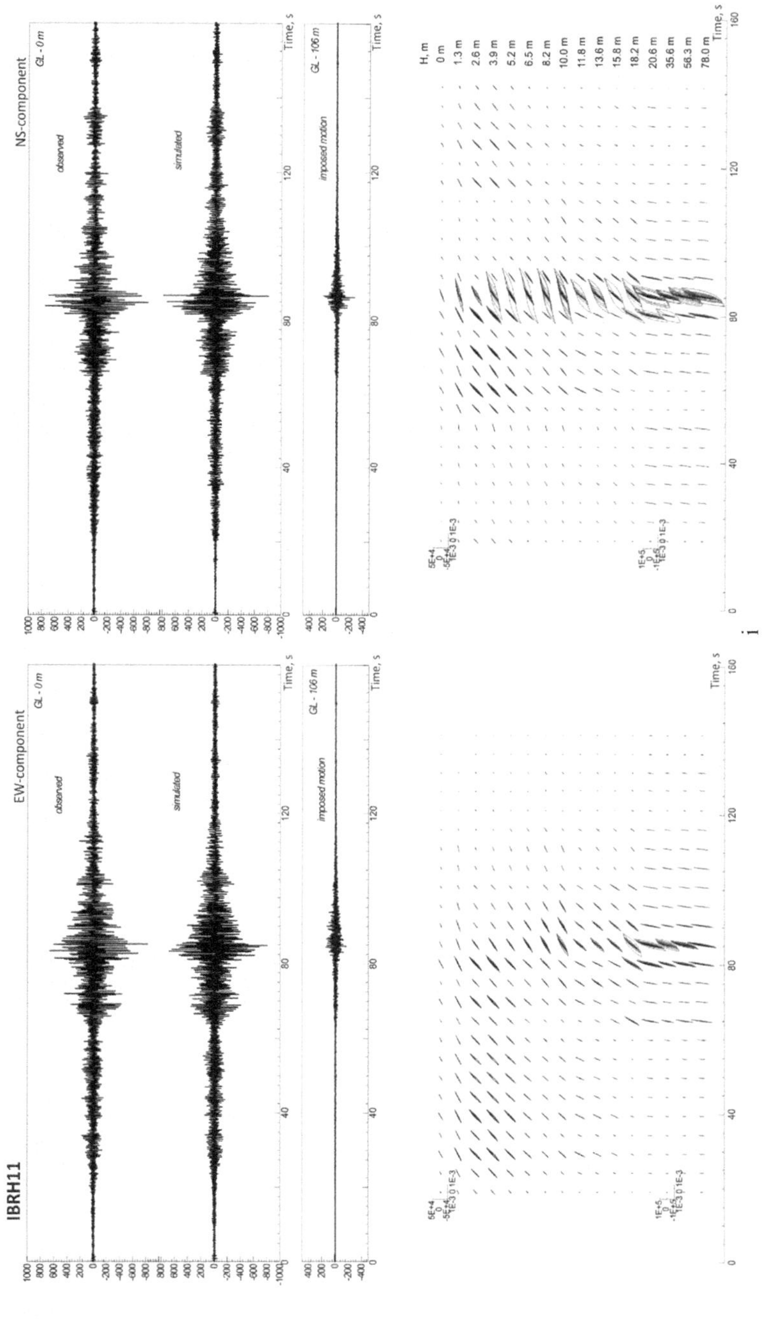

Fig. 4.3 (continued)

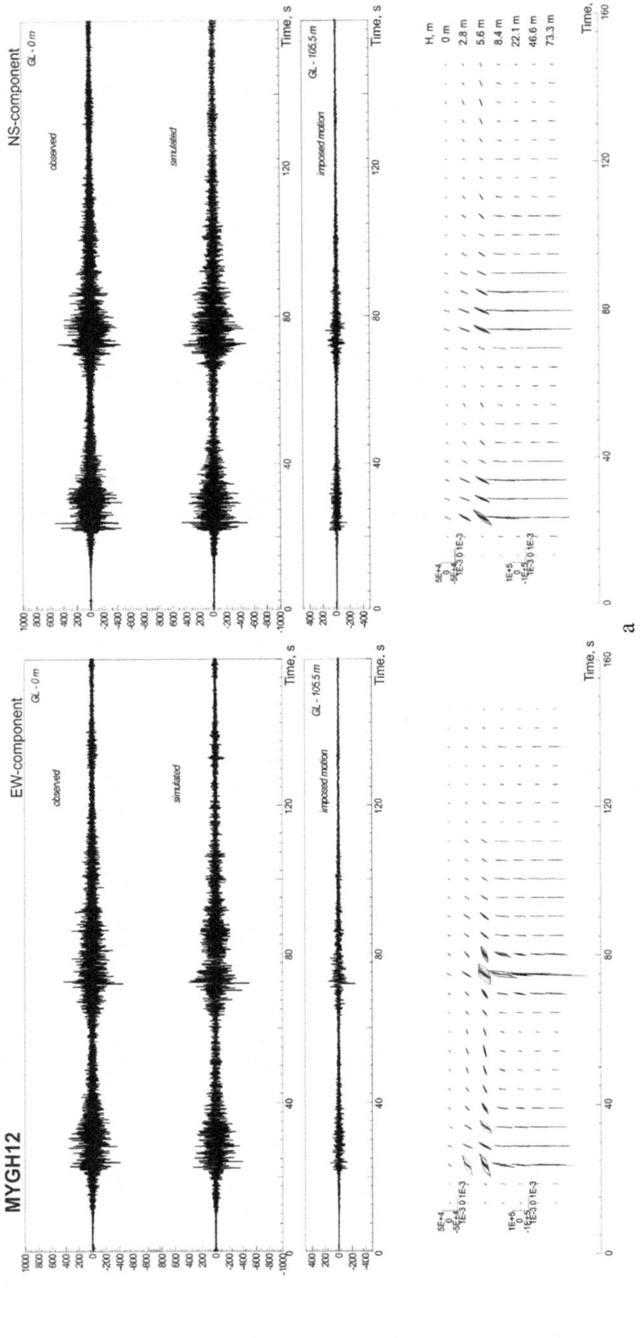

Fig. 4.4 Acceleration time histories of the Tohoku earthquake at northern stations, close to the source, observed and simulated, and estimated stress–strain relations in soil layers, changing with time during strong motion. Accelerations are given in cm/s^2; stresses are given in Pa, and strains in strain

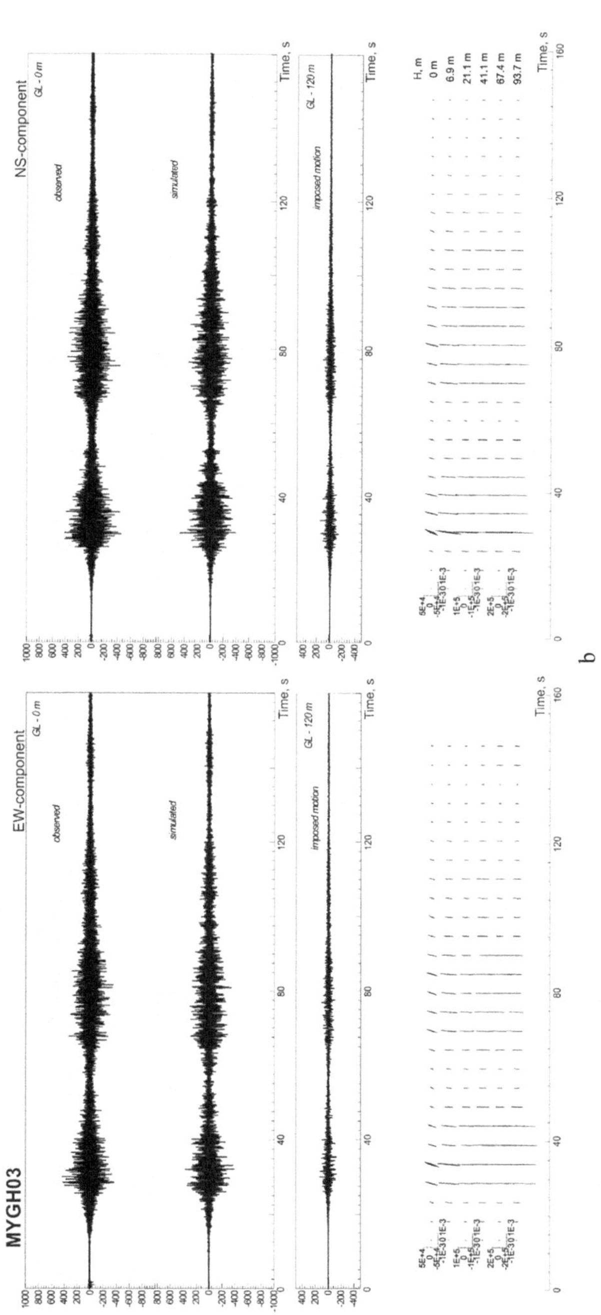

Fig. 4.4 (continued)

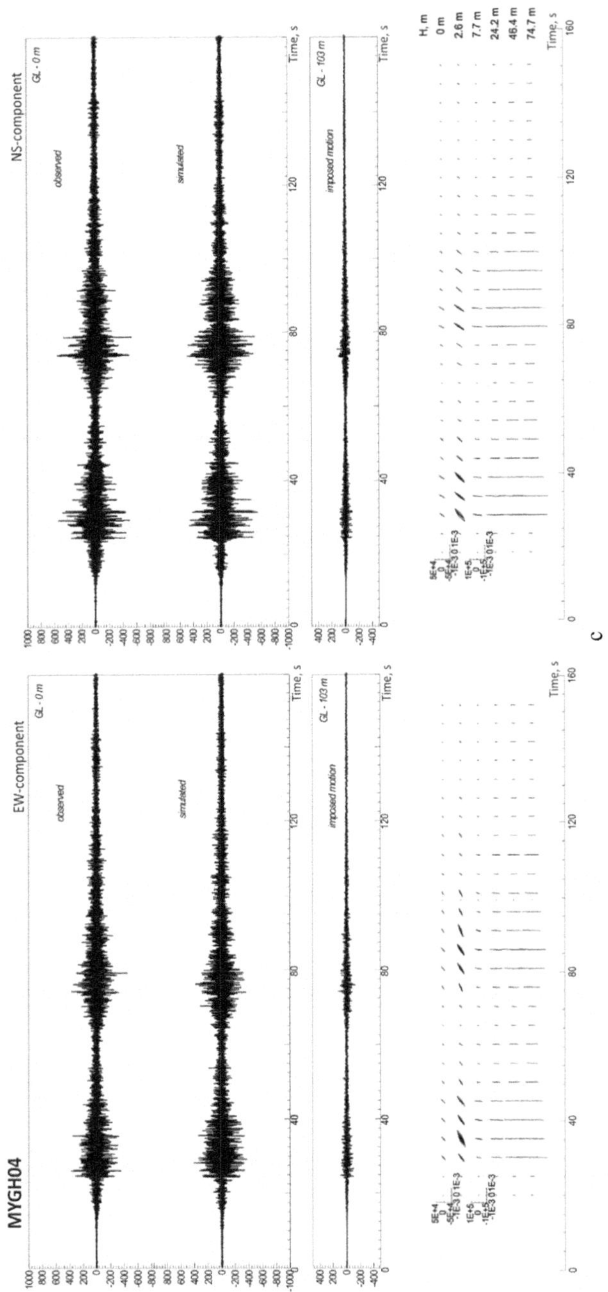

Fig. 4.4 (continued)

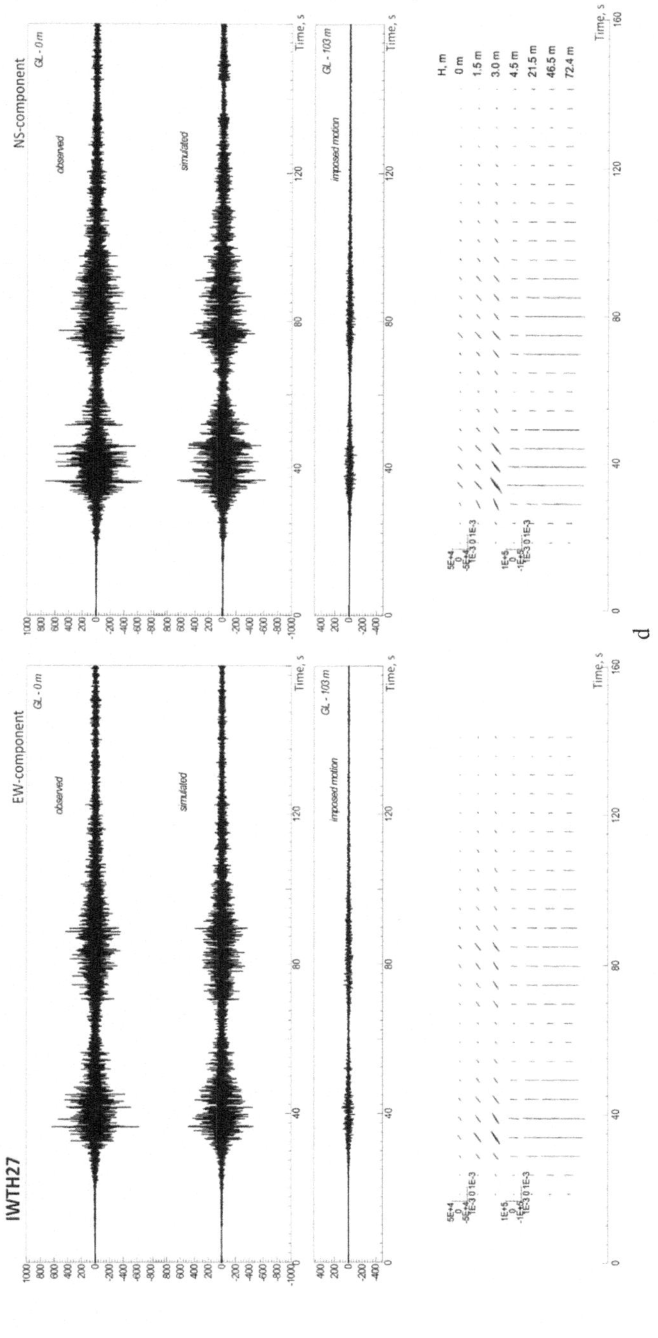

Fig. 4.4 (continued)

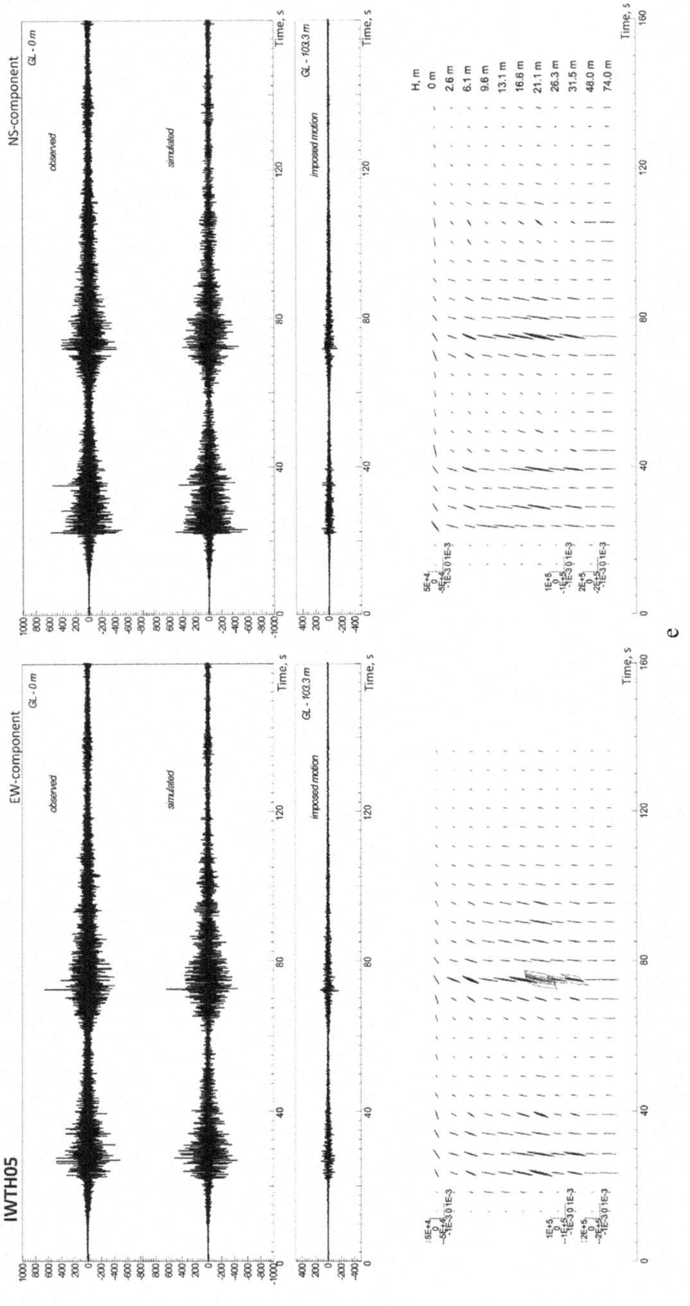

Fig. 4.4 (continued)

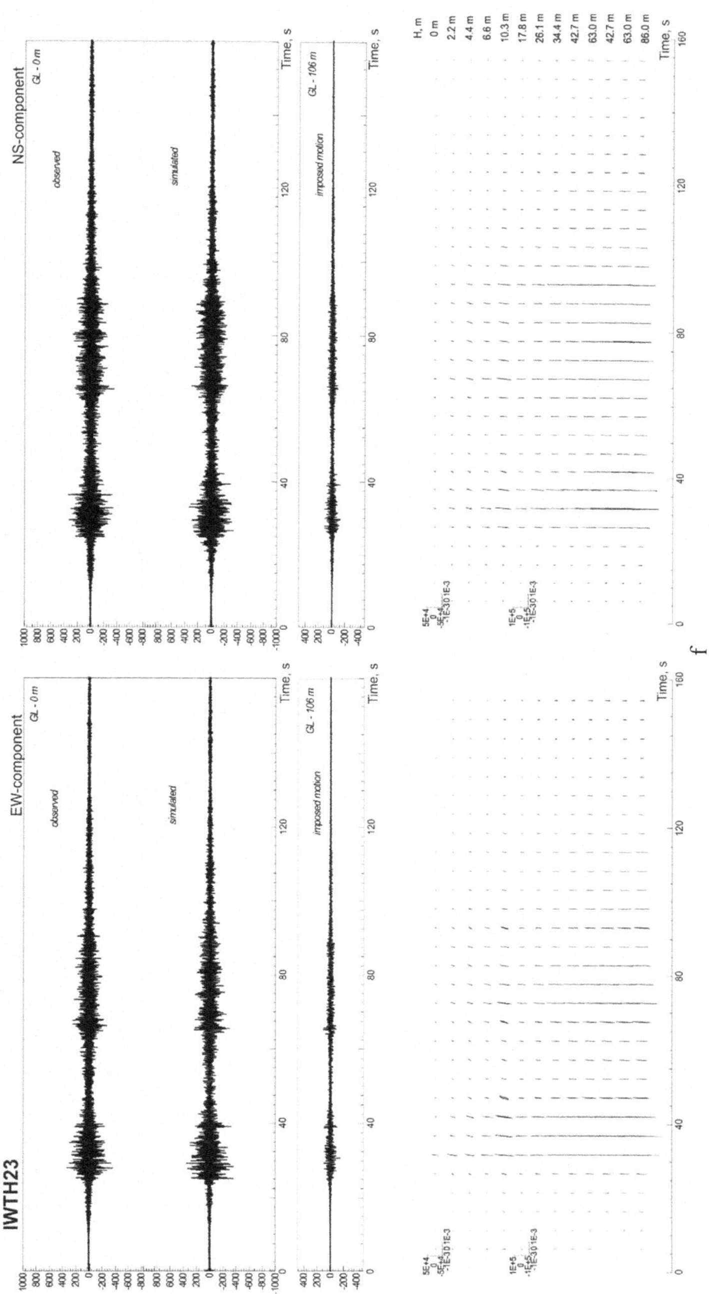

Fig. 4.4 (continued)

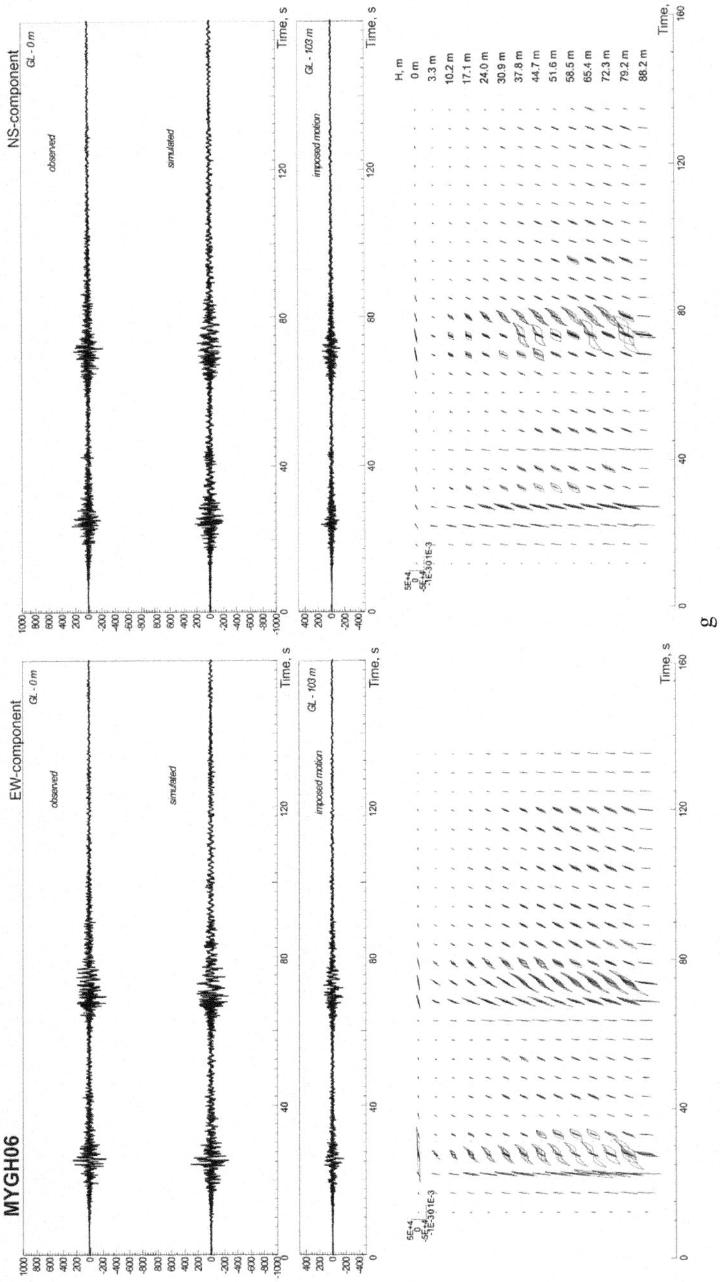

Fig. 4.4 (continued)

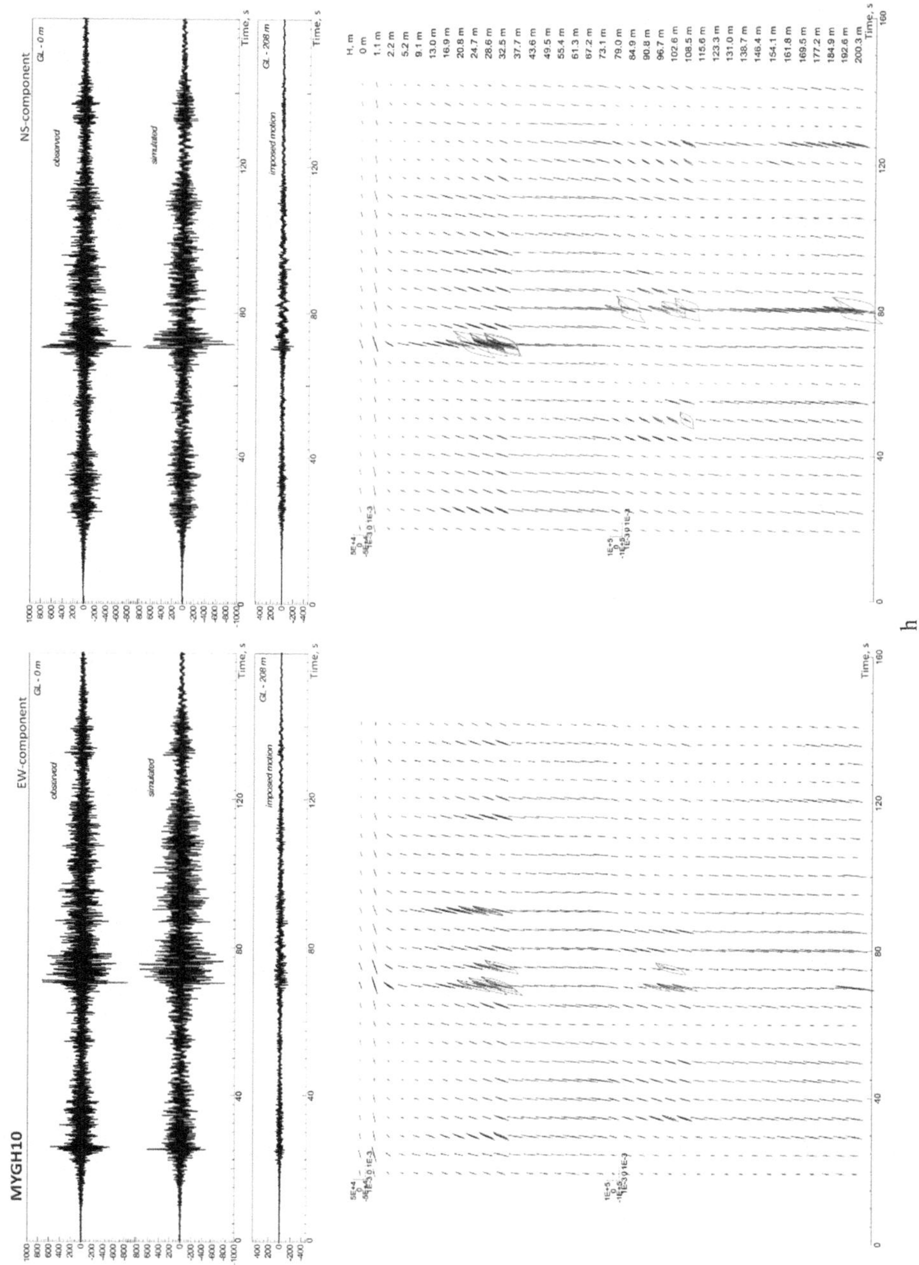

Fig. 4.4 (continued)

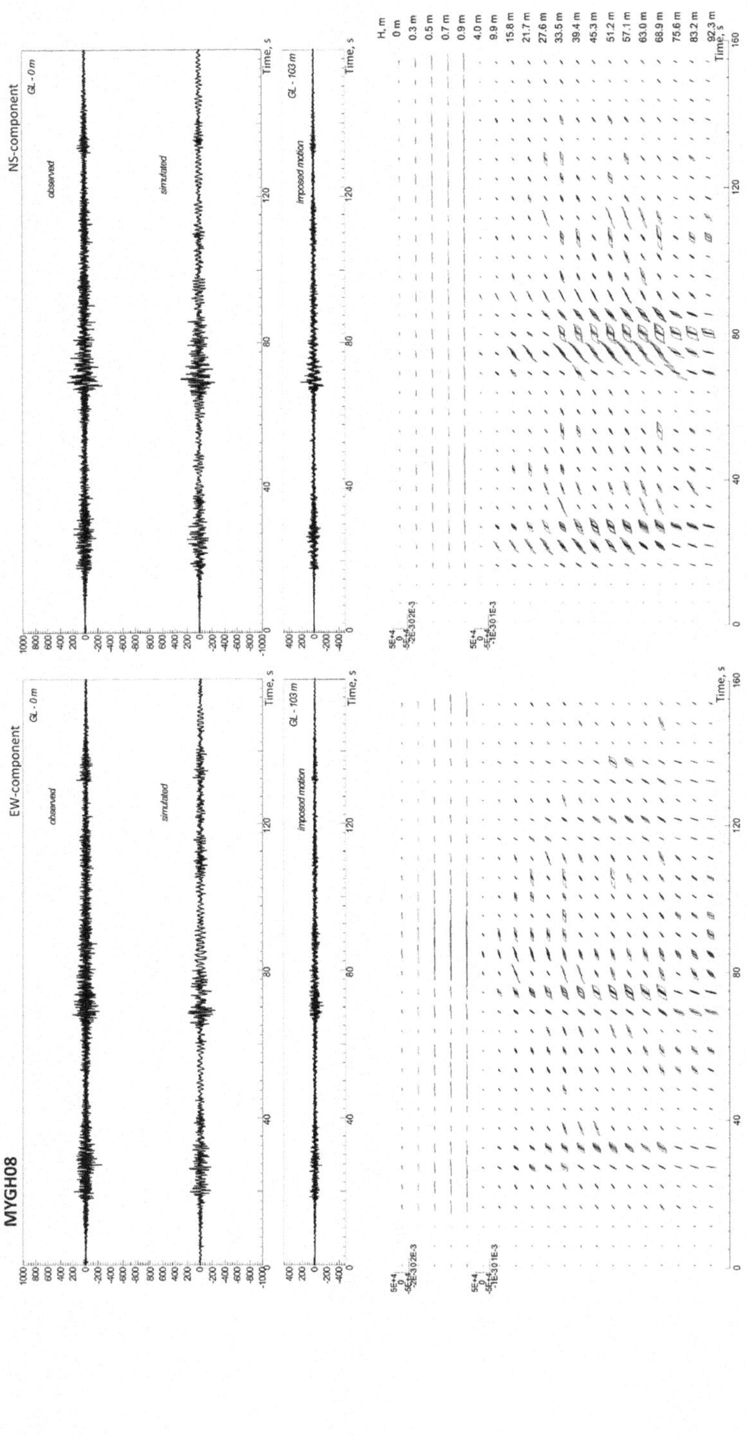

Fig. 4.4 (continued)

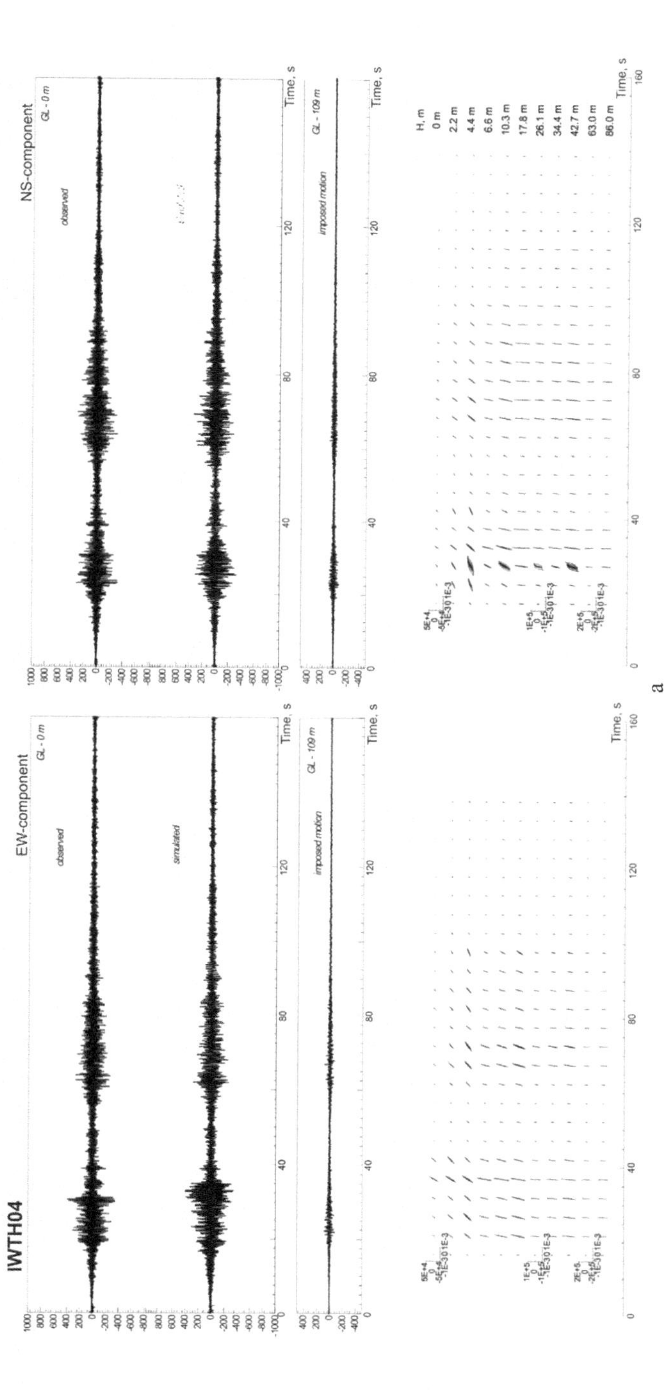

Fig. 4.5 Acceleration time histories of the Tohoku earthquake at northern stations, remote, observed and simulated, and estimated stress–strain relations in soil layers, changing with time during strong motion. Accelerations are given in cm/s^2; stresses are given in Pa, and strains in strain

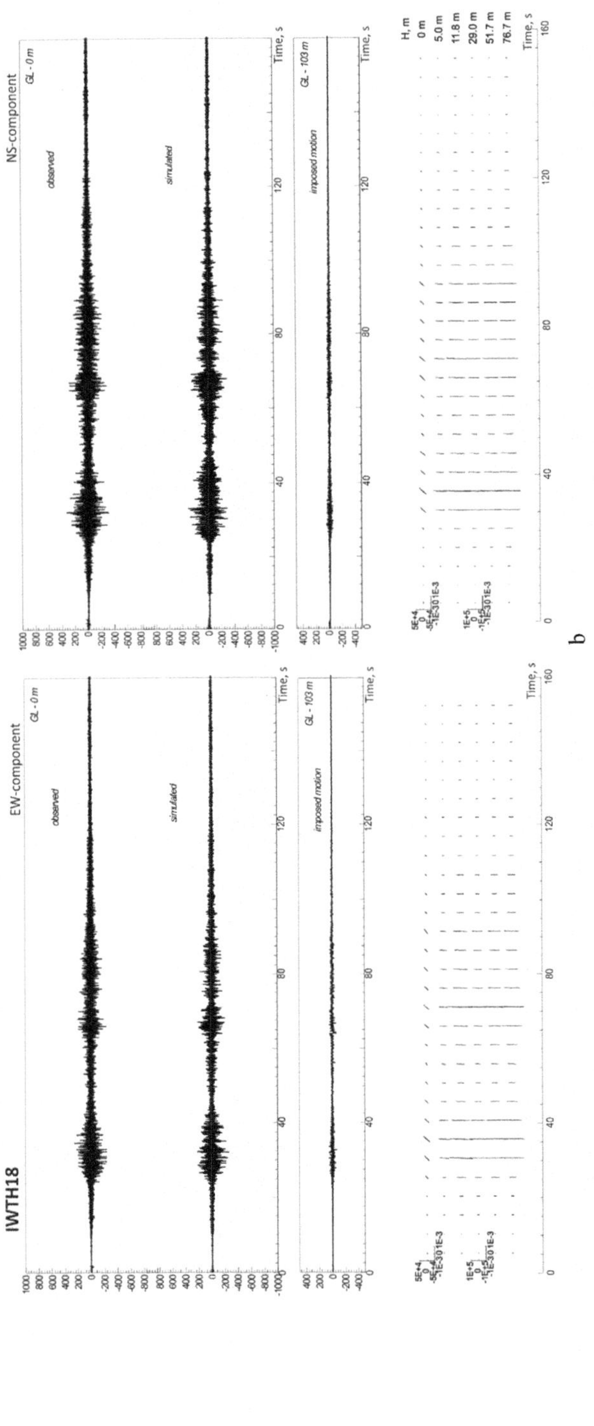

Fig. 4.5 (continued)

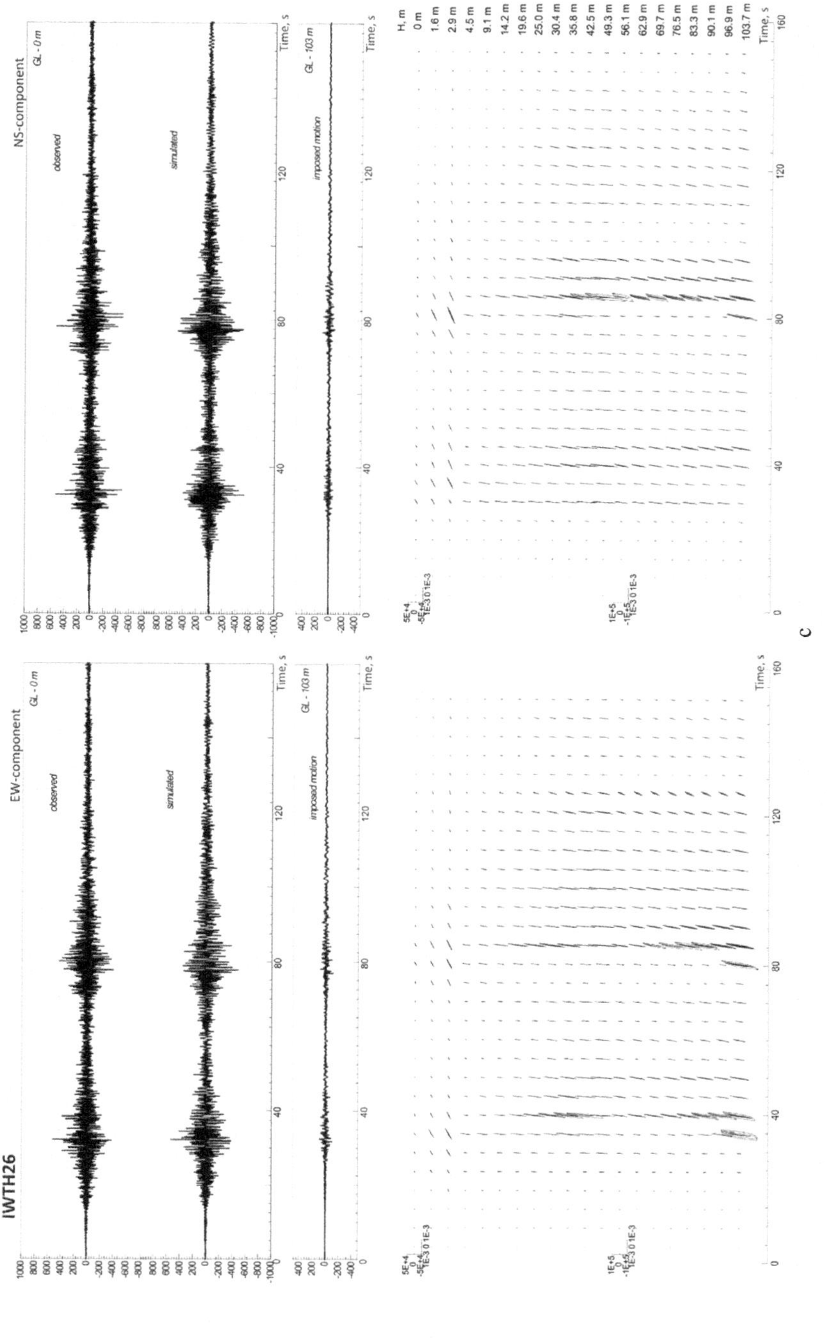

Fig. 4.5 (continued)

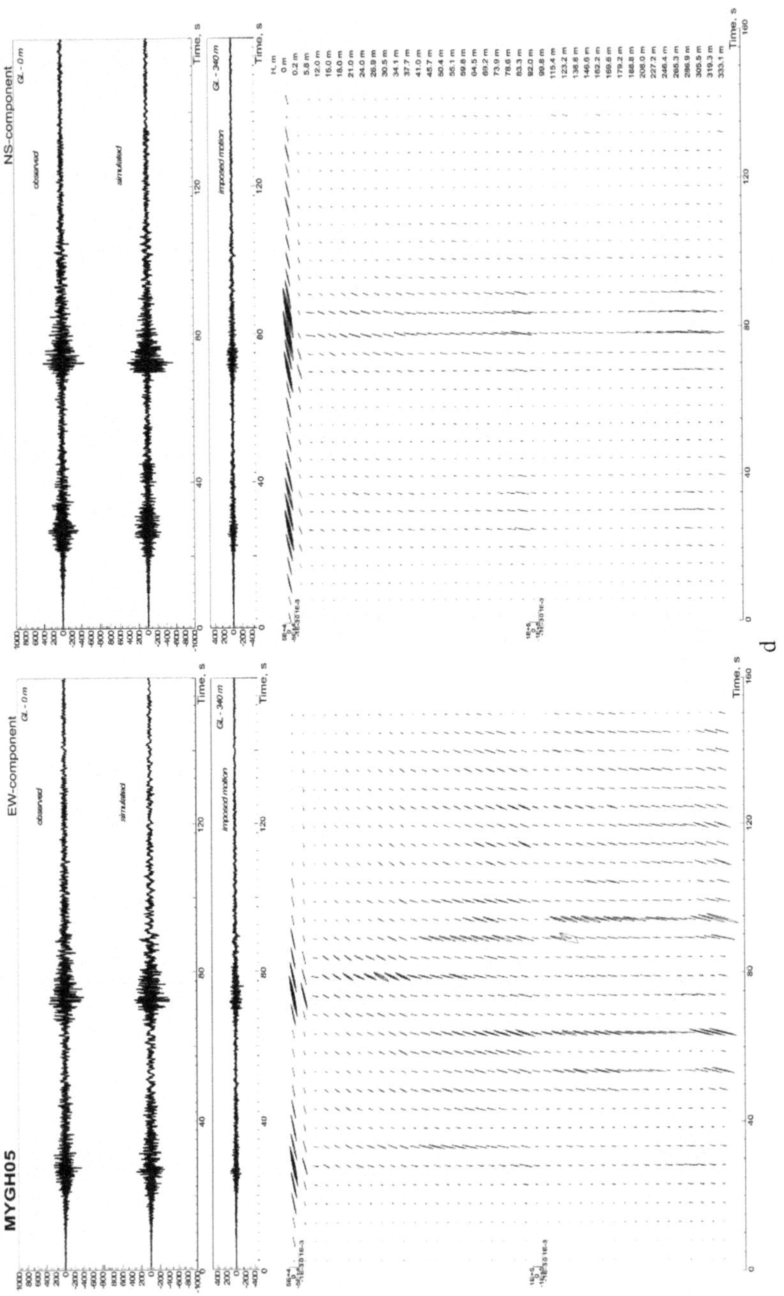

Fig. 4.5 (continued)

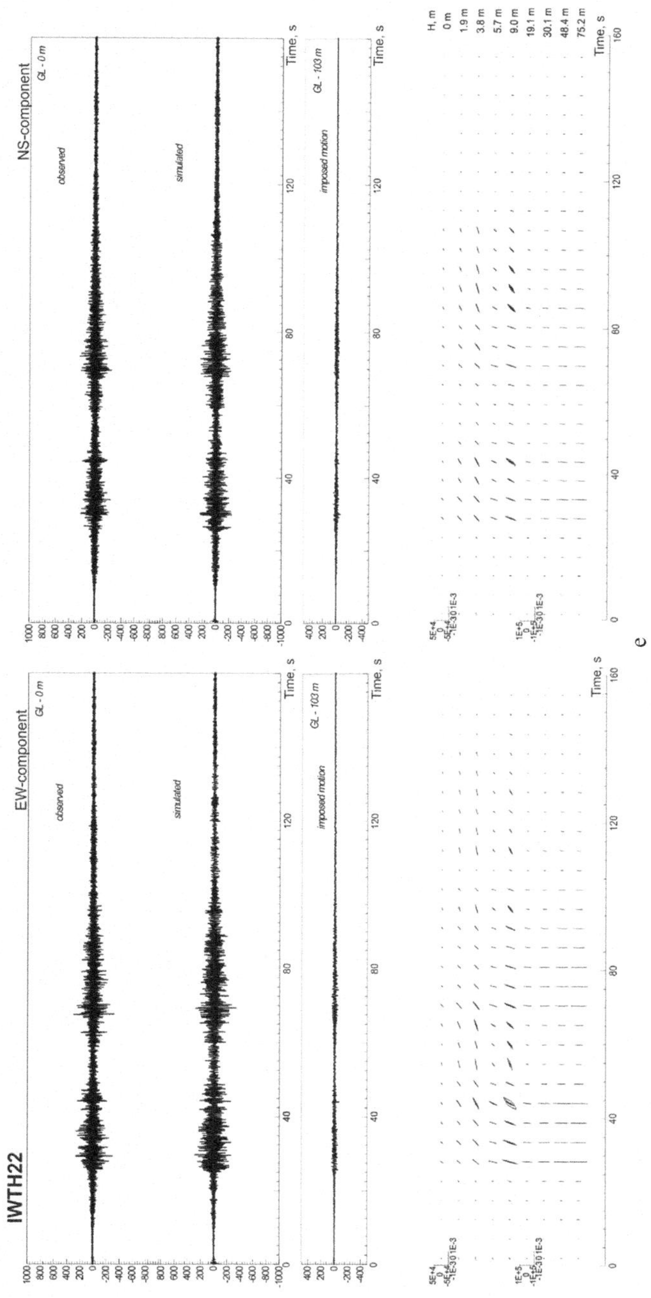

Fig. 4.5 (continued)

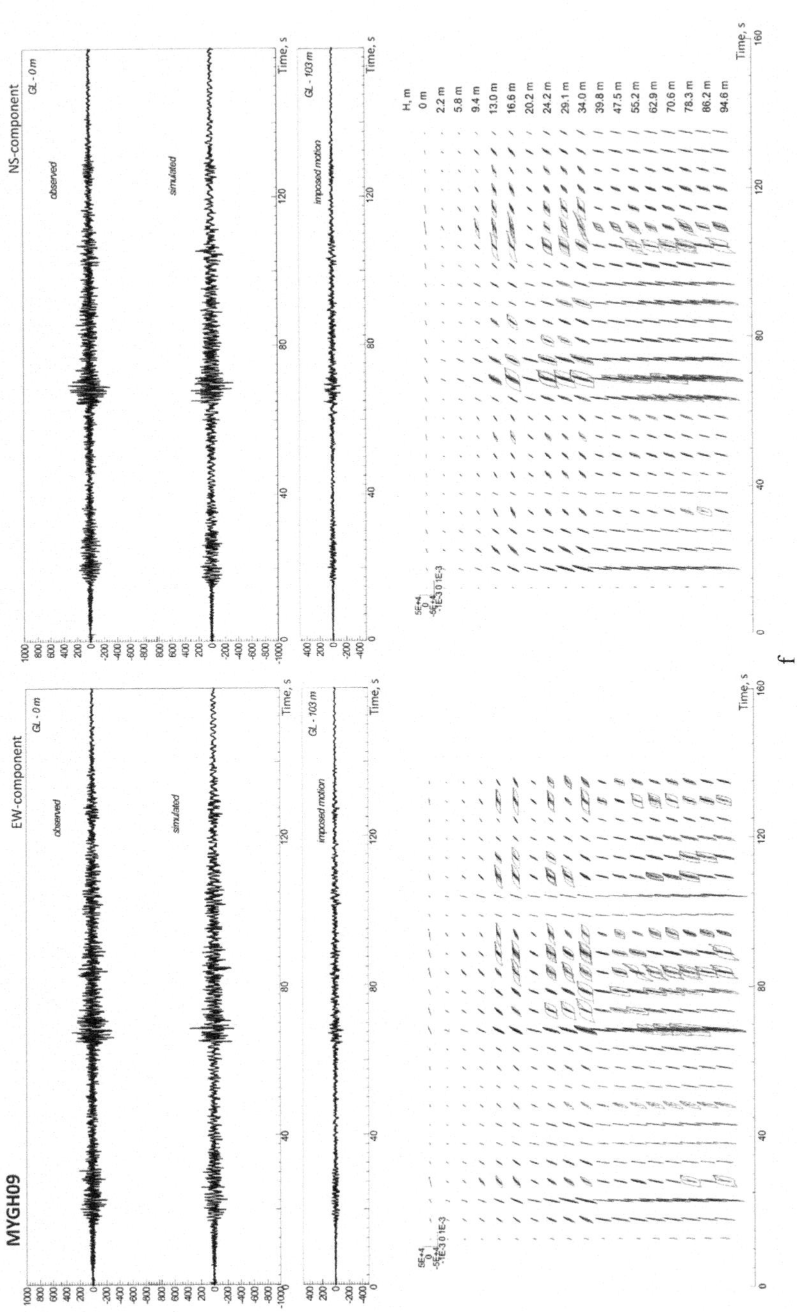

Fig. 4.5 (continued)

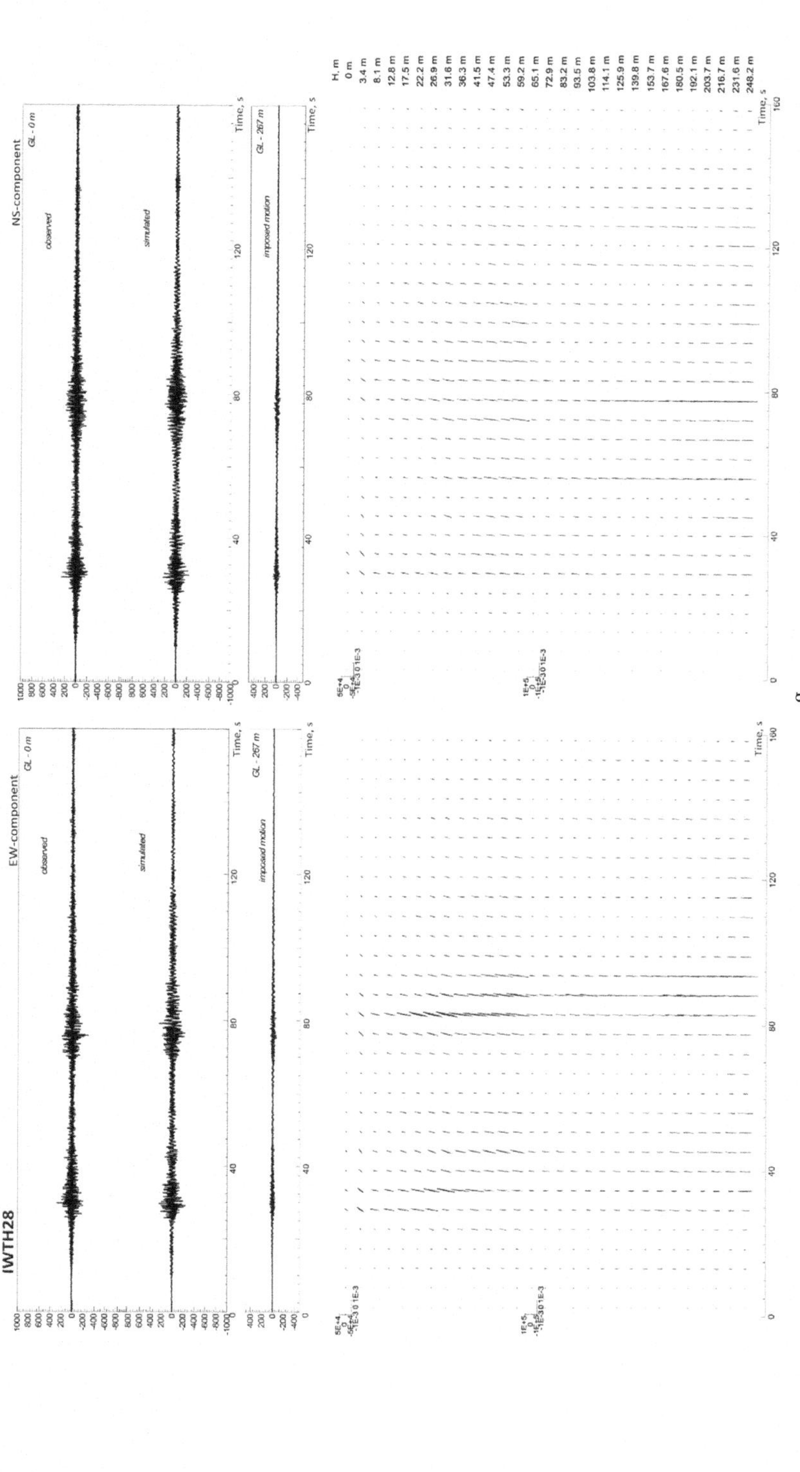

Fig. 4.5 (continued)

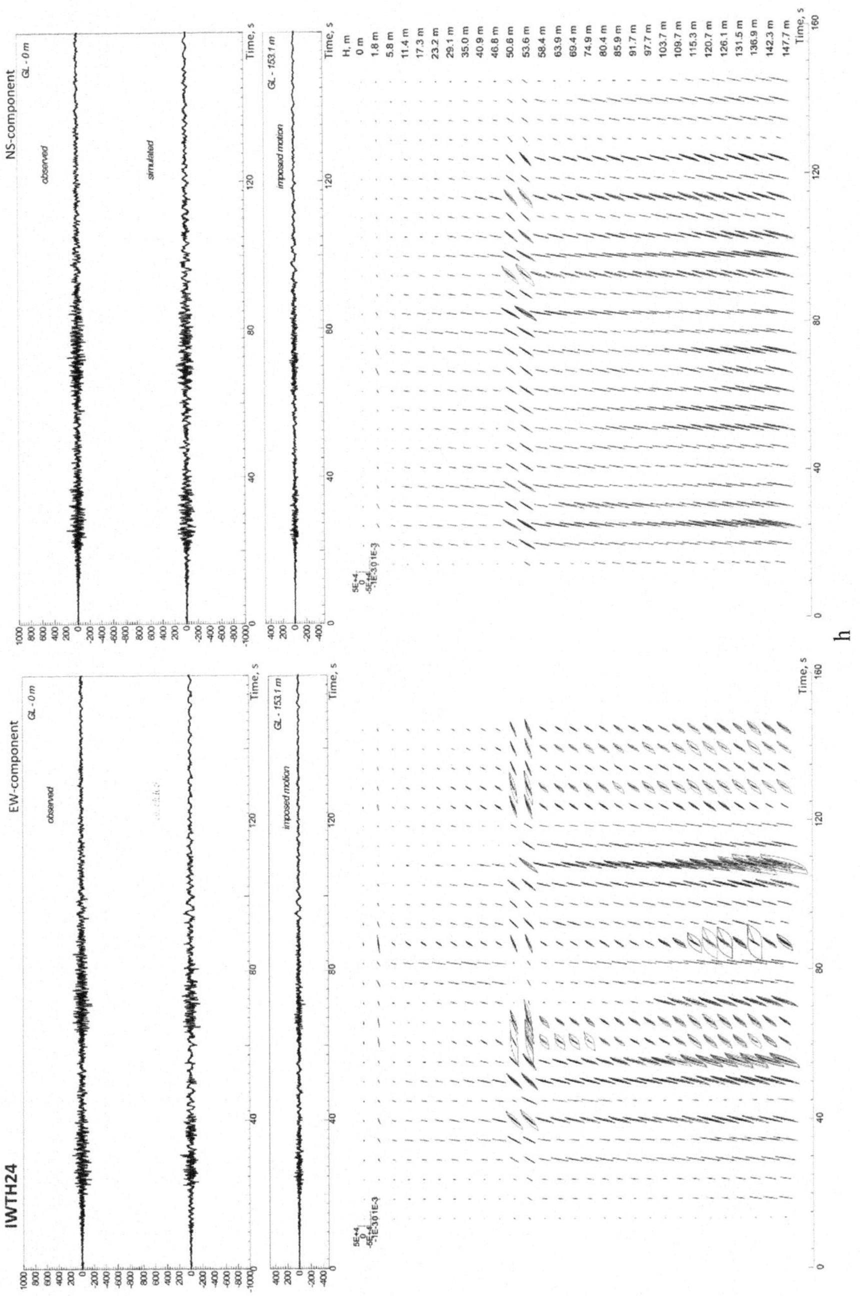

Fig. 4.5 (continued)

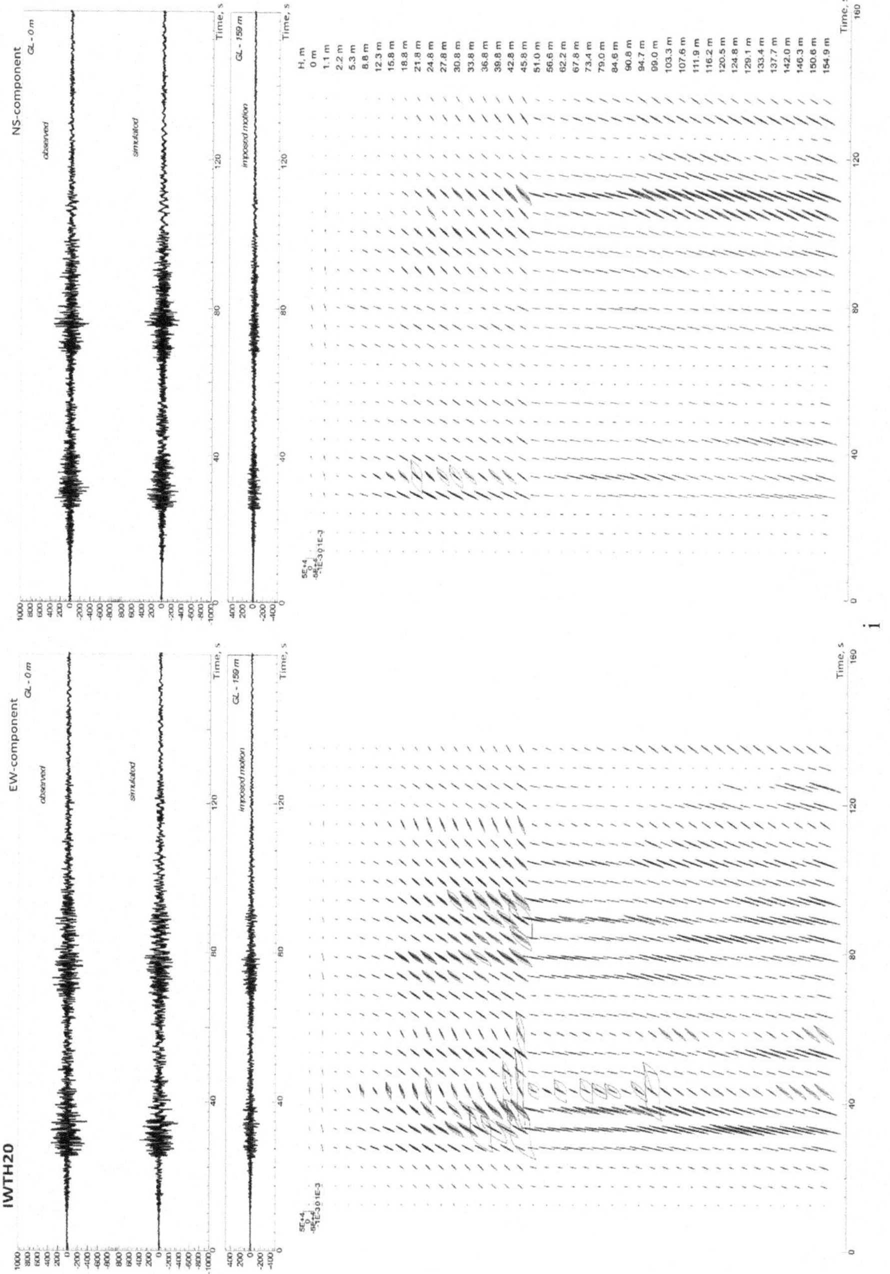

Fig. 4.5 (continued)

(at FKSH14, FKSH18, FKSH12, TCGH13, IBRH15, TCGH10, TCGH16, IRRH11, MYGH12, IWTH27, and MYGH08 sites).

Soil conditions at the studied KiK-net sites, such as, average *S*-wave velocities in the upper 30 m V_{S30}, thicknesses of the upper softer layers *d* and *S*-wave velocities V_S in these layers are described in Tables 4.1 and 4.2. If a soil profile contains soft soils in the upper 30 m (V_{S30} < 370 m/s), or rather thick upper soft layers (5 m and more with V_S < 300 m/s), such sites can be considered as soft soil sites, where nonlinear phenomena are likely to occur during strong motion.

To trace changes in soil behavior during strong motion, shear moduli in soil layers were estimated in successive time intervals (Figs. 4.8 and 4.9) as the ratios of mean normalized stresses to the mean normalized strains (normalization is described in Joyner and Chen 1975), averaged over all hysteretic curves within 5 s time intervals and over all the depths within the upper ~ 100 m of soil.

Thus, we do not see "typical" response of soft soils during the Tohoku earthquake, similar to that observed during past strong earthquakes with smaller magnitudes M_w ~ 6.7–7.6, i.e., soil softening and reduction of shear moduli at the beginning and recovery at the end of strong motion, though some stations show local drops of the shear moduli.

Soil behavior at southern KiK-net sites

The constructed models of soil behavior reveal the highest nonlinearity at FKSH20 and FKSH14 sites located in coastal near-source areas. At these sites, acceleration time histories contain strong low-frequency components and substantially nonlinear stress–strain relations in the middle and final parts of the strong motion (Fig. 4.2a, c). High acceleration peaks riding a low frequency carrier and "hard-type" nonlinear stress–strain relations indicate pore-pressure effects at FKSH20 and FKSH14 sites. To a smaller extent, similar effects are seen at FKSH12 site (Fig. 4.2e). At FKSH14 site, very sloping stress–strain relations and large strains in the upper ~ 8 m of water-saturated sandy soils testify to the liquefaction in these layers, which is in agreement with Roten et al. (2013).

At FKSH11 soft soil site, nonlinear stress–strain relations in soil layers (Fig. 4.2g) and small amplification of seismic waves in soil layers (Table 4.1) also show substantially nonlinear soil response. Low amplification of ~ 2.8–3.1 at FKSH20, FKSH14, and FKSH11 sites testify to the nonlinear soil response.

At the same time, at soft soil sites FKSH19 and IBRH13 located close to the fault, the soil response was virtually linear (Figs. 4.2b and 4.3a), shear moduli did not reduce during strong motion (Fig. 4.8), and the amplification of seismic waves in soil layers was rather high, ~ 5.8–6.0 (Table 4.1). This probably can be explained by the fact that only the upper 20–30 m are soft layers, and they are underlain by denser soils, while strong motion involves all the layers down to ~ 50–100 m in the near-fault zones, i.e., the behavior of all the layers should be similar. At FKSH19 site, PGA of the imposed motion (~ 350 Gal) is almost as high as at FKSH20 site (~ 356 Gal), however, the high PGA did not induce soil softening similar to that occurred at FKSH20 site. Instead, we observe soil hardening: soil behavior is described by very hard and quasi-linear stress–strain relations (Fig. 4.2b). Similar results are obtained

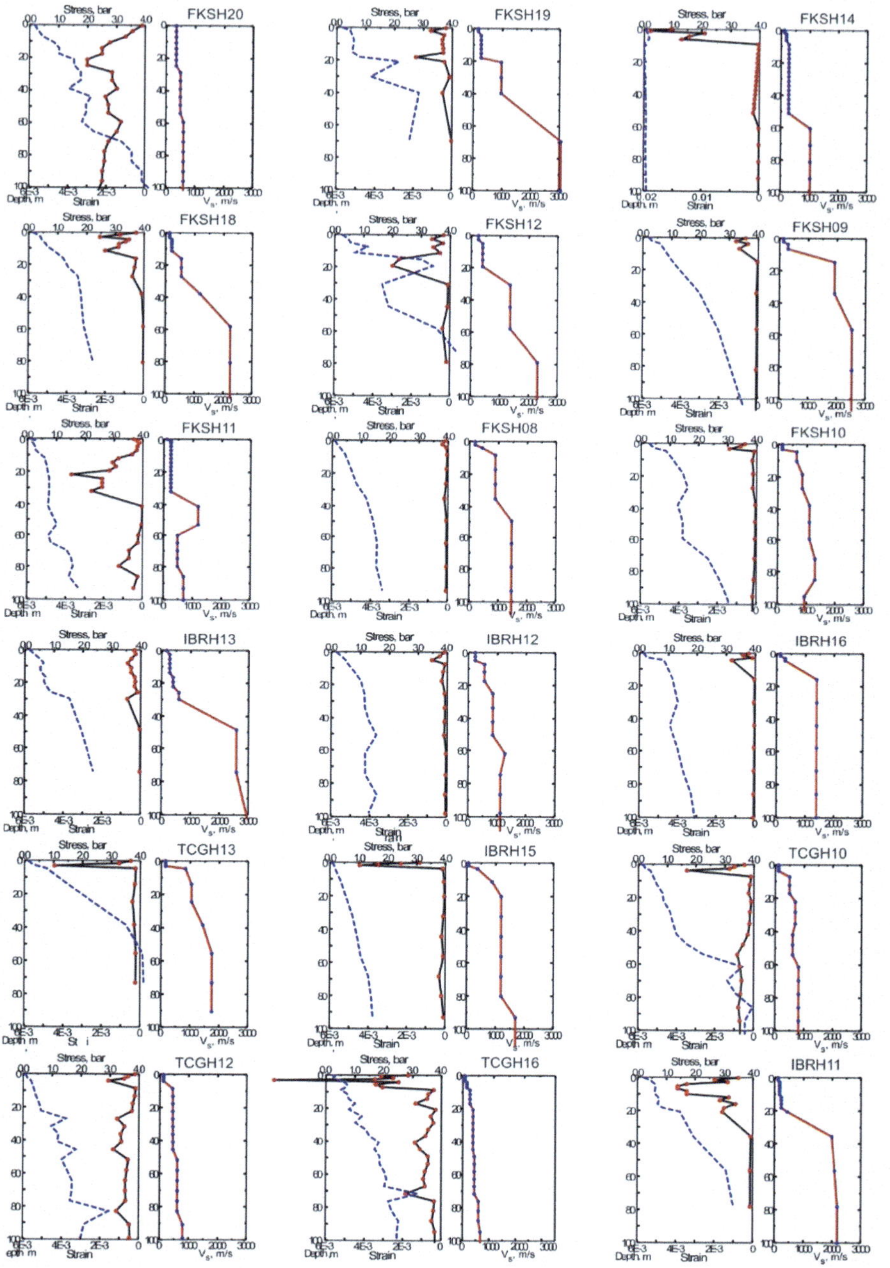

Fig. 4.6 The profiling data, V_S velocities (on the right), and the estimates of maximum stresses and strains (dash and solid lines) induced in the upper ~ 100 m by strong motion (on the left) at southern KiK-net sites. Stresses are given in Pa, and strains in strain

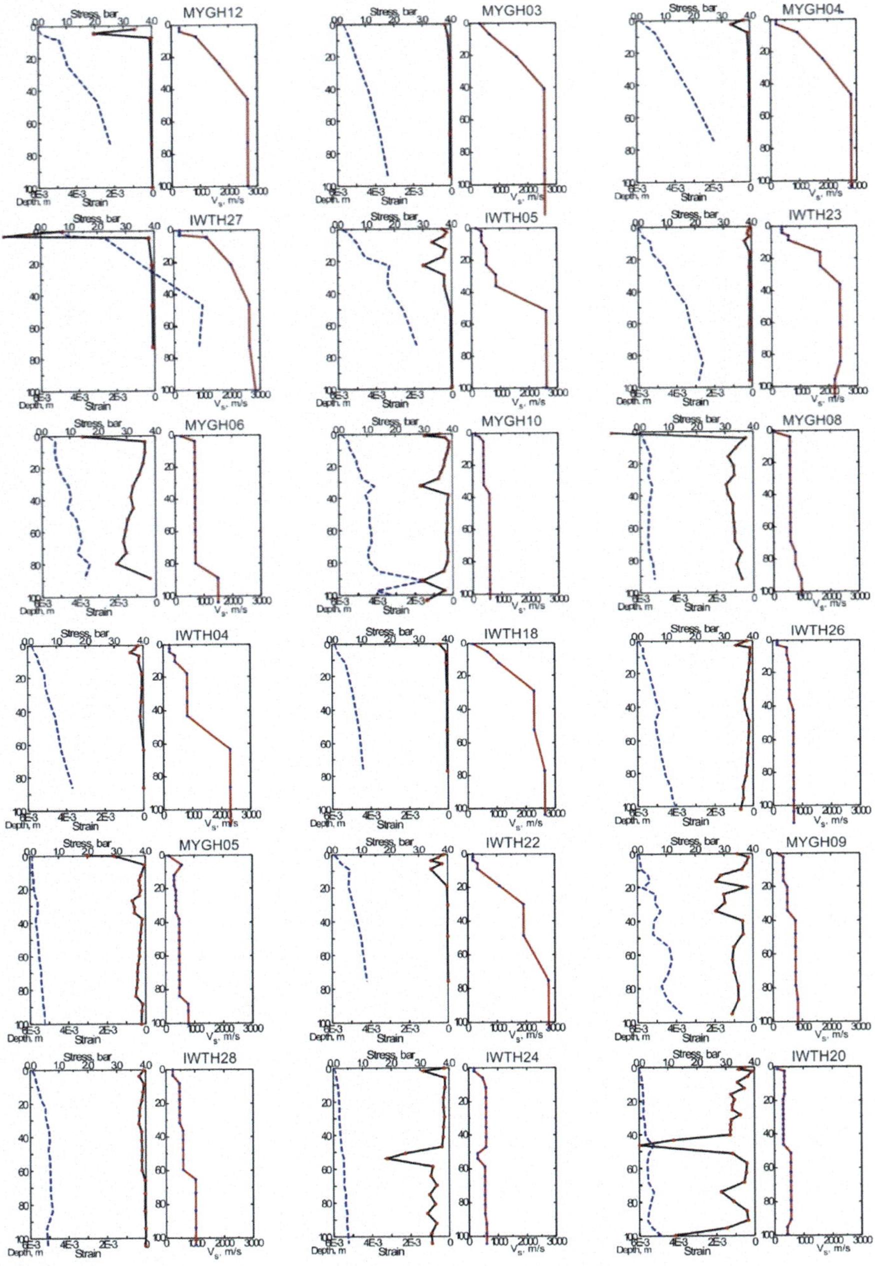

Fig. 4.7 The profiling data, V_S velocities (on the right), and the estimates of maximum stresses and strains (dash and solid lines) induced in the upper ~ 100 m by strong motion (on the left) at northern KiK-net sites. Stresses are given in Pa, and strains in strain

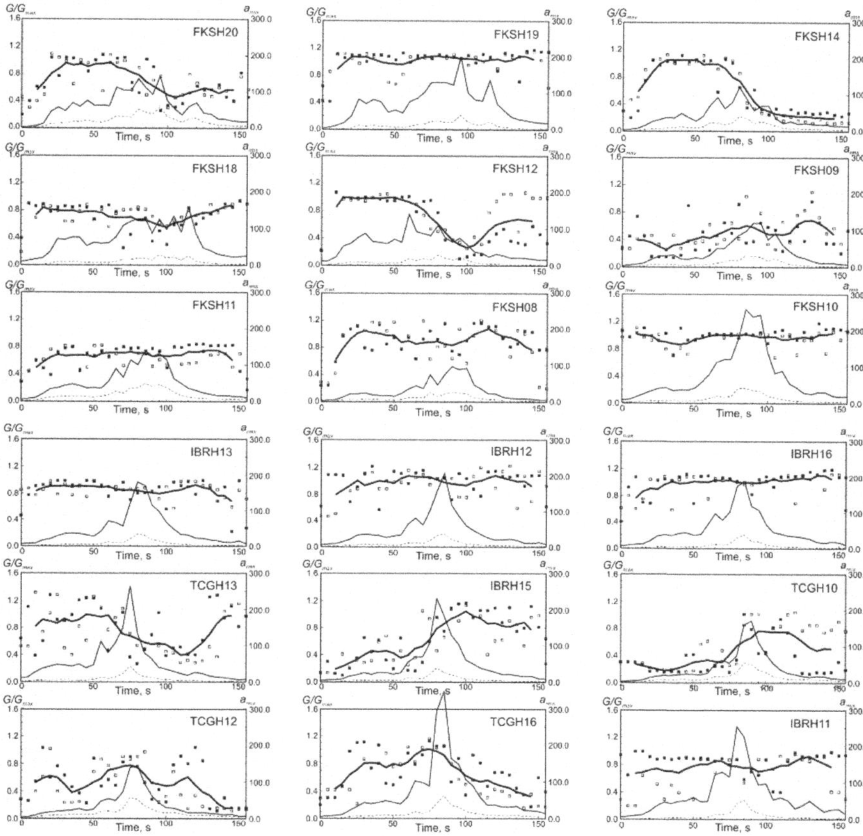

Fig. 4.8 Changes of shear moduli in soil layers at the southern KiK-net sites during the Tohoku earthquake. Points are the estimated shear moduli in successive time intervals at two horizontal components; thick lines show these values smoothed and averaged over two horizontal components. Thin lines are the intensities of motion on the surface (solid lines) and at depths of locations of the deep devices (dash lines)

for IBRH13 and FKSH10 sites: at the moments of the highest intensity of motion, the models of soil behavior indicate linear behavior and soil hardening (stress–strain relations decline to the stress axes).

At remote sites (at epicentral distances of ~ 265–309 km) recorded high PGA: FKSH10, TCGH13, IBRH15, TCGH10, TCGH12, TCGH16, and IBRH11, the intensity of the imposed motion was also high (at larger epicentral distances it drops sharply), and PGA at depth achieved 205–267 cm/s^2 (Table 4.1). Shear moduli reach their maxima and show soil hardening at the moments of the highest intensity of motion (Fig. 4.8); this is clearly seen on the constructed models of soil behavior, where stress–strain curves decline to the stress axes at the moments of the highest intensity of motion. At these sites soil conditions are diverse (Table 4.1; Fig. 4.6):

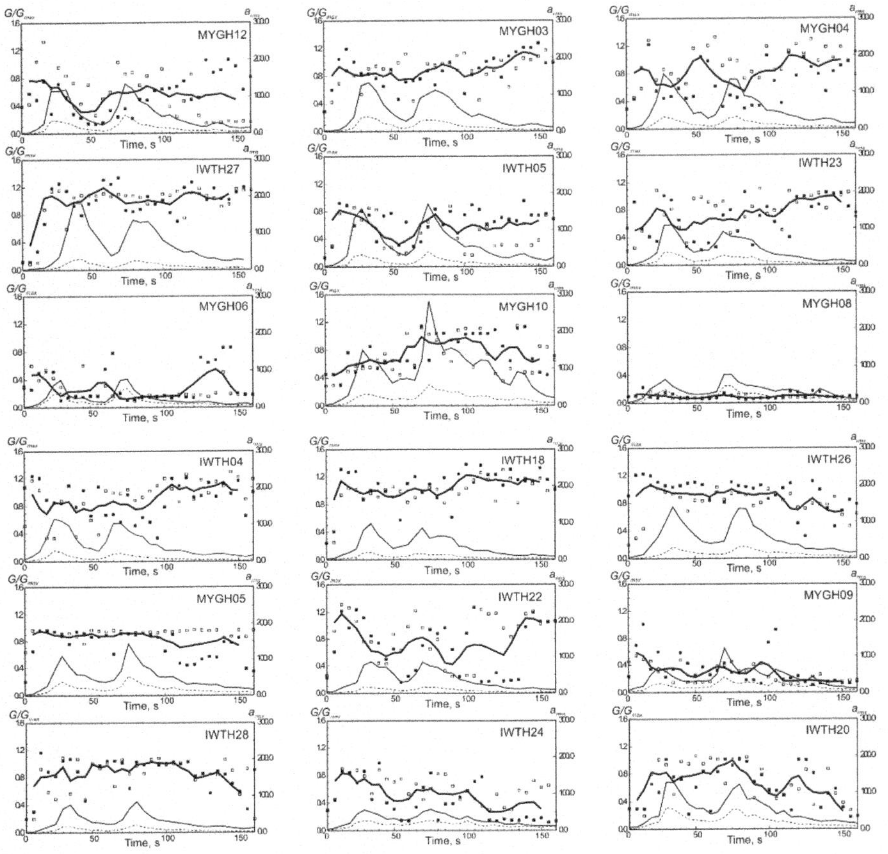

Fig. 4.9 Changes of shear moduli in soil layers at the northern KiK-net sites during the Tohoku earthquake. The notations are the same as in Fig. 4.8

dense soils at FKSH10, TCGH13, and IBRH15 sites (V_S ~ 840– 880 m/s at 4–5 m), softer soils at TCGH10 and TCGH12 sites (V_S gradually increase to ~ 800 m/s at ~ 60–70 m), and soft soils at TCGH16 and IBRH11 sites.

Nonlinear soil behavior was observed at soft soil sites TCGH16, IBRH11, TCGH10, and TCGH12 (Fig. 4.3), however, shear moduli behave atypically: they grow and reach their maxima at the moments of the highest intensity of motion (Fig. 4.8). Amplification of seismic waves in soil layers is rather high, ~ 5.5–6.5 on dense soils at FKSH10, TCGH13 and IBRH15, as well as on soft soils at TCGH16 and IBRH11.

An exciting feature of the acceleration time histories at soft soil sites FKSH14, TCGH16, and IBRH11 is that they are high-frequency ones, with the predominant frequencies of ~ 4–5 Hz till the moments of the highest intensity of motion. Then the frequency sharply falls (Fig. 4.10a), and shear moduli also fall. Similar behavior is observed at K-NET soft soil sites FKS012, IBR002, and IBR014 (Fig. 4.10b). The locations of K-NET sites are shown in Fig. 4.11.

Soil behavior at northern KiK-net sites

As seen from Figs. 4.4 and 4.5, the constructed models indicate quasi-linear soil behavior at many northern KiK-net sites, while nonlinear soil response is observed

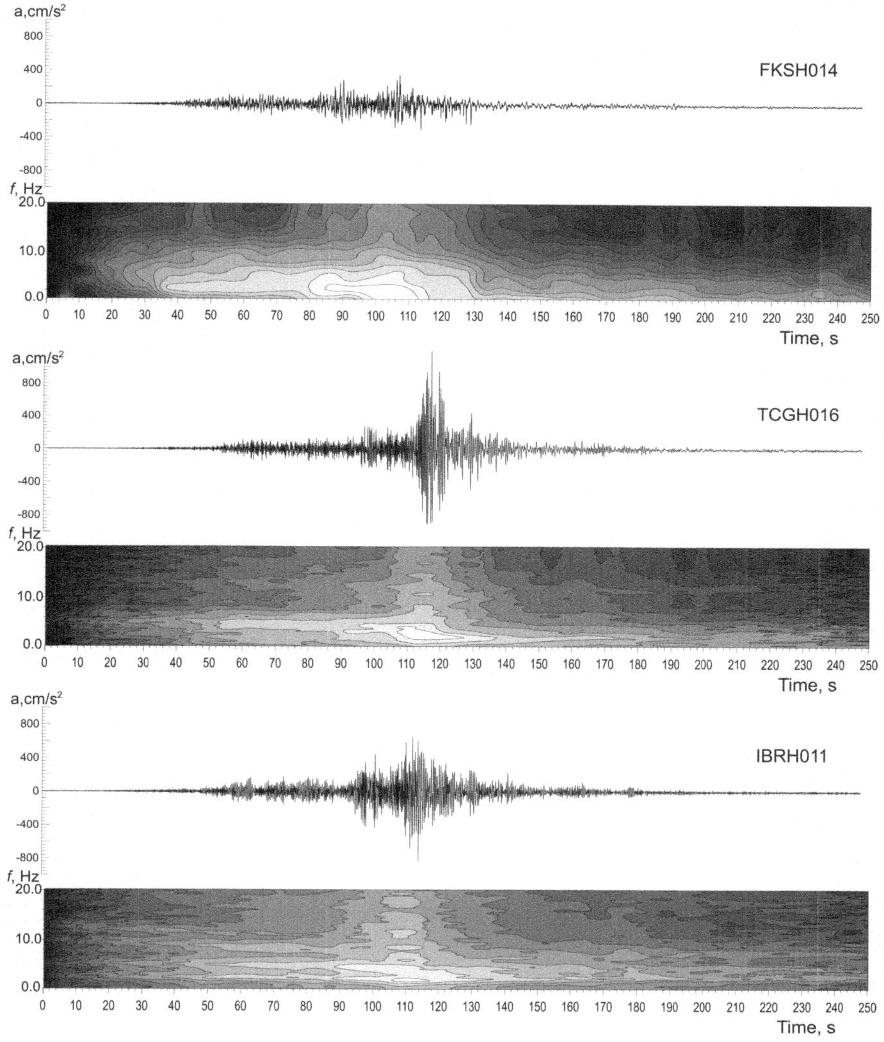

a

Fig. 4.10 Time–frequency diagrams of the acceleration time histories of the Tohoku earthquake at KiK-net (**a**) and K-NET (**b**) stations, at horizontal components

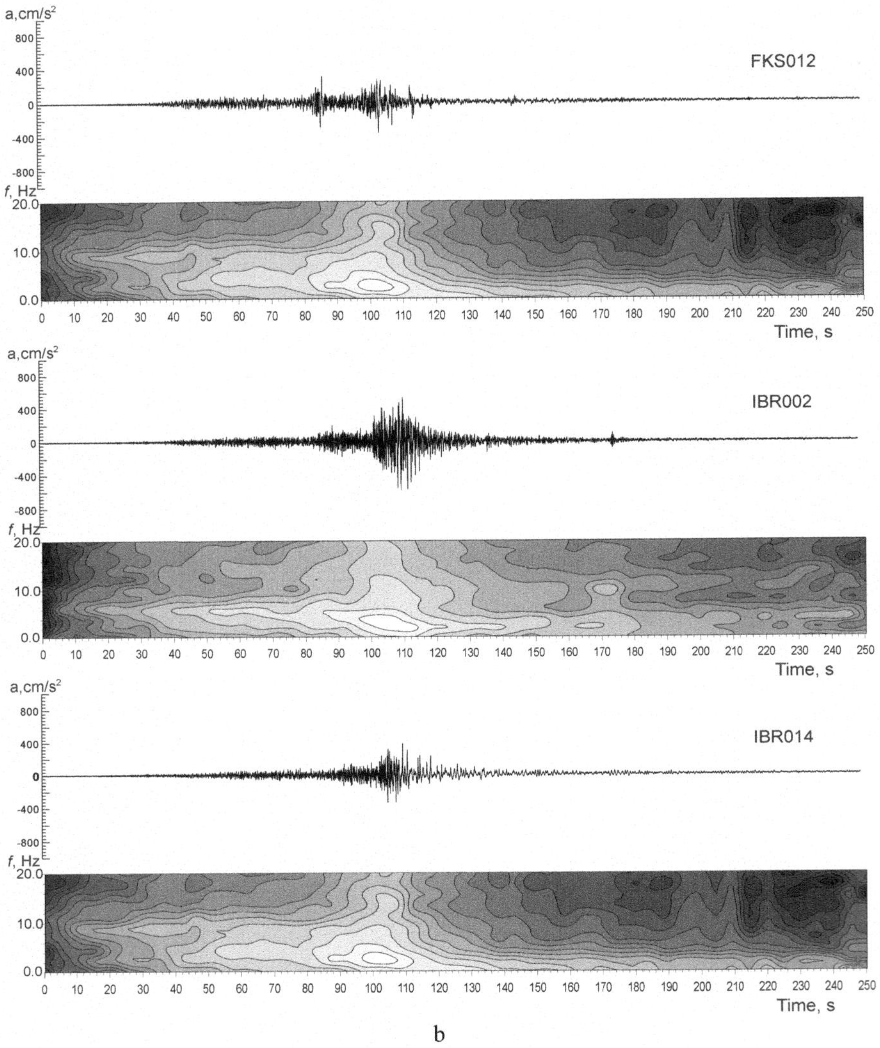

Fig. 4.10 (continued)

at coastal stations MYGH06, MYGH08, MYGH05, MYGH09 and at soft soil stations IWTH20 and IWTH24. At northern stations, shear moduli vary during strong motion in a more complicated way than at southern stations.

At sites MYGH12 and IWTH05 located closely to the fault (~ 137–156 km) and recorded high PGA ~ 568–791 Gal (Table 4.2), shear moduli correlate with the intensity of motion; they grow and reduce twice following two groups of intense

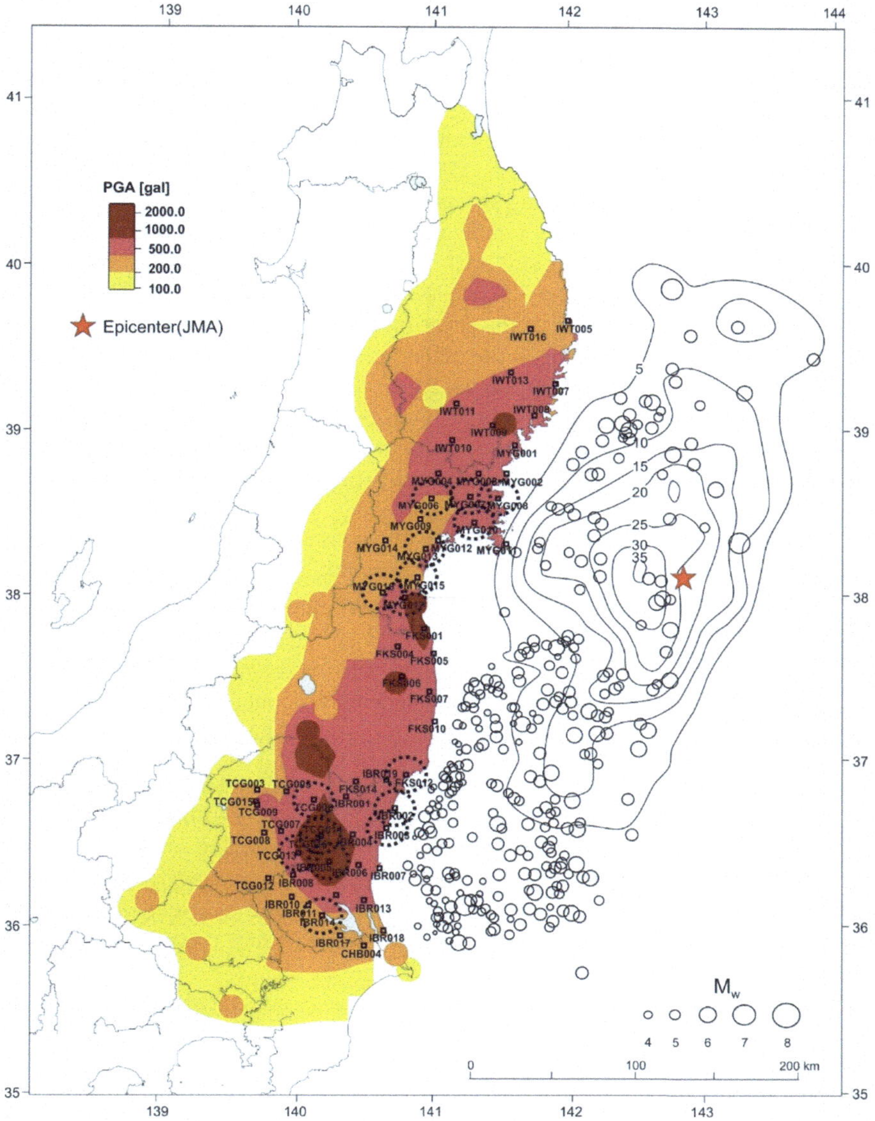

Fig. 4.11 Map showing the distribution of PGA recorded during the 2011 Tohoku earthquake (NIED 2011), locations of the mainshock (star) and aftershocks (circles) (M > 3) occurred in 24 h after the mainshock, and K-NET stations (squares). Contour lines depict slip distribution in meters according to the "joint source model" by Koketsu et al. (2011). Dashed circles mark sites with nonlinear soil response

waves (Fig. 4.9). At other sites (MYGH04, IWTH21), the correlation is negative (Fig. 4.9).

At sites MYGH10, IWTH20, and YMTH04, shear moduli grow at the beginning and fall at the end of strong motion (Fig. 4.9), whereas, at IWTH22 and IWTH04 sites we see the opposite: shear moduli reduce at the beginning and recover at the end of strong motion (Fig. 4.9). IWTH26, MYGH09, and IWTH24 sites show a slight decrease of the shear moduli during strong motion.

In the whole, the shear moduli remain rather high (~ 0.8–1.0) during the Tohoku earthquake at softer and denser soil sites (Fig. 4.9), except MYGH06, MYGH08, and MYGH09, where thick layers of very soft clays are deposited (NIED).

At many stations, variations of the shear moduli indicate soil softening at some time intervals and soil hardening at the others; generally, the behavior of softer and denser soils looks similar (Figs. 4.8 and 4.9).

Soil softening during strong ground motion is well known from laboratory and field experiments and in situ observations: soft soils usually soften and loosen in conditions of strong motion, i.e., seismic velocities and shear moduli fall in soft layers, while damping increases.

Soil hardening is an opposite process, i.e., it means solidification and compaction of the soil, inducing seismic velocities and shear moduli to grow during strong motion.

The constructed models of soil behavior at the studied KiK-net sites and the estimates of shear moduli show us a widespread soil hardening during the Tohoku earthquake, such as, shear moduli at many soft soil sites grow or remain stably high, and soil behavior is virtually linear with low damping.

Nonlinearity of soil behavior during the Tohoku earthquake

The degree of nonlinearity of soil response at the studied KiK-net sites during the Tohoku earthquake was estimated. The obvious way to estimate the degree of nonlinearity is to compare borehole transfer functions calculated for weak motions $G_W(f)$, where the soil response can be considered as linear, and for the Tohoku earthquake $G_T(f)$. This approach is similar to that proposed by Field et al. (1997), who compared linear and nonlinear amplification analyzing records of the 1994 Northridge earthquake.

The transfer functions and their confidence limits were calculated for series of weak earthquakes preceding the Tohoku earthquake. For each station, 20 records of weak earthquakes occurred in the 2000s were selected, for which PGA on the surface did not exceed 5–10 Gal, and soil behavior was virtually linear.

The method is based on our knowledge about nonlinear transformations of seismic waves in soil layers (described in detail in Chap. 5). As known, the propagation of intense seismic waves in soil layers is accompanied by nonlinear wave effects, such as, interaction of spectral components of the waves, nonlinear damping, higher harmonic generation, etc. The energy of the propagating waves is redistributed over the spectra so that spectral components at high and medium frequencies weaken, low-frequency components amplify, and power spectra of motion on the surface tend to take the form $E(f) \sim f^{-k}$. Thus, soil nonlinearity influences both the amplification and frequency content of seismic waves propagating in soil layers. The difference in the transfer functions calculated for weak and strong motions can be considered as the estimate of the degree of nonlinearity of the soil response during strong motion (another way to estimate the nonlinearity is described in Chap. 6).

Nonlinearity of soil response during strong motion leads to a decrease in the amplification of seismic waves propagating in soil layers. The decrease characterizes the degree of nonlinearity of soil response nl_1 and can be estimated as the difference between the integrals of the transfer functions calculated for weak $G_W(f)$ and strong motions $G_T(f)$ in seismic frequency range (f_1, f_2), normalized by the weak-motion transfer function $G_W(f)$:

$$nl_1 = \frac{\int_{f_1}^{f_2} (G_W(f) - G_T(f))df}{\int_{f_1}^{f_2} G_W(f)df}. \tag{4.1.1}$$

Nonlinearity of soil response also leads to changes in spectral composition of seismic waves propagating in soil layers, and these changes can also serve as a measure of the nonlinearity of soil response nl_2 and can be estimated by calculating the correlation between the transfer functions calculated for weak and strong motion.

Preliminary calculations have shown that the estimates nl_1 based on the differences in the amplification in weak and strong motion are more stable than the estimates nl_2 based on the correlation of the transfer functions in weak and strong motions, apparently because they are less dependent on the smoothness of the transfer functions, which affects nl_2. Realistic estimates of the nonlinearity of soil response nl can be obtained from the combination of nl_1 and nl_2: $nl = (nl_1 + nl_2)/2$, where nl_1 and nl_2 are calculated for frequencies below 12 Hz. The estimates based on the frequency components below 12 Hz actually seem to be more sensitive to nonlinear effects, because in seismic waves passed soft soil layers, low-frequency components are amplified. All the estimates of the nonlinearity of soil response in percent are given in the last columns in Tables 4.1 and 4.2.

Figure 4.12a shows the borehole transfer functions at the studied sites calculated for weak motions and for the Tohoku earthquake.

The technique for estimating nl was verified on records of the 2000 Tottori earthquake (M_w ~ 6.7), where TTRH02 and SMNH01 sites (~ 7 and ~ 8 km from the fault plane) showed strong nonlinearity of soil response (Chap. 3). At TTRH02 and SMNH01 sites, the estimates nl ~ 83% and nl ~ 69% were obtained. At remote sites HRSH06 and HRSH05 (~ 57 and ~ 80 km from the fault plane), the estimates of nonlinearity were much smaller: nl ~ 3% and ~ 9% (Fig. 4.12b).

The transfer functions shown in Fig. 4.12 clearly indicate high nonlinearity of soil response during the 2000 Tottori earthquake and rather low nonlinearity of soil response during the 2011 Tohoku earthquake.

As seen from Tables 4.1 and 4.2, the highest PGA at depth (~ 179–241 Gal) were recorded at the near-fault soft soil sites MYGH12, MYGH10, MYGH08, MYGH05, and IWTH20 (Fig. 4.1). However, the high PGA did not lead to substantial nonlinearity of soil response at these sites; the nonlinearity was rather low, nl ~ 1–11%. The strongest nonlinearity was observed at soft soil sites IWTH26 (nl ~ 18%) and

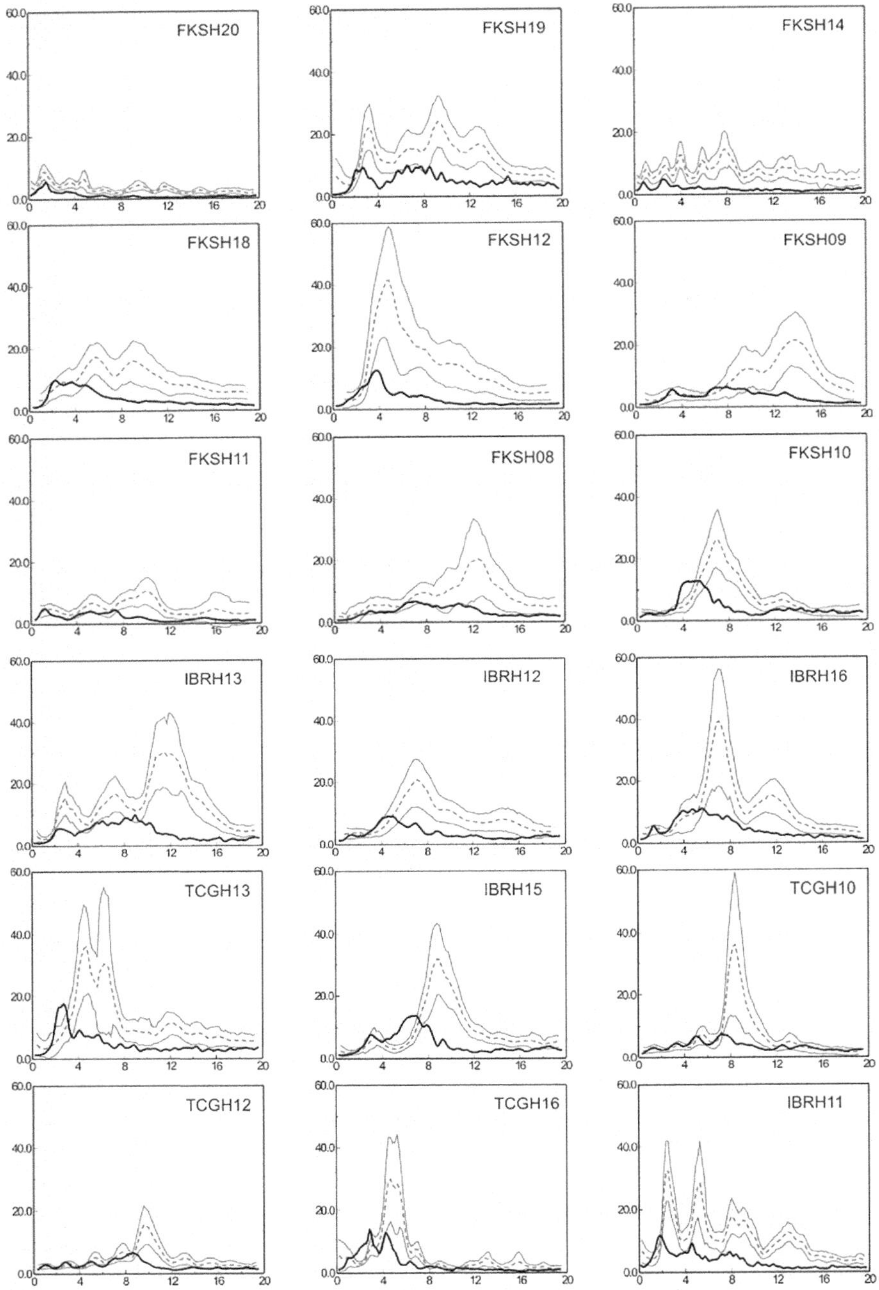

Fig. 4.12 Borehole transfer functions: **a**—at KiK-net sites calculated for the Tohoku earthquake $G_T(f)$ (black thick lines) and for weak motions $G_W(f)$ (thin dash lines—averages, thin solid lines—confidence limits); **b**—at sites in the near-fault zones of the 2000 Tottori earthquake

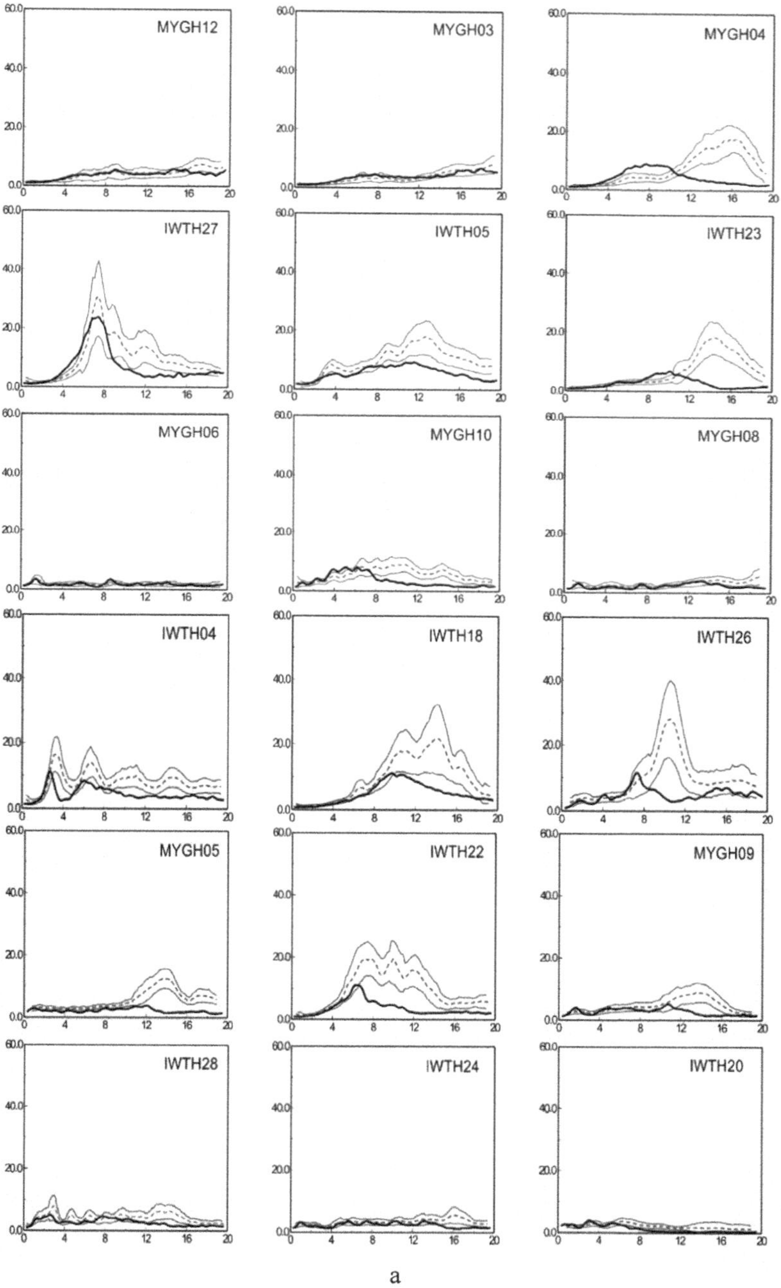

Fig. 4.12 (continued)

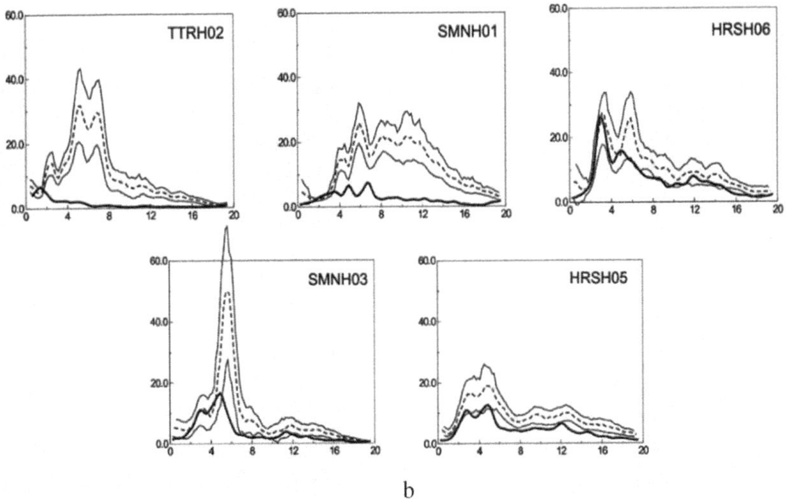

Fig. 4.12 (continued)

IWTH22 (*nl* ~ 16%). On the average, *nl* estimates were higher on softer (~ 8.4%) than on denser soils (~ 3.9%).

In the southern Tohoku and Kanto regions, PGA at depth are generally higher than in the northern Tohoku; they reach ~ 350–356 Gal, and the nonlinearity of soil response was ~ 27–33% (Table 4.1). The highest values (*nl* > 18%) are obtained at FKSH20, FKSH14, IBRH18, FKSH12, FKSH11, FKSH18, IBRH11, and IBRH17 sites (Fig. 4.1). Sites possessing higher nonlinearity are located on the coast near the source and at large epicentral distances; on the average, the nonlinearity is higher at sites with high PGA at depth. On soft soils *nl* ~ 14.9%, which is slightly higher than on denser soils, where *nl* ~ 12.8%.

Tables 4.3 and 4.4 present data on K-NET stations located in the near-fault zones of the Tohoku earthquake. Many of them are installed on soft soils, where the nonlinearity of soil response can be roughly evaluated based on the waveforms and spectra of the acceleration time histories on the surface (Figs. 4.13 and 4.14).

As shown above (Fig. 4.10b), at FKS012, IBR002 and IBR014 sites, a stepwise decrease of motion frequency is observed at the moments of the highest intensity of motion. At IBR003, IBR005, TCG006, TCG013, TCH014, MYG006, MYG007, MYG008, MYG010, MYG013, MYG015, MYG016 and MYG017 sites, in the time intervals of high intensity of motion low-frequency components increase and prevail, so that spectra of motion on the surface tend to take the form $E(f) \sim f^{-k}$, indicating substantial nonlinearity of soil response.

The KiK-net and K-NET sites showing nonlinear soil response are marked by dash lines in Figs. 4.1 and 4.11, respectively.

Actually, the obtained results demonstrate that the nonlinearity of soil behavior during the Tohoku earthquake was not widespread.

Table 4.3 Information on the southern K-NET stations

No.	Site code	Latitude (°)	Longitude (°)	V_{S20}, (m/s)	Epicentr. dist. (km)	PGA (Gal)
1	FKS012	36.91	140.79	199.3	226	387.0
2	IBR019	36.88	140.65		238	699.9
3	IBR002	36.71	140.71		245	683.7
4	FKS014	36.87	140.43	449.5	255	252.0
5	IBR003	36.59	140.65	238.7	258	1845.2
6	IBR001	36.78	140.36		266	454.0
7	IBR004	36.55	140.41	300.0	277	1311.9
8	IBR007	36.35	140.60		279	584.7
9	TCG006	36.76	140.13	413.9	284	436.0
10	IBR006	36.37	140.45	294.2	287	851.3
11	TCG014	36.54	140.18	348.0	294	1291.1
12	TCG016	36.53	140.16	454.7	297	401.1
13	TCG005	36.81	139.93	236.1	297	293.7
14	IBR005	36.39	140.24		301	996.0
15	IBR013	36.16	140.49	219.5	301	1762.3
16	IBR018	35.98	140.63	255.0	308	658.4
17	IBR012	36.19	140.29	211.6	312	365.3
18	TCG013	36.44	140.02	329.4	312	465.3
19	TCG003	36.82	139.72	321.9	312	306.8
20	TCG015	36.75	139.71	311.4	316	197.1
21	TCG009	36.73	139.72	201.8	317	1444.0
22	TCG008	36.56	139.77	237.8	323	346.4
23	CHB004	35.89	140.49	277.3	324	310.3
24	IBR008	36.31	139.98	476.2	324	298.5
25	IBR014	36.07	140.19	224.4	327	533.8
26	IBR017	35.95	140.32	234.0	329	553.4
27	IBR011	36.13	140.09	222.2	330	371.7
28	IBR010	36.18	139.97	190.2	334	436.1
29	TCH004	36.81	139.42	383.0	336	423.7
30	TCG012	36.29	139.80	318.5	338	430.0

At coastal sites FKSH20 and FKSH14, soil nonlinearity was rather high, while amplification of seismic waves in soil layers was low (Table 4.1). High PGA were evidently caused by the bursts from large slip patches and pore pressure effects, as concluded by Roten et al. (2013).

At remote sites FKSH10, IBRH15, TCGH16, IBRH11 and others recorded abnormally high accelerations, there should be another mechanism of generation of high PGA. At these sites, amplification of seismic waves in soils is high, stress–strain

Table 4.4 Information on the northern K-NET stations

No	Site code	Latitude (°)	Longitude (°)	V_{S20} (m/s)	Epicentr. dist. (km)	PGA (Gal)
1	MYG011	38.30	141.51	1255	120	939.2
2	MYG008	38.58	141.45	508.2	134	436.1
3	MYG002	38.73	141.51		137	772.4
4	MYG010	38.43	141.28	161.2	143	487.5
5	MYG001	38.90	141.57		143	430.9
6	IWT008	39.08	141.71		148	387.0
7	MYG007	38.59	141.25		151	695.0
8	MYG003	38.73	141.31		152	870.2
9	IWT007	39.27	141.86		156	741.6
10	IWT009	39.02	141.41		163	570.1
11	MYG012	38.32	141.02	786.1	163	2018.9
12	MYG013	38.27	140.93	228.9	170	1807.8
13	FKS005	37.64	140.99		173	741.7
14	MYG006	38.58	140.97	130.3	174	585.7
15	FKS001	37.79	140.92		174	682.8
16	MYG015	38.10	140.87	189.5	174	433.6
17	MYG004	38.73	141.02	210.7	175	2933.2
18	MYG009	38.45	140.89		177	553.5
19	IWT010	38.93	141.12	413.2	178	1225.8
20	IWT013	39.34	141.54	289.2	179	469.1
21	MYG017	37.98	140.78	110.1	183	357.9
22	FKS007	37.41	140.96		184	880.6
23	IWT011	39.15	141.15	250	188	359.8
24	IWT005	39.65	141.95		189	280.9
25	FKS010	37.23	141.00		190	1239.9
26	FKS004	37.68	140.73		193	687.5
27	IWT016	39.60	141.68		195	244.3
28	MYG014	38.32	140.64		196	535.6
29	FKS006	37.50	140.76		197	633.6
30	MYG016	38.01	140.62		197	426.0

relations in soil layers are close to linear ones and show soil hardening at the moments of the highest intensity of motion; high PGA exceeding 1 g are not related to pore-pressure effects.

Kawase et al. (2011), Irikura and Kurahashi (2012), and Roten et al. (2013) note the increased amplification of seismic waves in soil layers during the mainshock of the Tohoku earthquake compared to weak motion; at remote sites FKSH10, IBRH15,

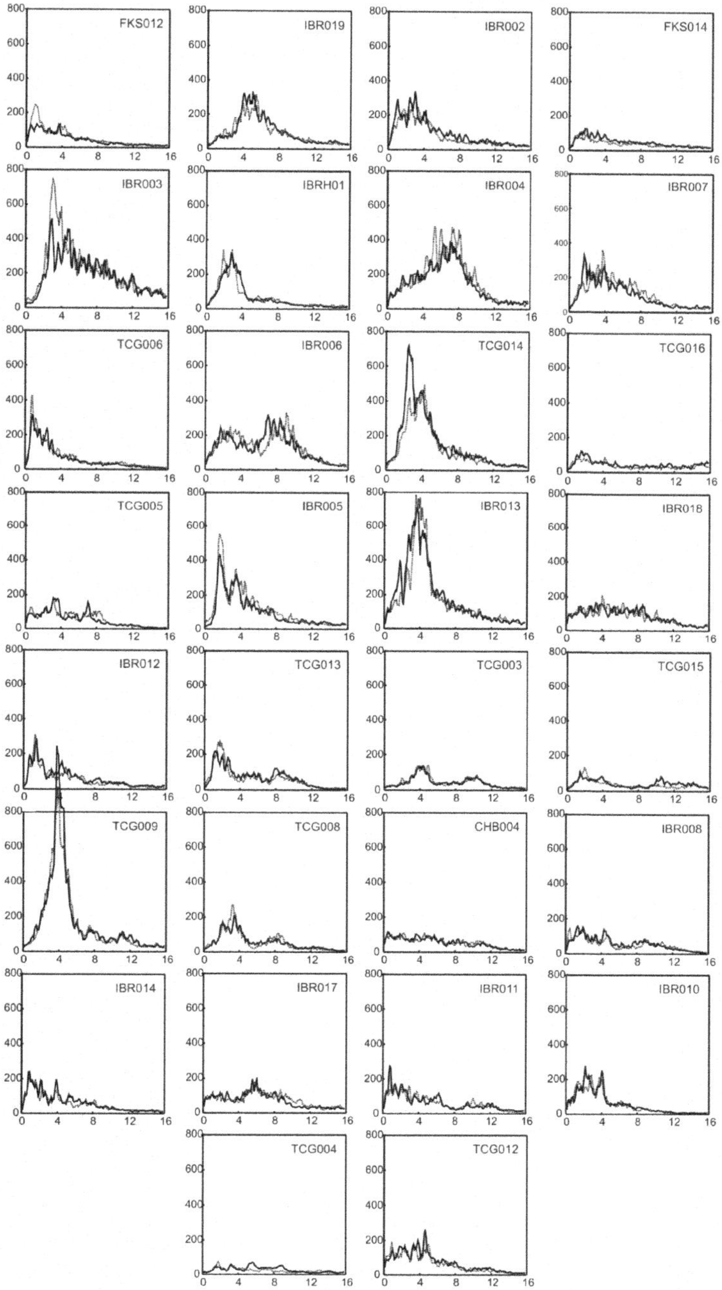

Fig. 4.13 Acceleration spectra of the Tohoku earthquake at southern K-NET stations at two horizontal components

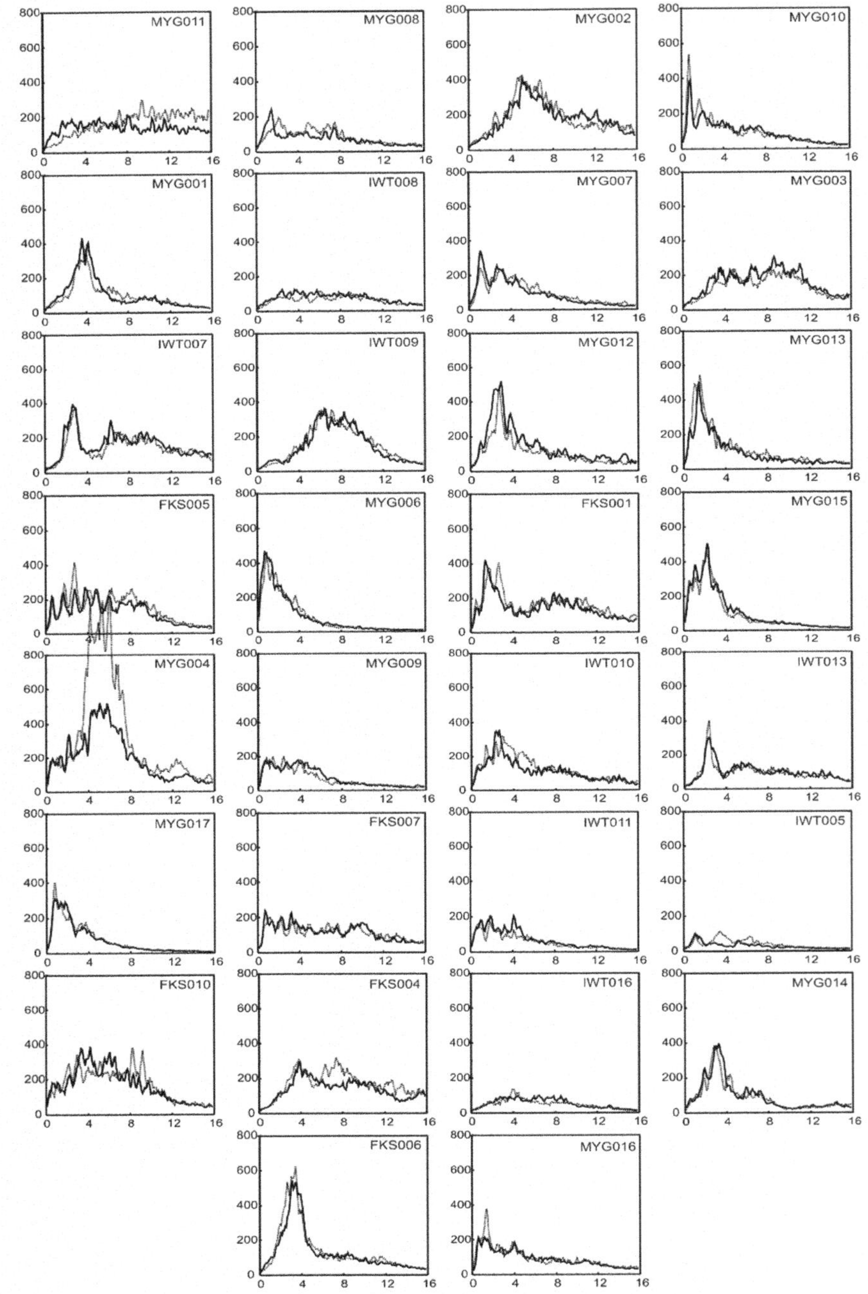

Fig. 4.14 Acceleration spectra of the Tohoku earthquake at northern K-NET stations at two horizontal components

TCGH16, IBRH11 and others recorded high accelerations, it can be explained by soil hardening at the moments of the highest intensity of motion.

The effects of soil hardening evidently contribute to the occurrence of high PGA during the earthquake. As seen from Tables 4.1 and 4.2, the highest PGA on the surface were recorded at sites showing the highest amplification of seismic waves in soil layers, such as, FKSH19, IBRH12, FKSH10, IBRH16, FKSH18, TCGH13, IBRH15, TCGH16, and IBRH11 sites.

Thus, the features of soil behavior in the near-fault zones of the Tohoku earthquake are: a rather widespread soil hardening (softening was also observed), high shear moduli at soft and dense soil sites, and relatively weak nonlinearity of soil response.

4.2 Mechanisms of Generation of Abnormally High PGA During the Tohoku Earthquake

As noted above, during the 2011 Tohoku earthquake nineteen K-NET and KiK-net stations reliably recorded PGA higher than 1 g, the highest reaching ~ 3 g. The distribution of PGA in the near-fault zones (Fig. 4.15) reveals a complicated pattern of seismic wave radiation and propagation from the extended earthquake source. In particular, the highest PGA in the southern Honshu (Fukushima-Ibaraki-Tochigi area) did not occur just near the fault, but at rather large distances from the fault.

The source process of the Tohoku earthquake was investigated by many researchers (Ide et al. 2011; Mitsui et al. 2012; Goto et al. 2012, etc.). Goto et al. (2012), Kozdon and Dunham (2013) and Huang et al. (2012, 2013) proposed 2-D rupture models, accounting for the heterogeneous stress distribution, depth-dependent material properties, curved fault geometry, and the seafloor geometry. Models and mechanisms of the 3-D dynamic ruptures accounting for the large slip and the propagation of the rupture over a wide region are suggested by Duan (2012), Mitsui et al. (2012) and Galvez et al. (2014). The models gave a physical interpretation of the depth-dependent frequency content of seismic radiation, variations of the rupture speed and slip distribution.

In the review paper by Tajima et al. (2013) source models for the 2011 Tohoku earthquake in published papers as of February 2012 are described. The models show notable differences depending on the sources of information employed, however, all the authors conclude that the source process of about 150 s included rupturing of an area of a very large slip (30–80 m, according to different authors) on the shallow part of the mega-thrust up-dip of the hypocenter. The area of the large slip produced large-amplitude low-frequency radiation, while the dominant high-frequency radiation was generated from deeper sources down-dip of the hypocenter.

Source models using tsunami data tend to give larger values of slip for the shallow portion of the fault near the trench (Tajima et al. 2013). Satake et al. (2013) identified some of the tsunami generating areas closely to the previous 869 Jogan earthquake and 1896 Sanriku earthquake models.

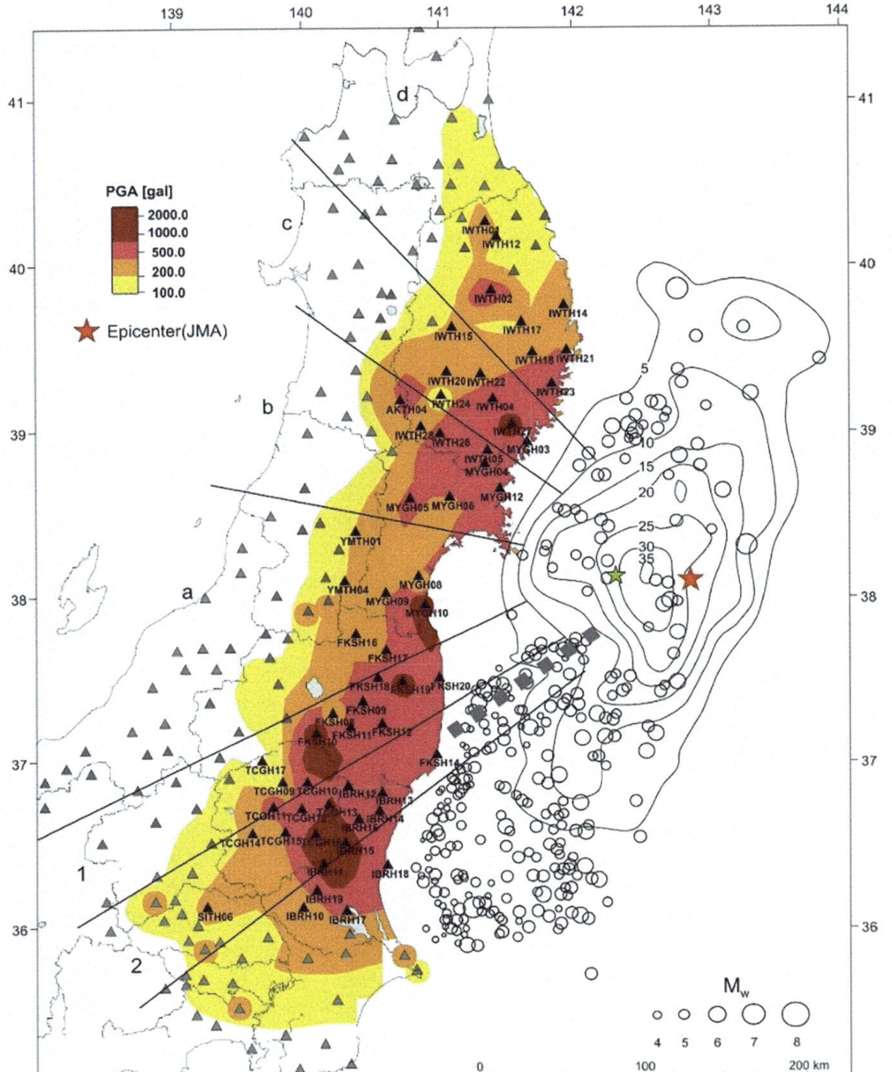

Fig. 4.15 The map showing the distribution of PGA recorded during the 2011 Tohoku earthquake, positions of the mainshock (big star), smaller event (aftershock) of 25 July 2011, $M_w = 6.3$ (small star), aftershocks (circles) ($M_w > 3$) occurred during 24 h after the mainshock, and KiK-net stations (triangles). Contour lines depict slip distribution in meters according to the "joint source model" by Koketsu et al. (2011). Dotted thick line, numbers 1, 2, letters a, b, c, d indicate the horizontal projection of the fault segment and sectors discussed in the text

The joint source model for the earthquake, based on the teleseismic, strong motion, geodetic, and tsunami data is presented by Koketsu et al. (2011). The authors believe that their "joint" model reflects all aspects of the Tohoku earthquake, and contour lines in Fig. 4.15 depict the slip distribution over the fault plane. The maximum slips gravitate to the epicenter, where they reach ~ 36 m and decrease to zero in the margins of the source plane.

As described above, during the Tohoku earthquake, no strong or widespread nonlinearity of soft soil response was observed. The lack of nonlinearity (lack of shear moduli reduction and soil softening) influenced the frequency content of motion on the surface at soil sites: the predominant frequencies remained rather high during strong motion. As a result, fortunately, the amount of shaking damage was not great, which can be partially attributed to relatively low amplitudes in the critical frequency range of 0.5–1 Hz (e.g. Furumura et al. 2011), combined with the high standards of building construction in Japan.

This is clearly seen in the southern Honshu, where acceleration time histories contain one group of intense waves. At soft soil sites shear moduli reduction after the moments of the highest intensity of motion was accompanied by a step-like drop of the predominant frequencies of motion on the surface (Fig. 4.10).

The increased amplification of seismic waves in soil layers during the mainshock compared to weak motion (Nagashima et al. 2012; Irikura and Kurahashi 2012; J-SHIS) is evidently related to the soil hardening and contributes to the generation of anomalously high PGA during the Tohoku earthquake.

Shock wave as a possible cause of abnormally high PGA recorded during the Tohoku earthquake

We can understand and explain the features of soil behavior, if we assume the presence of some additional compression acting on soils. As a result, the soils under compression harden and behave more linearly during strong motion. This compression, which led to soil hardening, could have been produced by a shock wave, generated in the earthquake source.

As known, shock waves may occur in different ways: by explosions, detonation, when bodies move at near- and super-sonic speeds. Shock waves represent the limiting case of accumulation of nonlinear distortions of finite-amplitude waves propagating in a nonlinear medium (sawtooth waves), when further distortions break down and transform them into small-amplitude waves (Zarembo and Krasil'nikov 1966).

Recent years publications appeared where seismologists discuss the generation of shock waves during strong earthquakes due to supershear crack propagation (Bouchon et al. 2001; Bouchon and Vallee 2003; Aagaard and Heaton 2004; Ellsworth et al. 2004; Robinson et al. 2006; Dunham and Bhat 2008; Vallee and Dunham 2012; Zhan et al. 2013).

The possibility of crack propagation at a supershear velocity was theoretically demonstrated in the 1970s. It was shown that shear cracks can either propagate at a sub-Rayleigh velocity ($V < V_R$) or at an intersonic velocity ($V_S < V < V_P$), though crack propagation at most of the intersonic velocities is unstable (Freund 1979).

Since that time, increasingly more observations of supershear crack propagation have appeared. During some strong earthquakes, the rupture velocity locally exceeded the *S*-wave velocity over parts of the faults and generated conical shock wave (or Mach wave) fronts. Archuleta (1984) showed strong indications that during the 1979 Imperial Valley earthquake in California, rupture propagated at supershear velocity over part of the fault. Such phenomena were observed during the 1999 Turkey earthquakes (Bouchon et al. 2001), 2001 Kunlunshan earthquake (Bouchon and Vallee 2003; Robinson et al. 2006), 2001 Kokoxili earthquake (Vallee and Dunham 2012), 2002 Denali (Alaska) earthquake (Ellsworth et al. 2004), and the aftershock of the 2013 Sea of Okhotsk earthquake ($M_w = 6.7$) (Zhan et al. 2013). It was also shown that "very long subduction zone thrust earthquakes such as the Sumatra earthquake could have a very long in-plane rupture front, which, in principle, could reach speeds as high as that for the Kunlun earthquake" (Robinson et al. 2006).

Since supershear crack propagation is usually accompanied by the substantial growth of peak accelerations at the Mach front, this issue is of critical importance to seismic hazard calculations (Dunham and Bhat 2008).

Rosakis et al. (1999) were first who successfully observed intersonic propagation of a shear crack in laboratory experiments, in conditions similar to those that occur during shallow earthquakes.

If a shock wave was generated by the source of the Tohoku earthquake, it can be traced by the shapes of the accelerograms and by their changes with distance from the fault. We studied the shapes of the acceleration time histories at KiK-net sites located within narrow sectors around the fault. Noticeable changes in the shapes were revealed in two sectors oriented in the southwest direction from the epicenter (sectors 1 and 2 in Fig. 4.15), where the highest PGA were recorded during the Tohoku earthquake (Fig. 4.16). The KiK-net stations were located at epicentral distances of ~ 178–316 km.

As seen from the figures, the intensity of motion and PGA do not noticeably change with distance within ~ 300 km from the fault, whereas the duration of strong motion decreases. At epicentral distances of ~ 270–310 km (at stations FKSH10, TCGH10, TCGH16, and IBRH11), the motion duration reaches its minimum, while PGA increase. At larger distances PGA and the intensity of motion sharply drop (Figs. 4.15 and 4.16).

The decrease of the duration of motion and the increase of PGA in sectors 1 and 2 (Fig. 4.16) are observed during the mainshock of the Tohoku earthquake. Evidently, these features refer to source effects, not path effects. In Fig. 4.17 similar graphs are shown for a smaller event, the aftershock of the Tohoku earthquake of 25 July 2011 ($M_w = 6.3$, $h \sim 46$ km) for the same KiK-net sites. The source of the event (small star in Fig. 4.15) was located just near the epicenter of the mainshock, and for this smaller event, there are no changes in the shapes of acceleration time histories with distance, similar to those shown in Fig. 4.16.

In cases of supershear crack propagation, shock waves are formed, because the distance between a seismic station and a crack tip decreases faster than *S*-waves propagate, and seismic waves radiated by the crack during sufficiently long time periods

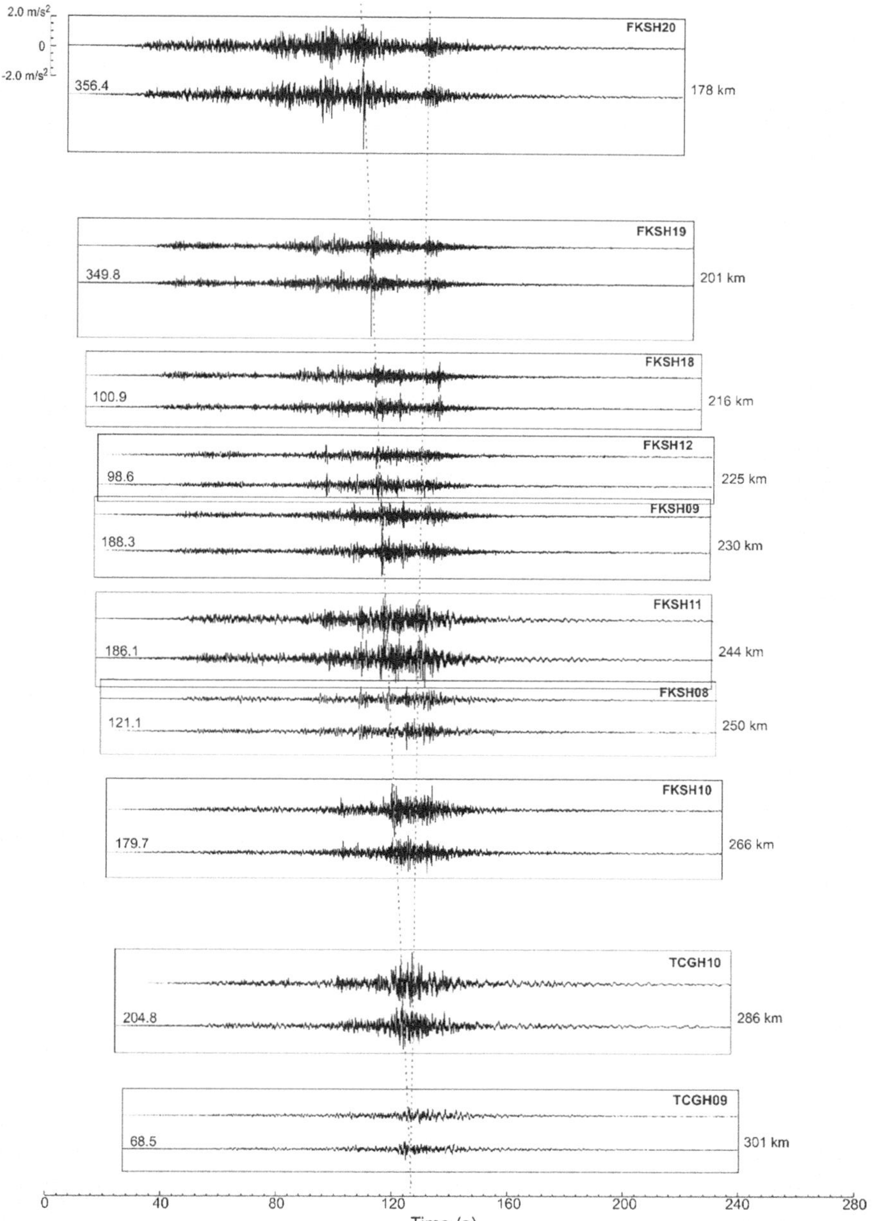

a

Fig. 4.16 The acceleration time histories (EW and NS components, depth records) of the 2011 Tohoku earthquake at KiK-net stations located in sectors 1 (**a**) and 2 (**b**). For each station, the epicentral distances (on the right) and PGA in cm/s^2 (on the left) are shown. Dotted lines show the decrease of the duration of strong motion with distance

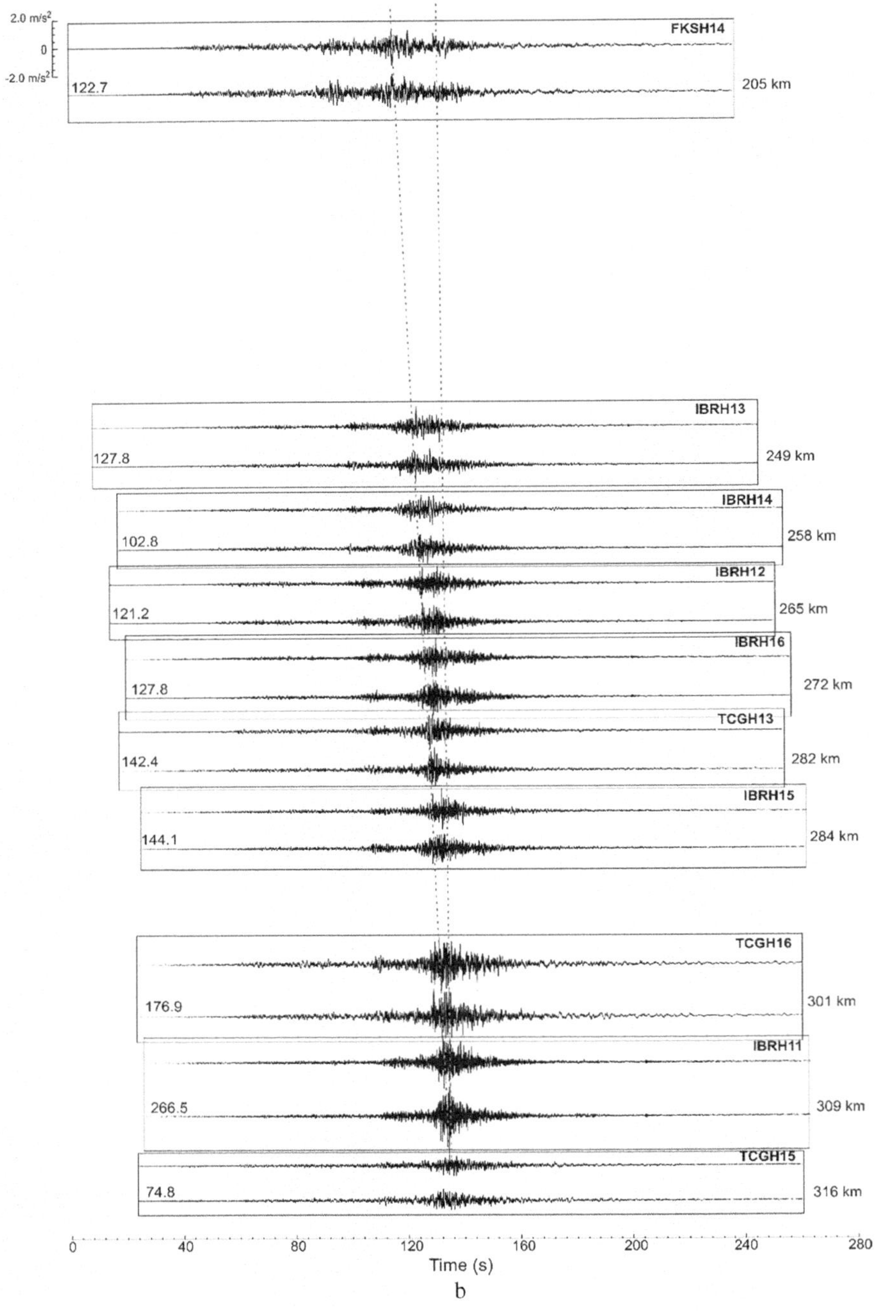

Fig. 4.16 (continued)

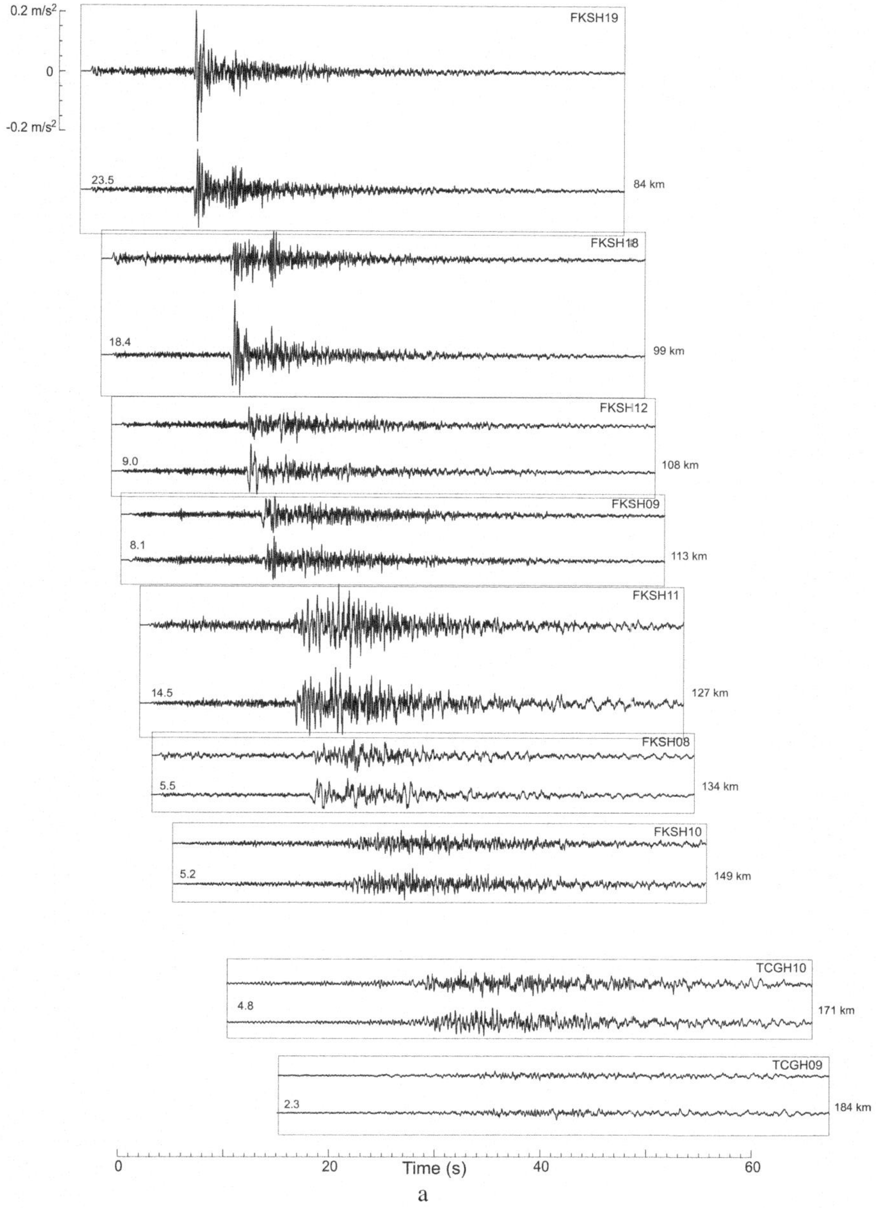

Fig. 4.17 The acceleration time histories (EW and NS components, depth records) of the aftershock of 25 July 2011, $M_w = 6.3$, at KiK-net stations located in sectors 1 (**a**) and 2 (**b**). For each station, the epicentral distances (on the right) and PGA in cm/s^2 (on the left) are shown

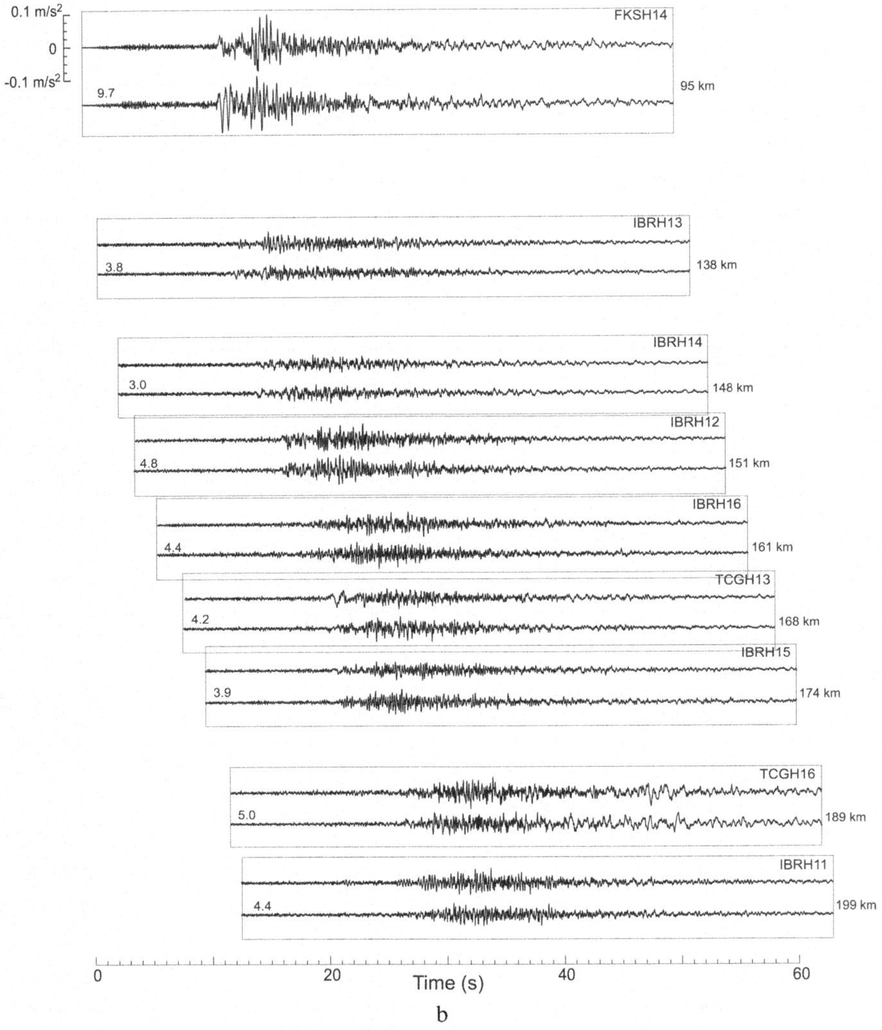

Fig. 4.17 (continued)

reach seismic stations almost simultaneously. Seismic waves interfere, overlap, and shock waves arise as a limiting case of the directivity effects.

During the Tohoku earthquake, the mechanism could be similar. The duration of strong motion reaches its minimum at the maximum intensity of motion at epicentral distances of ~ 290–310 km, where the shock wave arises. It disappears at larger distances: an abrupt decrease of PGA at ~ 300–310 km is clearly seen in the map (Fig. 4.15) and in Fig. 4.16 (at sites TCGH09 and TCGH15).

The cross-section of the fault plane of the earthquake and the locations of some KiK-net stations are schematically shown in Fig. 4.17. Simple calculations (Table 4.5) suggest that overlapping and interference of seismic waves could occur at the KiK-net sites during the Tohoku earthquake due to the orientation and dimensions of the fault plane, and the rupture velocity was not necessarily intersonic, it could have sub-Rayleigh values (0.7–0.9 V_S).

At sites located far from the epicenter (FKSH10, TCGH10, TCGH16, IBRH11, etc.), the distance between the station and the rupture tip decreased faster than shear waves propagated, and the waves radiated by the rupture during quite a long time arrived at the stations in shorter time intervals. The waves overlap, producing the shock wave front. At stations located closer to the fault plane (FKSH20, FKSH14) the distance between the station and the rupture tip did not decrease so fast, because it is largely dependent on the depth (Fig. 4.18). Seismic waves arrived to the stations in a usual order, and shock waves did not emerge.

Table 4.5 illustrates the calculations of *S*-wave propagation from the crack tip to three KiK-net stations: FKSH14, IBRH14 and IBRH11, located in forward direction. Travel times are calculated based on the distance between the crack tip and the stations and 4-layer velocity structure model approximated from the regional velocity model (Asano and Iwata 2012) (Fig. 4.18). The calculations were performed for time moments corresponding to the locations of the crack tip over successive time intervals of 7.2 s (points A, B, C, D, E) and the end point F. *S*-wave arrival times to the KiK-net stations were calculated as the sums of the moments of *S*-wave radiation by the crack tip and *S*-wave travel times from the crack tip to the stations.

As the calculations show, strong-motion durations that were observed at the KiK-net sites as results of crack propagation over a part of the fault plane (~ 13 s at FKSH14, ~ 6 s at IBRH14 and ~ 4 s at IBRH11) can be achieved with certain combinations of parameters describing the rupture length l (A–F in Fig. 4.18), rupture velocity at this part of the fault V, dip α, and the location and orientation of this part of the fault.

The part of the fault used in the calculations l is marked by thick dotted line in Fig. 4.15; it starts approximately at ~ 80 km from the epicenter and coincides with the area of high concentration of aftershocks, which is considered characteristic of areas of large fault slip, high stress drop, and the highest rupture velocity (Robinson et al. 2006). The calculations were performed with the following parameters selected in agreement with source models of the Tohoku earthquake: l ~ 120 km, α ~ 10°, and V ~ 4.0 km/s.

The estimates of strong-motion duration at the KiK-net sites located in sector 1 (~ 24 s at FKSH20, ~ 10 s at FKSH10, and ~ 5 s at TCGH10) also satisfy the observations, if we account for the locations of these sites several kilometers to the north from sector 2 and the direction of crack propagation. We assumed that the crack propagated towards sector 2 (Fig. 4.15), because the records of KiK-net stations located in sector 2 possess short duration and high intensity (Fig. 4.16b), and the shapes of the corresponding 3-component velosigrams and seismograms (Fig. 4.19) are similar to the shapes of the velosigrams and seismograms identified

Table 4.5 Duration of strong motion that was observed at KiK-net stations as a result of crack propagation over a certain part of the fault plane during the Tohoku event (l ~ 120 km, α ~ 10°, V ~ 4 km/s)

Time moment of *S*-wave radiation by a crack tip, sec	The projection of the distance between the crack tip and KiK-net stations on the *X* axis, km			Depth of the crack tip *h*, km	*S*-wave travel time from the crack tip to the KiK-net stations, sec			The time of *S*-wave arrival to the KiK-net stations, sec		
	FKSH14	IBRH14	IBRH11		FKSH14	IBRH14	IBRH11	FKSH14	IBRH14	IBRH11
0 (A)	115	168	219	30	30.4	42.4	53.9	30.4	42.3	53.9
7.2 (B)	86.5	139.5	190.5	35	24.3	36.1	47.5	31.5	43.3	54.7
14.4 (C)	58	111	162	40	18.4	29.9	41.3	32.8	44.3	55.7
21.6 (D)	29.5	82.5	133.5	45	14.0	23.9	35.0	35.6	45.5	56.6
28.8 (E)	1	54	105	50	12.8	18.7	29.1	41.6	47.5	57.9
30.2 (F)	– 4.7	48.3	99.3	51	13.2	18.0	27.8	43.4	48.2	58.0
					The duration of strong motion at FKSH14, IBRH14 and IBRH11:			13.0 sec	5.9 sec	4.1 sec

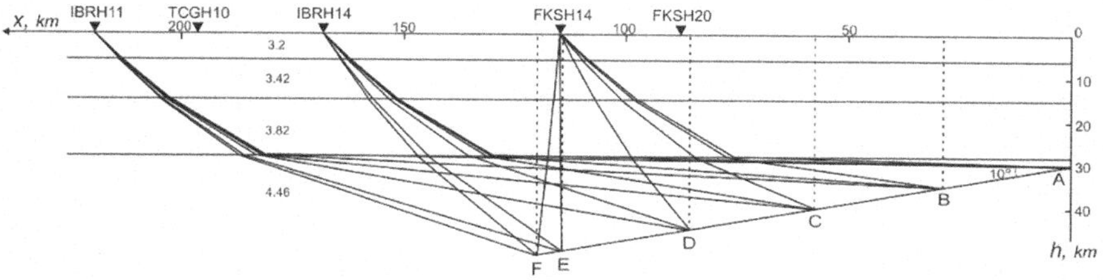

Fig. 4.18 The cross-section of the fault plane of the 2011 Tohoku earthquake, locations of KiK-net stations and seismic wave paths. The numbers indicate *S*-wave velocities in the layers in km/s; velocity structure model is taken from (Asano and Iwata 2012)

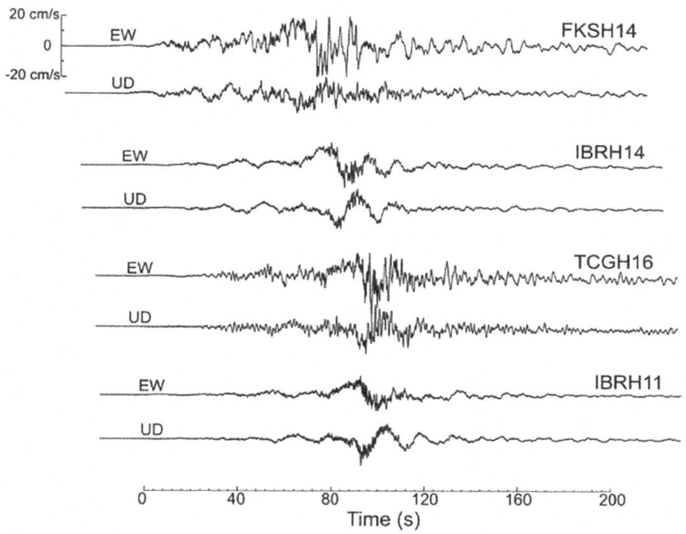

Fig. 4.19 Velocity time histories (EW and UD components, depth records) of the 2011 Tohoku earthquake at KiK-net stations FKSH14, IBRH14, TCGH16 and IBRH11

as shock waves during past earthquakes (Bouchon et al. 2001; Aagaard and Heaton 2004; Ellsworth et al. 2004; Zhan et al. 2013).

Source models for the Tohoku earthquake constructed by many researchers describe the source by a long-duration source process $\tau \sim 150$ s and a large rupture area of about 200 km × 500 km (Tajima et al. 2013). There are some indications that the rupture may have reversed itself and re-ruptured a portion of the fault (Tajima et al. 2013; Galvez et al. 2014). Obviously, the Tohoku earthquake had complex fault geometry, and even the slope of the fault varied with depth. Thus, mostly qualitative explanations of the mechanisms of high PGA generation are suggested, and the calculations shown in Table 4.5 illustrate the possibility of these mechanisms.

At the same time, some features of the source process are established more reliably and reinforce the explanations.

Based on various datasets, the researchers suggest patterns for the evolution of the source process, that give different results concerning the direction and rupture speed for the early portion of the earthquake; however, after about the first minute most results show unilateral rupture propagation toward the southwest (Tajima et al. 2013).

Ide et al. (2011) estimated the durations of the main phases of the source process: the initial phase, a deep rupture for up to 40 s, then extensive shallow rupture at 60 to 70 s, and continuing deep rupture lasting more than 100 s.

The isochrons in the source models by different authors show the spread of a large rupture in southwestern direction close to the dotted line in Fig. 4.16.

Early post-seismic events triggered by the Tohoku earthquake systematically propagated across southwestern Japan, associated with the strong seismic waves from the source (Miyazawa 2011). The propagation front travelled at 3.1–3.3 km/s and extended to a distance of 1350 km, suggesting dynamic triggering. There were much fewer observations of triggered earthquakes in the northern direction, where dynamic stresses were smaller.

According to many researchers, the fault plane of the 2011 Tohoku earthquake was distinctly separated into zones of high-frequency radiation down-dip and low-frequency radiation up-dip; therefore, it possessed substantial heterogeneity. The results of seismic tomography of the region point to noticeable variations of *P*- and *S*-wave velocities in the subducting slab and its environs in the epicentral area. The velocity variations evidently contributed to the nature of the subevents and the patterns of energy release during the Tohoku earthquake (Tajima et al. 2013).

Zhao et al. (2011) conclude that most of the largest foreshocks, aftershocks, and the mainshock of the Tohoku earthquake were located in the high-velocity patches or at the boundary between the low-velocity and high-velocity zones.

These considerable variations in physical properties of the source zone evidently produce variations in the velocities of rupture propagation in different parts of the fault. According to Ide et al. (2011), the deep rupture that occurred after about the first minute of the source process lasted more than 100 s; it propagated to a distance of ~ 250 km, i.e., about one half of the fault length. Then the average rupture speed (not accounting for heterogeneity of the fault plane) in the southwest direction was ~ 2.5 km/s, which agrees with Koketsu et al. (2011).

At the same time, Galvez et al. (2014) performed dynamic simulations of the 2011 Tohoku source process under the assumption that the medium is a half-space with *S*-wave velocity V_S ~ 3470 km/s, and obtained the estimates of the propagation speed of this rupture of ~ 3.5 km/s, which is close to V_S.

According to Zhao et al. (2011), the area of rupture propagation discussed here (A – F in Fig. 4.18) is located in the high-velocity zone. Accounting for the large size of the fault, the assumption that rupture velocity at a certain part of the fault (~ 120 km) was close to *S*-wave velocity (in agreement with Galvez et al. 2014) and reached ~ 4.0 km/s for ~ 30 s seems very likely. Past strong earthquakes that occurred in this part of the fault (Tajima et al. 2013; Satake et al. 2013) could weaken coupling, facilitate sliding, and accelerate the crack propagation.

Thus, the possible explanation of the anomalously high PGA and other features of the 2011 Tohoku earthquake is the following: seismic waves radiated by the crack tip during its propagation over a sufficiently long segment of the fault have all arrived to the remote stations TCGH10, TCGH16, IBRH11, and FKSH10 almost simultaneously, which resulted in the shock wave (amplification of motion at the bottoms of soil layers). The decrease in frequencies of motion observed at soft soil sites provides additional evidence for the compression and the following rarefaction of soft soils by the shock wave front.

Stations recorded the highest PGA experienced substantial amplification of strong motion at the bottoms of soil layers. At the same time, the shock wave hardened the soil layers and increased amplification of seismic waves in the layers. The recorded high PGA evidently result from the combination of these two factors.

This mechanism allows us to explain the anomalously high PGA observed during the Tohoku earthquake at sites with various (soft and dense) soil conditions and the other features of the Tohoku earthquake, such as, high PGA recorded far from the fault, sharp decrease of PGA at epicentral distances of 300–310 km in southwestern direction, soil hardening, observed changes in the shapes of acceleration time histories with increasing distance from the fault, step-like decrease of frequencies of motion at soft soil sites.

The waveforms of acceleration time histories in northern and central Honshu were also studied to check directivity effects similar to those found in southern Tohoku and Kanto.

Figures 4.20, 4.21, 4.22 and 4.23 show records of the Tohoku earthquake made by deep accelerometers of KiK-net stations located in sectors a, b, c, and d (Fig. 4.15). In sectors a and d, the intensity of motion and PGA values decrease with distance from the source, and the shapes of acceleration time histories are rather stable, whereas in sectors b and c, PGA do not definitely decrease with distance, and we can notice some decrease of the strong motion duration in the first group of waves in sector b. Thus, directivity effects are likely in northern Honshu in the first group of intense waves (Figs. 4.21 and 4.22).

The first group of intense waves is associated with one of the main ruptures on the fault plane. The isochrons shown in Fig. 4.15 reflect slip distribution on the fault plane according to the "joint" source model by Koketsu et al. (2011), constructed through joint inversions of teleseismic, strong motion, geodetic, and tsunami datasets. The authors conclude that the earthquake consisted of three main ruptures corresponding to the zones of large slips on the fault plane. After a small rupture in the initial 50 s, the first rupture expanded at a speed of ~ 1.8 km/s to the northeast and east; 20 s later the second rupture began to propagate westward at a speed of ~ 1.5 km/s and became dominant with the final slip of 36 m. The third rupture then played the leading role, propagating southward at a speed of 2.5 km/s.

The large second rupture on the fault plane propagating in western direction is responsible for the first group of intense waves recorded by northern stations. This rupture is clearly seen from tsunami waveform data as a "very large (~ 25 m) slip

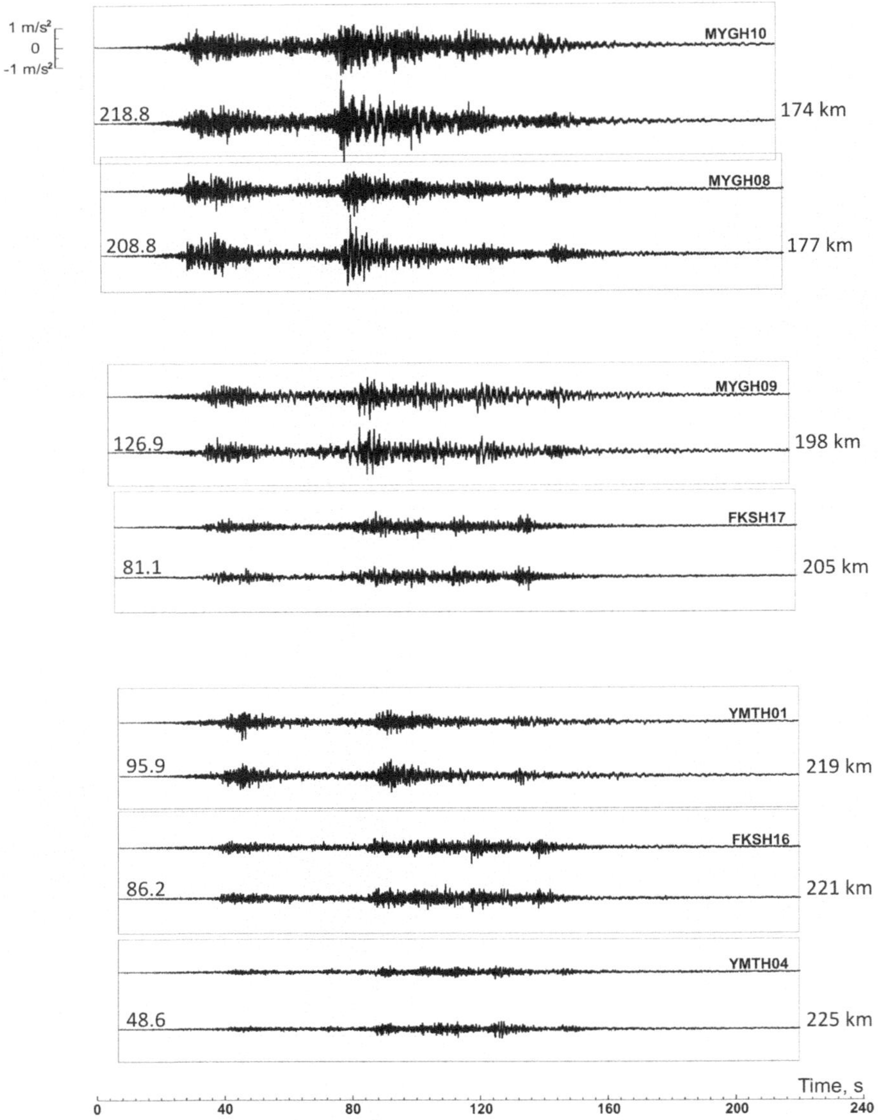

Fig. 4.20 The acceleration time histories (EW and NS components, depth records) of the 2011 Tohoku earthquake at KiK-net stations located in sector a (Fig. 4.15). For each station, the epicentral distances (on the right) and PGA in cm/s^2 (on the left) are shown

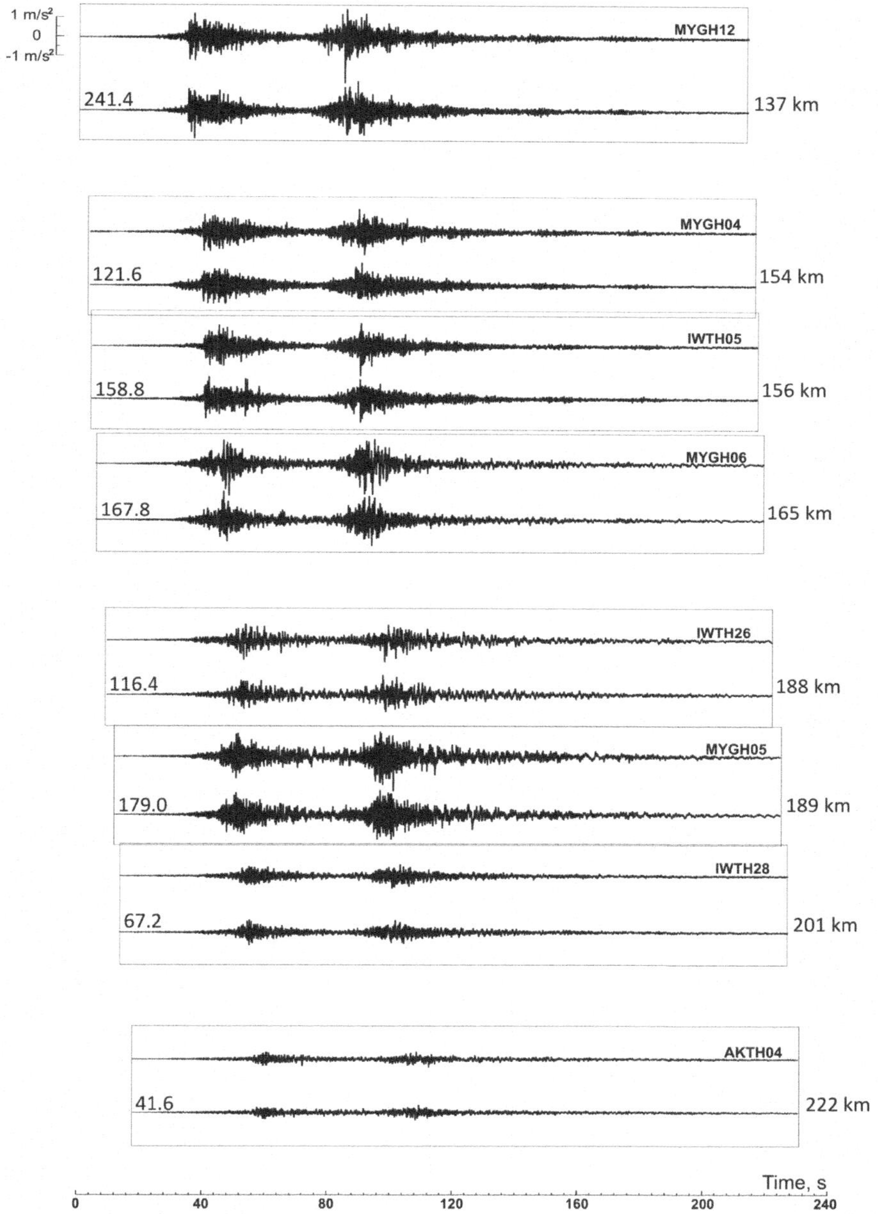

Fig. 4.21 The acceleration time histories (EW and NS components, depth records) of the 2011 Tohoku earthquake at KiK-net stations located in sector b (Fig. 4.16)

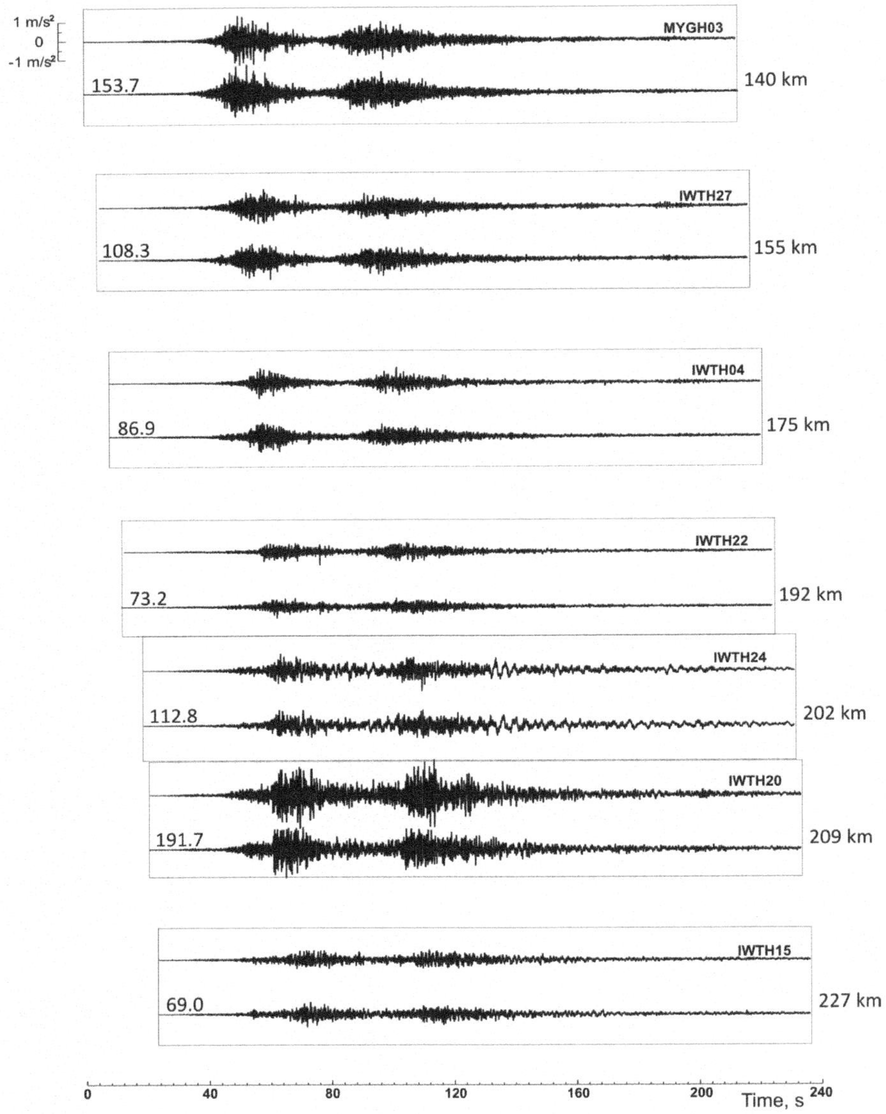

Fig. 4.22 The acceleration time histories (EW and NS components, depth records) of the 2011 Tohoku earthquake at KiK-net stations located in sector c (Fig. 4.15)

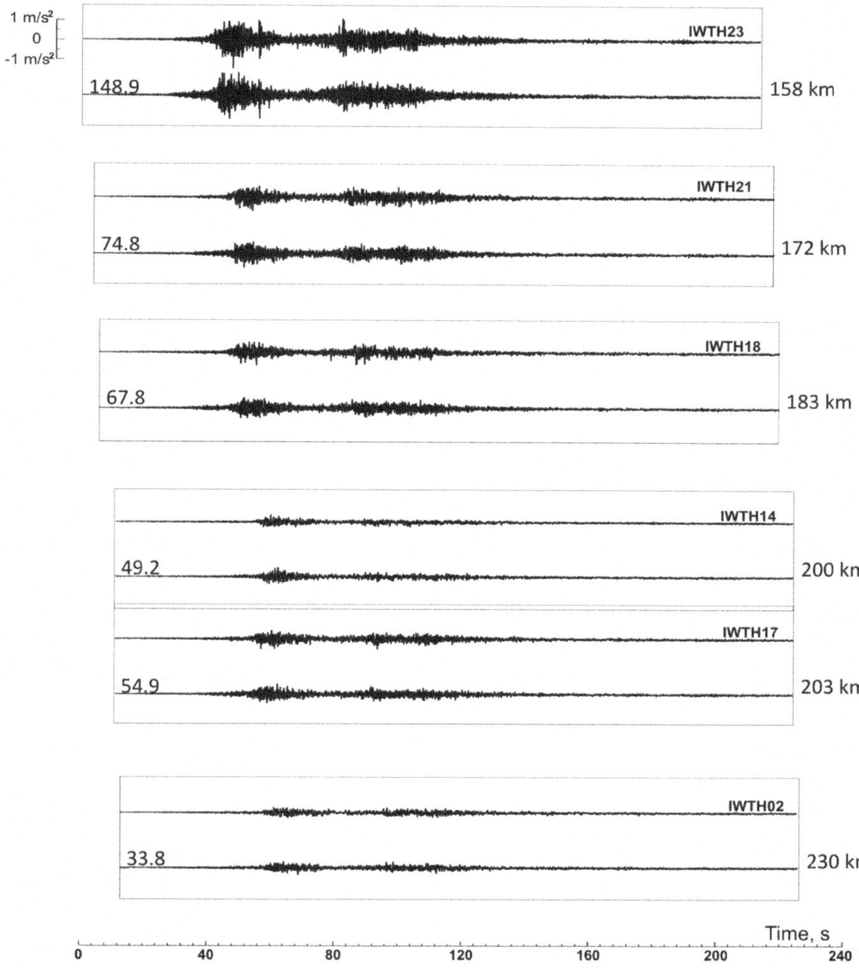

Fig. 4.23 The acceleration time histories (EW and NS components, depth records) of the 2011 Tohoku earthquake at KiK-net stations located in sector a (Fig. 4.15)

on the deep plate interface near Miyagi Prefecture" (Satake et al. 2013), and the directivity effects (seen in sector b) can be associated with this rupture.

These effects could lead to soil hardening during the passage of the first group of waves in sector b at sites MYGH12, MYGH04, IWTH05, IWTH26, MYGH05, where shear moduli in soil layers were high (Fig. 4.9).

However, high values of shear moduli at many sites in the near-fault zones during the passages of the two groups of intense waves suggest that there may be other mechanisms that lead to soil compaction and hardening.

The mega-thrust Tohoku earthquake caused long-duration ground motions, when soils experienced long-term dynamic loadings. Compaction and hardening can be the soil response to these long-lasting dynamic loadings.

As known, compaction is considered as a typical response of non-cohesive soils to dynamic loadings of moderate stress amplitudes (Voznesensky 1999), which is due to fundamental physical laws associated with the tendency of loose media to reduce their volumes. Stresses induced in soil layers by the strong motion range from units to hundreds of kPa. Loading stresses of units to tens of kPa correspond mostly to soil compaction mode, whereas higher stresses reaching hundreds of kPa usually correspond to soil loosening mode (Asaoka et al. 2012). Thus, vibrational compaction in soils possibly took place in the near-fault zones during the Tohoku earthquake. Soil hardening could occur during the passage of the second group of intense waves associated with the large rupture on the fault plane moving in southern direction, at sites IWTH04, MYGH03, MYGH04, IWTH27, IWTH23, IWTH21, and IWTH18, where shear moduli grew till the end of strong motion, showing continuous soil hardening (Fig. 4.9).

Thus, we should admit that, despite the success in studying in situ soil response in strong ground motion, numerous laboratory and field experiments, there still remain insufficiently studied issues concerning soil behavior in conditions of loading.

We cannot always correctly predict the behavior of certain soils in certain loading conditions; in particular, the behavior of dry loose sands, or dense water-saturated sands under long-term dynamic loadings, the behavior of dense water-saturated sands in drained conditions, etc. (Voznesensky 1999). Probably, data on strong motion from future strong earthquakes will help us in solving these problems.

4.3 Models of Soil Behavior and Generation of Abnormally High PGA During the 2003 Tokachi-Oki Earthquake

The subduction-type Tokachi-oki earthquake occurred off the coast of Hokkaido, Japan, on September 26, 2003, with the moment magnitude M_w ~ 8.0–8.3 and the focal depth of ~ 18–42 km. According to various authors, the maximum slip on the fault plane varied from ~ 4.3 m to ~ 13 m (Robinson and Cheung 2010). The earthquake caused extensive damage, landslides, soil liquefaction and tsunami up to 4 m in height.

The distribution of PGA in the near-fault zones shows similar mosaic pattern (Fig. 4.24) as observed during the Tohoku earthquake. The highest accelerations were recorded close to the epicenter, at HKD100 (988.4 Gal), HKD086 (800.6 Gal), and HKD092 (672.7 Gal) sites, as well as at remote sites KSRH03 (884.4 Gal), KSRH10 (613.0 Gal), NMRH02 (607.2 Gal), HKD066 (591.1 Gal), and HKD075 (560.8 Gal). Yagi (2004) and Koketsu et al. (2004) attribute some of these high accelerations to the effects of directivity.

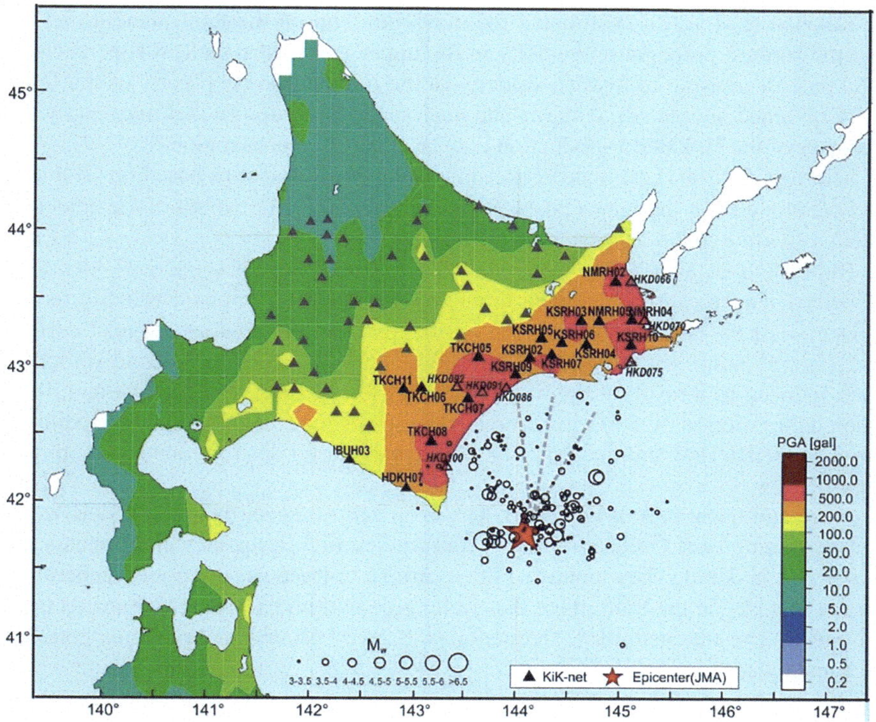

Fig. 4.24 Map showing the distribution of PGA recorded during the 2003 Tokachi-oki earthquake (NIED data), locations of the mainshock (star), aftershocks (circles) ($M_w > 3$) occurred in 24 h after the mainshock (JMA data), and KiK-net stations (triangles). Dotted lines show horizontal projections of the rupture propagation towards KSRH09, KSRH06, and KSRH10 sites

The source process was studied by many research groups. Yamanaka and Kikuchi (2004) estimated source parameters based on teleseismic *P*- and *SH*- waves and concluded that the rupture propagated northward from a shallow to a deep region. Their results suggest that the 2003 Tokachi-oki earthquake was a recurrent event of the 1952 Tokachi-oki earthquake ($Ms \sim 8.2$).

Yagi (2004), Koketsu et al. (2004) and Honda et al. (2004) also estimated source parameters based on teleseismic body waves and strong motion records (Yagi 2004), strong motion records (Honda et al. 2004), geodetic data and strong motion records (Koketsu et al. 2004). The authors used strong motion records at K-NET and KiK-net sites located in the near-fault zones, at 12–15 sites. Yagi (2004) found that a rupture front velocity of 4.5 km/s in the middle and deep parts of the fault shows the best fit to the observations. Honda et al. (2004) identified three main slip areas: around the hypocenter, in the northwest part of the fault (with a maximum slip of 5.9 m), and in the northeast part of the fault, the longest in time.

Koketsu et al. (2004) estimated the maximum slip on the fault plane as 7.1 m, and the rupture propagation velocity in the upper part of the fault as supershear—4.3 km/s, decreasing to *S*-wave velocity in the middle and deep parts of the fault. They obtained simple slip patterns and near-supershear rupture that may imply the maturity of the Hokkaido subduction zone around the source region.

Miura et al. (2004) estimated coseismic and postseismic slips based on GPS data and concluded that the major postseismic slip extended from south of the epicenter about 180 km to the northeast.

Rubinstein et al. (2007) based on repeating earthquake sequences identified velocity changes caused by the earthquake. They found significant velocity reductions up to 0.3% close to the receivers, evidently caused by damage to near-surface materials by nonlinear strong motion. The second region where seismic velocities were reduced as a result of the mainshock was the rupture zone. Hatayama (2008) studied long-period (4–8 s) waves induced by strong motion in the Yufutsu sedimentary basin (IBUH03 station area, Fig. 4.24) that caused destructions of oil storage tanks.

Nozu and Irikura (2008) inverted the waveforms of the earthquake at 39 KiK-net sites with empirical Green's functions and studied high-frequency components (up to ~ 1 Hz) in detail. They identified the locations of three strong motion generation areas (SMGA) on the fault plane that well agree with the asperities identified from the inverted results using low-frequency (< 0.2 Hz) ground motions plus geodetic data and tsunamis.

Robinson and Cheung (2010) reviewed source parameters obtained by many researchers and estimated the parameters themselves based on broadband *SH*-wave seismograms; they concluded that the rupture occurred at a depth of ~ 18 km, in agreement with Yagi (2004), and the maximum slip was about 12 m.

As we can see from the review, the researchers mainly concerned the source mechanism, they did not comment the occurrence of high PGA at remote sites KSRH03, KSRH10, NMRH02 and others. The only work by Nozu and Irikura (2008) discusses the records at KSRH03 site. The simulations give lower amplitudes at KSRH03 site than those observed, and to correct this, the authors introduce SMGA at the edge of the fault just near KSRH03, which slightly corrected the situation.

Here we try to explain the distribution of PGA in the near-fault zones of the 2003 Tokachi-oki earthquake. To do this, we study soil behavior at KiK-net sites and changes of shear moduli in soil layers during strong motion and then we interpret the results taking into account the effects of source directivity.

Many KiK-net sites in the near-fault zones are located on soft soils, and the records provide us information on soft soil behavior during long-lasting strong motion.

Figure 4.24 shows the locations of the mainshock, aftershocks occurred during the first 24 h after the mainshock, and 18 KiK-net stations, at which soil behavior was studied. Table 4.6 provides information on the studied sites: the coordinates, epicentral distances, soil conditions (V_{S30}, the thickness of surface softer layers d and *S*-wave velocities in the layers V_S), recorded PGA, borehole depth, amplification. As seen from the Table, KiK-net stations recorded the highest PGA are located mostly on soft soils.

Table 4.6 Information on KiK-net stations, for which models of soil behavior during the 2003 Tokachi-oki earthquake were constructed

Site code	Latitude(°)	Longitude (°)	d/V_S	V_{S30} (m/s)	Epic. dist (km)	PGA surface (Gal)	PGA depth (Gal)	Bor. depth (m)	Ampification
TKCH08	42.49	143.15	4/130	353	80	539.8	131.9	103	4.0
HDKH07	42.13	142.92	2/190	495	84	207.7	94.9	104	2.3
TKCH07	42.81	143.52	38/120	140	89	414.0	170.5	103	2.7
KSRH09	42.99	143.98	10/118	230	102	454.6	164.4	103	2.6
TKCH06	42.89	143.06	2/130	300	115	192.6	91.3	230	2.0
KSRH02	43.11	144.12	18/162	219	116	462.0	173.3	108	2.5
TKCH05	43.12	143.62	6/140	337	118	497.4	128.3	103	3.6
KSRH07	43.14	144.33	14/125	204	122	538.0	219.7	225	2.1
TKCH11	42.87	142.88	1/150	459	124	316.2	77.7	103	3.5
KSRH06	43.22	144.43	2/90	326	133	394.5	126.1	240	3.1
KSRH05	43.26	144.23	4/100	389	133	347.6	106.6	333	3.1
KSRH04	43.21	144.68	30/189	189	140	362.2	163.9	243	2.3
KSRH03	43.38	144.63	16/192	250	155	884.4	161.1	110	3.7
KSRH10	43.21	145.12	36/224	213	159	613.0	134.2	258	5.2
NMRH05	43.39	144.80	20/174	209	161	406.6	139.7	223	2.8
NMRH04	43.40	145.12	146/258	168	175	518.3	155.1	219	3.0
IBUH03	42.65	141.86	76/174	111	182	377.9	58.9	156	5.9
NMRH02	43.68	144.96	10/181	315	195	607.2	114.8	106	5.2

Models of soil behavior, such as, vertical distributions of stresses and strains in soil layers changing with time during strong motion are constructed using the methods described in Chap. 3.

At all the studied sites soil behavior was described by hard-type stress–strain relations down to the locations of the deep devices; the best-fit nonlinear stress–strain relations were selected from the series of 250 parametric curves. Records of two horizontal components were analyzed, and calculations were performed in successive time intervals of 1.5 s duration.

Thus, soil behavior at the KiK-net sites during the earthquake was simulated in 50 1.5 s time intervals (75 s of strong motion), i.e., stress–strain relations in soil layers were found that reflect the features of soil behavior during the earthquake.

The constructed models of soil behavior at sites with softer and denser soils are shown in Figs. 4.25a–l and 4.26a–f. We consider sites with $V_{S30} > 300$ m/s (Table 4.6) as denser soil sites; these sites possess mostly gravelly soils in the upper layers. Softer soil sites ($V_{S30} < 300$ m/s) contain silts, sands, clays, and volcanic ash in the upper layers.

As seen from the figures, a fairly good agreement was obtained between the calculations and observations. Differences in the behavior of softer and denser soils are observed only at sites located close to the epicenter. TKCH07, KSRH09 and TKCH06 sites indicate nonlinear behavior of soft soils (Fig. 4.25a–c), whereas, TKCH08, HDKH07, TKCH05, TKCH11 sites demonstrate virtually linear behavior of denser soils (Fig. 4.26a–d).

Soft soil behavior becomes more linear with increasing distance from the fault (Fig. 4.25e–l). At TKCH07 and KSRH09 sites, there are signs of liquefaction of sandy water-saturated soils in the upper 14 m (at TKCH07) and in the upper 6 m (at KSRH09) and nonlinear behavior of sandy soils in deeper layers (Fig. 4.25a, b). Quasi-linear behavior of soft soils at remote sites means low nonlinear damping, which agrees with high accelerations recorded at some of them.

At KSRH07, KSRH03, KSRH10, and NMRH02 sites at the moments of the highest intensity of motion, soil behavior is described by quasi-linear stress–strain relations indicating soil hardening (Fig. 4.25e, g, h, l). Similar soft soil behavior was observed during the 2011 Tohoku earthquake.

The estimates of shear moduli in soil layers changing with time during strong motion (75 s) at softer and denser soil sites are shown in Fig. 4.27a, b. The estimates should be considered as approximate ones, because they are based on the models of soil behavior, approximating real soil behavior; however, we can notice some clear regularities: shear moduli grow and fall during strong motion, i.e., soft and dense soils experience hardening and softening; each cycle lasts ~ 10–15 s. At the majority of sites, we observe soil softening after reaching the highest intensity of motion, which is clearly seen at sites KSRH09, KSRH02, KSRH07, KSRH03, KSRH10, TKCH08 and others (Fig. 4.27a, b).

At the beginning of strong motion, shear moduli estimation is more difficult, because the motion is weak (and several stress–strain relations can fit the records); fairly reliable estimates can be obtained after ~ 15–20 s of motion. At these time intervals shear moduli are usually high (~ 0.8), and they reduce down to ~ 0.2 at the final parts of strong motion.

At remote soft soil sites recorded high PGA, such as, KSRH04, KSRH10, NMRH05, NMRH04 (Fig. 4.24; Table 4.6), shear moduli variations are rather simple: they are high (~ 0.8–1.0) at the beginning of motion and reduce to ~ 0.2 after reaching the highest intensity of motion (Fig. 4.27a). Similar behavior was observed at remote soft soil sites recorded high PGA during the 2011 Tohoku earthquake.

During the Tohoku earthquake, such remote soft soil sites showed a stepwise decrease of predominant frequencies of motion on the surface from ~ 4 to ~ 2 Hz after reaching the highest intensity of motion.

During the Tokachi-oki earthquake many KiK-net sites located on soft soils also show a decrease of the predominant frequencies of motion from 6–8 to ~ 1–2 Hz (Fig. 4.28) after reaching the highest intensity of motion.

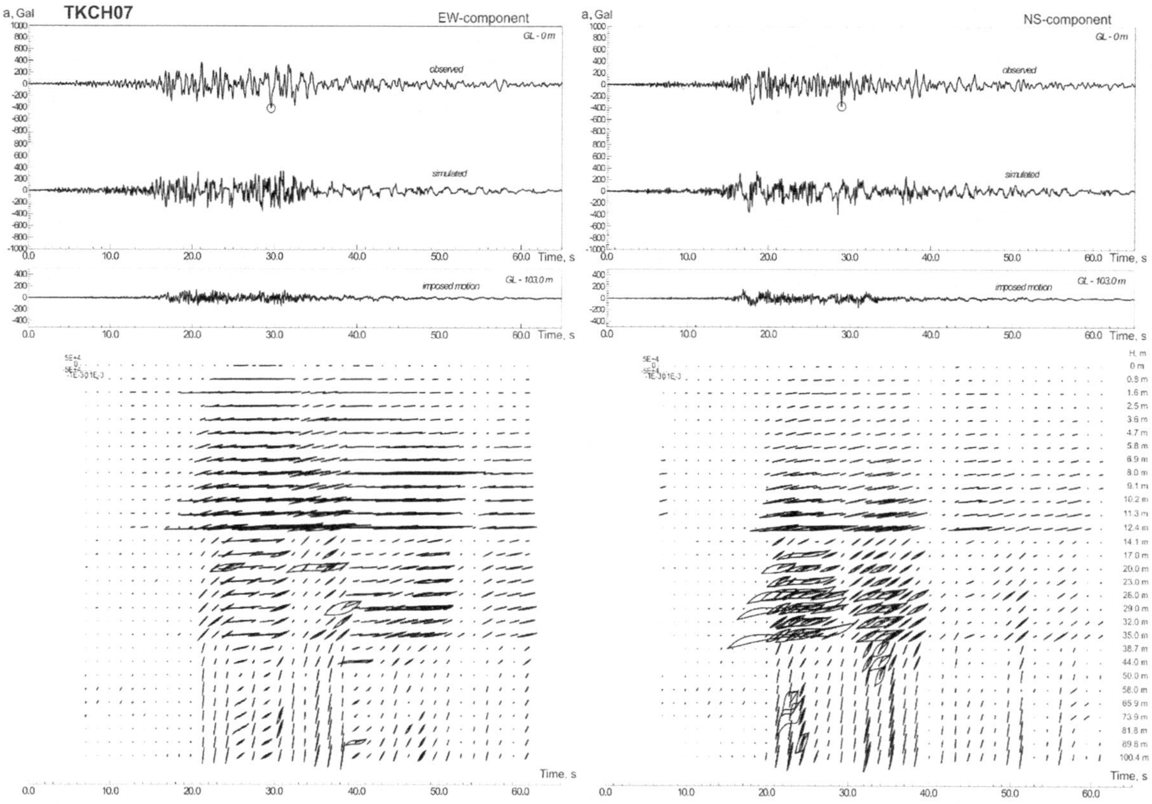

a

Fig. 4.25 Acceleration time histories of the 2003 Tokachi-oki earthquake, observed and simulated at soft soil sites and estimated stress–strain relations in soil layers changing with time during strong motion. Stresses are given in Pa, strains in strain

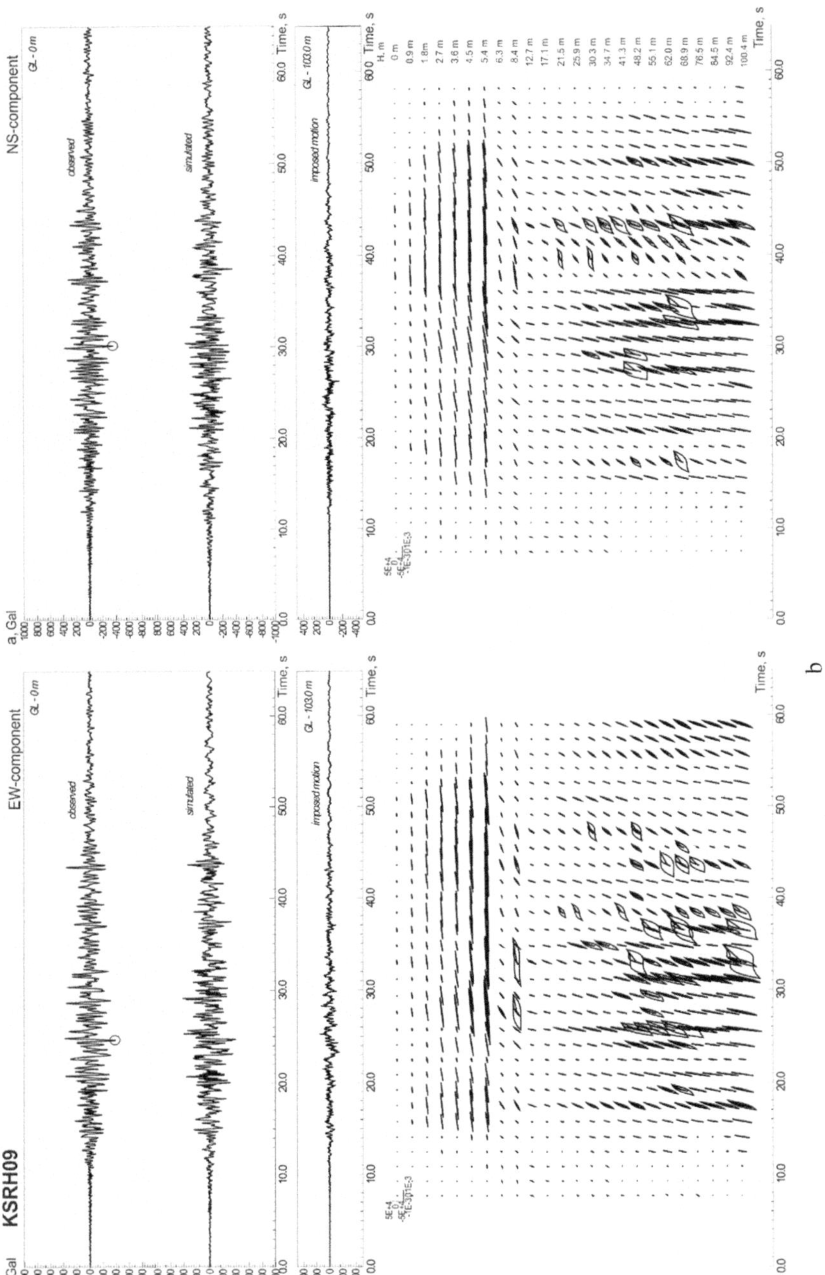

Fig. 4.25 (continued)

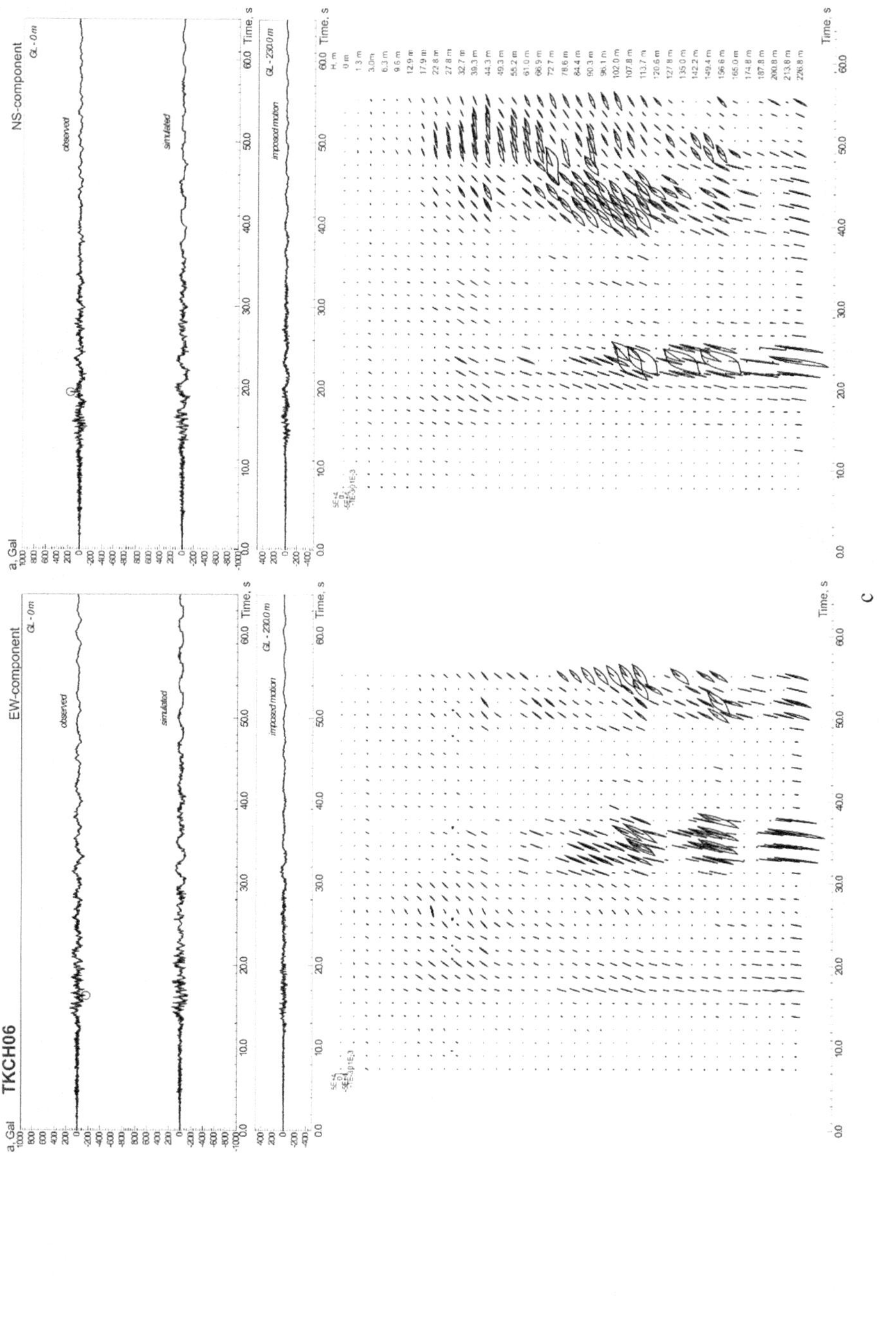

Fig. 4.25 (continued)

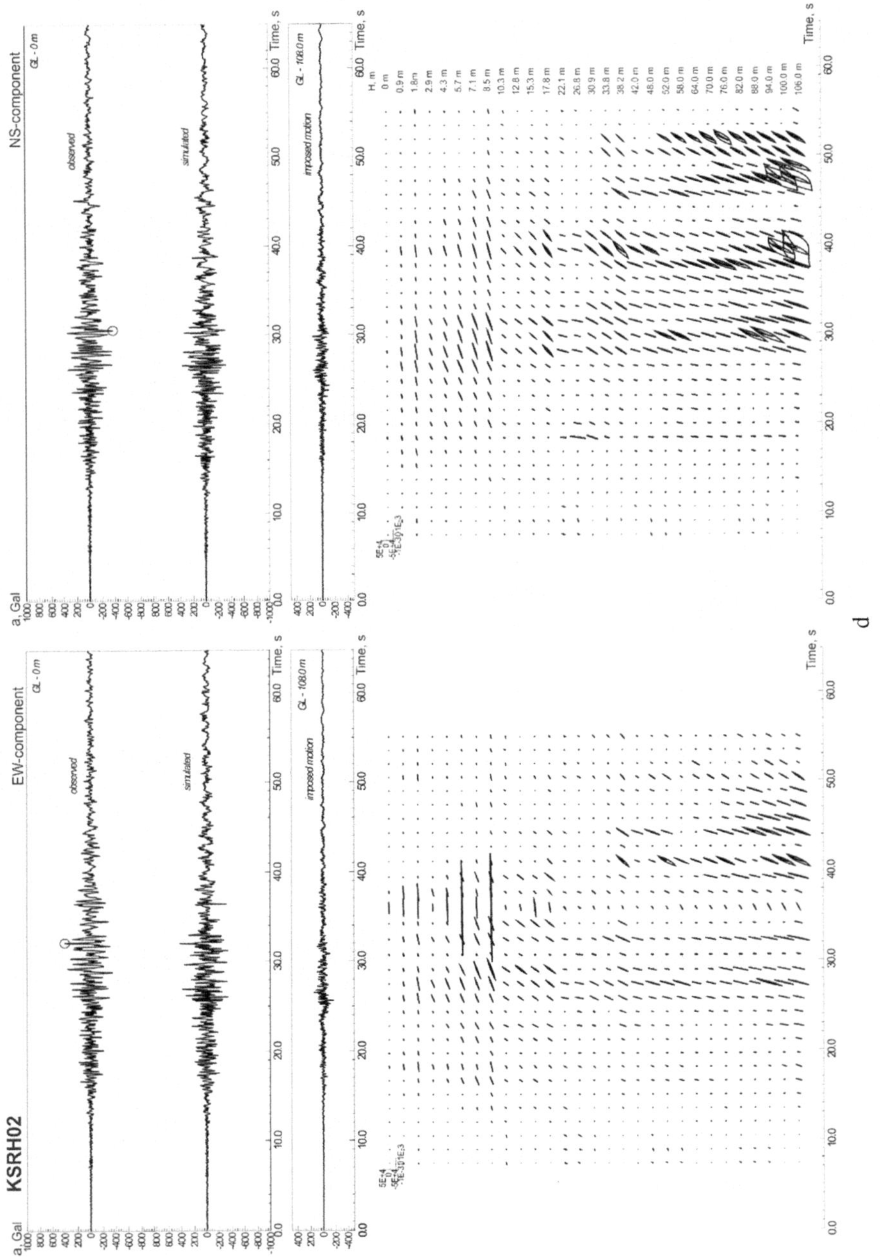

Fig. 4.25 (continued)

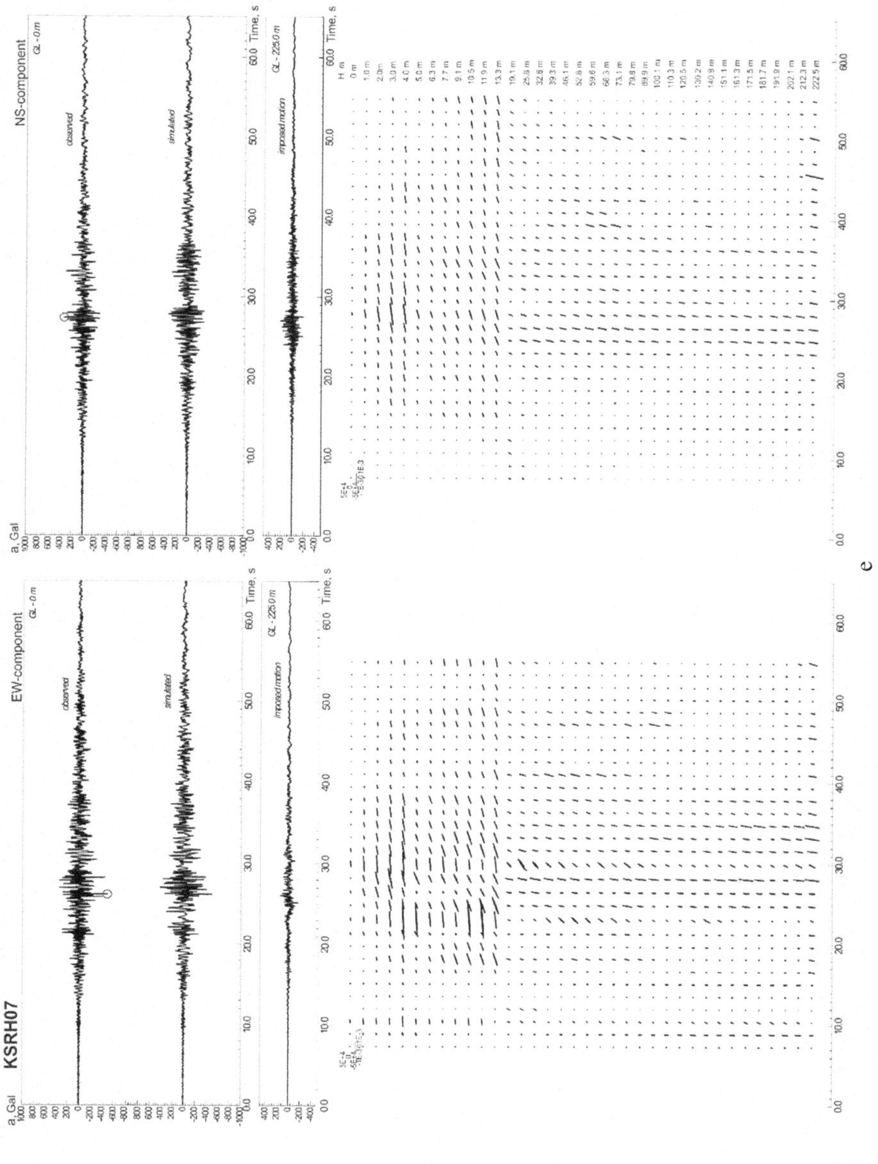

Fig. 4.25 (continued)

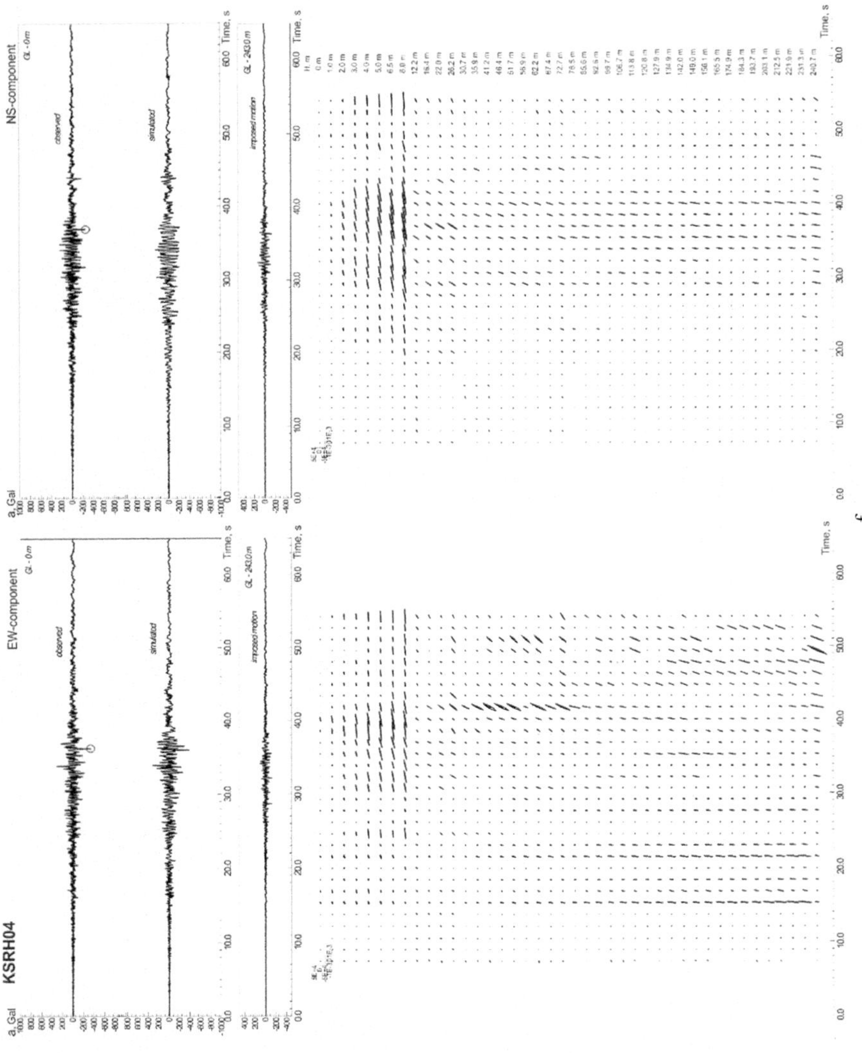

Fig. 4.25 (continued)

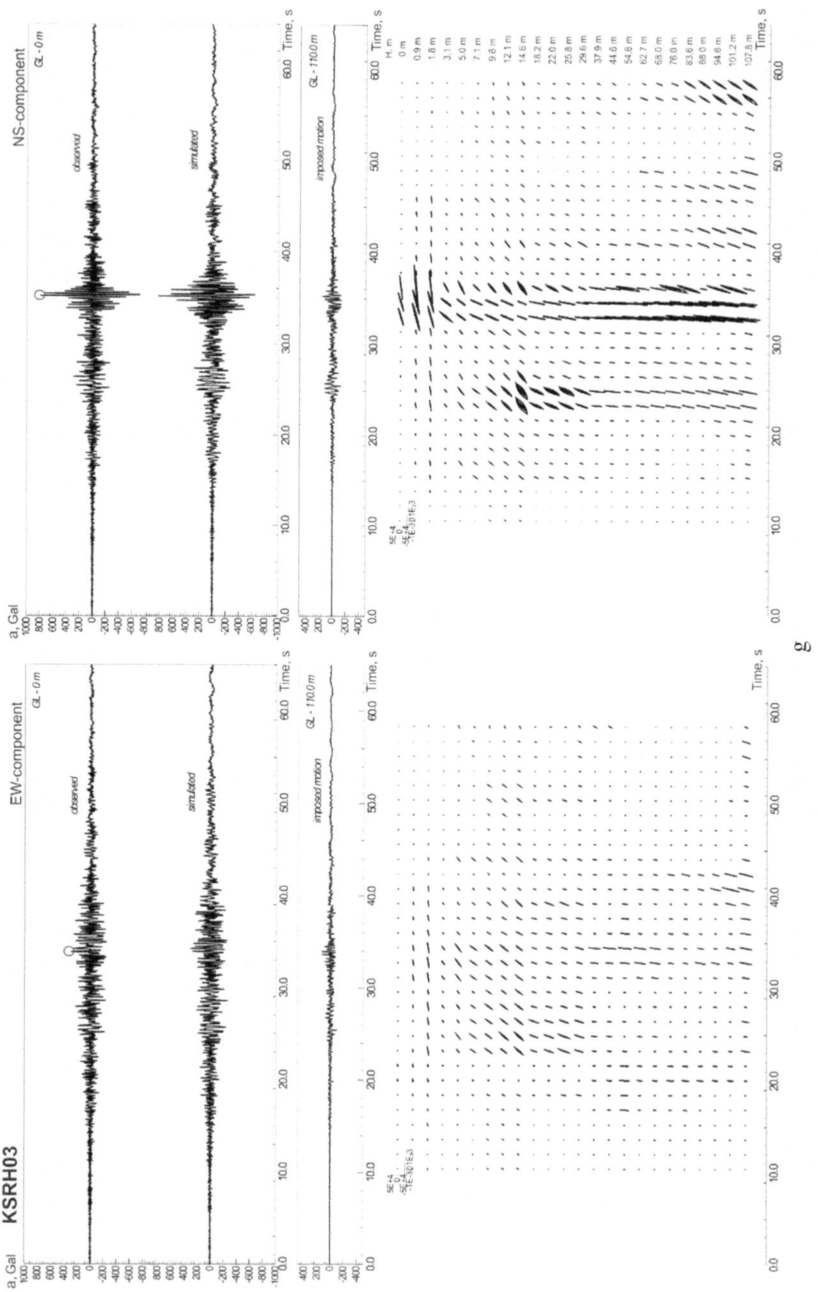

Fig. 4.25 (continued)

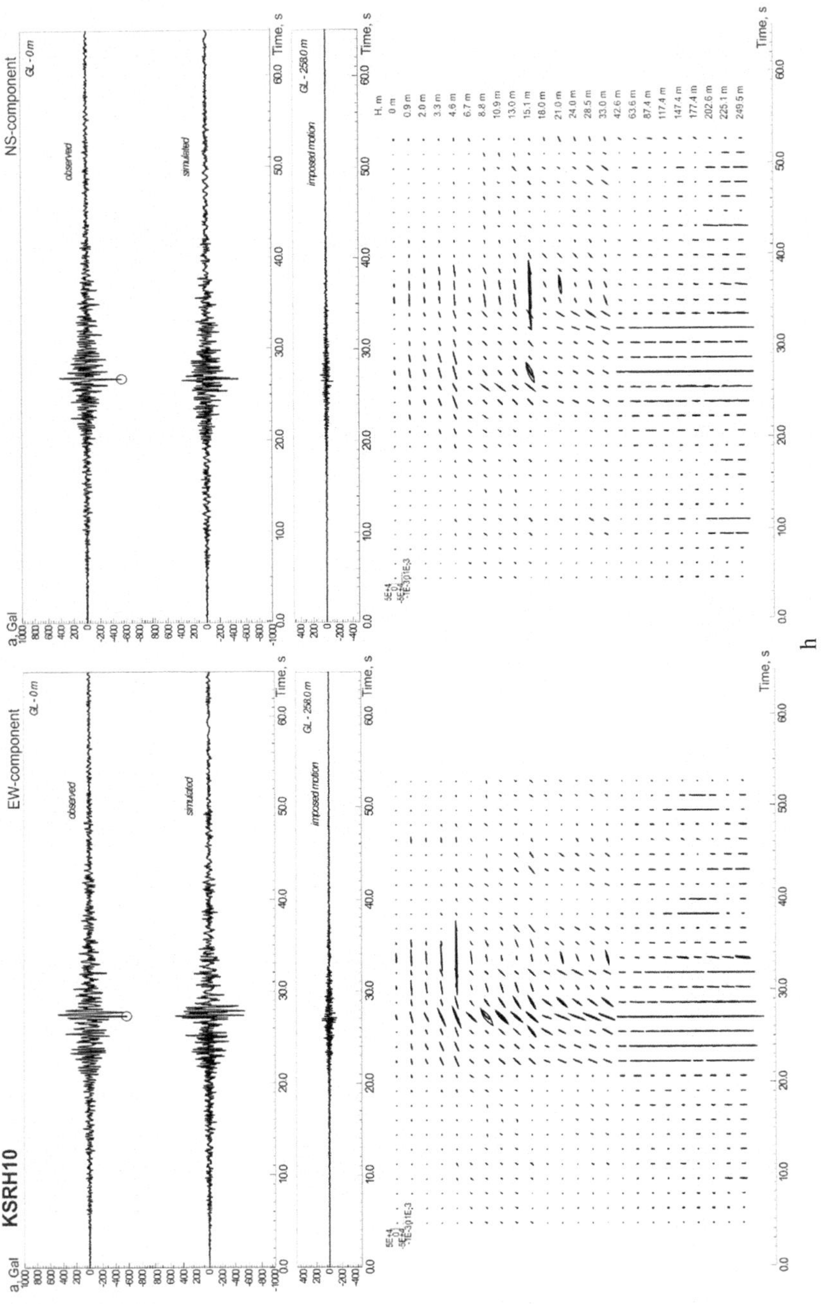

Fig. 4.25 (continued)

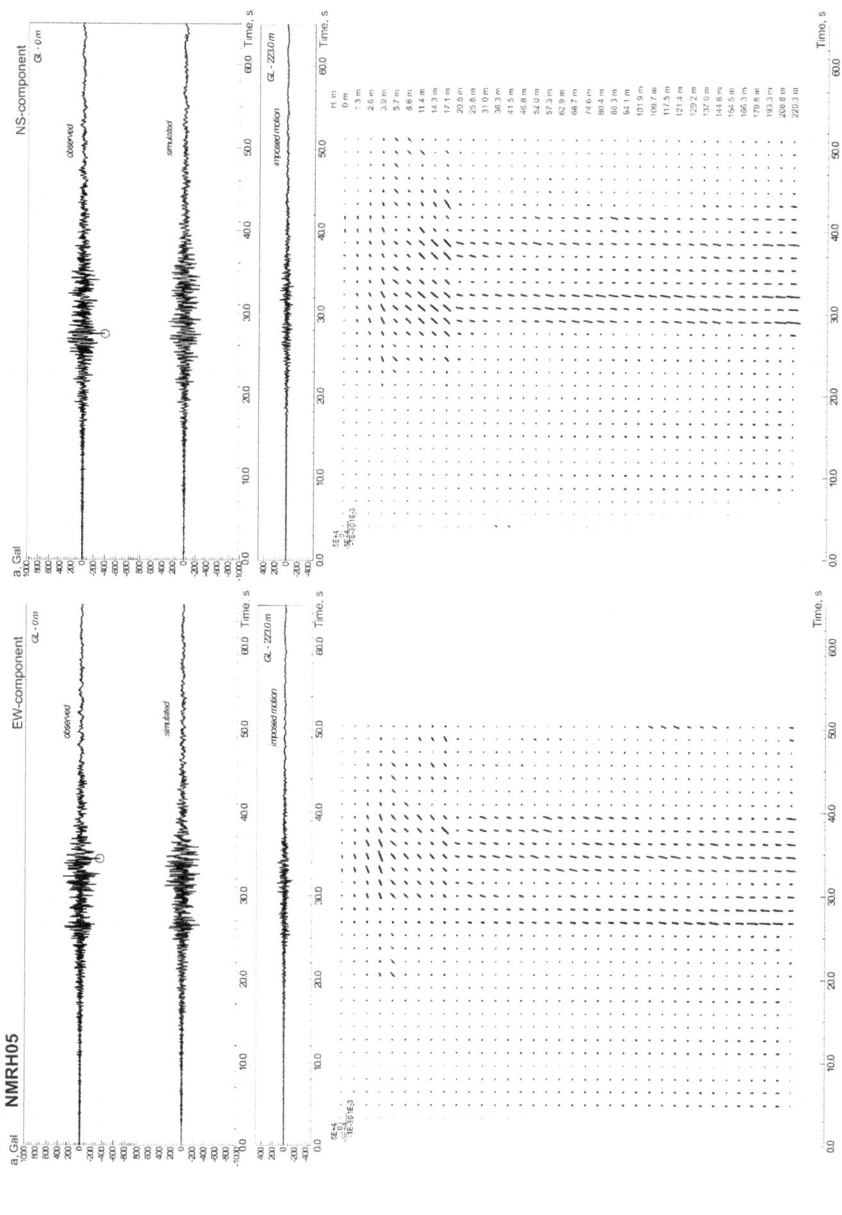

i

Fig. 4.25 (continued)

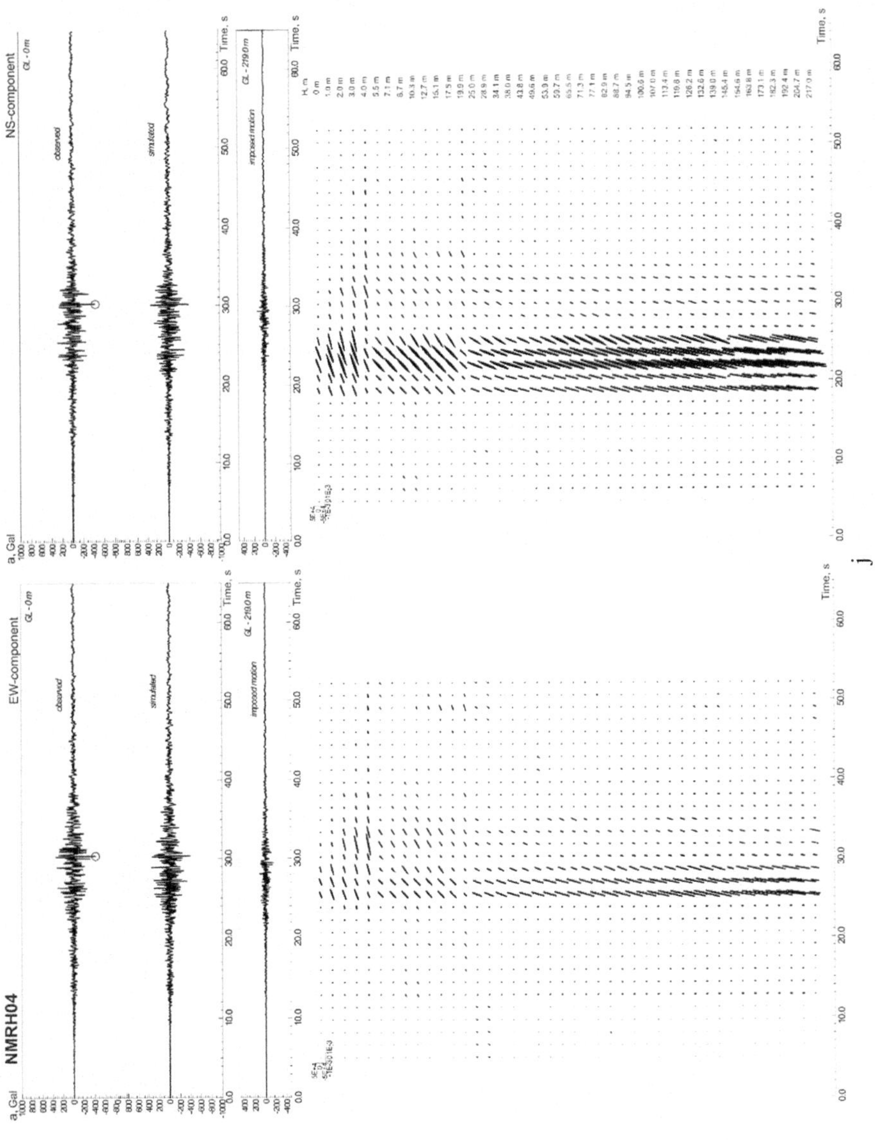

Fig. 4.25 (continued)

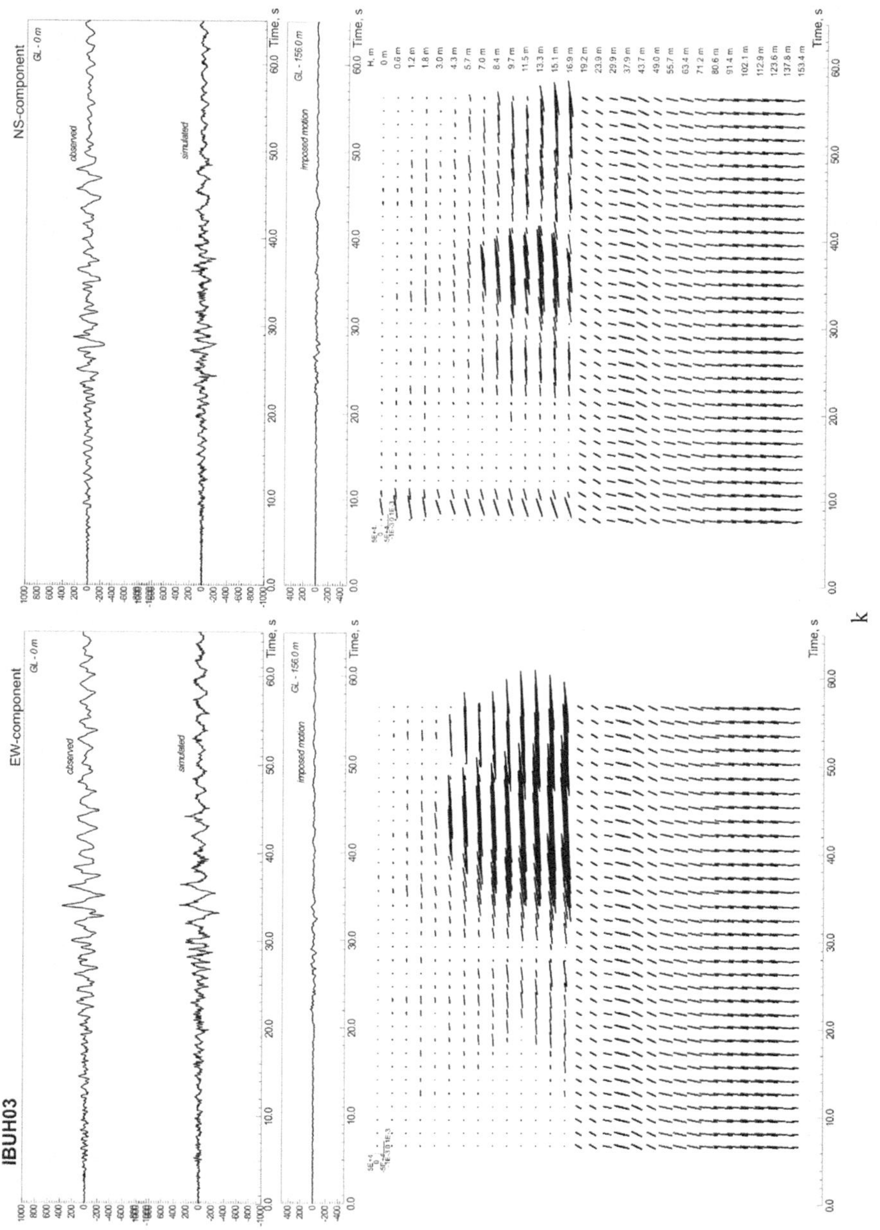

Fig. 4.25 (continued)

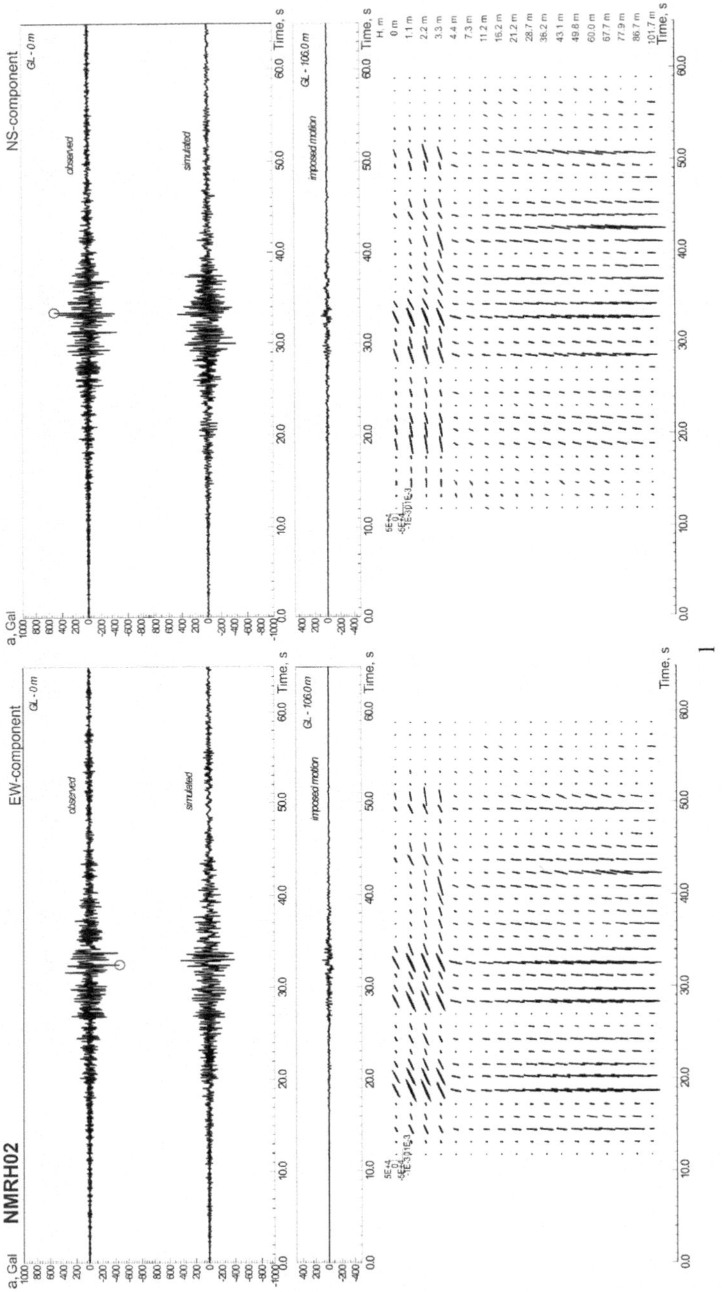

Fig. 4.25 (continued)

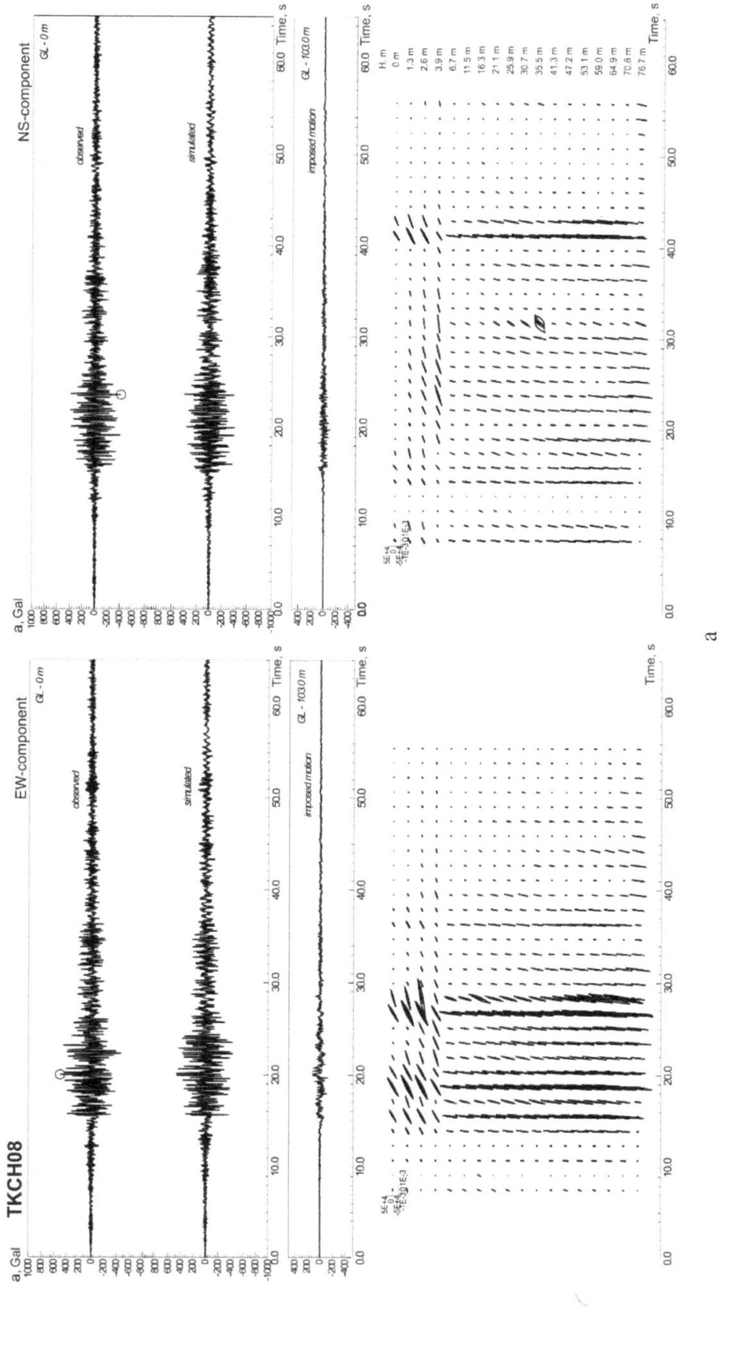

a

Fig. 4.26 Acceleration time histories of the 2003 Tokachi-oki earthquake, observed and simulated at denser soil sites and estimated stress–strain relations in soil layers, changing with time during strong motion. Stresses are given in Pa, strains in strain

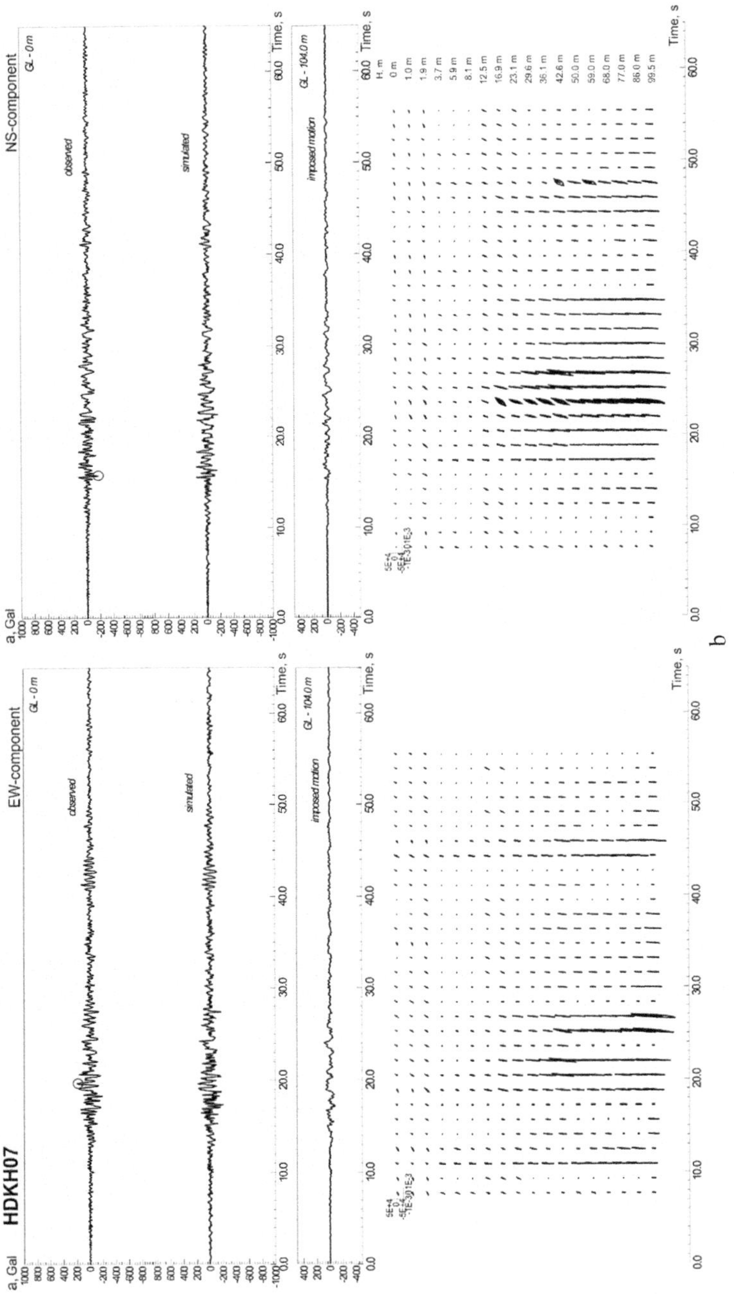

Fig. 4.26 (continued)

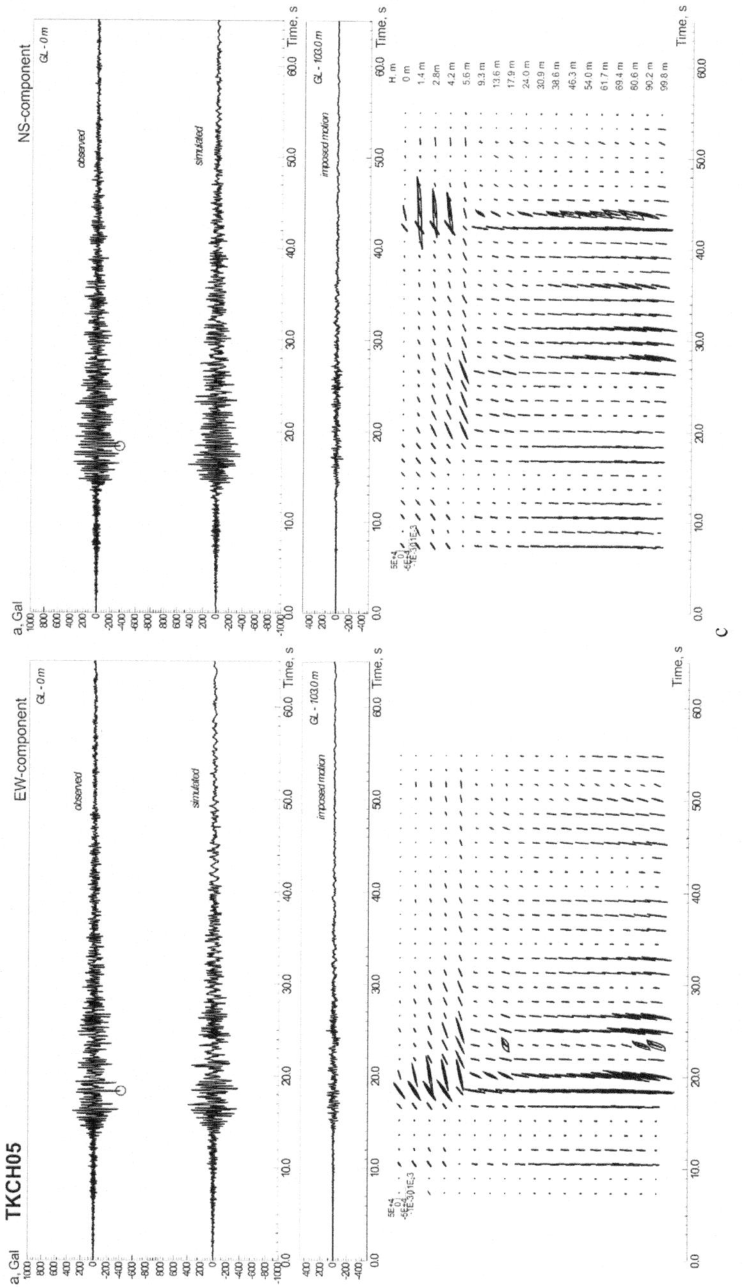

Fig. 4.26 (continued)

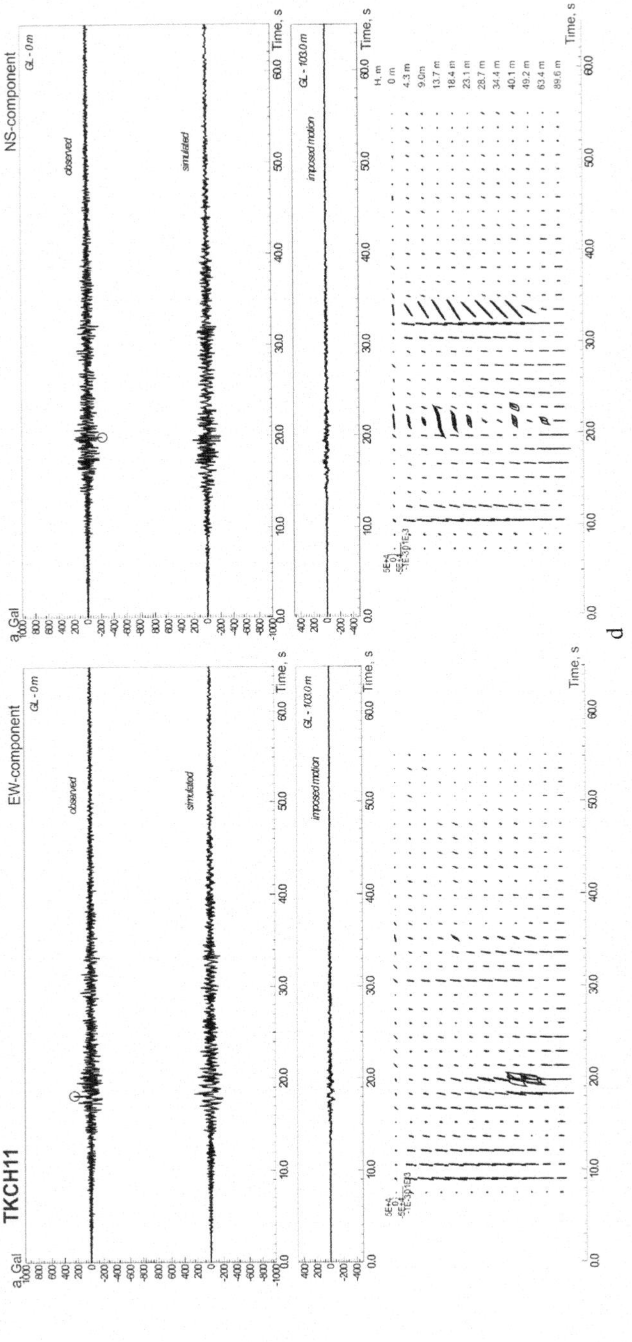

Fig. 4.26 (continued)

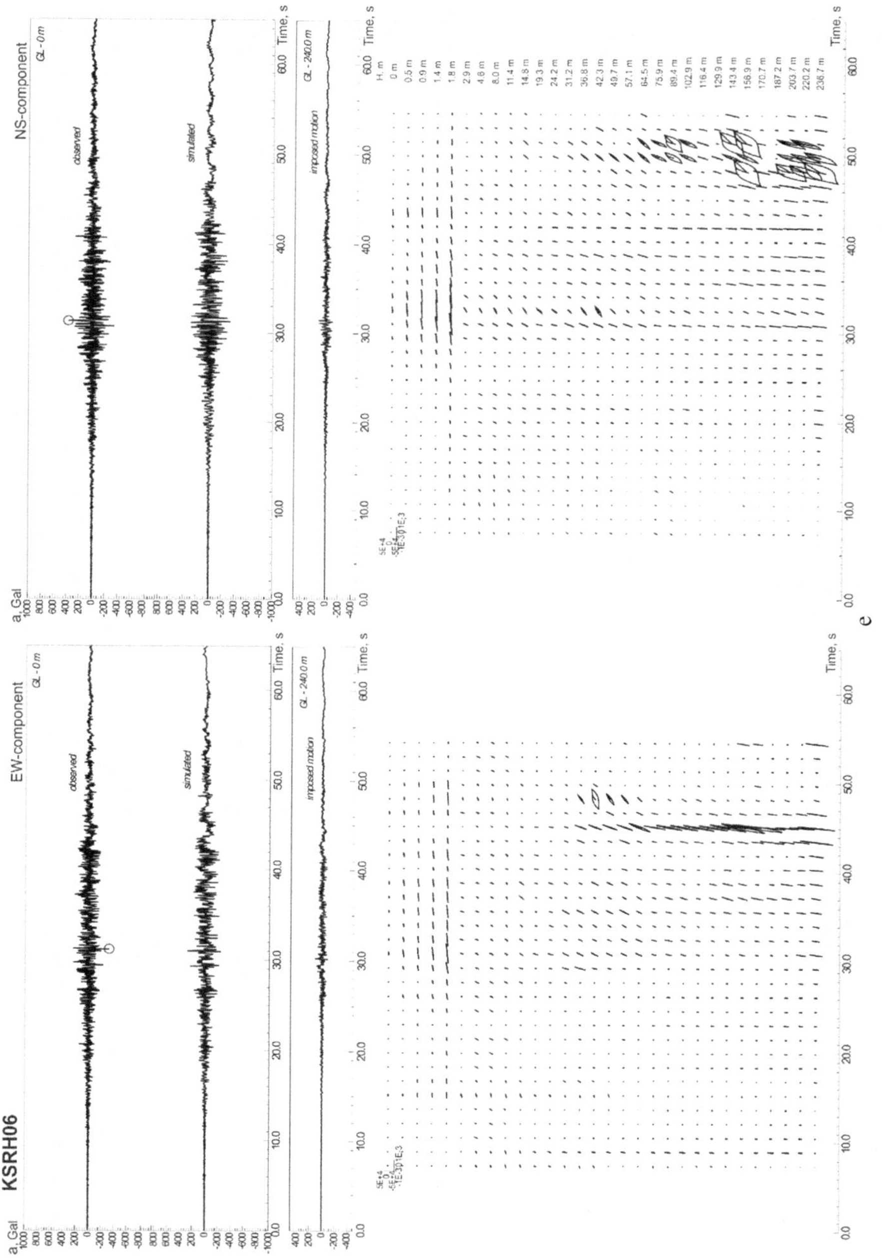

Fig. 4.26 (continued)

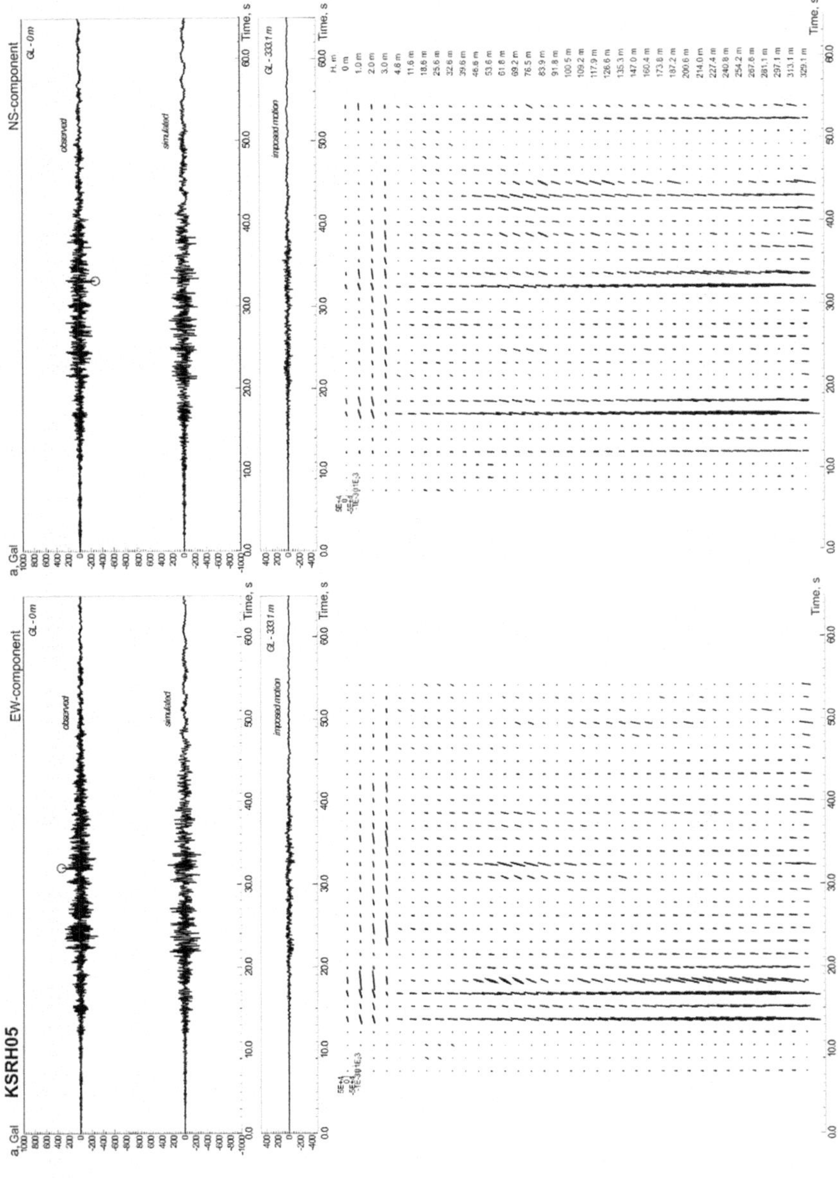

Fig. 4.26 (continued)

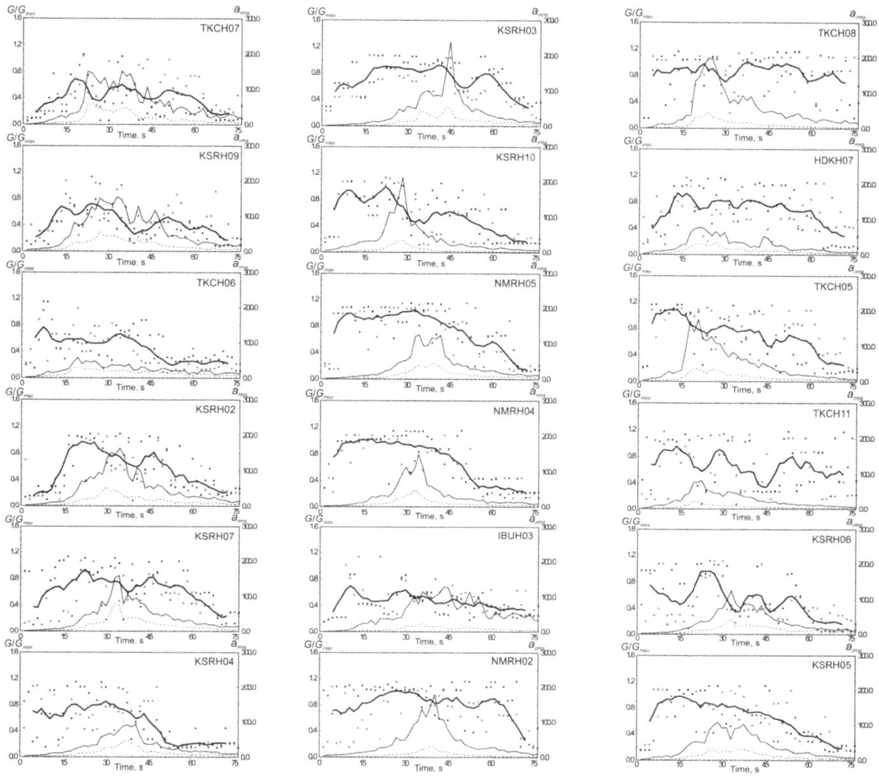

Fig. 4.27 Changes of shear moduli in soil layers at the studied KiK-net sites during the 2003 Tokachi-oki earthquake (75 s of strong motion): **a**—at sites with softer subsurface soils, **b**—at sites with denser subsurface soils. Points indicate estimated shear moduli in successive time intervals at two horizontal components (empty and filled circles mark EW and NS components), and thick lines show these values smoothed and averaged over two horizontal components. Thin lines are the intensities of motion on the surface (solid lines) and at depths of locations of deep devices (dash lines)

Information on the K-NET stations recorded high accelerations during the 2003 Tokachi-oki earthquake is given in Table 4.7; many of the sites are located on soft soils. Figure 4.29 shows the acceleration spectra on two horizontal components that allow a rough evaluation of the degree of nonlinearity of soil response at these sites.

As seen from Fig. 4.28, at remote K-NET sites HKD066 and HKD070 we observe high-frequency vibrations before the moments of the highest intensity of motion and stepwise drop of the motion frequency after the maximum of PGA. At HKD091 and HKD086 sites located closer to the fault, low-frequency components prevail in the strong motion so that spectra of motion on the surface tend to take the form $E(f) \sim f^{-k}$, testifying to the nonlinearity of soil response.

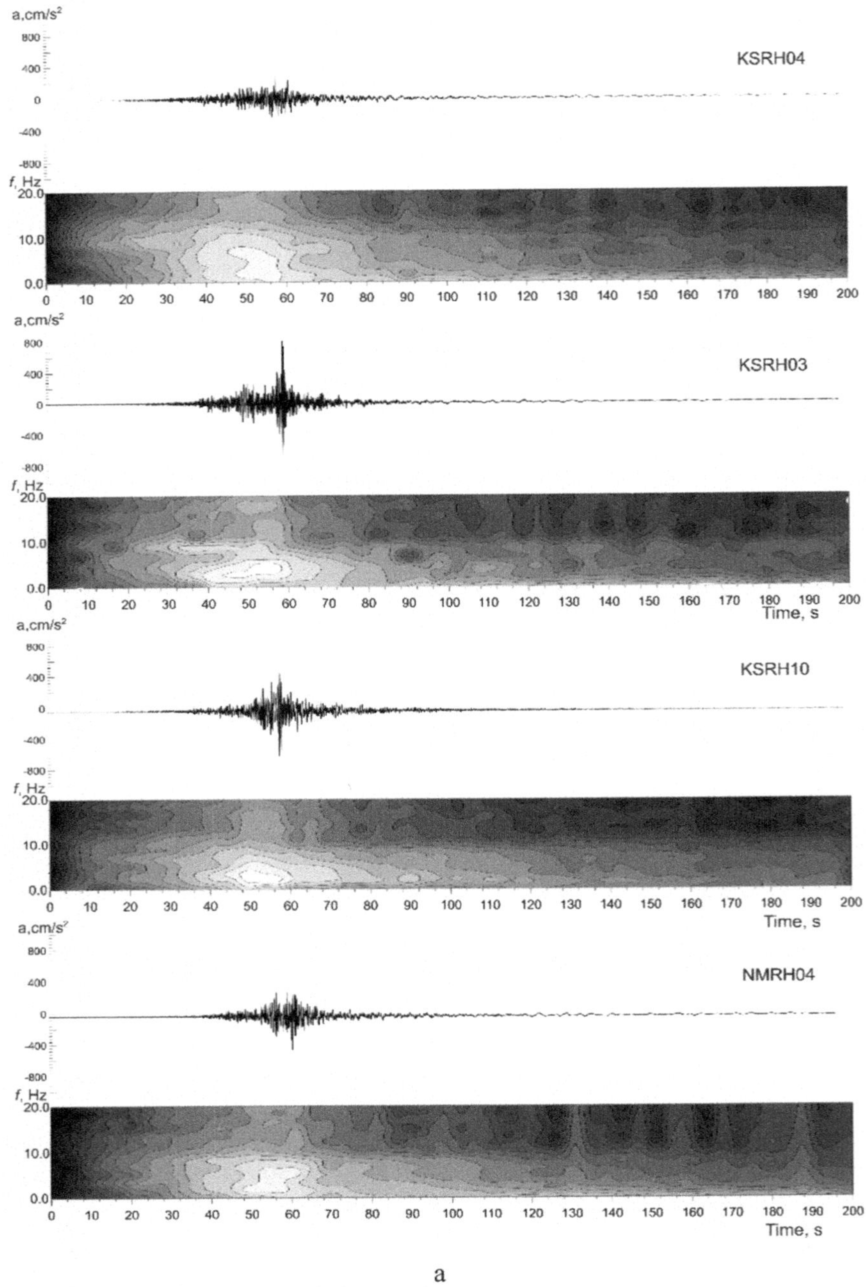

Fig. 4.28 Time–frequency diagrams of the acceleration time histories of the 2003 Tokachi-oki earthquake at KiK-net sites KSRH04, KSRH03, KSRH10, and NMRH04 (**a**) and at K-NET sites HDK066 and HDK070 (**b**) at one of the horizontal components

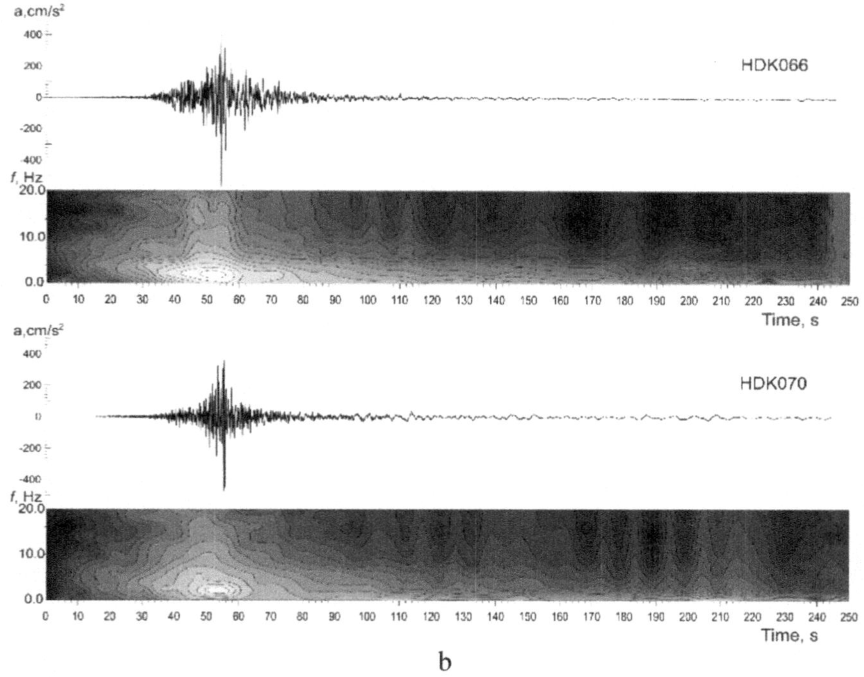

Fig. 4.28 (continued)

Table 4.7 Information on the K-NET stations located in the near-fault zones of the earthquake

Site code	Latitude (°)	Longitude (°)	V_{S20} (m/s)	Epic. distance (km)	PGA on surface (Gal)
HKD100	42.29	143.31		84	988.4
HKD091	42.81	143.66	76.3	119	455.0
HKD086	42.85	143.86	160.6	120	800.6
HKD092	42.93	143.45		138	672.7
HKD090	43.12	143.62	Hard	154	485.0
HKD075	43.08	145.13	345.4	168	560.8
HKD070	43.39	145.28	131.7	204	481.6
HKD066	43.66	145.13	116.1	226	591.1

Thus, during the 2003 Tokachi-oki earthquake, we observe phenomena similar to those observed during the 2011 Tohoku earthquake, such as, nonlinear behavior of soft soils at sites closest to the epicenter, soft soil hardening at remote sites recorded high PGA and decrease of frequencies of motion after the moments of the highest intensity of motion at these sites.

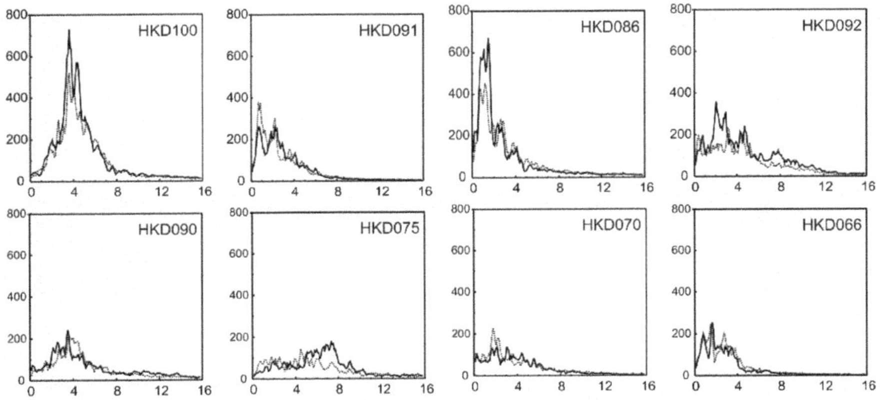

Fig. 4.29 Spectra of acceleration time histories at K-NET sites (two horizontal components) in the near-fault zones of the 2003 Tokachi-oki earthquake

In the previous Sect. 4.2, an explanation is suggested that such behavior of soft soils at remote sites during the Tohoku earthquake was due to the influence of a shock wave as a limiting case of directivity effects that occurred during the propagation of a crack in an extended earthquake source over a rather long distance toward the remote stations. These effects can be traced by changes in the waveforms of the acceleration time histories with increasing distance from the source.

Similarly, during the 2003 Tokachi-oki earthquake directivity effects were essential, as noted, for example, by Koketsu et al. (2004) and Honda et al. (2004). High PGA were recorded in coastal areas to the northwest of the epicenter: at HKD100, HKD086, and HKD092 (Table 4.7) sites. Honda et al. (2004) concluded that directivity effects of the largest asperity might have increased the amplitudes of seismic waves incident to the Yufutsu sedimentary basin, where oil storage tanks were severely damaged. Koketsu et al. (2004) found that the rupture propagated at a supershear speed on the upper part of the fault plane near the hypocenter, and this supershear ($V > V_S$) or near-supershear ($V \sim V_S$) situation resulted in directivity effects on *P*- and *S*-waves. The effects are confirmed by long-period pulses in the observed velosigrams at sites located to the north of the epicenter: TKCH08, TKCH10, TKCH11, and ABSH04.

At the same time, high PGA ~ 1 g were recorded at remote sites KSRH03, KSRH10, NMRH02, HKD066, and HKD075 (Tables 4.6 and 4.7) located northeast of the epicenter, where they can also be associated with directivity effects.

Figure 4.30 shows horizontal components of acceleration time histories of the 2003 Tokachi-oki earthquake (records of deep devices) at KiK-net sites located north and northeast of the source at epicentral distances of ~ 134–223 km. Similar to the Tohoku earthquake, we observe a decrease of the strong motion duration with distance, while PGA are not changed significantly. The minimal duration of strong motion is observed at sites KSRH10, KSRH03 (second group of intense waves), and NMRH02, where the highest PGA were recorded (Fig. 4.30; Table 4.6). At KSRH03 and the neighboring NMRH05 sites, intense motion is seen before the analyzed peaks, which may be related to some internal heterogeneities.

To check the influence of the directivity effects on the records at sites located to the northeast of the epicenter, simple calculations were performed. The propagation of seismic waves radiated by the crack tip moving from the hypocenter towards three stations: KSRH09, KSRH06, and KSRH10 is calculated. Figure 4.31 schematically shows the cross-section of the fault plane and the locations of the stations, and Tables 4.8, 4.9 and 4.10 present the calculations illustrating seismic wave propagation to the three stations.

Superposition of seismic waves could occur at KSRH10 site, i.e., seismic waves radiated by the crack tip during a certain time period, came to the site almost simultaneously, whereas they came to KSRH09 site, located closer to the fault, sequentially, and no superposition occurred. KSRH06 site is an intermediate case.

In the calculations, parameters of the fault plane are used taken from the paper by Koketsu et al. (2004), which partially coincide with the estimates by Yamanaka and Kikuchi (2004): the depth of the hypocenter of ~ 25 km and the orientation of the fault plane (230°, 20°). The location of the hypocenter was taken according to JMA: 41.78N, 144.078E. The locations of the three KiK-net sites, KSRH09, KSRH06, and KSRH10, relative to the fault plane (Fig. 4.24) do not allow studying directivity effects in narrow sectors as in case of the Tohoku earthquake; the calculations were performed for three directions shown by dotted lines in Fig. 4.24.

The researchers, who studied the source process of the 2003 Tokachi-oki earthquake, emphasize a simple model of slip distribution over the fault plane and radial propagation of ruptures from the hypocenter. Thus, we assume a radial rupture front in the sectors marked by dotted lines (Fig. 4.24).

Travel times to the stations (Tables 4.8, 4.9 and 4.10) were calculated based on the distances between the crack tip and the stations, with 6-layer velocity structure model by Koketsu et al. (2004) (Fig. 4.31). The calculations were performed for time moments corresponding to the crack tip positions over 20 km intervals (points A, B, C, D, E, F); the total path 100 km.

According to Koketsu et al. (2004), the crack started near the hypocenter in the area of high concentration of the aftershocks (Fig. 4.24). *S*-wave arrival times to KiK-net stations were calculated as the sums of the time moments of *S*-wave radiation by the crack tip and the travel times to the stations. The crack propagation velocity was also taken from Koketsu et al. (2004): supershear, 4.3 km/s in the upper part of the fault (A – B – C – D) and decreasing to *S*-wave velocity, 4.0 km/s in the middle and deep parts of the fault (D – E – F). The estimates of strong motion duration at the stations are: ~ 8 s at KSRH09, ~ 4.5 s at KSRH06 and ~ 2 s at KSRH10 (Tables 4.8, 4.9 and 4.10), which agrees with the observations.

Thus, the distance between the crack tip and KSRH10 station decreased faster than *S*-waves propagate, and seismic waves radiated by the crack tip over a long time period reach KSRH10 site almost simultaneously. The waves overlap, and PGA increase while the duration of motion decrease.

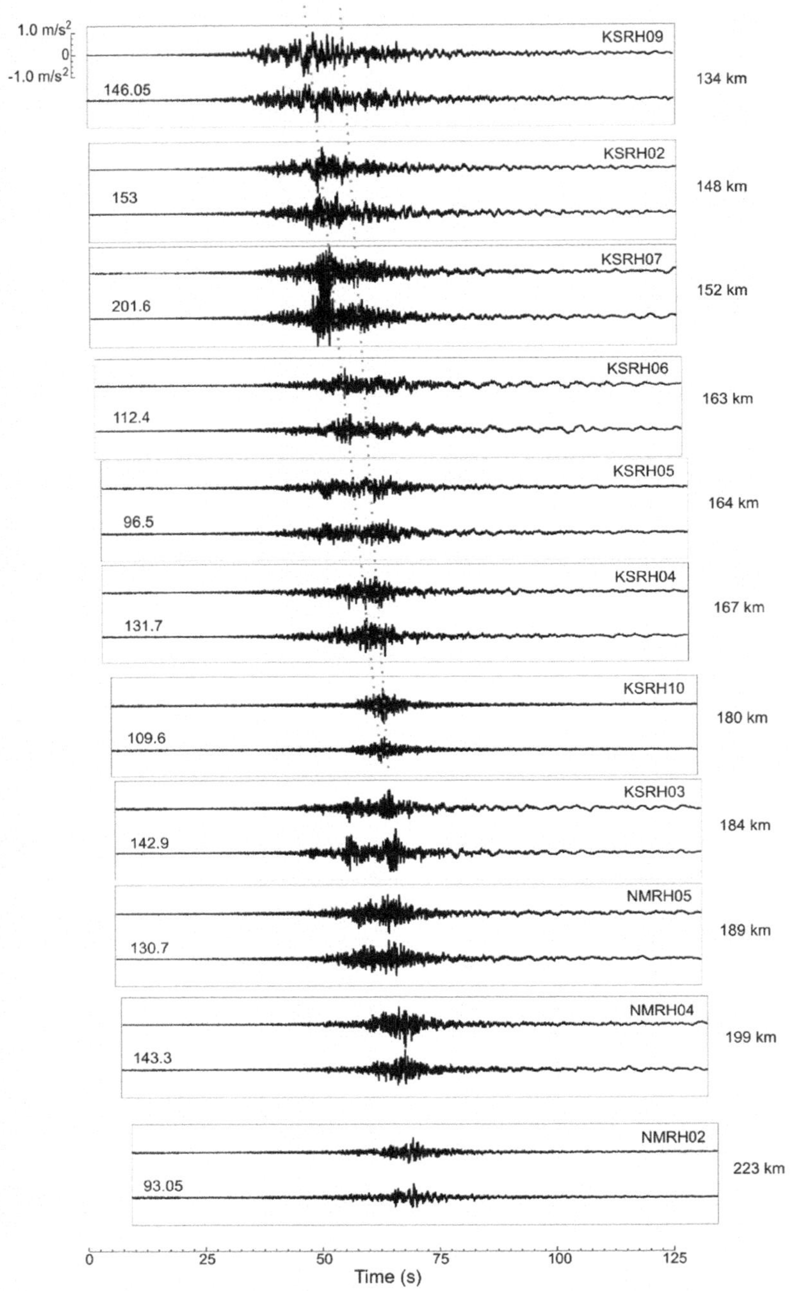

Fig. 4.30 Acceleration time histories (EW and NS components, depth records) of the 2003 Tokachi-oki earthquake at KiK-net sites located to the north of the epicenter. For each site, the epicentral distances (on the right) and PGA in cm/s^2 (on the left) are shown. Dotted lines show the decrease of strong motion duration with distance

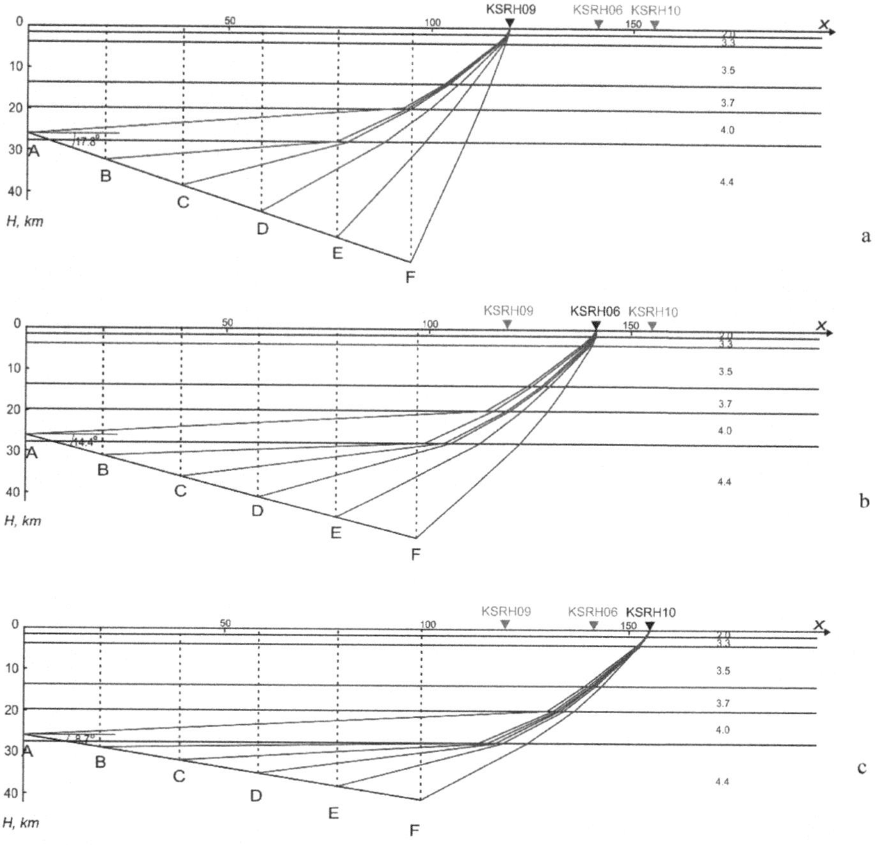

Fig. 4.31 The cross-section of the fault plane of the 2003 Tokachi-oki earthquake, locations of KiK-net sites and seismic wave paths. The numbers indicate *S*-wave velocities in the layers in km/s; velocity structure model is taken from Koketsu et al. (2004). *S*-wave propagation: **a**—to KSRH09 site, **b**—to KSRH06 site, **c**—to KSRH10 site

The model (Fig. 4.31) and the calculations (Tables 4.8, 4.9 and 4.10) should be considered as approximate ones, illustrating directivity effects of the finite-fault source, leading to high PGA at remote sites.

According to various authors, the fault plane of the 2003 Tokachi-oki earthquake was 160 km × 140 km on average (the estimates vary from 90 km × 70 km to 200 km × 160 km). The researchers agree that the fault had a fairly simple slip distribution that may imply the maturity of the subduction zone around the source region and a near-supershear rupture propagation in the hypocentral area as estimated by Koketsu et al. (2004). High values of rupture propagation velocity, 4.5 km/s in north direction, were obtained by Yagi (2004). Earthquakes have repeatedly occurred in this area in the past; evidently, they smoothed out asperities and facilitated sliding on the fault.

Table 4.8 Duration of strong motion that was observed at KSRH09 site as a result of crack propagation over a certain part of the fault plane during the 2003 Tokachi-oki event (l ~ 100 km, α ~ 17.8°, V ~ 4.3–4.0 km/s)

Time moment of *S*-wave radiation by a crack tip (s)	The projection of the distance between the crack tip and KiK-net station on X axis (km)	Depth of the crack tip *h* (km)	*S*-wave travel time from the crack tip to the KiK-net station (s)	Time of *S*-wave arrival to the KiK-net station (s)
0 (A)	141.6	26	38.63	38.63
4.65 (B)	122.23	30.97	32.46	37.11
9.3 (C)	102.86	35.95	28.11	37.41
13.95 (D)	83.49	40.92	24.02	37.97
18.95 (E)	64.11	45.90	20.34	39.29
23.95 (F)	44.74	50.87	17.42	41.37

The duration of strong motion: 7.52 sec

Table 4.9 Duration of strong motion that was observed at KSRH06 site as a result of crack propagation over a certain part of the fault plane during the 2003 Tokachi-oki event (l ~ 100 km, α ~ 14.4°, V ~ 4.3–4.0 km/s)

Time moment of *S*-wave radiation by a crack tip (s)	The projection of the distance between the crack tip and KiK-net station on X axis (km)	Depth of the crack tip *h* (km)	*S*-wave travel time from the crack tip to the KiK-net station (s)	Time of *S*-wave arrival to the KiK-net station (s)
0 (A)	141.6	26	38.63	38.63
4.65 (B)	122.23	30.97	32.46	37.11
9.3 (C)	102.86	35.95	28.11	37.41
13.95 (D)	83.49	40.92	24.02	37.97
18.95 (E)	64.11	45.90	20.34	39.29
23.95 (F)	44.74	50.87	17.42	41.37

The duration of strong motion: 4.26 sec

As known, high concentration of the aftershocks is usually associated with the areas of large slip, high stress drop, and high rupture velocity, as noticed by Robinson et al. (2006). Thus, the distribution of aftershocks occurred during the first days and months after the mainshock was studied. The highest aftershock concentrations were observed to the northwest of the hypocenter, where Koketsu et al. (2004) and Yagi (2004) found near-supershear rupture propagation and directivity effects, as well as to the northeast, to KSRH10 site. The major postseismic slip extended from south of the epicenter about 180 km to the northeast.

Table 4.10 Duration of strong motion that was observed at KSRH10 site as a result of crack propagation over a certain part of the fault plane during the 2003 Tokachi-oki event (l ~ 100 km, α ~ 8.7°, V ~ 4.3–4.0 km/s)

Time moment of S-wave radiation by a crack tip (s)	The projection of the distance between the crack tip and KiK-net station on X axis (km)	Depth of the crack tip h (km)	S-wave travel time from the crack tip to the KiK-net station (s)	Time of S-wave arrival to the KiK-net station (s)
0 (A)	155.4	26	42.07	42.07
4.65 (B)	135.63	29.03	35.43	40.08
9.3 (C)	115.86	32.05	30.94	40.24
13.95 (D)	96.09	35.08	26.56	40.51
18.95 (E)	76.32	38.10	22.28	41.23
23.95 (F)	56.55	41.13	18.32	42.27

The duration of strong motion: 2.19 sec

Considering all these facts, it seems quite likely that the rupture spread at a speed of ~ 4.3–4.0 km/s radially to the north-northeast from the hypocenter and produced directivity effects at remote sites KSRH10, KSRH03, NMRH02 and some others.

Seismic waves radiated by the crack tip during its propagation over a rather long segment (~ 100 km) of the fault came to remote sites almost simultaneously, (it is seen from decreased duration and increased PGA at the inputs to soil layers). Soil layers experienced hardening at the moments of high intensity of motion; amplification of seismic waves increased, and high PGA appeared on the surface. Similar mechanisms worked during the 2011 Tohoku earthquake. During future large earthquakes directivity effects may also be significant, and we may expect generation of high PGA in the near-fault zones.

******

Note the different behavior of soft soils during earthquakes with magnitudes M_w ~ 6.5–7.0 and stronger earthquakes with magnitudes M_w ~ 8.0 and higher.

In the first case, the duration of strong motion in the near-fault zones does not exceed ~ 15–20 s, soft soil behavior is substantially nonlinear near the source, and nonlinearity gradually decreases with distance. Shear moduli in soil layers reduce at the beginning and recover at the end of strong motion (Chap. 3).

During stronger earthquakes with magnitudes M_w ~ 8.0 and higher, strong motion duration reaches tens to hundreds of seconds, so that soil layers experience several stages of softening—hardening, reduction and recovering of the shear moduli. On the whole, the behavior of soft soils can become rather linear in the near-fault zones. During the 2003 Tokachi-oki earthquake, shear moduli at sites with softer and denser soils finally reduced at the end of strong motion and did not recover.

During large earthquakes with magnitudes $M_w > 8.0$ and extended sources, directivity effects can play an important role in the distribution of PGA in the near-fault zones. When a crack in the earthquake source propagates at a high speed along a sufficiently long section of the fault plane (which is generally typical for large earthquakes, since they occur regularly on the same faults), seismic waves radiated during a certain time period come to seismic stations, located in forward direction, almost simultaneously; in this case, source effects (directivity) and local site effects (soil hardening) lead to high PGA at these sites. Soil hardening contributes to the generation of high PGA during strong earthquakes, because consolidated soils behave more linearly and effectively amplify strong motions.

Thus, soil behavior during large earthquakes with magnitudes $M_w > 8.0$ differs from soil behavior during weaker events with magnitudes $M_w \sim 6.5–7.6$, and apparently, we can expect similar effects of soil hardening and generation of high PGA in future strong earthquakes.

References

Aagaard, B.T., Heaton, T.H.: Near-source ground motions from simulations of sustained intersonic and supersonic fault ruptures. Bull. Seism. Soc. Am. **94**, 2064–2078 (2004)

Archuleta, R.J.: A faulting model for the 1979 imperial valley earthquake. J. Geoph. Res. **89**, 4559–4585 (1984)

Archuleta, R.J., Hartzell, S.H.: Effects of fault finiteness on near-source ground motion. Bull. Seism. Soc. Am. **71**, 939–957 (1981)

Asano, K., Iwata, T.: Source model for strong ground motion generation in the frequency range 0.1–10 Hz during the 2011 Tohoku earthquake. Earth Plan. Space **64**(12), 1111–1123 (2012)

Asaoka, A., Sawada, Y., Noda, T., Yamada, S., Shimizu, R.: An attempt to replicate the so-called "trampoline effect" in computational geomechanics. In: Proceedings of 15th World Conference Earthquake Engineering Lisbon Portugal (2012)

Bonilla, L.F., Tsuda, K., Pulido, N., R´egnier J., Laurendeau A.: Nonlinear site response evidence of K-NET and KiK-net records. Earth. Plan. Space **63**, 785–789 (2011)

Bouchon, M., Bouin, M.-P., Karabulut, H., Toksoz, M.N., Dietrich, M., Rosakis, A.J.: How fast is rupture during an earthquake? New Insights from the 1999 Turkey Earthquakes. Geoph. Res. Lett **28**, 2723–2726 (2001)

Bouchon, M., Vallee, M.: Observation of long supershear rupture during the magnitude 8.1 Kunlunshan Earthquake. Science **301**, 824–826 (2003)

Duan, B.: Dynamic rupture of the 2011 Mw 9.0 Tohoku-Oki earthquake: roles of a possible subducting seamount. J. Geoph. Res. **117**, B05311 (2012). https://doi.org/10.1029/2011JB009124

Dunham, E.M., Bhat, H.S.: Attenuation of radiated ground motion and stresses from three-dimensional supershear ruptures. J. Geoph. Res, **113**, B08319 (2008). https://doi.org/10.1029/2007JB005182

Ellsworth, W.L., Celebi, M., Evans, J.R., Jensen, E.G., Kayen, R., Metz, M.C., Nyman, D.J., Roddick, J.W., Spudich, P., Stephens, C.D.: Near-field ground motion of the 2002 denali fault, Alaska, Earthquake recorded at pump station 10. Earthq. Spectra **20**, 597–615 (2004)

Field, E.H., Johnson, P.A., Beresnev, I.A., Zeng, Y.: Nonlinear ground-motion amplification by sediments during the 1994 Northridge earthquake. Nature **390**(6660), 599–602 (1997)

Freund, L.B.: The mechanisms of dynamic shear crack propagation. J. Geoph. Res. **84**, 2199–2209 (1979)

Furumura, T., Takemura, S., Noguchi, S., Takemoto, T., Maeda, T., Iwai, K., Padhy, S.: Strong ground motions from the 2011 off-the Pacific-Coast-of-Tohoku, Japan ($M_W = 9.0$) earthquake obtained from a dense nationwide seismic network. Landslides **8**, 333–338 (2011)

Galvez, P., Ampuero, J.-P., Dalguer, L.A., Somala, S.N., Nissen-Meyer, T.: Dynamic earthquake rupture modelled with an unstructured 3-D spectral element method applied to the 2011 M9 Tohoku earthquake. Geoph. J. Int. **198**, 1222–1240 (2014)

Goto, H., Yamamoto, Y., Kita, S.: Dynamic rupture simulation of the 2011 off the Pacific coast of Tohoku Earthquake: multi-event generation within dozens of seconds. Earth. Plan. Space **64**, 1167–1175 (2012)

Hatayama, K.: Lessons from the 2003 Tokachi-oki, Japan, earthquake for prediction of long-period strong ground motions and sloshing damage to oil storage tanks. J. Seism. **12**, 255–263 (2008)

Honda, R., Aoi, S., Morikawa N., Sekiguchi H., Kunugi, T., Fujiwara, H.: Ground motion and rupture process of the 2003 Tokachi-oki earthquake obtained from strong motion data of K-NET and KiK-net. Earth, Planet. Space **56**, 317–322 (2004)

Huang, Y., Meng, L., Ampuero, J.-P.: A dynamic model of the frequency-dependent rupture process of the 2011 Tohoku-Oki earthquake. Earth, Plan. Space **64**, 1061–1066 (2012)

Huang, Y., Ampuero, J.-P., Kanamori, H.: Slip-weakening models of the 2011 Tohoku-Oki earthquake and constraints on stress drop and fracture energy. Pure Appl. Geoph. (2013). https://doi.org/10.1007/s00024-013-0718-2

Ide, S., Baltay, A., Beroza, G.C.: Shallow dynamic overshoot and energetic deep rupture in the 2011 M_W 9.0 Tohoku-Oki earthquake. Science **332**, 1426–1429 (2011)

Irikura, K., Kurahashi, S.: High acceleration motions generated from the 2011 Pacific coast off Tohoku, Japan, earthquake. In: Proceedings of 15th World Conference Earthquake Engineering, Lisbon, Portugal (2012)

Joyner, W.B., Chen, T.F.: Calculation of nonlinear ground response in earthquakes. Bull. Seism. Soc. Am. **65**(5), 1315–1336 (1975)

Kawase, H., Sanchez-Sesma, F.J., Matsushima, S.: Application of the H/V spectral ratios for earthquake and microtremor ground motions. In: 4th IASPEI/IAEE International Symposium Effects of Surface Geology on Seismic Motion (2011)

Koketsu, K., Yokota, Y., Nishimura, N., Yagi, Y., Miyazaki, S., Satake, K., Fujii, Y., Miyake, H., Sakai, S., Yamanaka, Y., Okada, T.: A unified source model for the 2011 Tohoku earthquake. Earth Planet. Sci. Lett. **310**, 480–487 (2011)

Koketsu, K., Hikima, K., Miyazaki, S., Ide, S.: Joint inversion of strong motion and geodetic data for the source process of the 2003 Tokachi-oki, Hokkaido, earthquake. Earth, Planet. Space **56**, 329–334 (2004)

Kozdon, J.E., Dunham, E.M.: Rupture to the trench: dynamic rupture simulations of the 11 March 2011 Tohoku earthquake. Bull. Seism. Soc. Am. **103**(2B), 1275–1289 (2013)

Mitsui, Y., Iio, Y., Fukahata, Y.: A scenario for the generation process of the 2011 Tohoku earthquake based on dynamic rupture simulation: role of stress concentration and thermal fluid pressurization. Earth, Plan. Space **64**, 1177–1187 (2012)

Miura, S., Suwa, Y., Hasegawa, A., Nishimura, T.: The 2003 M8.0 Tokachi-Oki earthquake—how much has the great event paid back slip debts? Geoph. Res. Lett. **31**, L05613 (2004)

Miyazawa, M.: Propagation of an earthquake triggering front from the 2011 Tohoku-Oki earthquake. Geoph. Res. Lett. **38**, L23307 (2011)

Nagashima, F., Kawase, H., Matsushima, S., Sanchez-Sesma, F. J., Hayakawa, T., Satoh, T., Oshima, M.: Application of the H/V spectral ratios for earthquake ground motions and microtremors at K-NET sites in Tohoku Region, Japan to Delineate Soil Nonlinearity. In: Proceedings of 15th World Conference Earthquake Engineering, Lisbon, Portugal (2012)

NIED – National Research Institute for Earth Science and Disaster Resilience, Japan. https://www.bosai.go.jp

Nozu, A., Irikura, K.: Strong-motion generation areas of a great subduction-zone earthquake: waveform Inversion with empirical Green's functions for the 2003 Tokachi-oki earthquake. Bull. Seism. Soc. Am. **98**, 180–197 (2008)

Robinson, D.P., Cheung, L.T.: Source process of the M_W8.3, 2003 Tokachi-Oki, Japan earthquake and its aftershocks. Geoph. J. Int. **181**, 334–342 (2010)

Robinson, D.R., Brough C., Das, S.: The M_W7.8, 2001 Kunlunshan earthquake: Extreme rupture speed variability and effect of fault geometry. J. Geoph. Res, **111**, B08303 (2006). https://doi.org/10.1029/2005JB004137

Rosakis, A.J., Samudrala, O., Coker, D.: Cracks faster than the shear wave speed. Science **284**, 1337–1340 (1999)

Roten, D., Fah, D., Bonilla, L.F.: High-frequency ground motion amplification during the 2011 Tohoku earthquake explained by soil dilatancy. Geoph. J. Int. **193**, 898–904 (2013)

Rubinstein, J.L., Uchida, N., Beroza, G.C.: Seismic velocity reductions caused by the 2003 Tokachi-Oki earthquake. J. Geoph. Res. **112**, B05315 (2007)

Satake, K., Fujii, Y., Harada, T., Namegaya, Y.: Time and space distribution of coseismic slip of the 2011 Tohoku earthquake as inferred from tsunami waveform Data. Bull. Seism. Soc. Am. **103**, 1473–1492 (2013).https://doi.org/10.1785/0120120122

Somerville, P., Smith, N.F., Graves, R.W., Abrahamson, N.A.: Modification of empirical strong ground motion attenuation relations to include the amplitude and duration effects of rupture directivity. Seism. Res. Lett. **68**, 199–222 (1997)

Tajima, F., Mori, J., Kennett, B.L.N.: A review of the 2011 Tohoku-Oki earthquake (M_W 9.0): Large-scale rupture across heterogeneous plate coupling. Tectonoph. **586**, 15–34 (2013)

Vallee, M., Dunham, E.M.: Observation of far-field Mach waves generated by the 2001 Kokoxili supershear earthquake. Geoph. Res. Lett. **39**, L05311 (2012). https://doi.org/10.1029/2011GL050725

Voznesensky, E.A.: Dynamic Instability of Soils. Editorial URSS, Moscow (1999). (in Russian)

Yagi, Y.: Source rupture process of the 2003 Tokachi-oki earthquake determined by joint inversion of teleseismic body wave and strong ground motion data. Earth, Planet. Space **56**, 311–316 (2004)

Yamanaka, Y., Kikuchi, M.: Source process of the recurrent Tokachi-oki earthquake on September 26, 2003, inferred from teleseismic body waves. Earth, Planet. Space **55**, e21–e24 (2004)

Zarembo, L.K., Krasilnikov, V.A.: Introduction to Nonlinear Acoustics. Nauka, Moscow (1966). (in Russian)

Zhan, Z., Helmbergar, D.V., Kanamori, H., Shearer, P.M.: Supershear rupture in a M_W6.7 aftershock of the 2013 Sea of Okhotsk earthquake. Science **345**, 204–207 (2013)

Zhao, D., Huang, Z., Umino, N., Hasegawa, A., Kanamori, H.: Structural heterogeneity in the megathrust zone and mechanism of the 2011 Tohoku-Oki earthquake (M_W 9.0). Geoph. Res. Lett. **38**, L17308 (2011). https://doi.org/10.1029/2011GL048408

Chapter 5
Nonlinear Components in Soil Response

The constructed models of soil behavior allow us to study soil profiles as nonlinear systems converting input motion to soil layers into motion on the surface. To estimate nonlinear components in soil response, nonlinear identification methods are used based on the calculation of high-order impulse characteristics of nonlinear systems. Under conditions of stationarity, analyticity, and finite memory of a nonlinear system, the response of the system can be represented as a Volterra-Wiener series, i.e. a sum of multiple integrals of the input signal. Knowing the input and output signals, we can conclude about the types and quantitative characteristics of the system's nonlinearity. The Chapter describes nonlinear identification of the soil profiles and estimation of nonlinear components in the soil response in the near-fault zones of the studied earthquakes. It was obtained that the estimates of nonlinear components of the soil response are comparable to the estimates of shear moduli reduction in the layers during strong motion.

Subsurface soils possess mostly odd-type nonlinearities (cubic and other higher-order odd-type nonlinearities), because hysteretic stress–strain relations are essentially odd functions. Nonlinearities of even types are usually weak, but become significant in cases when loading and unloading curves of the stress–strain relations acquire noticeable even components (as in liquefied soils or at the moments of most intense motion).

5.1 Nonlinear System Identification by the White-Noise Method

Let's consider nonlinear corrections to Hooke's law during the propagation of seismic waves in a nonlinear elastic medium. In one-dimensional case, motion of a particle along x-axis after its deviation from the equilibrium is described by the equation of motion:

O. V. Pavlenko, *Seismic Waves in Soil Layers*,
https://doi.org/10.1007/978-3-031-52338-0_5

$$m\ddot{x} = f(x), \tag{5.1.1}$$

where *m* is the particle mass, *f(x)* is the component of the force acting on the particle located at point *x*. If the deviations from the equilibrium position are small, function *f(x)* can be represented as the Taylor series:

$$f(x) = f(0) + xf'(0) + \frac{x^2}{2!}f''(0) + \frac{x^3}{3!}f'''(0) + \ldots. \tag{5.1.2}$$

Evidently, $f(0) = 0$, if $x = 0$ is the equilibrium point. Usually in real physical systems the term $xf'(0)$ is nonzero and the main member in the expansion. $f'(0) < 0$ if x is the equilibrium point, and for small deviations from the equilibrium, the equation of motion describes harmonic oscillations:

$$m\ddot{x} = xf'(0) = -kx, \tag{5.1.3}$$

where $k = -f'(0) > 0$. Almost all physical systems behave as linear oscillators at small deviations from the equilibrium.

If the deviations of particles from the equilibrium cannot be considered small, in expression (5.1.2) we should take into account the term $\frac{x^2}{2!}f''(0)$ and the following terms.

In case of nonlinear oscillations of systems with many degrees of freedom, the Lagrange function defined up to third-order terms, accounting for quadratic nonlinearity, takes the form (Landau and Lifshitz 1960):

$$L = \frac{1}{2}\sum_{i,k}(m_{ik}\dot{x}_i\dot{x}_k - k_{ik}x_ix_k) + \frac{1}{2}\sum_{i,k,l}n_{ikl}\dot{x}_i\dot{x}_kx_l - \frac{1}{3}\sum_{i,k,l}l_{ikl}x_ix_kx_l, \tag{5.1.4}$$

where m_{ik}, k_{ik}, n_{ikl}, and l_{ikl} are constant coefficients. The second and third terms appeared in the expression as a result of taking into account the values of the third order of smallness in the expressions for kinetic and potential energy, respectively. In normal coordinates Q_α, we get:

$$L = \frac{1}{2}\sum_{\alpha}(\dot{Q}_\alpha^2 - \omega_\alpha^2 Q_\alpha^2) + \frac{1}{2}\sum_{\alpha,\beta,\gamma}\lambda_{\alpha\beta\gamma}\dot{Q}_\alpha\dot{Q}_\beta Q_\gamma - \frac{1}{3}\sum_{\alpha,\beta,\gamma}\mu_{\alpha\beta\gamma}Q_\alpha Q_\beta Q_\gamma. \tag{5.1.5}$$

$\lambda_{\alpha\beta\gamma}$ and $\mu_{\alpha\beta\gamma}$ are constant coefficients. The corresponding equations of motion take the form:

$$\ddot{Q}_\alpha + \omega_\alpha^2 Q_\alpha = f_\alpha(Q, \dot{Q}, \ddot{Q}), \tag{5.1.6}$$

where f_α are second-order homogeneous functions of coordinates Q and their time derivatives. By the method of successive approximations, we are looking for the solution in the form:

$$Q_\alpha = Q_\alpha^{(1)} + Q_\alpha^{(2)}, \tag{5.1.7}$$

where $Q_\alpha^{(2)} \ll Q_\alpha^{(1)}$, and functions $Q_\alpha^{(1)}$ satisfy the non-perturbed equations:

$$\ddot{Q}_\alpha^{(1)} + \omega_\alpha^2 Q_\alpha^{(1)} = 0, \tag{5.1.8}$$

i.e., represent ordinary harmonic oscillations

$$Q_\alpha^{(1)} = A_\alpha \sin(\omega_\alpha t + \alpha_\alpha). \tag{5.1.9}$$

Keeping in the right of Eq. (5.1.6) only the terms of the second order of smallness, we obtain for $Q_\alpha^{(2)}$:

$$\ddot{Q}_\alpha^{(2)} + \omega_\alpha^2 Q_\alpha^{(2)} = f_\alpha(Q^{(1)}, \dot{Q}^{(1)}, \ddot{Q}^{(1)}), \tag{5.1.10}$$

where expressions (5.1.9) should be used in the right part. As a result, we get linear inhomogeneous differential equations, the right parts of which can be converted to sums of simple periodic functions, for example,

$$\begin{aligned} Q_\alpha^{(1)} Q_\beta^{(1)} &= A_\alpha A_\beta \cos(\omega_\alpha t + \alpha_\alpha)\cos(\omega_\beta t + \alpha_\beta) \\ &= \frac{1}{2} A_\alpha A_\beta \{\cos[(\omega_\alpha + \omega_\beta)t + \alpha_\alpha + \alpha_\beta] + \cos[((\omega_\alpha - \omega_\beta)t + \alpha_\alpha - \alpha_\beta)]\} \end{aligned} \tag{5.1.11}$$

Thus, in the right parts of Eqs. (5.1.10) there are terms corresponding to oscillations with frequencies equal to the sums and differences of the fundamental frequencies of the system, and we should look for the solutions containing such periodic multipliers. Additional oscillations with combinational frequencies $\omega_\alpha \pm \omega_\beta$, including doubled frequencies and a zero frequency (constant component) are superimposed on the normal oscillations of the system.

In the following approximations (third, fourth, etc.), when taking into account higher-order terms in the expansion of the Lagrange function, combinational oscillations arise with frequencies that are sums and differences of a larger number of frequencies ω_α . Thus, on a set of basic waves with frequencies $\omega_1, \omega_2, \omega_3, \omega_4 \ldots$, and wave vectors $\vec{k}_1, \vec{k}_2, \vec{k}_3, \vec{k}_4, \ldots$, waves of smaller intensities with combinational frequencies $\omega_1 \pm 2\omega_2$, $2\omega_1 \pm \omega_2$, $\omega_1 \pm \omega_2 \pm \omega_3$, $2\omega_1 \pm \omega_3$, $\omega_1 \pm 2\omega_3$ (in case of cubic nonlinearity), or $\omega_1 \pm 3\omega_2$, $3\omega_1 \pm \omega_2$, $2\omega_1 \pm \omega_2 \pm \omega_3$, $2\omega_1 \pm 2\omega_3$, $\omega_1 \pm \omega_2 \pm \omega_3 \pm \omega_4$, etc. (in case of 4th-order nonlinearity), and so on, with the corresponding wave vectors, are superimposed. The amplitudes of the combinational oscillations are proportional to the products of the amplitudes of the corresponding normal oscillations.

The right parts of Eqs. (5.1.10) contain the products of normal coordinates in the form (5.1.11) and also the products of their derivatives, so the amplitudes of combinational waves of summary frequencies are proportional to the sums of the wave vectors, and the amplitudes of combinational waves of difference frequencies are proportional to the differences of the wave vectors of the interacting waves. Thus, in the absence of absorption, low-frequency combinational harmonics grow slower than high-frequency ones.

Nonlinearities of all orders are the base (or condition) for generation of higher harmonics and mixing frequencies, and odd-order nonlinearities produce frequencies coinciding with the original ones. As a result, the attenuation of the propagating waves depends on their intensities.

Combinational waves can be considered as some kind of the correction effect, because they cannot exist by themselves. Generally speaking, the relations between the combinational frequencies and the corresponding wave vectors, similar to those that occur between the frequencies and wave vectors in ordinary monochromatic waves, are not satisfied. However, at some specially selected frequencies and wave vectors the relations are fulfilled. In such cases, the superposition of two or more monochromatic waves leads to the appearance of secular terms at summary and difference frequencies, which amplitudes increase with time. In particular, in an isotropic solid, the resonance is possible only in the two following cases (Landau and Lifshitz 1970): (1) the interacting waves are S-waves, and the combinational wave is a P-wave; (2) one of the interacting waves is an S-wave, the other is a -wave, and the resulting wave is a P-wave. If the interacting waves propagate in one direction, the resonance is possible in cases when all the waves are P-waves or S-waves.

The nonlinearity of a soil response is defined by the nonlinearity of its hysteretic stress–strain relation. Only initial parts of the stress–strain curves, i.e., the range of small strains, are close to linear relations, and weak seismic waves propagating in soils keep their shapes, while intense waves suffer nonlinear distortions.

The mathematical description of the response of a nonlinear system can be obtained in different ways. Some of the approaches assume available information on the internal structure of the system and the possibility to conduct experiments, when we can study the system by tracing the passage of signals from the input to the output. The constructed models of soil behavior (Chaps. 3 and 4) allow such experiments.

Another approach is described in this chapter. A studied system is presented as a black box with an input and output, and we look for connections between input and output signals. Under the conditions of stationarity, analyticity and finite memory of the system (usually satisfied for real physical systems), the output signal can be represented by the Volterra series, as a sum of multiple integrals of the input signal (Marmarelis and Marmarelis 1978):

$$y(t) = k_0 + \int_0^\infty k_1(\tau)x(t-\tau)d\tau + \int_0^\infty\int_0^\infty k_2(\tau_1,\tau_2)x(t-\tau_1)x(t-\tau_2)d\tau_1 d\tau_2$$

$$+ \int_0^\infty\int_0^\infty\int_0^\infty k_3(\tau_1,\tau_2,\tau_3)x(t-\tau_1)x(t-\tau_2)x(t-\tau_3)d\tau_1 d\tau_2 d\tau_3 + \ldots.. \quad (5.1.12)$$

where $x(t)$ is the input signal, $y(t)$ is the output signal, k_0, $k_1(\tau)$, $k_2(\tau_1, \tau_2)$, $k_3(\tau_1, \tau_2, \tau_3)$ are the Volterra kernels or impulse characteristics of the system of zero-, first-, second- and third- orders. The Volterra series is a generalization of the Taylor series for a function of many variables. In the limiting case, the function becomes the functional, the variables become continuous functions, and the coefficients of the Taylor series become the kernels of the Volterra series.

If a system is linear, the first term of the Volterra series, defining the linear part of the system response, and one kernel $k_1(\tau)$ are sufficient to describe it. To describe a nonlinear system, several terms of the Volterra series are needed. Most physical systems can be satisfactorily described by the first few terms; the series converge fairly quickly, which is explained, on one hand, by the smoothness of the functionals, and on the other hand, by the linearizing effect of noise (Marmarelis and Marmarelis 1978).

The first-order Volterra kernel can be considered as the model accounting for the influence of past values of the input signal on the output at present, and higher–order kernels can be considered as the model describing the dependence of the output signal at present on the interactions of the input signals in the past. The n-th order interaction can be described by not only the n-th order Volterra kernel, but also by the higher-order kernels.

The principle of superposition is not fulfilled in a nonlinear system, and the response of the system to the sum of two input signals generally differs from the sum of the responses to each of the signals separately.

Therefore, sinusoidal, stepwise, pulse, etc., signals are uninformative as testing signals: we should test the nonlinear system with all possible types of signals and make a catalog of its responses. Noise-like signals are most effective in this case, since they contain components of various frequencies and amplitudes: if the Gaussian noise is applied as the input to the system for a sufficiently long time, finally, any form of disturbance will be represented in this signal (Marmarelis and Marmarelis 1978).

For nonlinear system identification, the Wiener series are widely used, consisting of functionals, which are linear combinations of the Volterra-type functionals. The terms of the Wiener series are orthogonal with respect to the input signal in the form of the Gaussian white noise, which allows us to determine the Wiener kernels independently of each other. On the other hand, application of input signals in the form of the Gaussian white noise is the most effective way to test nonlinear systems.

The dependence of the system response $y(t)$ on the input signal takes the form (Marmaris and Marmarelis 1978):

$$y(t) = \sum_{m=0}^{\infty} G_m\left[h_m(\tau_1, \ldots, \tau_m); x(t'), t' \le t\right], \tag{5.1.13}$$

where G_m are orthogonal functionals, if $x(t)$ is the Gaussian white noise with a zero mean, $\{h_m\}$ is the sequence of the Wiener kernels; each kernel being a symmetric function. The first four Wiener functionals take the form:

$$G_0[h_0; x(t)] = h_0, \tag{5.1.14}$$

$$G_1[h_1; x(t)] = \int_0^{\infty} h_1(\tau)x(t-\tau)d\tau, \tag{5.1.15}$$

$$G_2[h_2; x(t)] = \int\int_0^{\infty} h_2(\tau_1, \tau_2)x(t-\tau_1)x(t-\tau_2)d\tau_1 d\tau_2 - P\int_0^{\infty} h_2(\tau_1, \tau_1)d\tau_1, \tag{5.1.16}$$

$$G_3[h_3; x(t)] = \int\int\int_0^{\infty} h_3(\tau_1, \tau_2, \tau_3)x(t-\tau_1)x(t-\tau_2)x(t-\tau_3)d\tau_1 d\tau_2 d\tau_3$$

$$-3P\int\int_0^{\infty} h_3(\tau_1, \tau_2, \tau_2)x(t-\tau_1)d\tau_1 d\tau_2, \tag{5.1.17}$$

where P is the intensity of the white noise, not depending on frequency.

The orthogonality of the terms in the Wiener series relative to the Gaussian white noise at the input provides us significant advantages, since: (1) effective methods based on cross-correlation functions can be used for evaluating kernels; (2) any partial sum of the Wiener series gives the best approximation to the response of the system (minimum root-mean-square error) (Marmarelis and Marmarelis 1978).

Similarly to the impulse response of a system $h(t)$, a set of kernels $\{h_m\}$ can be considered as a generalized "combined" impulse response of the system. The first–order kernel describes the linear part of the system response, and higher-order kernels describe nonlinear corrections to the linear part of the response, due to quadratic nonlinearity, cubic nonlinearity, and nonlinearities of higher orders. The Gaussian (normal) distribution of the amplitudes is not necessary in the white-noise approach, but it allows us to represent autocorrelation functions of even orders through the second-order autocorrelation function, which noticeably simplifies the expressions for the Wiener functionals and facilitates the calculations (Marmarelis and Marmarelis 1978).

A quantitative measure of the agreement of the model with the real system's response is the root-mean-square error. The constant component corresponding to zero-order kernel $\{h_0\}$ provides zero-approximation: we estimate the deviation of the system response from this line and normalize it so that it was 100%. Then we estimate the deviation of the system response from the model based on the first-order kernel $\{h_0, h_1\}$ and normalize it by the same value; then we calculate the deviations of the system response from the responses of the nonlinear models predicted by kernels $\{h_0, h_1, h_2\}$ and $\{h_0, h_1, h_2, h_3\}$ and normalize the results by the same value.

These calculations allow us to conclude about the contributions of the second $\{h_2\}$ and third $\{h_3\}$ -order kernels to the nonlinear component of the system response, i.e., about the types and quantitative characteristics of the system's nonlinearity. If kernel $\{h_2\}$ provides the largest contribution, the system has predominantly quadratic nonlinearity, if $\{h_3\}$—cubic nonlinearity, and so on. If the contribution of high-order kernels is negligible, the system is linear (Marmarelis and Marmarelis 1978).

Inaccuracies in the estimates of kernels and functionals and methods for their correction are described in (Marmarelis and Marmarelis 1978). The inaccuracies may be caused by the finite duration and boundedness of the spectral range of the Gaussian noise, by truncation of tails of the normal distribution of the amplitudes and by inaccuracies in the calculations. The choice of parameters for the nonlinear system identification should be aimed at reducing the errors.

When choosing the parameters, we should first evaluate the complexity of the system, i.e., the system's memory, bandwidth, and the degree of nonlinearity. The memory is estimated as a time period, during which the influence of the input on the output virtually disappears. The frequency band of the testing noise should overlap the bandwidth of the system, but should not be too wide, to reduce errors. In practice, the upper limit of the frequency band of the testing noise is defined by the highest frequency of the system response that can be detected on the background of the system's noise.

The degree of nonlinearity of the system can be estimated based on the system response to a sinusoidal signal; the Volterra kernels can be approximately estimated by the higher harmonics. However, the results may significantly depend on the frequency, since nonlinear elements have memory. We should estimate the degree of the system's nonlinearity in order to determine the number and types of kernels that should be calculated during the nonlinear system identification.

Approximately, variations of the estimates of kernels are inversely proportional to the duration of testing signals; and the duration of testing signals required to achieve a sufficient accuracy of the estimates depends on the memory of the system, the degree of nonlinearity, and the type and intensity of the system's noise. Usually it is limited only by our computational capabilities (Marmarelis and Marmarelis 1978).

The Wiener functionals and kernels depend on the intensity of testing noise according to (5.1.16) and (5.1.17), and thus, they differ from the Volterra kernels, which are system's invariants. The correct choice of the intensity of testing signals is important for obtaining adequate estimates of the Wiener kernels and functionals.

A set of Wiener kernels completely describes the system, i.e., if we know the kernels and the intensity of testing white noise, we can predict the system's response to any input signal.

The Wiener and Volterra series represent expansions of the same function, and they are related by an unambiguous analytical dependence: the Wiener kernels of even (odd) orders are polynomials of the white noise intensity P with coefficients depending on all even (odd) higher-order Volterra kernels:

$$h_{2n}(\sigma_1, \ldots, \sigma_{2n}) = \sum_{m=n}^{\infty} \frac{2m!P^{m-n}}{2n!(m-n)!2^{m-n}}$$

$$\int_0^{\infty} \ldots \int_0^{\infty} k_{2m}(\tau_1, \tau_1, \ldots, \tau_{m-n}, \tau_{m-n}, \sigma_1, \ldots, \sigma_{2n}) d\tau_1 \ldots d\tau_{m-n},$$

$$h_{2n+1}(\sigma_1, \ldots, \sigma_{2n+1}) = \sum_{m=n}^{\infty} \frac{(2m+1)!P^{m-n}}{(2n+1)!(m-n)!2^{m-n}}$$

$$\int_0^\infty \ldots \int_0^\infty k_{2m+1}(\tau_1, \tau_1, \ldots, \tau_{m-n}, \tau_{m-n}, \sigma_1, \ldots, \sigma_{2n+1}) d\tau_1 \ldots d\tau_{m-n}, \quad (5.1.18)$$

As a result, the first-order Wiener functional of a nonlinear system generally differs from the linear description of the system by the first-order Volterra functional and depends on higher-order Volterra kernels, i.e., it reflects nonlinear properties of the system. For any signal, the Wiener series converges faster than the Volterra series, and a partial sum of the Wiener series provides the best truncated model of the system, which can be expanded by introducing higher-order terms without changing lower-order terms (Marmarelis and Marmarelis 1978). The Wiener and Volterra kernels coincide only for systems, in which all the kernels of orders higher than three are zero (Marmarelis and Marmarelis 1978).

For better understanding of the results of nonlinear identification of the soil response, let us consider some methodological issues.

The choice of the intensity of testing signals. In nonlinear system identification, we should choose the intensity of testing Gaussian white noise so that the content of nonlinear components in the soil response to the Gaussian white noise would be (approximately) the same as in the soil response to the input signal—the earthquake.

Since the nonlinearity of the soil response is defined by nonlinear stress–strain relations in soil layers, we should choose the intensity of testing noise signals so that the average working intervals of the stress–strain relations during the propagation of testing signals and seismic waves from the earthquake in soil layers coincide.

Figures 5.1 and 5.2 present the results of two numerical experiments illustrating the choice of the intensity of testing signals.

Let us consider a nonlinear filter possessing quadratic and cubic nonlinearities:

$$y(t) = x + 0.2 \cdot x^2(t) - 0.2 \cdot x^3(t) \quad (5.1.19)$$

Harmonic signals with a frequency of 2 Hz (the results do not depend on frequency, because the filter has no memory) and the Gaussian white noise were used as testing signals. The amplitudes of the harmonics and the intensity of the Gaussian white noise were chosen so that the average intensities of the outputs coincide, which provides an approximate coincidence of the working intervals of the nonlinear characteristics of the filter.

Figures 5.1 and 5.2 illustrate two calculations with lower and higher intensities of the input signals. Harmonics that have passed the nonlinear filter were enriched with the second- and third- higher harmonics of the frequency 2 Hz, representing nonlinear quadratic and cubic components of the filter response. The intensities of the nonlinear components were estimated in percent of the output signal intensity (Figs. 5.1a and 5.2a). When the input signal was the Gaussian white noise, the nonlinear quadratic and cubic components of the output were evaluated as the Wiener functionals.

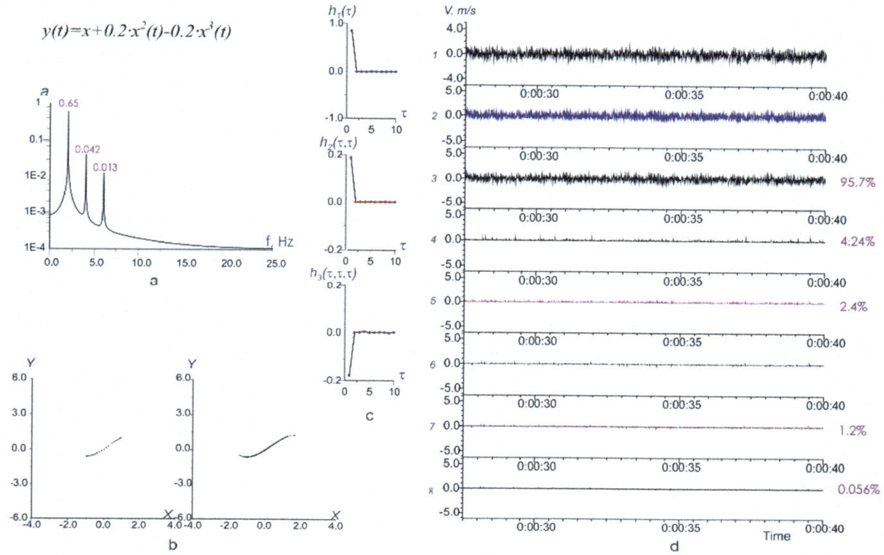

Fig. 5.1 A test example with lower-intensity input signals: **a**—the output spectrum, when the input was a harmonic at $f = 2$ Hz; **b**—the intensity of the output in cases when the input was a harmonic (left) and the Gaussian white noise (right); **c**—impulse characteristics $h_1(\tau)$, $h_2(\tau_1, \tau_2)$, and $h_3(\tau_1, \tau_2, \tau_3)$; **d**—1 is the input Gaussian white noise, 2 is the filter response, 3 is the response predicted by the linear model, 4 is the difference between the filter response and the response predicted by the linear model, 5 is the nonlinear correction associated with quadratic nonlinearity of the filter response, 6 is the difference between the filter response and the response predicted by the linear model with nonlinear quadratic correction, 7 is the nonlinear correction associated with cubic nonlinearity of the filter response, 8 is the difference between the filter response and the response predicted by the linear model with nonlinear quadratic and cubic corrections. The intensities of the quadratic and cubic nonlinear components of the filter response are expressed in percent of the total intensity of the response

Figures 5.1c and 5.2c show the first-order impulse characteristic $h_1(\tau)$ and diagonal values of the second- and third-order impulse characteristics of the filter $h_2(\tau_1, \tau_2)$ and $h_3(\tau_1, \tau_2, \tau_3)$. Since the Wiener and Volterra kernels coincide for the third-order filter, the extreme values of the impulse characteristics $h_1(\tau)$, $h_2(\tau_1, \tau_2)$ and $h_3(\tau_1, \tau_2, \tau_3)$ at zero time delays are equal to the coefficients of the corresponding terms in expression (5.1.19). The average intensities of the harmonics and the Gaussian white noise that have passed the filter are close to each other (Figs. 5.1b and 5.2b).

The ratio of the amplitudes of the 2nd and 3rd harmonics of the filter response and their ratios to the amplitude of the main frequency harmonic depend on the intensity of the input signal, as well as the ratios of the nonlinear and linear components in the output in case of the Gaussian white noise as an input. Also, the ratio of the amplitudes of the 2nd and 3rd harmonics corresponds to the ratio of the intensities

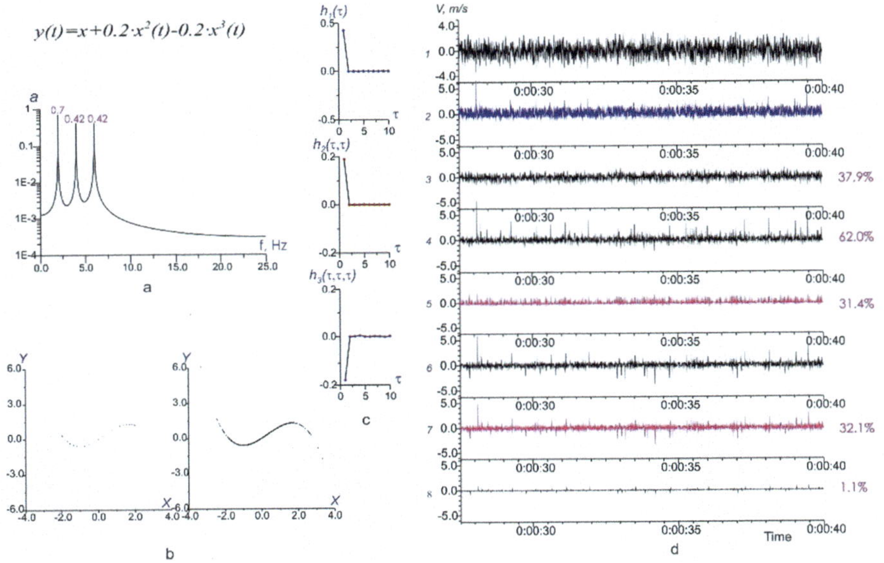

Fig. 5.2 A test example with higher-intensity input signals. The designations are the same as in Fig. 5.1

of the quadratic and cubic nonlinear components in the filter response in case when the input signal was the Gaussian white noise.

Thus, the contents of nonlinear components in the response of a nonlinear system to two different input signals is approximately the same, when the average working intervals of the nonlinear characteristics of the system are similar.

Description of a soil thickness as a nonlinear filter; the connection of $h_2(\tau_1, \tau_2)$ *and* $h_3(\tau_1, \tau_2, \tau_3)$ *with the filter coefficients.* As shown in Chap. 3, the nonlinearity of the soil response in the near-fault zones can be rather high. Numerical experiments show us that if harmonic signals with amplitudes corresponding to strong motion in the near-fault zones of strong earthquakes propagate in soft soil layers, higher harmonics up to the 11th–13th-orders of the main frequency are generated. Then, the stress–strain relations in soil layers can be considered as nonlinear filters in the form

$$\sigma(\varepsilon) = \alpha_1\varepsilon + \alpha_2\varepsilon^2 + \alpha_3\varepsilon^3 + \alpha_4\varepsilon^4 + \alpha_5\varepsilon^5 + \ldots + \alpha_{11}\varepsilon^{11}, \tag{5.1.20}$$

where σ is stress, ε is strain, and $\alpha_1, \alpha_2, \alpha_3, \alpha_4, \alpha_5, \ldots$ are constant coefficients.

The soil thickness can be represented as a nonlinear system consisting of the sequence of a linear unit with an impulse characteristic $g(\tau)$ and a set of nonlinear units. The output of such a system is a series with the terms up to 11th–13th orders:

$$y(t) = a_1x(t) + a_2x^2(t) + a_3x^3(t) + a_4x^4(t) + a_5x^5(t) + \dots$$
$$= \int_0^\infty a_1g(\tau)x(t-\tau)d\tau + \int_0^\infty\int_0^\infty a_2g(\tau_1)g(\tau_2)x(t-\tau_1)x(t-\tau_2)d\tau_1d\tau_2$$
$$+ \int_0^\infty\int_0^\infty\int_0^\infty a_3g(\tau_1)g(\tau_2)g(\tau_3)x(t-\tau_1)x(t-\tau_2)x(t-\tau_3)d\tau_1d\tau_2d\tau_3$$
$$+ \int_0^\infty\int_0^\infty\int_0^\infty\int_0^\infty a_4g(\tau_1)g(\tau_2)g(\tau_3)g(\tau_4)x(t-\tau_1)x(t-\tau_2)x(t-\tau_3)x(t-\tau_4)$$
$$d\tau_1d\tau_2d\tau_3d\tau_4 + \dots \quad (5.1.21)$$

where coefficie0nts a_1, a_2, a_3, a_4, a_5,... are connected to coefficients $\alpha_1, \alpha_2, \alpha_3, \alpha_4, \alpha_5, \dots$ by some linear relations. The Volterra kernels of such a system take the form:

$$k_1(\tau) = a_1g(\tau), \quad (5.1.22)$$

$$k_2(\tau_1, \tau_2) = a_2g(\tau_1)g(\tau_2), \quad (5.1.23)$$

$$k_3(\tau_1, \tau_2, \tau_3) = a_3g(\tau_1)g(\tau_2)g(\tau_3), \quad (5.1.24)$$

$$k_4(\tau_1, \tau_2, \tau_3, \tau_4) = a_4g(\tau_1)g(\tau_2)g(\tau_3)g(\tau_4), \quad (5.1.25)$$

$$k_5(\tau_1, \tau_2, \tau_3, \tau_4, \tau_5) = a_5g(\tau_1)g(\tau_2)g(\tau_3)g(\tau_4)g(\tau_5), \text{ and so on} \quad (5.1.26)$$

If *x(t)* is the Gaussian white noise, the Wiener kernels can be determined by the method of cross-correlation functions. They differ from the Volterra kernels, because the system possesses nonlinearities of orders higher than three. Using the Wiener kernels $h_2(\tau_1, \tau_2)$ and $h_3(\tau_1, \tau_2, \tau_3)$, , we obtain a truncated model of the series (5.1.20):

$$y(t) = b_1x(t) + b_2x^2(t) + b_3x^3(t). \quad (5.1.27)$$

The first term of the series, corresponding to $b_1\varepsilon$, is associated with the linear part of the effective average stress–strain relation of the soil thickness and with the first-order Wiener kernel $h_1(\tau)$, so that the maximum of $h_1(\tau)$ (corresponding to time τ of the propagation of a seismic wave from the "input" to soil layers to the surface) is the estimate of coefficient b_1.

The second- and third- order Wiener kernels $h_2(\tau_1, \tau_2)$ and $h_3(\tau_1, \tau_2, \tau_3)$ are associated with the quadratic and cubic nonlinear corrections in expression (5.1.27), and the extremums of $h_2(\tau_1, \tau_2)$ and $h_3(\tau_1, \tau_2, \tau_3)$, achieved at the same time delays as the maximum of $h_1(\tau)$ are the estimates of coefficients b_2 and b_3, respectively.

As follows from the properties of the Wiener and Volterra kernels, coefficients b_1, b_2 and b_3 depend on the intensity of the testing noise, and the weaker the nonlinearity of the system (or the smaller coefficients a_4, a_5, a_6,.....), the closer b_1, b_2 and b_3 to coefficients a_1, a_2, and a_3, which are invariants with respect to the input signals.

Coefficient b_1 depends not only on a_1, but also on the nonlinear part of the response, and coefficients b_2 and b_3 approximate the contributions of even and odd coefficients of the series (5.1.21), i.e., a_4, a_6, a_8, ... and a_5, a_7, a_9, ..., respectively.

Estimating coefficients b_1, b_2 and b_3, we obtain a mathematical description of the response of the soil thickness (5.1.21) with the minimum number of terms, and we can say that the estimates of coefficients b_1, b_2 and b_3 (as the extremums of the Wiener kernels $h_1(\tau)$, $h_2(\tau_1, \tau_2)$ and $h_3(\tau_1, \tau_2, \tau_3)$ in testing soil layers with rather weak noise signals) approximate the estimates of coefficients a_1, a_2, and a_3, which are nonlinear characteristics of soil layers measured in field experiments as parameters of the quadratic and cubic soil nonlinearity (Chap. 2).

5.2 Nonlinear Identification of Soil Profiles in the Near-Fault Zones During the 1995 Kobe and 2000 Tottori Earthquakes

Port Island, SGK, and TKS Sites, 1995 Kobe Earthquake (Mw ~6.8)

Let's apply nonlinear identification methods to estimate the degree of nonlinearity of the soil response at Port Island, SGK and TKS sites during the 1995 Kobe earthquake. The locations of the sites, epicenters of the mainshock and the aftershocks are shown in Fig. 2.10, the ground profiles are presented in Fig. 3.1. Stress–strain relations in soil layers were estimated from the surface down to the location of the deep device (Sect. 3.1). The observed and simulated acceleration time histories of the mainshock (horizontal components) are shown in Fig. 3.2, where dotted lines mark time intervals used in the nonlinear identification.

To define the intensity of testing noise signals, average working intervals of the stress–strain relations in soil layers were estimated. Figure 5.3 illustrates the stress–strain relations, averaged within the groups of the upper layers at PI, SGK and TKS sites.

In our calculations, the stress–strain relations are used in their normalized form, as suggested by Hardin and Drnevich (1972), so that the same relations describe the behavior of different layers within a group. Averaging stress–strain relations within a group means finding average working intervals of the normalized stress–strain relations.

The degree of nonlinearity of the soil response was estimated by testing the soil profiles by harmonic signals. As shown above, the intensity of testing signals should be chosen so that the working intervals of stress–strain relations in the layers during

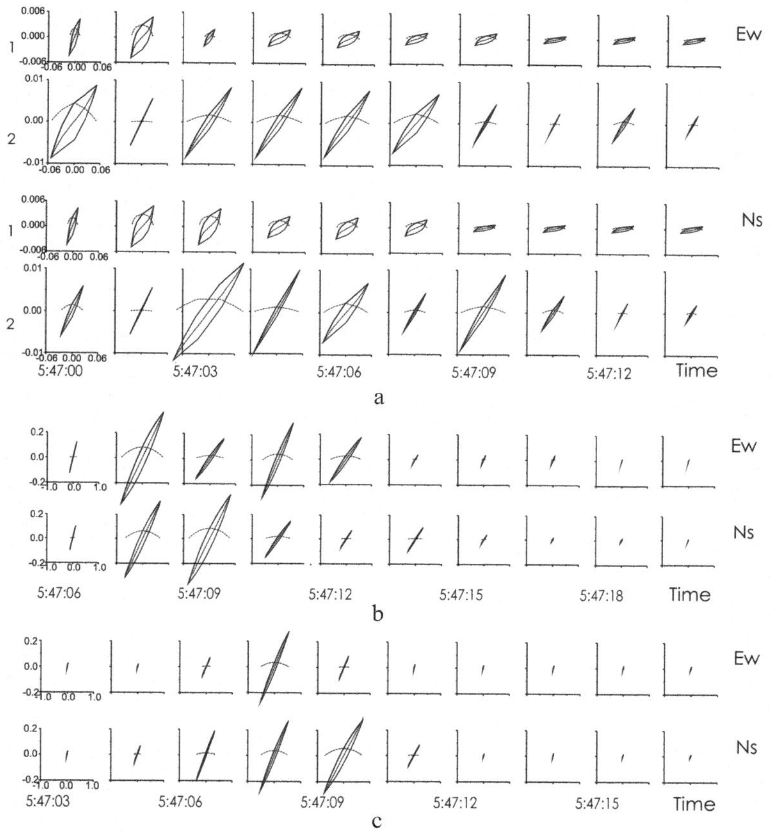

Fig. 5.3 Estimated stress–strain relations in soil layers during the 1995 Kobe earthquake, averaged within the groups of layers at **a**—PI (1: 0–13 m and 2: 13–17.5 m), **b**—SGK (0–11 m) and **c**—TKS (0–14 m), in 10 successive time intervals (each of 1.5 s duration). Scales in relative units are the same for all time intervals

the propagation of testing signals coincide with those during strong ground motion. In this case, the soils exhibit similar degree of nonlinearity, and with testing signals we can estimate the contents of nonlinear components in the soil response to strong motion.

However, in a layered medium the propagation of seismic waves of different spectral compositions is different, and generally we cannot achieve a coincidence of the working intervals of stress–strain relations during the propagation of testing signals and seismic waves from the earthquake.

Therefore, selecting the intensities of testing signals, we try to adjust the controlling parameters, such as, mean-square stresses, strains, and slopes of the stress–strain curves within the groups of layers. Two groups of layers were considered for SGK and TKS sites and five groups of layers for PI (Fig. 3.1).

Figure 5.4 shows the results of testing the obtained models of soil behavior (EW components) at the three sites during the 1995 Kobe earthquake by harmonic signals. To calculate the propagation of the harmonics in the soil profiles, stress–strain relations in soil layers during the Kobe earthquake (Sect. 3.1) were used.

Spectra of input harmonic signals are shown in lower rows at each site, and above them, there are spectra of simulated signals at depths of locations of the recording devices in 10 successive time intervals. As seen from the figure, the signals propagating up to the surface acquire higher harmonics of the main frequencies. A

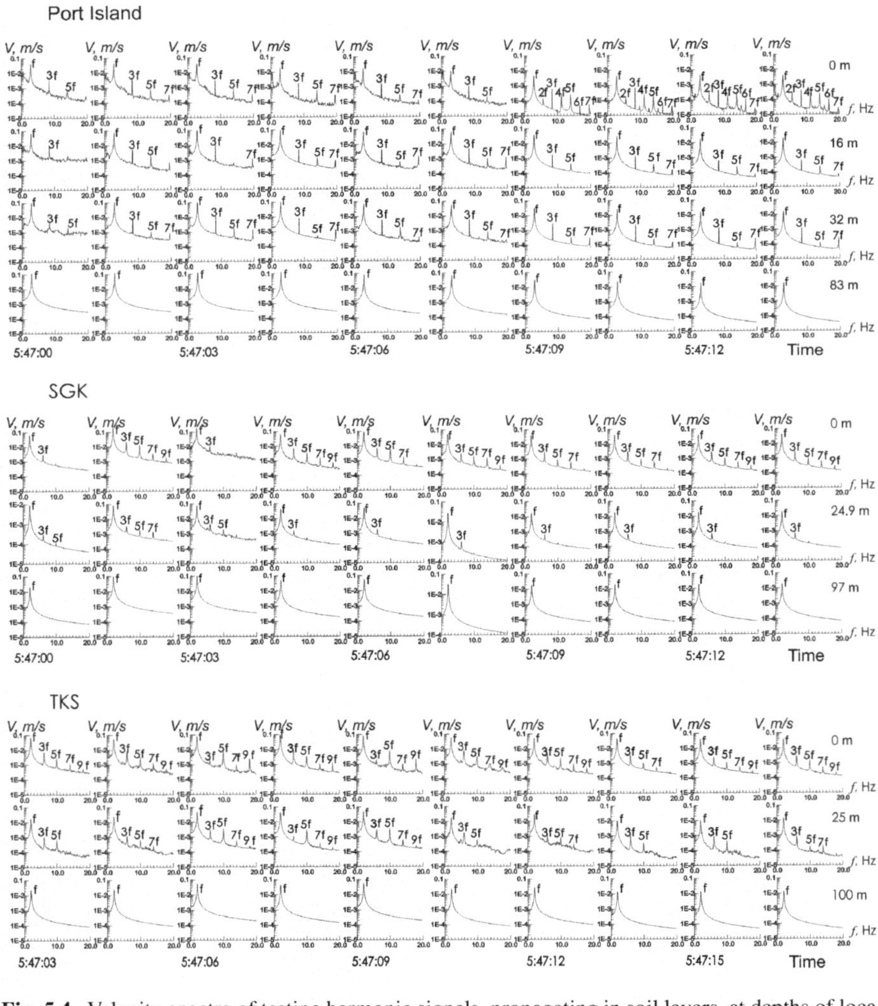

Fig. 5.4 Velocity spectra of testing harmonic signals, propagating in soil layers, at depths of locations of the recording devices at PI (0, 16, 32 and 83 m), SGK (0, 24.9 and 97 m) and TKS (0, 25 and 100 m) in 10 successive time intervals (15 s of strong motion)

large number of higher harmonics at PI and SGK sites testify to a strong nonlinearity of the soil response at these sites. At TKS site, fewer higher harmonics are generated, indicating weaker nonlinearity of the soil response.

Generation of the third, fifth, seventh and other odd-order harmonics testifies to odd-order nonlinearities of soil behavior. Even-order higher harmonics, such as, the second, fourth, etc., are generated only in liquefied soils at PI, in final parts of strong motion, and at SGK at the moments of the highest intensity of motion (Fig. 5.4).

Since nonlinear hysteretic stress–strain relations define transformations of input seismic motion into the ground response, generation of higher harmonics depends on the shapes of the stress–strain curves. Loading and unloading parts of the curves can be represented as sums of even and odd functions (any function defined within the interval $(-x; x)$ can be represented as a sum of an even function $0.5 * (f(x) + f(-x))$ and an odd function $0.5 * (f(x) - f(-x))$).

In Fig. 5.3, the odd components of the loading curves are shown as diagonal lines inside the hysteretic curves, and the even components are shown as convex arcs crossing the curves. The amplitudes of the odd components are higher than the amplitudes of the even components, except the latter intervals in the upper 13 m at Port Island, where the stress–strain curves decline to horizontal axis, due to liquefaction (Fig. 5.3). In these intervals, the soil response contains both odd and even nonlinear components (Fig. 5.4). Even-order higher harmonics are also generated at the moments of the highest intensity of motion at SGK, where the stress–strain relations acquire significant even components.

Thus, the hysteretic nature and the shapes of stress–strain relations cause the predominance of odd-order nonlinearities in the soil response, whereas even-order nonlinearities appear in cases when the loading and unloading parts of the hysteretic curves possess significant even components.

Figures 5.5, 5.6, and 5.7 show the results of testing the soil profiles by the Gaussian white noise in the range 0.1–100 Hz in 10 successive time intervals. Total nonlinear components of the soil response and nonlinear quadratic and cubic nonlinear components are shown in percent of the whole intensity of the response. The Wiener functionals were calculated by the method of cross-correlation functions (Marmarelis and Marmarelis 1978). Estimates of the zero-, first-, second- and third-order kernels are based on input and output signals of long duration (~40 min). The kernels were used to calculate the response of the linear models and nonlinear corrections accounting for quadratic and cubic nonlinearities of the response.

At Port Island site, liquefaction occurred in the upper layers, and five time intervals, corresponding to different stages of liquefaction are shown (Fig. 5.5). Figures 5.6 and 5.7 present examples of the nonlinear identification of the soil response at the beginning, middle and final parts of strong motion at SGK and TKS sites.

A noticeable difference in the shapes of input and output signals at PI (Fig. 5.5) indicates high nonlinearity of the soil response, which increases with developing liquefaction. Substantial nonlinear components (~40%) appear in the soil response during the first seconds of strong motion; they increase to ~60% with developing liquefaction (Fig. 5.5). As seen from the figure, at the beginning of strong motion in non-liquefied soils, the nonlinear cubic component of the response substantially

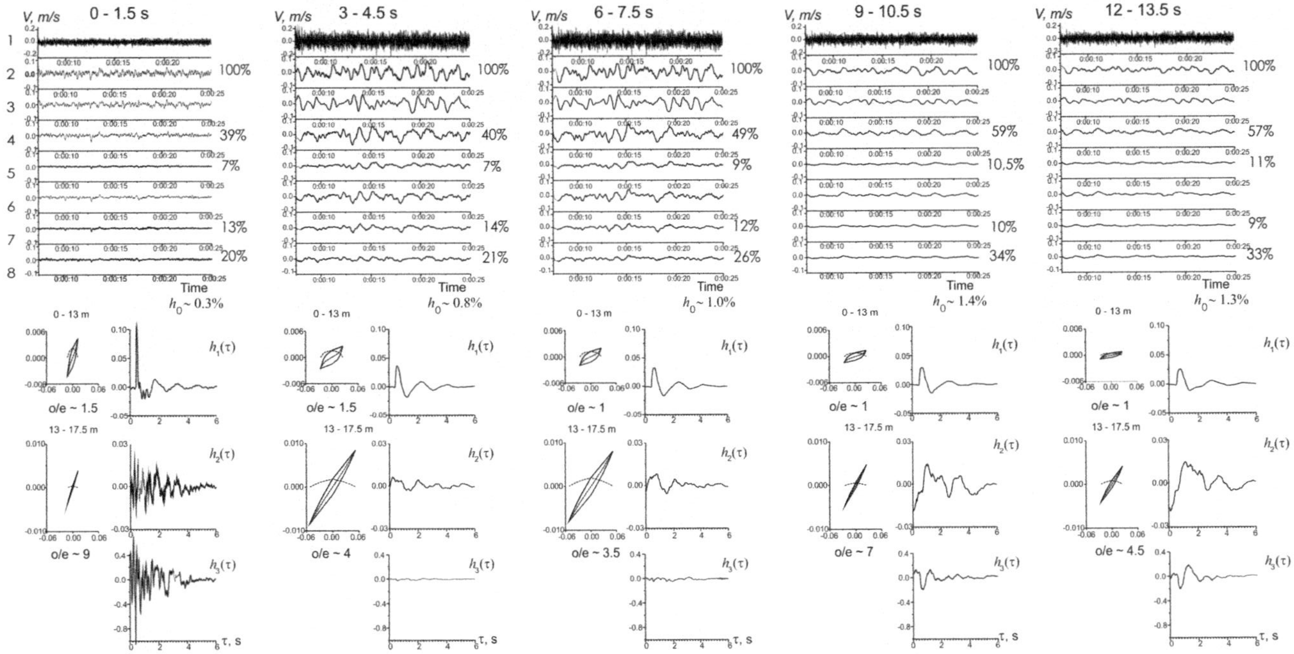

Fig. 5.5 Testing of soil layers at PI site by the Gaussian white noise (NS component). On the top, time passed from the beginning of strong motion is shown. 1 is the input, 2 is the output, 3 is the response predicted by the linear model, 4 is the deviation of the soil response from the response predicted by the linear model, 5 is the nonlinear correction due to quadratic nonlinearity predicted by kernel $\{h_2\}$, 6 is the deviation of the soil response from the response predicted by model $\{h_0, h_1, h_2\}$, 7 is the nonlinear correction due to cubic nonlinearity predicted by kernel $\{h_3\}$, 8 is the deviation of the soil response from the response predicted by model $\{h_0, h_1, h_2, h_3\}$. Below corresponding average stress–strain relations for the groups of soil layers are shown, represented as even and odd functions (o/e—are their ratios), and the diagonal values of kernels of the first, second and third orders. Zero-order kernels h_0 increase with developing liquefaction

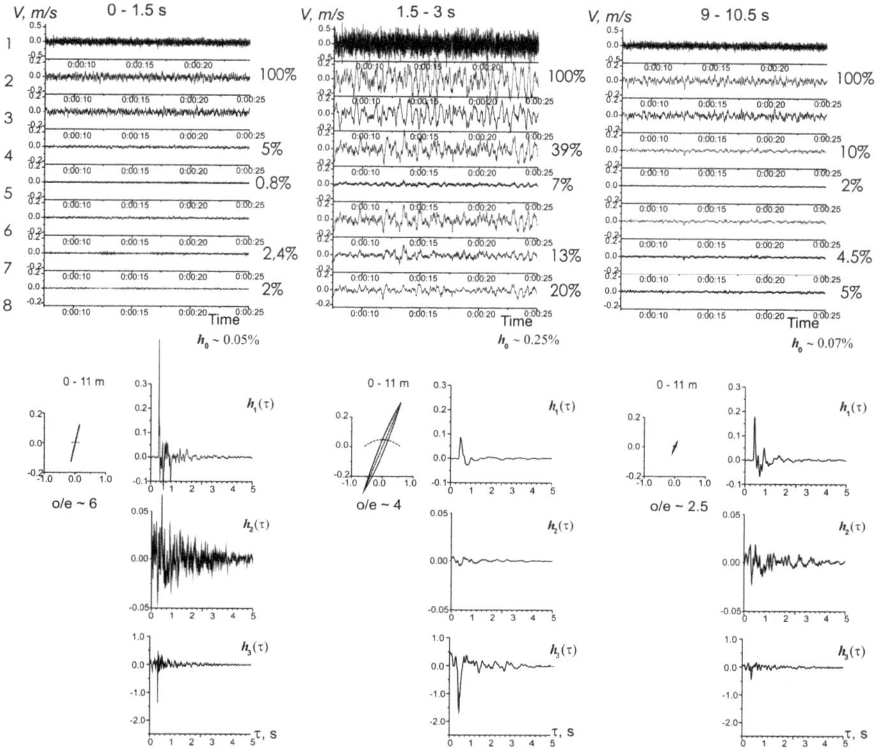

Fig. 5.6 Testing of soil layers at SGK site by the Gaussian white noise (NS component). The designations are the same as in Fig. 5.5

exceeds the nonlinear quadratic component. As liquefaction develops, the nonlinear quadratic component increases and approaches the nonlinear cubic component (Fig. 5.5).

The residual, i.e., the deviation of the soil response from the response predicted by model $\{h_0, h_1, h_2, h_3\}$, is due to higher-order nonlinearities and inaccuracies in the estimates of the Wiener kernels and functionals. The residual increases with developing liquefaction, which is obviously due to the increase of the total nonlinearity of the soil response.

To evaluate inaccuracies in the obtained estimates, numerical simulations were performed with linear models, where linear stress–strain relations were used instead of nonlinear hysteretic ones. For chosen duration of testing white noise signals (~40 min) the estimated nonlinear components of the response did not exceed 0.8–1% of the intensity of the response; these values characterize the inaccuracies in the obtained estimates of linear and nonlinear components of the soil response.

At SGK site, PGA recorded during the Kobe earthquake were high, and nonlinear components in the soil response reached ~40%, while at TKS site they did not

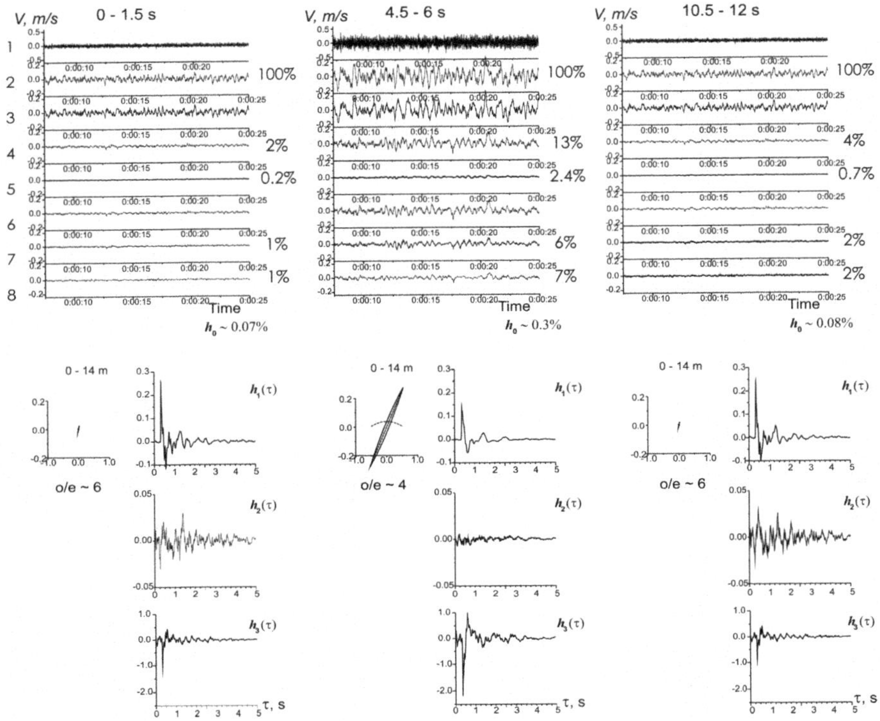

Fig. 5.7 Testing of soil layers at TKS site by the Gaussian white noise (EW component). The designations are the same as in Fig. 5.5

exceed ~13%. As seen from Figs. 5.6 and 5.7, the parts of nonlinear components in the soil response at SGK and TKS sites change with time, depending on the intensity of motion: nonlinear components are small at the beginning and at the end of strong motion, and they grow up and reach their maxima at the moments of the highest intensity of motion. The quadratic and cubic nonlinear components and the residual part of the response show similar trends, and nonlinear cubic components are substantially higher than nonlinear quadratic components.

Nonlinear components of even orders appear in the soil response at SGK at the moments when the intensity of motion increases to its maximum values (Fig. 5.6). At TKS, the intensity of motion is not so high, and the loading and unloading parts of the stress–strain relations are mainly odd functions; there are no even-order nonlinear components in the soil response (Fig. 5.7).

Zero-order kernels h_0 represent quasi-static deformations on the surface, that is, non-zero h_0 indicate slow shifting of surface layers to one or the other side during strong motion. This can be interpreted as accumulation of residual shear deformations on the surface (Zvolinsky 1982).

According to our estimates, at PI and SGK sites, quasi-static deformations reach ~15–20 cm, and at TKS site ~2–3 cm. At PI, quasi-static deformations increase with developing liquefaction (Fig. 5.5).

Figures 5.5, 5.6, and 5.7 show the first-order Wiener kernels $h_1(\tau)$, diagonal values of the second- and third-order Wiener kernels $h_2(\tau_1, \tau_2)$ and $h_3(\tau_1, \tau_2, \tau_3)$, and stress–strain relations averaged in the upper layers that correspond to the propagation of the Gaussian white noise in the soil profiles. Temporal changes of the Wiener kernels during strong motion are significant: at PI, maxima of $h_1(\tau)$ decrease and shift to larger time delays with developing liquefaction; i.e., amplification of seismic waves and shear moduli in the upper layers decrease, while the propagation time increases. Also, $h_1(\tau)$ kernels are smoothed and widened, and low-frequency components of motion on the surface increase. Second-order kernels $h_2(\tau_1, \tau_2)$ also increase and smooth with developing liquefaction.

At SGK and TKS sites, changes of the Wiener kernels during strong motion are also significant, and increase in the intensity of strong motion leads to smoothing the Wiener kernels and decreasing their amplitudes. The extreme values of $h_3(\tau_1, \tau_2, \tau_3)$ are usually higher than the extreme values of $h_2(\tau_1, \tau_2)$, especially at small strains. The first-order kernels $h_1(\tau)$ are more stable than $h_2(\tau_1, \tau_2)$ and $h_3(\tau_1, \tau_2, \tau_3)$, and the extreme values of $h_3(\tau_1, \tau_2, \tau_3)$ can be tens to hundreds times higher than that of $h_1(\tau)$ (Figs. 5.6 and 5.7).

As shown above (Sect. 5.1), the extreme values of $h_2(\tau_1, \tau_2)$ and $h_3(\tau_1, \tau_2, \tau_3)$ provide the estimates of the coefficients of the nonlinear filter that describes the behavior of the soil column and is defined by nonlinear stress–strain relations. At small strains, their values are close to the coefficients at the quadratic and cubic terms of the Volterra series.

Obviously, these coefficients and their ratios to the coefficients at linear terms were treated as characteristics of nonlinearity of subsurface soils in field experiments with seismic vibrators described in Sect. 2.2. Strains induced in soil layers during the 1995 Kobe earthquake at Port Island and SGK sites reached the values of ~10^{-3} in surface layers and ~10^{-5} at depth; at TKS site the strains reached the values of ~10^{-4} at the surface and ~10^{-5} at depth. This is comparable to strains excited in subsurface soils in the near field of a seismic vibrator.

The results of numerical simulations explain high values of effective parameters of nonlinearity in subsurface soils obtained in field experiments, ~$10^2 \div 10^4$ (Gushchin and Shalashov 1981; Shalashov 1984; Groshkov et al. 1991), and large scattering of the estimates made by various authors in their experiments. The latter is due to the established high sensitivity of the nonlinear coefficients to the shapes of the stress–strain relations. Since soils change their rheological properties during strong motion, large variations of the parameters of nonlinearity obtained in field experiments can be easily explained. Variations in the parameters of nonlinearity are associated with local variations of the subsurface soils and their rheological properties during the dynamic loading, with seasonal fluctuations of temperature and humidity.

The connection between the shapes of stress–strain relations in fractured rocks and manifestations of their nonlinearity was studied by Van Den Abeele and Johnson

(1996) and Kadish et al. (1996). "Hysteretic effects" in rocks were studied analytically and in laboratory experiments. Hysteretic loops were approximated by parallelograms, and if the parallelograms were narrow, the model predicted "zero spectral density at even-order harmonics of the source frequency" and "amplitude coupling" at odd-order harmonics.

Spectra of testing signals of the Gaussian white noise at Port Island, SGK and TKS sites calculated at depths of locations of the devices of the vertical arrays, are shown in Fig. 5.8. Transformations of spectra of seismic waves propagating in soil layers are clearly seen: spectra of motion on the surface tend to take the smoothed form $E(f) \sim f^{-k}$.

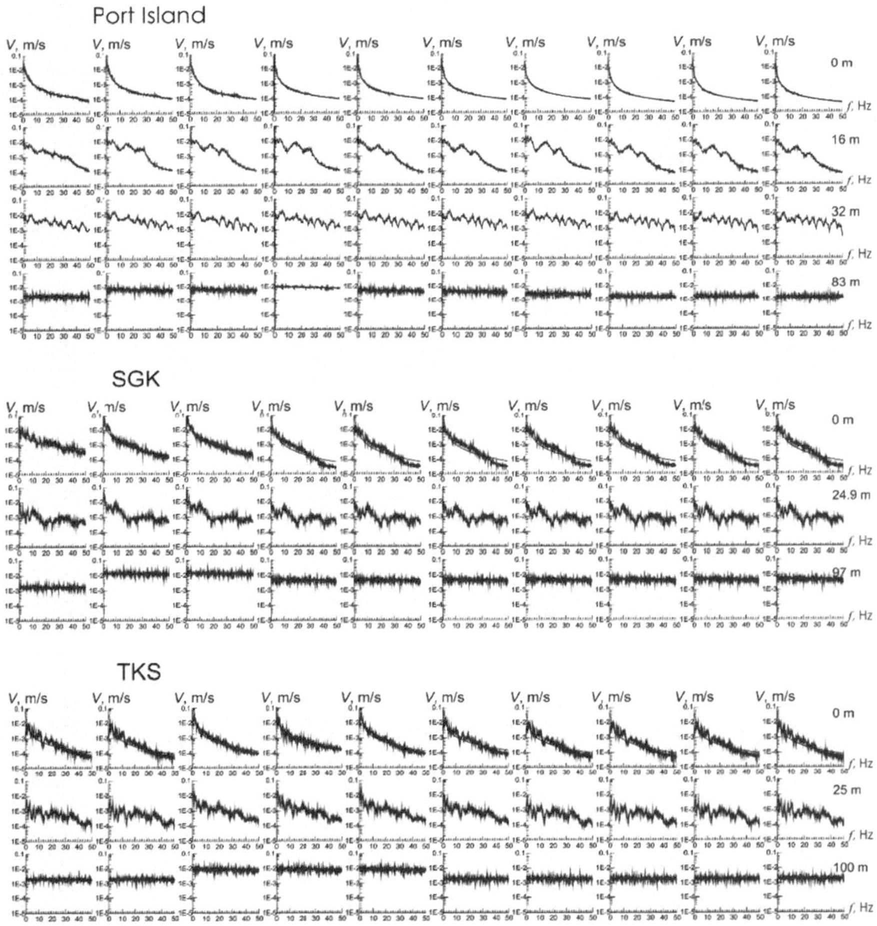

Fig. 5.8 Testing of soil layers at Port Island, SGK, and TKS sites by the Gaussian white noise: velocity spectra of input motion (at depths 83, 97, and 100 m) and spectra of motion at depths of locations of the devices of the vertical arrays in 10 successive time intervals (15 s of strong motion)

In soil layers at depths of 16 and 32 m at Port Island, 24.9 m at SGK and 25 m at TCS, along with nonlinear interactions of spectral components and generation of combination-frequency harmonics in low-frequency and high-frequency domains, resonant amplification occurs in some spectral ranges. Finally, the energy of propagating seismic waves is redistributed over the spectral bands so that spectral peaks smooth, and the resulting spectra on the surface tend to take the form $E(f) \sim f^{-k}$ (Fig. 5.8) (in Chap. 6, this is described in more detail).

Acceleration spectra of the 1995 Kobe earthquake at Port Island, SGK and TKS are presented in Fig. 5.9, and the trends described above are clearly seen: low-frequency components increase, and spectral peaks smooth with approaching the surface. The spectral shapes on the surface and at depth (~100 m) noticeably differ, due to strong nonlinearity of the soil response.

Parameter k can be estimated in numerical simulations or based on real acceleration time histories. In cases, when we can approximate spectra of motion on the surface by the relation $E(f) \sim f^{-k}$, k values are close to ~1.3–1.4. In numerical simulations, k ~1.13–1.44 at Port Island, and k ~1.08–1.51 at SGK at the highest intensity of motion.

Stations of KiK-net Strong Motion Network, Near-Fault Zones of the 2000 Tottori Earthquake (Mw ~6.7)

Nonlinear identification of soil behavior was performed for KiK-net sites located in the near-fault zones of the 2000 Tottori earthquake. Figures 3.15 and 3.16 show the locations of TTRH02, SMNH01, HRSH06, SMNH03 and HRSH05 sites in the near-fault zones, the epicenters of the mainshock and aftershocks, and the soil profiles. For the nonlinear identification, the constructed models of soil behavior during strong

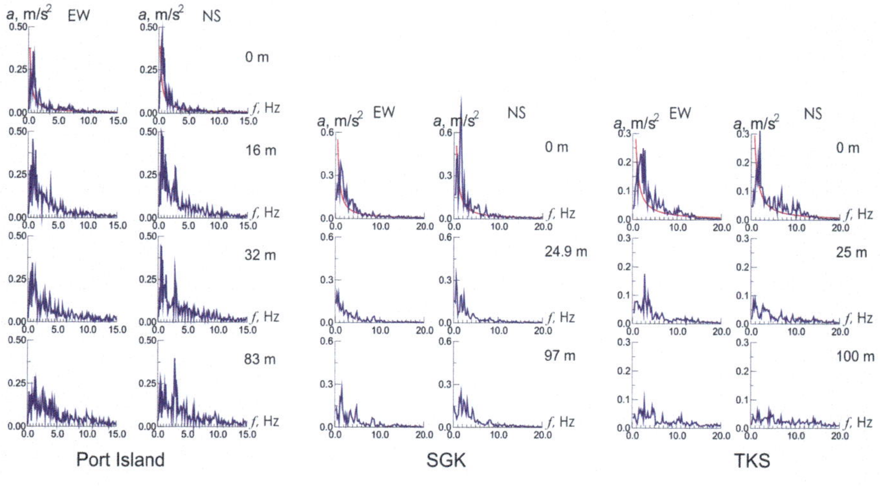

Fig. 5.9 Acceleration spectra of the mainshock of the 1995 Kobe earthquake at PI, SGK and TKS sites

motion were used (Figs. 3.17–3.26). The intensity of testing signals was chosen so that the root-mean-square (*rms*) stresses, strains, the ratios of *rms* stresses to *rms* strains and the root-mean-square slopes of the stress–strain relations during the propagation of seismic waves from the earthquake and testing signals most closely fit each other.

To estimate the degree of nonlinearity of the soil response, the soil profiles were tested by harmonic signals at frequency *f* ~3 Hz (Fig. 5.10). At TTRH02 and SMNH01 sites, soil behavior was substantially nonlinear, i.e., odd-order harmonics of the main frequency were generated in soil layers, and even-order harmonics were also generated, when the intensity of motion was rather high. At TTRH02 site, subharmonics of the main frequency were also generated at the moments of the highest intensity of motion. At remote sites HRSH06, SMNH03 and HRSH05 the nonlinearity of the soil response was rather low (Fig. 5.10).

The results of testing the soil profiles at TTRH02, SMNH01, HRSH06, SMNH03 and HRSH05 sites by the Gaussian white noise are shown in Figs. 5.11, 5.12, and 5.13. The Wiener kernels are estimated by the method of cross-correlation functions (Marmarelis and Marmarelis 1978); the estimates are based on long-duration (~40 min) input and output signals. The response of the linear model and the nonlinear

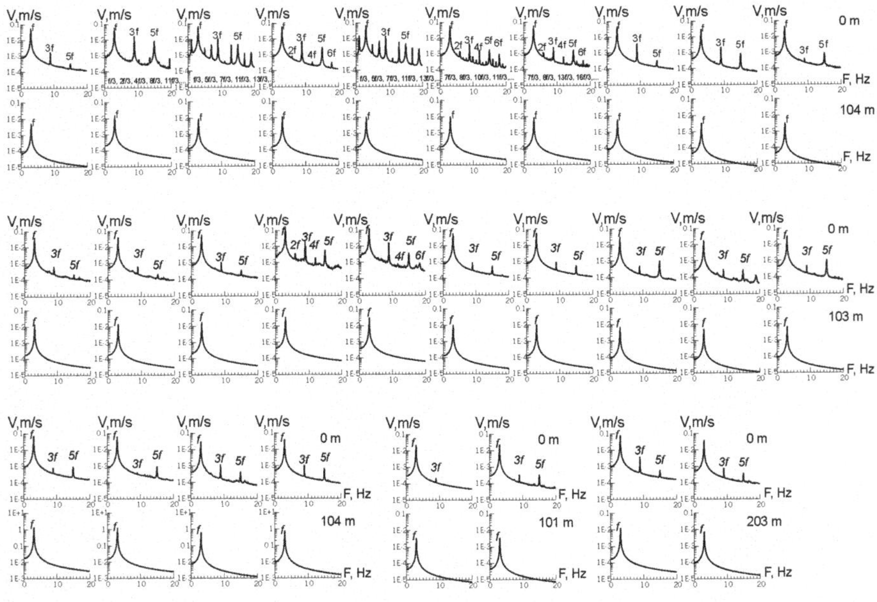

Fig. 5.10 Testing of the soil profiles at KiK-net sites by harmonic signals: velocity spectra of input motion at depths of locations of deep devices and spectra of motion on the surface. Upper row—TTRH02 (EW component), middle row—SMNH01 (EW component), lower row—HRSH06 (EW and NS components), SMNH03 (EW component), and HRSH05 (EW component). At TTRH02 and SMNH01 sites—in 10 successive time intervals; at HRSH06, SMNH03, and HRSH05 sites—at the moments of the highest intensity of motion

corrections due to quadratic, cubic and higher-order nonlinearities of the soil response are estimated based on the Wiener kernels.

As seen from the figures, the soil behavior at TTRH02 and SMNH01 sites was significantly nonlinear. At TTRH02 site located at the epicentral distance of ~7 km, surface soil layers are soft, with *S*-wave velocities increasing from 210 m/s at the surface to 350 m/s at ~20 m (Fig. 3.16), and the content of nonlinear components in the soil response was ~60%.

At SMNH01 site located at ~8 km from the epicenter, the nonlinearity of the soil response was lower, ~40–45%, evidently because of the soil composition: surface layers contain gravel, and V_S ~290 to 550 m/s in the upper 20 m (Fig. 3.16).

At HRSH06, SMNH03 and HRSH05 sites located at epicentral distances of ~57–80 km, the nonlinearity of the soil response did not exceed ~10%. The percentage of nonlinear components in the soil response approximately coincides with shear moduli reduction during strong motion, expressed in percent (Sect. 3.2).

Odd-order nonlinear components predominate (Figs. 5.11, 5.12, and 5.13); nonlinearities of even types are seen only at TTRH02 and SMNH01 sites at the moments of the highest intensity of motion. Zero-order kernels h_0 characterizing quasi-static surface deformations were also evaluated: at TTRH02, h_0 ~10–15 cm, at SMNH01 and other sites, h_0 ~2–3 cm.

The first order Wiener kernels $h_1(\tau)$ and the diagonal values of the second- and third-order kernels $h_2(\tau_1, \tau_2)$ and $h_3(\tau_1, \tau_2, \tau_3)$ change with time during strong motion, and the extreme values of $h_3(\tau_1, \tau_2, \tau_3)$ are substantially higher than the extreme values of $h_2(\tau_1, \tau_2)$, especially at the beginning of strong motion (Figs. 5.11, 5.12, and 5.13).

At TTRH02 and SMNH01 sites, maxima of $h_1(\tau)$ decrease with increasing the intensity of motion and shift towards larger time delays, indicating a decrease in amplification of seismic motion and reduction of shear moduli in the upper layers. Also, $h_1(\tau)$ kernels are smoothed and widened, and low-frequency components of motion on the surface increase. The second- and third-order kernels $h_2(\tau_1, \tau_2)$ and $h_3(\tau_1, \tau_2, \tau_3)$ decrease, smooth and obtain low-frequency components with increasing the intensity of motion. Oscillations of functions $h_1(\tau)$, $h_2(\tau, \tau)$ and $h_3(\tau, \tau, \tau)$ indicate multiple reflections of seismic waves from the boundaries of soil layers.

Figure 5.14 shows spectra of testing signals of the Gaussian white noise at KiK-net sites in 10 successive time intervals at depths of locations of the devices of the vertical arrays. Transformations of spectra of seismic waves propagating in soil layers are clearly seen: at a sufficiently high intensity of input motion, spectra of motion on the surface take the form $E(f) \sim f^{-k}$. Figure 5.15 shows acceleration spectra of the Tottori earthquake at KiK-net sites located in the near-fault zones.

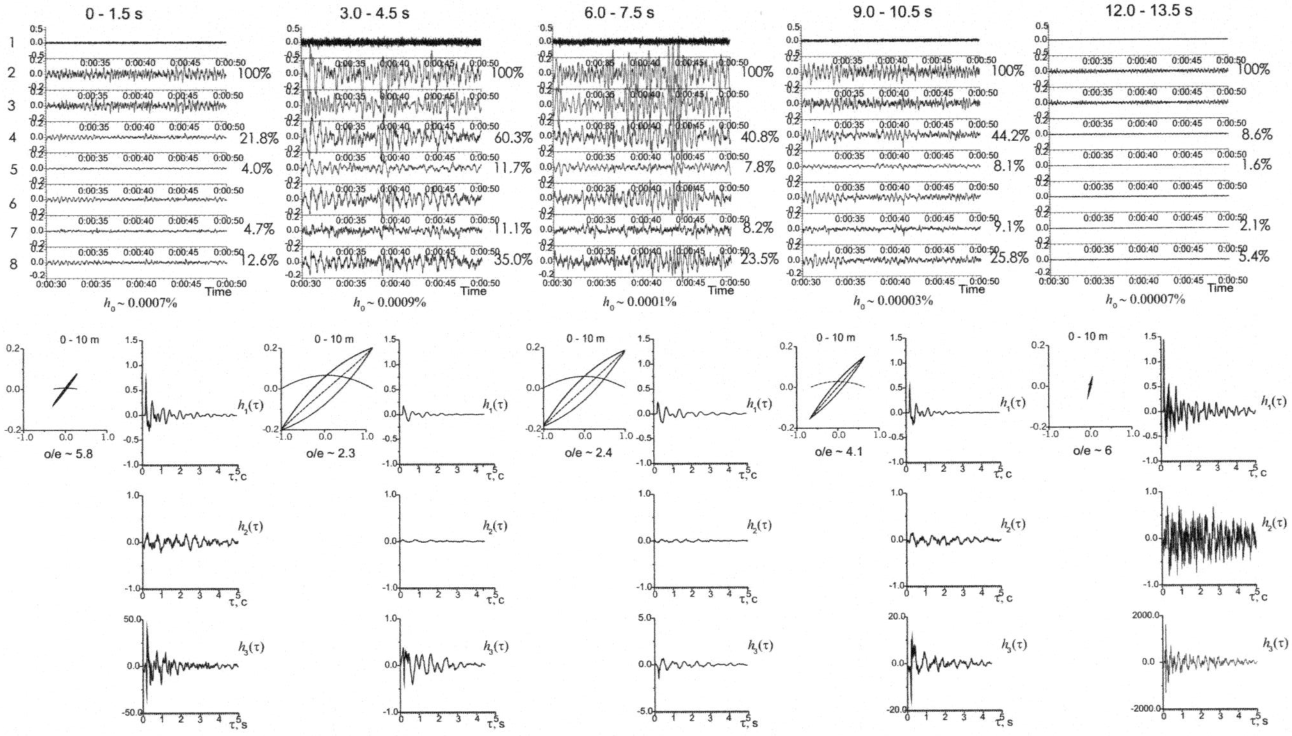

Fig. 5.11 Testing of soil layers at TTRH02 site by the Gaussian white noise (EW component). On the top, time passed from the beginning of strong motion is shown. 1 is the input, 2 is the output, 3 is the response predicted by the linear model, 4 is the deviation of the soil response from the response predicted by the linear model, 5 is the nonlinear correction due to quadratic nonlinearity predicted by kernel $\{h_2\}$, 6 is the deviation of the soil response from the response predicted by model $\{h_0, h_1, h_2\}$, 7 is the nonlinear correction due to cubic nonlinearity predicted by kernel $\{h_3\}$, 8 is the deviation of the soil response from the response predicted by model $\{h_0, h_1, h_2, h_3\}$. Below corresponding average stress–strain relations for the groups of soil layers are shown, represented as even and odd functions (o/e—are their ratios), and the diagonal values of kernels of the first, second and third orders

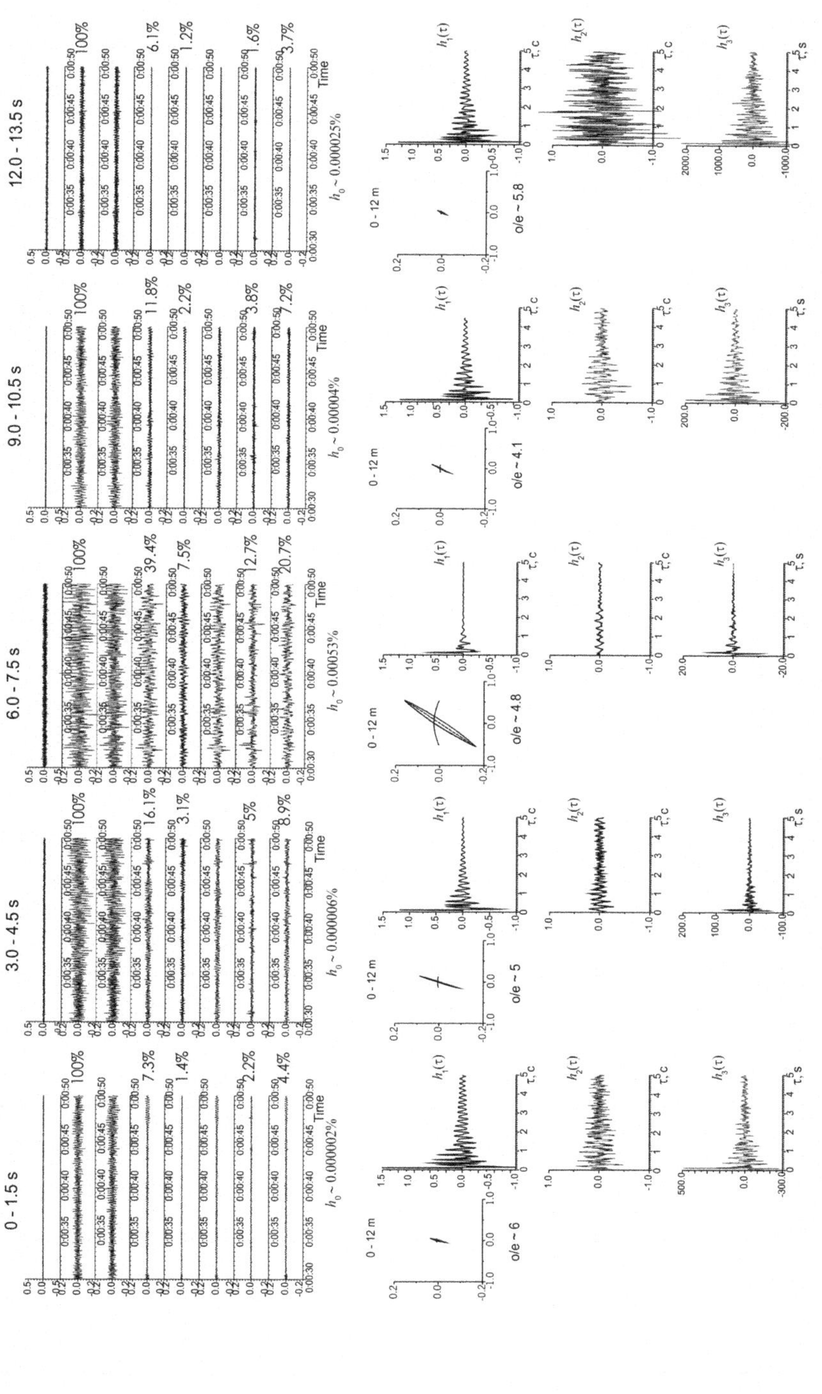

Fig. 5.12 Testing of soil layers at SMNH01 site by the Gaussian white noise. The designations are the same as in Fig. 5.11

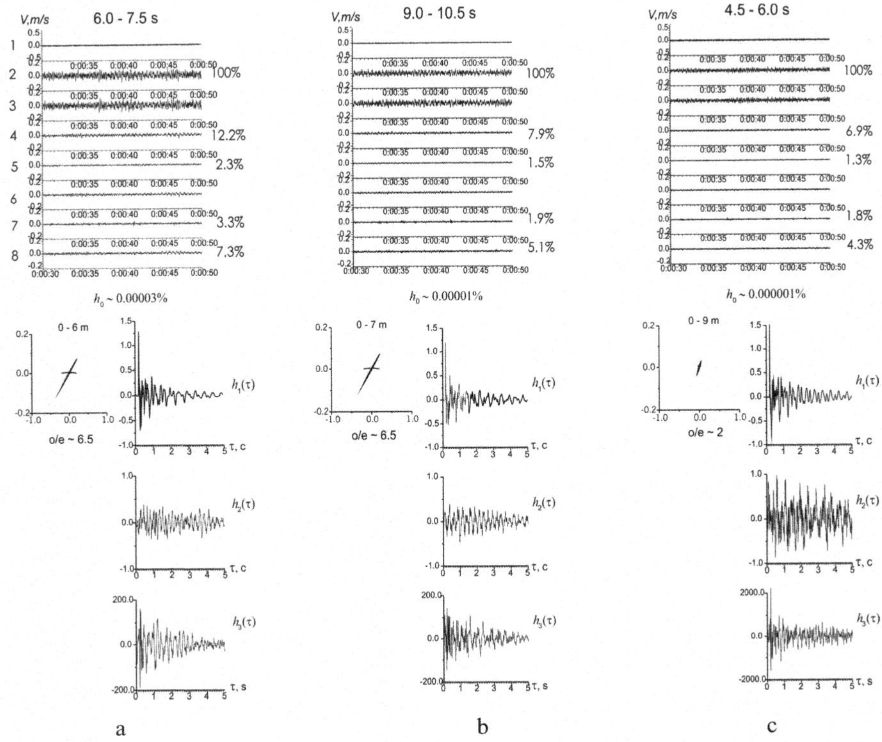

Fig. 5.13 Testing of soil layers at HRSH06 (**a**), SMNH03 (**b**) and HRSH05 (**c**) sites by the Gaussian white noise. The designations are the same as in Fig. 5.11

Spectral transformations to the form $E(f) \sim f^{-k}$ are clearly seen in cases of strong nonlinearity, i.e., at TTRH02 and SMNH01 sites. At other sites, the nonlinearity is not so strong, and resonant phenomena prevail, or spectral transformations caused by nonlinearity are not noticeable, because of thin soft soil layers. In cases when the spectral shape is close to $E(f) \sim f^{-k}$ (at TTRH02 and SMNH01 sites), parameter k is ~1.3–1.5, which is close to k estimates during the 1995 Kobe earthquake.

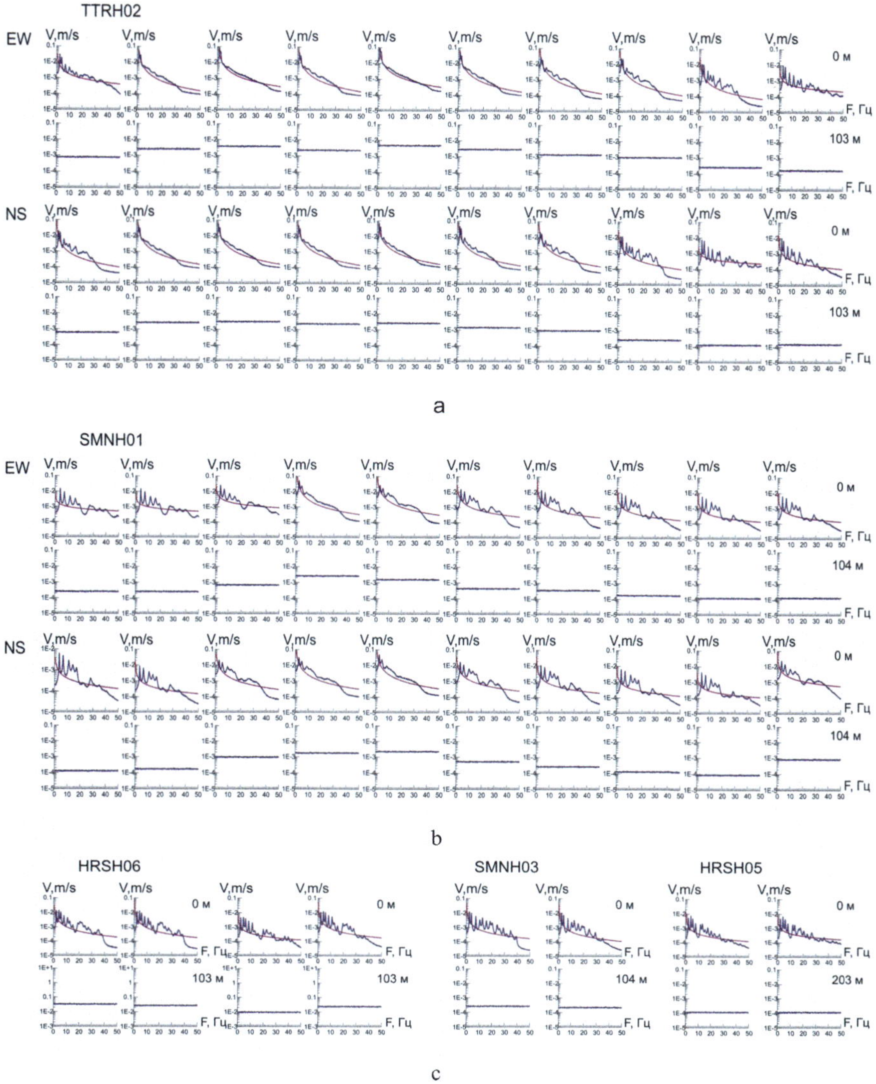

Fig. 5.14 Testing of the soil profiles at KiK-net sites by the Gaussian white noise: velocity spectra of input motion at depths of locations of deep devices and spectra of motion on the surface. **a**—TTRH02 (EW and NS components), **b**—SMNH01 (EW and NS components), **c**—HRSH06 (EW and NS components), SMNH03 (EW component), and HRSH05 (EW component). At TTRH02 and SMNH01 sites—in 10 successive time intervals; at HRSH06, SMNH03, and HRSH05 sites—at the moments of the highest intensity of motion

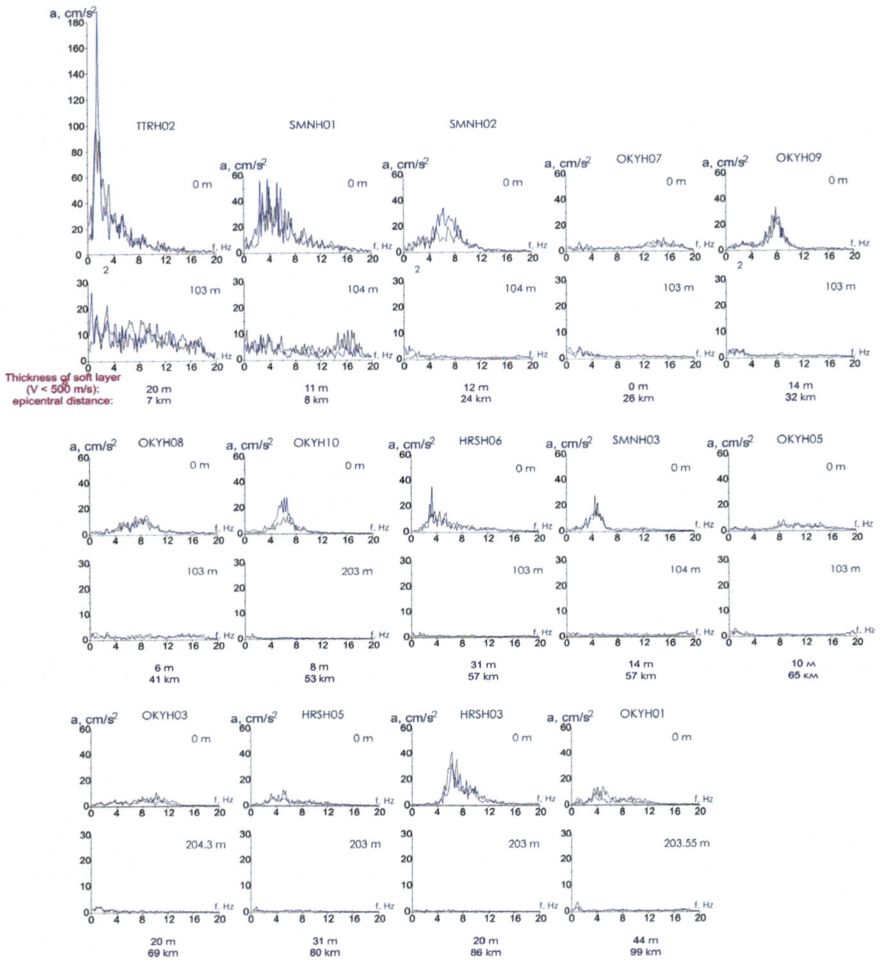

Fig. 5.15 Acceleration spectra of the 2000 Tottori earthquake at KiK-net sites located in the near-fault zones

5.3 Characteristics of the Soil Response in the Near-Fault Zones During the 1999 Chi-Chi Earthquake

The Chi-Chi earthquake ($M_w = 7.6$) that occurred in central Taiwan on September 21, 1999 was recorded by more than 400 strong motion devices island-wide. The majority of near-fault records were obtained at soil sites, and they apparently represent a rather complete database allowing a study of soil behavior during a strong earthquake in the near-fault zones.

Models of soil behavior are constructed for 31 soil sites located at various distances from the fault plane within a wide range of azimuthal directions (Sect. 3.3), and they allow us to obtain a general representation of the soil behavior during the Chi-Chi earthquake. With these models, we can investigate some regularities in soil behavior during a strong earthquake at various distances from the fault.

Based on the constructed models, parameters characterizing the soil response are estimated. Four groups of parameters were studied, such as, (1) characteristics of amplification of seismic waves by soil layers; (2) stresses and strains induced in soil layers by the Chi-Chi earthquake; (3) reduction of shear moduli in soil layers due to strong motion; and (4) contents of nonlinear components in the soil response. The distribution of these parameters around the fault plane is analyzed. Soft soil behavior during the Chi-Chi earthquake was essentially nonlinear in the near-fault zones, and in this section, the influence of nonlinearity of the soil response on the studied parameters is discussed.

The parameters of the soil response depend on the distance of a site from the fault plane. All the studied soil sites were located near the fault, and their distances from the fault plane are comparable with the fault length. Generally speaking, two values can be considered as the distance of a site from the earthquake source, such as, the shortest distance of a site from the fault plane and the average distance of a site from all subfaults of the fault plane (or the weighted average distance, with weights being proportional to subfault radiation intensities). It was found that the dependencies of the studied parameters on their distance from the source can be approximated by power functions, and the approximations are more exact (scattering of the estimates is smaller), when we define "the distance from the source" as the shortest distance of a site from the fault plane. Therefore, later on distance from the fault r designates the shortest distance of a site from the fault plane.

The locations of the studied soil sites, identified liquefaction zones, and the area of basin-induced surface waves are shown in Fig. 5.16.

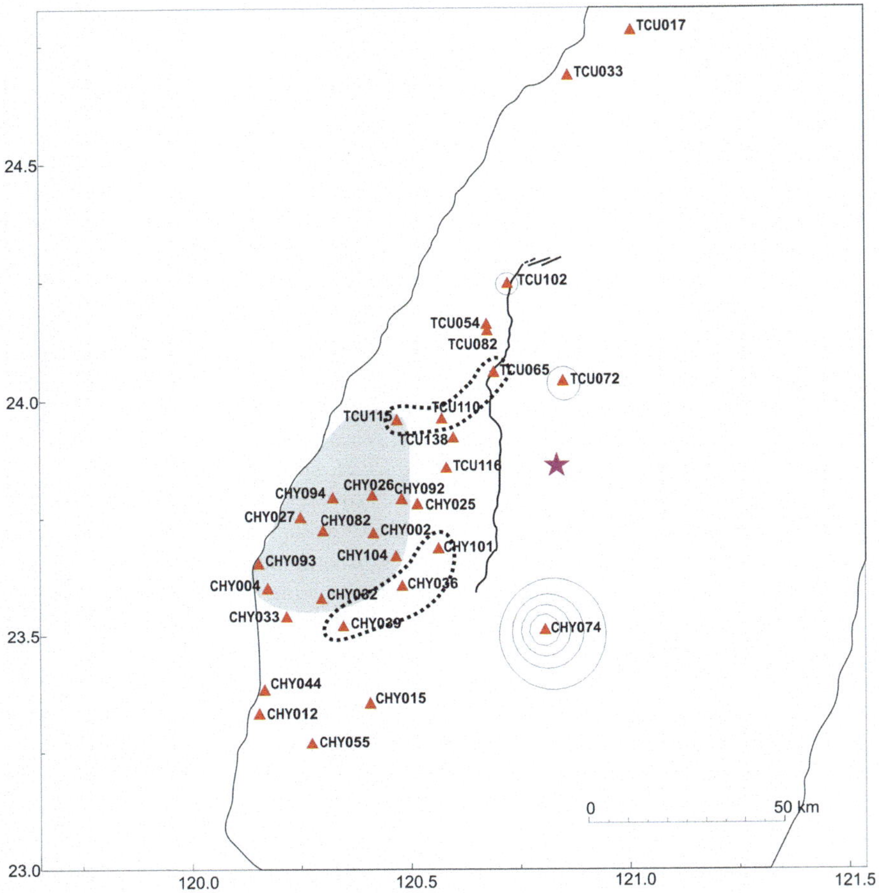

Fig. 5.16 Locations of the studied soil sites around the fault of the Chi-Chi earthquake. Dotted lines mark areas showing signs of soil liquefaction; area of basin-induced surface waves is marked by gray color

Table 5.1 summarizes information of the studied soil sites (closest distances to the fault plane r, average S-wave velocities in the upper 30 m of soil, V_{S30}, site classification according to Lee et al. (2001) and the obtained estimates of the soil response parameters, such as, amplification factors for acceleration and velocity F_a and F_v, average stresses τ_{30}, average strains γ_{30}, and shear moduli reduction in the upper 30 m of soil $\Delta G/G_{max30}$, as well as the contents of nonlinear components in the soil response. As seen from the Table, the majority of sites (22 sites) are class D and E sites, possessing soft soils in the upper 30 m, with $V_{S30} < 300$ m/s. Other 9 sites possess denser subsurface soils, where V_{S30} varies within 448–735 m/s, i.e., class C sites. It was found reasonable to consider separately two these groups of sites, such as sites with "softer" and "denser" subsurface soils.

Table 5.1 Characteristics of the soil response in the near-fault zones during the Chi-Chi earthquake

Station code	*r*, km	V_{S30}, m/s	F_a	F_v	τ_{30}, Pa	γ_{30}, 10^{-5}	$\Delta G/G_{max30}$, %	*NI*, %	Site class (Lee et al. 2001)
CHY101	1.9	260	1.5	2.0	1817	18.0	81	75	D
TCU110	2.4	207	0.7	2.9	1013	32.0	46	57	E
TCU116	3.4	483	1.0	1.8	852	0.3	32	16	E
TCU138	3.8	605	1.1	1.6	2140	0.54	10	24	D
TCU082	5.7	478	1.0	1.7	1162	0.3	15	15	D
CHY025	5.9	277	1.2	2.7	1487	11.5	41	62	E
TCU054	6.1	454	1.1	1.7	1494	0.8	21	11	D
TCU065	6.9	300	2.5	2.9	1677	9.3	63	54	D
CHY036	7.5	282	2.4	2.1	1995	9.6	73	61	D
TCU102	7.8	735	1.4	2.2	2325	3.4	51	56	D
CHY092	9.5	260	1.3	2.5	440	1.2	37	11	E
CHY104	9.5	225	2.2	2.9	1435	5.7	13	28	E
TCU115	11.4	190	1.4	3.1	654	17.0	34	50	E
CHY074	12.8	546	1.6	2.1	1824	3.5	44	25	C
TCU072	14.9	563	2.7	2.1	3481	6.2	37	48	D
CHY002	15.0	231	1.9	2.6	420	0.7	16	28	E
CHY026	16.1	226	1.2	2.0	530	8.0	22	33	E
TCU033	16.9	448	2.3	1.9	1642	0.5	12	22	D
CHY039	20.8	198	2.0	2.2	450	5.8	86	27	E
CHY094	24.9	227	1.6	2.7	1048	4.1	9	16	E
CHY032	25.6	202	3.1	3.2	780	2.9	9	33	E
CHY015	25.9	234	2.2	2.6	361	1.0	19	31	D
CHY0 82	26.6	209	2.2	3.0	211	0.8	13	25	E
TCU017	29.7	601	2.2	2.3	227	0.2	28	8	?
CHY027	32.1	221	3.9	3.4	645	4.9	13	24	E
CHY033	33.6	194	3.0	3.6	970	5.3	7	13	E
CHY004	38.2	279	2.8	3.0	1342	3.5	7	11	E
CHY093	41.2	199	2.8	3.4	1145	4.7	5	11	E
CHY055	41.5	240	3.7	3.0	1234	3.8	5	12	E
CHY044	43.0	193	3.0	2.9	576	2.2	5	30	E
CHY012	46.6	202	3.5	2.8	276	1.0	2	28	E

Amplification of Seismic Waves in Subsurface Soils in the Near-Fault Zones During the Chi-Chi Earthquake

Coefficients of amplification of seismic waves by soil layers represent the ratio of the intensity of an observed ground motion to a reference value of that intensity for a particular site condition (intact rock or rock-average). Here amplification factors were estimated separately for acceleration and velocity (from accelerograms and velocigrams, F_a and F_v) as the ratios of root mean-square (*rms*) accelerations and velocities on the surfaces and at the bottoms of soil layers. Values of *rms* accelerations and velocities, a_{rms} and v_{rms}, were calculated for 50-s time intervals of strong motion; estimates obtained for NS and EW components averaged. Amplification was estimated for the same acceleration time histories that were previously used for constructing models of soil behavior (which were taken as one of the series of accelerograms calculated by the stochastic method).

Accelerograms of the Chi-Chi earthquake possess a fairly uniform intensity during strong motion, and *rms* values should give us rather good estimates of the square roots of the ratios of intensities of oscillations on the surfaces and at the bottoms of soil layers. The selected 50-s time intervals correspond to the largest duration of records observed at southern stations (as mentioned in Sect. 3.3, directivity effects are clearly seen as decreased durations of acceleration time histories in forward northern directions and increased durations in backward southern directions; eastern and western directions represent intermediate cases).

At stations located to the north, east and west of the fault plane, the duration of records is smaller, however these differences (as well as the increase of the duration with distance) can only slightly influence amplification estimates, because the ratios are taken of *rms* values of oscillations on the surfaces and at the bottoms of soil layers. Generation of surface basin waves does not change the estimates of amplification, because the estimates are based on the simulated records, and basin waves are not simulated in the one-dimensional calculations.

Since amplification factors are considered to be functions of the amplitude of shaking, their dependencies on the distance from the fault plane (i.e., on the level of input motion) are studied. The estimates are shown in Fig. 5.17 and in Table 5.1.

As seen from the figure, a_{rms} and v_{rms} values at the bottoms of soil layers decrease with distance from the fault plane approximately as $\sim r^{-1}$ (Fig. 5.17), which corresponds to theoretical decrease of amplitudes due to geometrical spreading.

Amplification factors F_a and F_v vary from 1 to 4 for virtually all the sites (Fig. 5.17, Table 5.1). As seen from the figures, amplification increases with distance, i.e., with decreasing the amplitudes of "inputs" to soil layers; it is clearer for acceleration than for velocity. Linear functions approximate the dependencies; however, scattering is rather large, which can be due to various factors.

As known, three main mechanisms of seismic wave transformation in soil layers are: (1) increase of the intensity of motion on the surface due to transition of seismic waves from the earthquake sources to upper soil layers, possessing smaller seismic velocities and densities, (2) resonant phenomena in the upper softer soil layers also

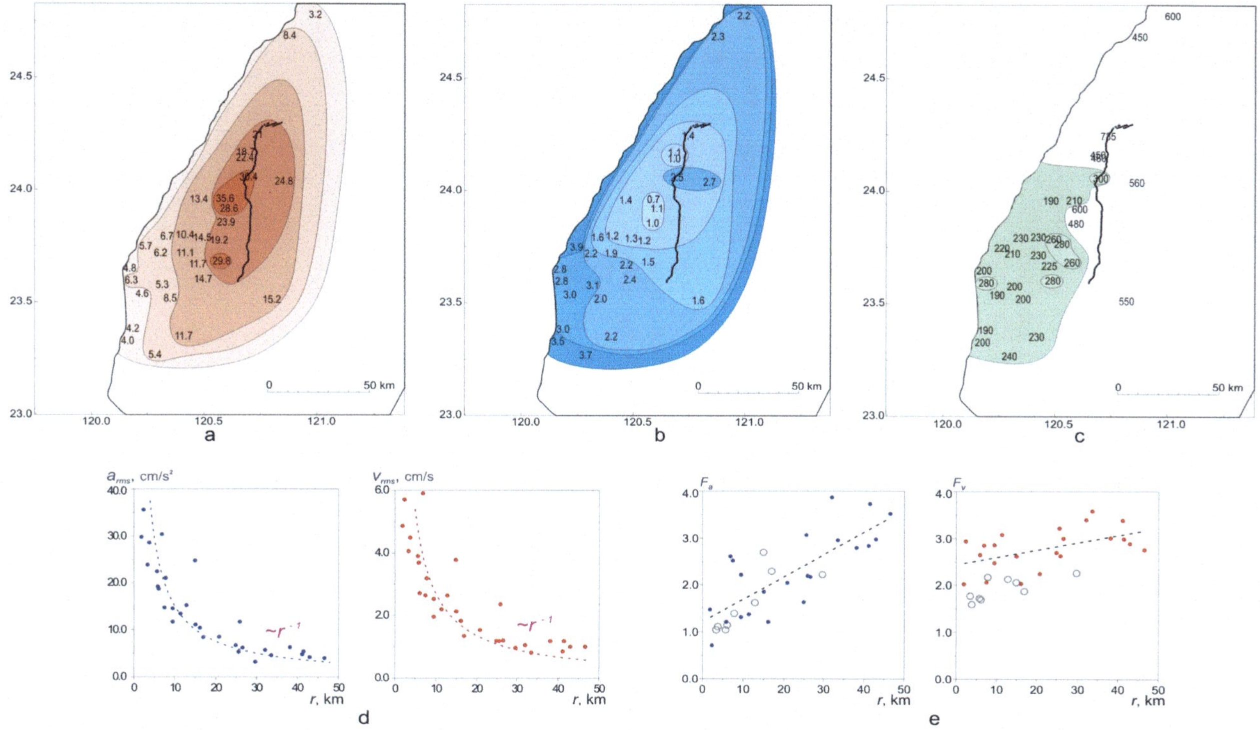

Fig. 5.17 Upper raw: distribution around the fault plane of: **a**—"input" motion to soil layers (*rms* accelerations at the bottoms of soil layers) a_{rms}, **b**—amplification of seismic waves by soil layers for acceleration F_a, **c**—average *S*-wave velocities in the upper 30 m, V_{S30}. Lower raw: **d**—"input" motion to soil layers a_{rms}, and v_{rms}, **e**—amplification for acceleration and velocity F_a and F_v as functions of the distance from the fault. Dots and circles show F_a and F_v at stations with softer and denser subsurface soils, respectively

leading to amplification of seismic motion on the surface, and (3) nonlinearity of the soil response, often leading to de-amplification of motion on the surface.

To evaluate the contribution of each mechanism to amplification of seismic waves in soil layers during the Chi-Chi earthquake, let's consider F_a and F_v separately at sites with softer and denser subsurface soils. As seen from Fig. 5.17, amplifications for acceleration F_a at softer and denser sites are close to each other and can be approximated by the same function of the distance from the fault (or of the intensity of input motion), whereas, amplifications for velocity F_v are noticeably smaller on denser than on softer subsurface soils. This agrees with seismological observations that "peak ground velocity and displacement show higher amplifications at soil sites than at rock sites, while peak ground acceleration is roughly independent of the site classification" (Aki and Irikura 1991) (that is, does not show higher amplification at soft soil sites).

The obtained dependencies of amplification on the level of input motion to soil layers is evidently due to nonlinearity of the soil response, which is higher at the near-fault sites. At rather large distances from the fault (~40–50 km), amplification factors F_a and F_v are close to each other (Fig. 5.17), whereas at small distances nonlinearity of the soil response noticeably decreases amplification for acceleration and to a lesser extent, for velocity. This can be due to nonlinear transformations of seismic waves propagating in soil layers: in cases of strong nonlinearity, their spectra tend to take the form $E(f) \sim f^{-k}$, when high- and medium-frequency spectral components decrease, while low-frequency components remain virtually at the same level or slightly increase. Obviously, the effect is stronger for accelerations than for velocities or displacements.

As concluded in Sect. 3.3, at the majority of sites, soil behavior during the Chi-Chi earthquake was defined by resonant oscillations induced by strong motion in the upper 40–60 m of soil, and by nonlinearity of soil response. At near-fault sites, subsurface water-saturated sandy soils possess hard-type nonlinearity, when de-amplification of seismic motion due to nonlinear damping can be rather small, because (1) stresses rapidly increase at large strains due to pore-pressure built-up (therefore, accelerations also increase) and (2) nonlinear damping (defined by the areas inside hysteretic curves) can be rather small, as, for example, at TCU065 site (Fig. 3.31).

As a consequence, in the near-fault zones, in sandy water-saturated soils possessing hard-type nonlinearity, de-amplification of seismic motion due to nonlinearity cannot compensate its amplification stipulated by two other mechanisms; as a whole, we observe amplification of motion on the surface. This is clearly seen, for example, in vertical array records of the 1995 Kobe earthquake at SGK site and at near-fault sites TCU065 and TCU072, where amplification factors F a are rather large, ~2.5–2.7. Thus, amplification can be rather large even in cases of strong nonlinearity.

Amplification of motion on the surface resulting from resonant phenomena in subsurface soils and hard-type soil nonlinearity is also observed at other near-fault sites, such as, CHY101 (F_a ~1.5), CHY036 (F_a ~2.4), TCU102 (F_a ~1.4), CHY104 (F_a ~2.2), TCU115 (F_a ~1.4), CHY074 (F_a ~1.6), and CHY002 (F_a ~1.9). Hard-type nonlinearity of the soil response virtually disappears at distances of more than

~12–15 km from the fault, because of the decrease of the intensity of input motion to soil layers. At rather large distances from the fault, amplification of seismic motion is due to two linear mechanisms described above, whereas nonlinear effects decrease with distance and can no longer effectively deamplify seismic motion.

Stresses and Strains Induced in Soil Layers in the Near-Fault Zones During the Chi-Chi Earthquake

Analyzing stresses and strains induced in soil layers during the Chi-Chi earthquake, we will also distinguish sites with softer ($V_{S30} \leq 300$ m/s) and denser ($V_{S30} \geq 450$ m/s) subsurface soils. Figure 5.18 presents estimated maximum stresses and strains induced in soil layers at the studied sites during the Chi-Chi earthquake. To obtain the estimates, at each site, for each soil layer, maximum (during the strong motion) stresses and strains were found and averaged over two horizontal components.

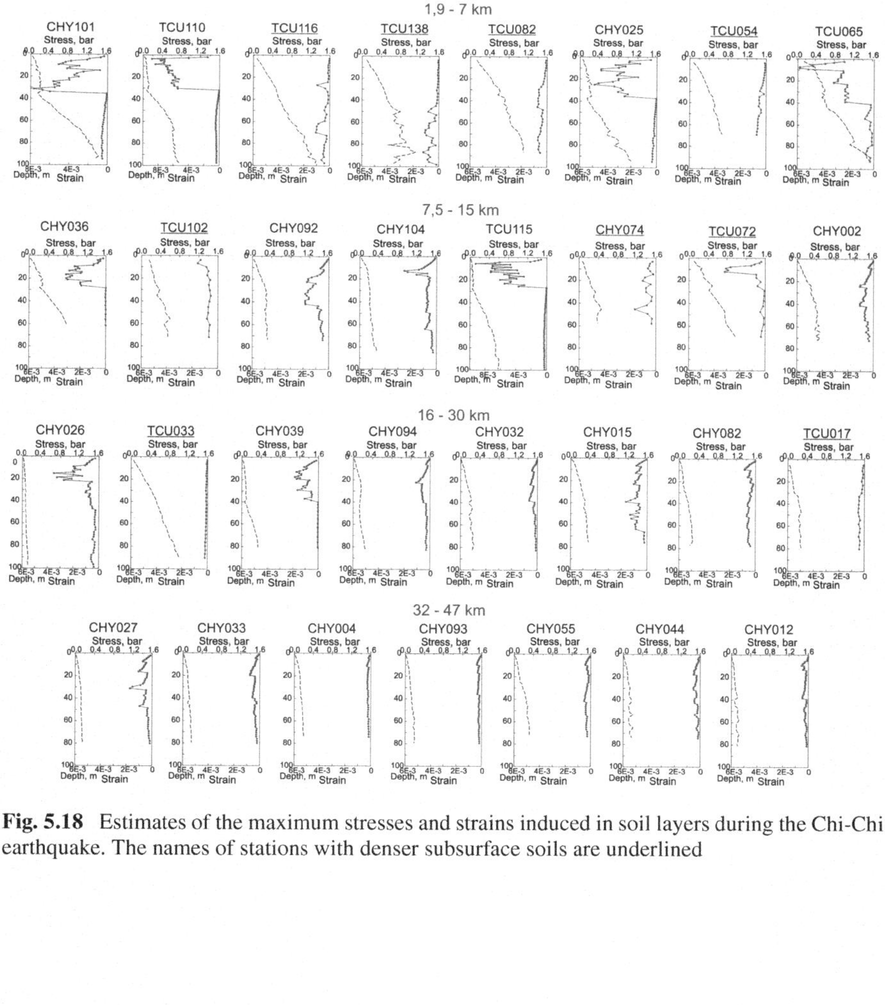

Fig. 5.18 Estimates of the maximum stresses and strains induced in soil layers during the Chi-Chi earthquake. The names of stations with denser subsurface soils are underlined

Resonant phenomena were observed at sites closest to the fault (within ~20 km from the fault) in the upper softer layers, down to ~40–60 m. Stresses in soil layers increased with depth, whereas maximum strains usually occurred in the upper ~15–40 m (Fig. 5.18). At sites, where soft surface layers with $V_{S30} \leq 300$ m/s were underlain by denser layers and impedance contrast was higher, resonant phenomena were more pronounced. Maximum strains were achieved at sites closest to the fault, such as, TCU065, TCU110, TCU115, CHY101 (~0.6–0.8%), CHY025, and CHY036 (~0.4–0.6%), where signs of soil liquefaction were observed. At other sites maximum strains in soil layers did not exceed ~0.1–0.4%.

The obtained estimates of average stresses and strains induced in the upper 30 m of soil during the strong motion, τ_{30} and γ_{30}, are given in Table 5.1, and the distributions of these parameters around the fault plane are shown in Fig. 5.19a, b. To obtain the average values, absolute values of limiting stresses and strains corresponding to the end points of loading and unloading cycles of the hysteretic curves were averaged; then the averages of the values obtained for NS and EW components for the upper 30 m of soil were taken as the final estimates.

As seen from the figures, areas of softer subsurface soils with $V_{S30} \leq 300$ m/s in bajada to the west of the fault correspond to zones of decreased stresses and increased strains. At sites with softer subsurface soils, average stresses and strains decrease with distance, and increase with the level of input motion; however, scattering of the estimates around approximating curves is rather large. At sites possessing denser subsurface soils, average stresses and strains also deviate from power functions with appreciably higher scattering. As expected, at these sites average stresses are higher, and average strains are lower than at sites with softer subsurface soils.

Evidently, average stresses and strains in the upper 30 m of soil depend on the distance from the fault (i.e., on the level of input motion) and on the mechanical properties of soils (*S*-wave velocities and densities). Stresses and strains naturally decrease with distance from the fault; at the same time, stresses τ_{30} increase, and strains γ_{30} decrease with increasing V_{S30}. Decreased stress values at some sites in the vicinity of the fault (for example, CHY092 and CHY002) are obviously related to decreased V_{S30} values in the upper layers at these sites, whereas decreased strain values at some other sites (for example, TCU116, TUC138, TCU082, and TCU054) are due to relatively high V_{S30} in the upper soil layers.

Thus, according to the obtained estimates, at sites possessing softer soils, average strains decrease with distance from the fault faster than average stresses, probably because stresses in soil layers are defined to a larger extent by radiation from the earthquake source (level of input motion to soil layers) than by mechanical properties of soils, whereas strains depend more on the soil properties, i.e., on resonant phenomena in the upper layers.

Shear Moduli Reduction in Soil Layers in Near-Fault Zones During the Chi-Chi Earthquake

At sites closest to the fault, soil behavior during the Chi-Chi earthquake was substantially nonlinear, and the nonlinearity can be characterized by the reduction of shear

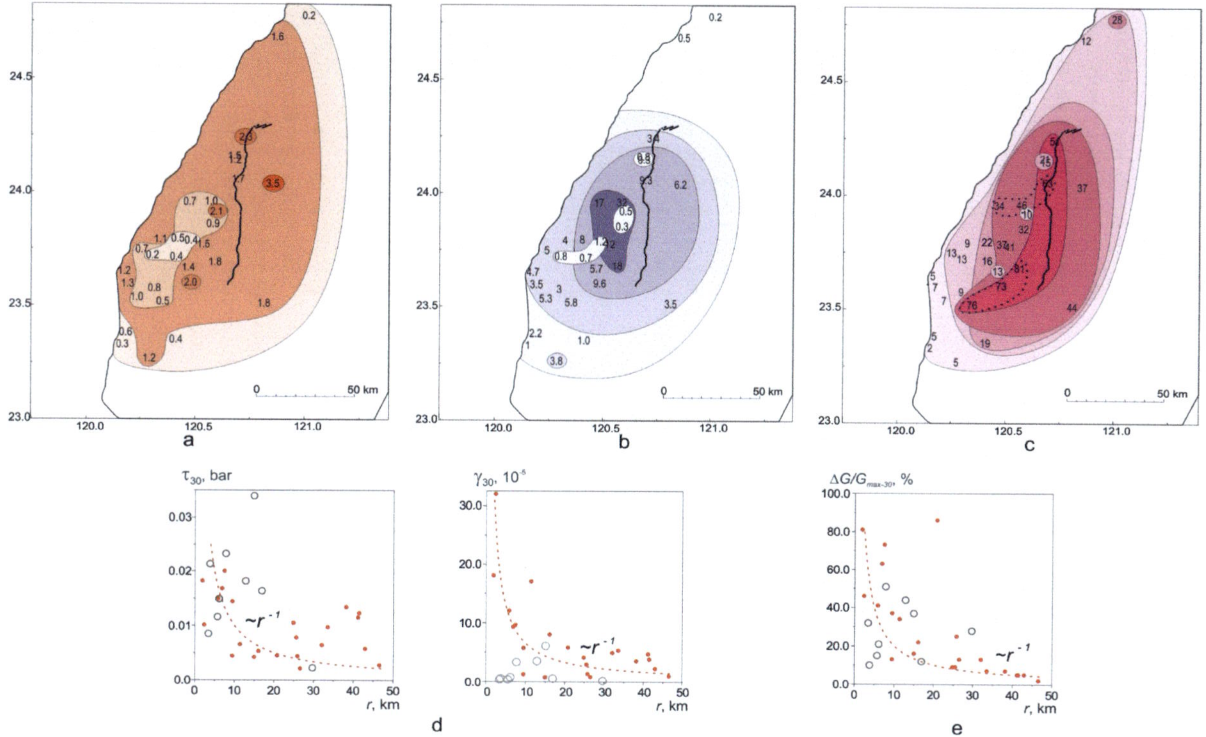

Fig. 5.19 Upper raw: distribution around the fault plane of: **a**—average stresses in the upper 30 m of soil τ_{30}, **b**—average strains in the upper 30 m of soil γ_{30}, **c**—estimates of shear moduli reduction in the upper 30 m of soil $\Delta G/G_{\max 30}$ (dotted lines indicate areas showing signs of soil liquefaction). Lower raw: **d**—average stresses τ_{30} and average strains γ_{30}, **e**—estimates of shear moduli reduction $\Delta G/G_{\max 30}$ as functions of the distance from the fault. The values of τ_{30}, γ_{30} and $\Delta G/G_{\max 30}$ for stations with softer and denser subsurface soils are shown by dots and circles, respectively

moduli in soil layers. Thus, the constructed models of soil behavior were used to estimate shear moduli reduction in the upper 30 m of soil.

For each site, average shear moduli were calculated as the ratios of stresses and strains averaged in time (during strong motion), depth (over the upper 30 m), and two horizontal components; stresses and strains were taken as their absolute values corresponding to the final points of loading and unloading cycles of the hysteretic curves. Shear moduli reduction $\Delta G/G_{\max 30}$ was calculated as the difference between the maximum values of shear moduli, calculated within small time intervals at the beginning of strong motion, and the average values, normalized by the average values and expressed in percent. As mentioned in Sect. 3.3, the obtained estimates of shear moduli reduction could be slightly underestimated, due to averaging and smoothing of the models of soil behavior. At sites where soil behavior was not adequately simulated in the middle and final parts of strong motion because of basin effects (Fig. 5.16), estimates of shear moduli reduction should be treated as approximate.

The obtained $\Delta G/G_{\max 30}$ estimates are presented in Table 5.1 and in Fig. 5.19; they testify to rather high nonlinearity of the soil response during the Chi-Chi earthquake. In the closest vicinity of the fault, at sites TCU110, CHY025, TCU065, TCU102, and CHY074, shear moduli reduction in the upper 30 m achieved ~40–50%, and the highest $\Delta G/G_{\max 30}$ estimates were obtained at CHY101, CHY036, and CHY039 sites, where the constructed models of soil behavior indicate soil softening and probably, liquefaction in surface layers. This area is marked by a dotted line in Figs. 5.16 and 5.19.

High values of shear moduli reduction $\Delta G/G_{\max 30}$ ~30–50% were obtained in the closest vicinity of the fault, within ~25 km from the surface rupture, at soft soil sites TCU102, TCU110, TCU115, TCU072, CHY074, TCU116, CHY025, and CHY092 (Fig. 5.19). At sites TCU138, TCU082, TCU054, and TCU033, possessing denser soils (V_{S30} ~450–605 m/s), lower values were obtained, $\Delta G/G_{\max 30}$ ~10–20%. At CHY104, CHY002, CHY026, CHY094, and CHY032 sites, reduction of shear moduli can be underestimated because of basin-induced surface waves (Fig. 5.16).

With increasing distance from the fault, $\Delta G/G_{\max 30}$ values decrease to ~10% at 35–50 km and to ~5% at 50–70 km from the fault (Fig. 5.19), where the soil response becomes virtually linear. At softer sites, the estimates of shear moduli reduction decrease with distance approximately as ~r^{-1}. The points deviating from the approximation with increased $\Delta G/G_{\max 30}$ values correspond to CHY039 and CHY036 sites, where the constructed models of soil behavior show signs of liquefaction. As seen from the plots, $\Delta G/G_{\max 30}$ estimates at denser sites substantially increase scattering of points around the approximating relationships (Fig. 5.19).

As mentioned above, resonant phenomena in the upper softer layers occurred during the Chi-Chi earthquake at many sites, and since soil conditions are similar at soil sites located mostly to the south and south-west of the fault (Fig. 5.16), average strains in the upper 30 m are proportional to the level of input motion to soil layers, and their decrease with distance from the fault can be approximately described as ~r^{-1}. Accordingly, shear moduli reduction values depend on distance and the level of input motion in the same manner.

At sites with denser subsurface soils, resonant phenomena were not observed in the upper 30 m (maximum strains usually correspond to depths of ~40 m and more, Fig. 5.18), and $\Delta G/G_{\max 30}$ estimates depend, more than for softer soils, on the profiling data, which are very diverse at these sites, dispersed over a large area around the fault. Consequently, shear moduli reduction values can be quite different, even at sites located equidistantly from the fault.

Nonlinear Components of Soil Response in the Near-Fault Zones During the Chi-Chi Earthquake

The contents of nonlinear components in the soil response were estimated by nonlinear identification methods, i.e., by testing the models of soil behavior at the studied sites (Sect. 3.3) by the Gaussian white noise. Linear and nonlinear components of the soil response were distinguished, and their intensities estimated in percent of the total intensity of the response.

The estimates of the zero-, first-, second-, and third-order Wiener kernels were constructed based on input and output signals of 500,000 points of duration (more than 40 min). Based on the estimated Wiener kernels, the response of the linear model and nonlinear corrections accounting for quadratic and cubic nonlinearities of the soil response were constructed. Nonlinear components were estimated in percent of the intensity of the soil response.

The whole nonlinear component of the soil response was estimated as the deviation of the real soil response from the response of the linear model constructed by zero- and first-order kernels. The nonlinear quadratic and cubic components were estimated as nonlinear corrections due to quadratic and cubic nonlinearities predicted by kernels $\{h_2\}$ and $\{h_3\}$, respectively. Then the residual part of the soil response was estimated, which is related to higher-order (4th-, 5th-, etc.) nonlinearities, as well as to inaccuracies in kernel estimates (Marmarelis and Marmarelis 1978).

The obtained estimates are presented in Table 5.1, and Fig. 5.20. Soft soils in the closest vicinity of the fault possess the highest nonlinearity of the response: nonlinear components of the soil response *Nl* achieve ~60–80%, and nonlinear quadratic and cubic components were up to ~12–16% of the intensity of the response. At the same time, nonlinear residual components of the soil response related to higher-order nonlinearities (4th order and higher) were also high, up to ~30–40% of the intensity of the response. Evidently, this indicates very high nonlinearity of the soil response in the near-fault zones of the Chi-Chi earthquake.

With increasing distance from the fault, nonlinear components of the soil response decrease, and their dependence on distance can be approximated as $\sim r^{-1}$. Accounting for sites possessing denser subsurface soils increases scattering of the estimates around the approximation curve. Note that nonlinear quadratic and cubic components are almost equal in the closest vicinity of the fault, in conditions of high nonlinearity, but with increasing distance from the fault, nonlinear cubic components become predominant (Fig. 5.20).

At two closest to the fault plane sites possessing denser subsurface soils, TCU072 and TCU102, nonlinearity of the soil response was also high, evidently because of pronounced resonant phenomena in the upper softer layers. At TCU072 site, resonant

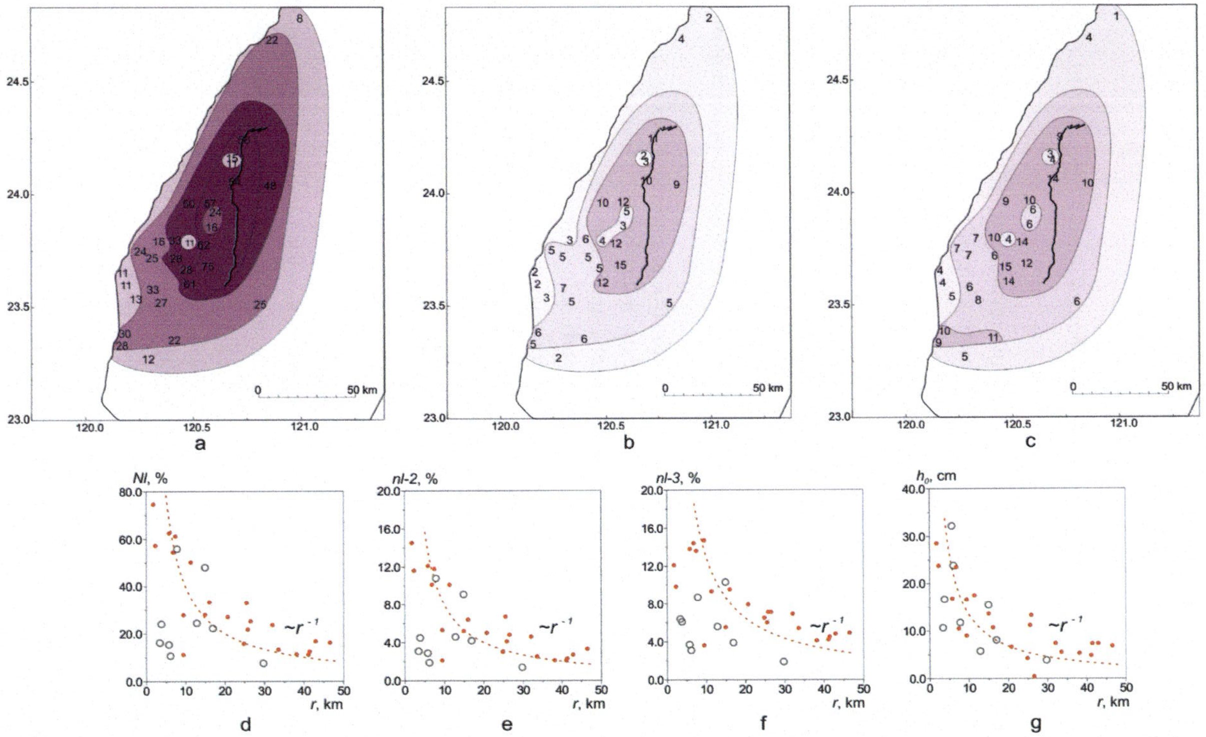

Fig. 5.20 Upper raw: distribution around the fault plane of: **a**—the whole nonlinear components of the soil response *Nl*, **b**—nonlinear quadratic components *nl*-2, **c**—nonlinear cubic components *nl*-3. Lower raw: **d**—the whole nonlinear components of the soil response *Nl*, **e**—nonlinear quadratic components *nl*-2, **f**—nonlinear cubic components *nl*-3, **g**—quasi-static deformations of the surface h_0 as functions of the distance from the fault. Values obtained for sites with softer and denser subsurface soils are shown by points and circles, respectively

amplification of seismic motion occurred in the upper ~11 m of colluvium ($V_S \approx$ 250 m/s), and at TCU102 site, resonant phenomena were related to the upper ~4–6 m of silty soils, which were softened during strong motion. The behavior of these upper layers was substantially nonlinear and is described by very sloping stress–strain relations (Sect. 3.3). Since soft surface layers were underlain by hard breccia (at TCU072) and gravel (at TCU102), the average *S*-wave velocities in the upper 30 m were rather high, 563 and 735 m/s, respectively, and the sites were assigned as possessing denser subsurface soils, however, nonlinear components at these sites satisfy regularities obtained for sites possessing softer subsurface soils (Fig. 5.20).

Quasi-static shear deformations on the surface h_0 achieve 20–35 cm at sites closest to the fault, according to our estimates (Fig. 5.20). They decrease approximately as $\sim r^{-1}$ with increasing distance from the fault, and the estimates for softer and denser sites possess similar regularities. At distances of ~40–50 km from the fault (~1/2 of the fault length), nonlinear components of the soil response decrease to ~10–15% (Fig. 5.20).

Changes in the contents of linear and nonlinear components in the soil response with increasing distance from the fault can be traced by impulse characteristics of soil layers (Fig. 5.21), calculated based on the simulated acceleration time histories on the surface and at the bottoms of soil layers (Sect. 3.3).

Impulse characteristics represent estimates of the first-order Wiener kernels, that is, they describe mostly linear parts of the soil response. At sites located at large distances from the fault, they possess oscillating character, indicating an essential role of resonant phenomena in the soil response. With approaching the fault, nonlinear damping increases, and the oscillations decrease; impulse characteristics become smoothed and low-frequency ones, which indicates substantial changes in spectral contents of seismic waves propagating in soil layers, noticeable decrease of linear components and increase of nonlinear components in the soil response.

Thus, amplification of seismic waves in soil layers, average stresses and strains induced in soil layers by the strong motion, shear moduli reduction and characteristics of nonlinearity of the soil response in the near-fault zones during the Chi-Chi earthquake are estimated. In spite of approximate character of the estimates, they allow us to draw some conclusions.

Amplification of seismic waves in subsurface soils estimated for accelerations and velocities increase approximately in proportion to the distance from the fault, $\sim r$. Other parameters decrease with distance, or with decreasing the level of input motion to soil layers.

Since the estimates possess a rather large scattering, their behavior was approximated by simple functions. Note a slower (than other parameters) decrease of average stresses in soil layers with distance and a large scattering of the amplification factors, average stresses and strains, and nonlinear components in the soil response around the approximating curves, whereas, scattering of $\Delta G/G_{\max 30}$ estimates is smaller.

Let's consider possible causes of these observations. As shown in Sect. 3.3, in the near-fault zones of the Chi-Chi earthquake, resonant phenomena in subsurface soils and nonlinearity of soil response were most important. Resonant effects involved thick soil layers, down to ~40–60 m, they noticeably increased strains and shear

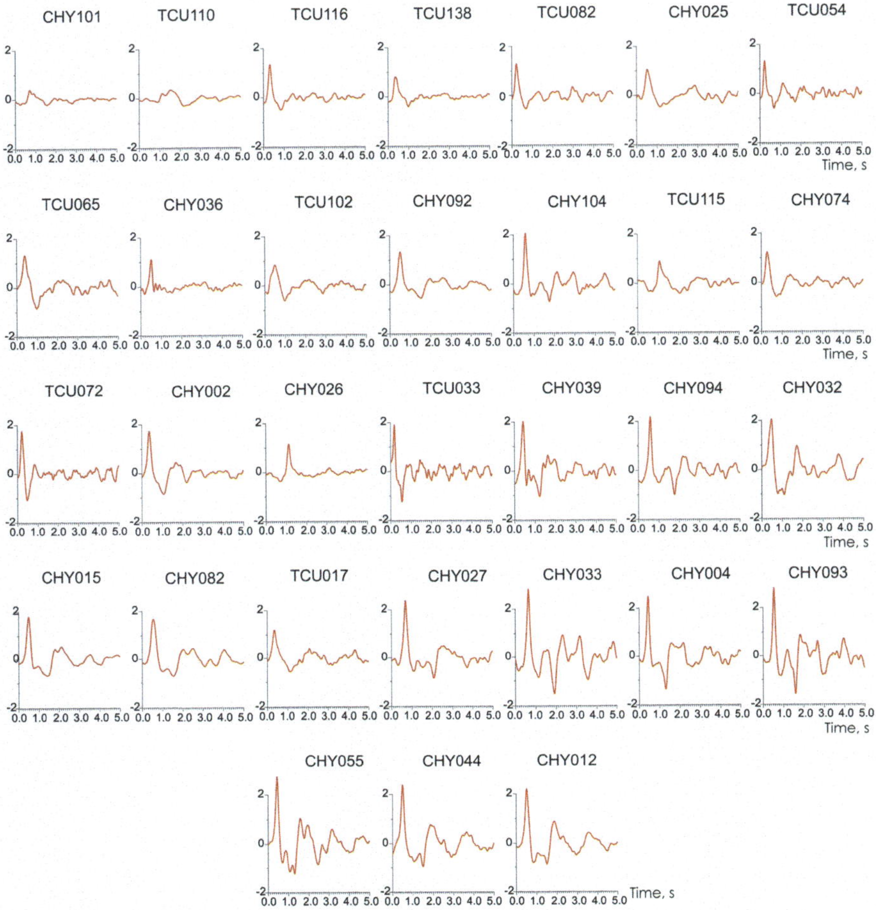

Fig. 5.21 Impulse characteristics of the soil response during the Chi-Chi earthquake at the studied soil sites arranged according to the distance from the fault

moduli reduction in soil layers, but only slightly influenced stresses. Therefore, these three parameters decreased with distance in different ways.

A large scattering in the obtained estimates testifies to the variety of manifestations of the soil response, depending on the profiling data and the proximity of a site to the fault plane. In horizontally layered soft soils, resonant amplification of seismic waves occurred, and the soil response depended on the distance from the fault (i.e., on the level of input motion). Whereas the response of denser soils was more diverse and cannot be easily approximated by functions of the distance or the level of input motion. Scattering of parameters describing the response of soft soils is related to the level of input motion. At sites closest to the fault plane, in water-saturated sandy soils, a noticeable amplification of seismic motion occurs at rather large strains due to pore pressure built-up and hard-type nonlinear soil behavior (as described in Chap. 3).

The estimates of nonlinear components of the soil response characterize the whole soil thickness down to ~70–80 m. According to the theory of nonlinear interactions, the amplitudes of combinational-frequency harmonics generated on system's nonlinearity are proportional to the products of the amplitudes of the interacting waves (Zarembo and Krasil'nikov 1966). Thus, the estimates of quadratic and cubic nonlinear components should be proportional to the second and third powers of the amplitudes of input motion. The whole nonlinear component, which includes, among quadratic and cubic, higher-order nonlinearities, should also be proportional to some power of the level of input motion; the power is higher than 2 and depends on the content of higher-order nonlinear components in the soil response. Here "input motion" means testing Gaussian white noise used for estimating nonlinear components in the soil response.

The intensities of the testing signals (that influence the estimates of nonlinear components) were selected so that the average stresses and strains induced in soil layers by the earthquake and by the testing noise were close to each other. The dependencies of the obtained estimates of nonlinear components in the soil response on the amplitudes of testing signals agree with the theoretical predictions. However, plots in Fig. 5.20 show other dependencies, such as, the dependencies of the estimated nonlinear components in the soil response on the level of preliminary calculated input motion from the Chi-Chi earthquake to soil layers, and these dependencies can be roughly approximated by linear functions, where a large scattering around the approximating lines can be partially due to inaccuracies of the estimations.

Thus, the estimates of nonlinear components in the soil response show the same regularities as shear moduli reduction values, which also characterize nonlinearity of the soil response during the earthquake. For sites closest to the fault plane, both estimates achieve ~60–80% (Table 5.1). However, shear moduli reduction values decrease with distance faster than the estimates of nonlinear components in the soil response: at ~35–50 km from the fault (~1/2 of the fault length) shear moduli reduction and nonlinear components were ~2–7% and ~10–15%, respectively. Partially this can be due to the fact that shear moduli reduction was estimated for the upper 30 m of soil, whereas nonlinear components relate to the whole soil thickness (~70–80 m).

Based on the estimates of shear moduli reduction and the contents of nonlinear components in the soil response, we can distinguish the area of strong nonlinearity within ~20–25 km from the fault (~1/4 of its length). In this area, shear moduli in the upper 30 m of soil reduced by ~40–50% of their initial values during strong motion, and the contents of nonlinear components in the soil response were also high, ~30–80% of the whole intensity of the response. This area includes, in particular, CHY101, TCU110, TCU116, CHY025, TCU065, CHY036, CHY092, CHY104, and TCU115 soft-soil sites. At these sites strong motion induced changes in rheological properties of the upper layers, and spectra of motion on the surface were close to the form $E(f) \sim f^{-k}$.

******

Records of recent strong earthquakes, such as, the 1995 Kobe earthquake (M_w ~ 6.8), 2000 Tottori earthquake (M_w ~ 6.7), and 1999 Chi-Chi earthquake (M_w ~ 7.6) showed significant nonlinearity of soil response in the near-fault zones. Nonlinear components of the soil response were estimated by nonlinear identification methods, i.e., by testing the constructed models of soil behavior by the Gaussian white noise. The obtained results lead us to the following conclusions.

In the near-fault zones of the three earthquakes, soil response contains significant nonlinear components, rapidly decreasing with distance from the fault.

The shapes of stress–strain relations in soil layers (mainly odd functions) define the types of nonlinearity of the soil response. Soils possess mainly odd-order nonlinearities, i.e., cubic nonlinearity, nonlinearities of the 5th, 7th, 9th, etc. orders, whereas nonlinearities of even orders are usually weakly expressed, but become significant in cases when loading and unloading parts of the stress–strain relations acquire significant even components: in liquefied soils, at the moments of high intensity of motion.

Based on the estimates of shear moduli reduction and nonlinear components in the soil response, we can identify areas of strong nonlinearity of soil behavior in the vicinity of the fault plane.

During the Kobe and Tottori earthquakes (M_w ~6.7–6.8), nonlinear components of the soil response reached 40–60% of the intensity of the response within ~7–8 km from the fault plane (~1/4 of its length); at ~15–20 km (~1/2 of the length of the fault plane) nonlinear components decreased to ~10–12%, and at ~50–80 km (2–3 lengths of the fault plane) they did not exceed ~2–5% of the intensity of the soil response.

During the Chi-Chi earthquake, the content of nonlinear components in the soil response was ~30–80% of the intensity of the response within ~20–25 km from the fault plane (~1/4 of its length). At ~35–50 km from the fault (~1/3–1/2 of its length) nonlinear components decreased to ~10–30% of the intensity of the response.

Thus, the area of strong nonlinearity of soil response, where changes in rheological properties of soft soils occur and spectra of motion on the surface approach $E(f) \sim f^{-k}$, is about 1/4–1/3 of the length of the fault plane.

Reduction of shear moduli expressed in percent is close to the content of nonlinear components in the soil response.

If we compare soil behavior in the near-fault zones during the 1999 Chi-Chi earthquake ($M_w = 7.6$), 1995 Kobe earthquake ($M_w = 6.8$), and 2000 Tottori earthquake ($M_w = 6.7$), we notice that the majority of the studied sites (except Port Island) showed hard type of nonlinear soil behavior: SGK and TKS sites in case of Kobe earthquake; TTRH02 and SMNH01 sites in case of Tottori earthquake; TCU065, TCU102, and TCU110 sites in case of Chi-Chi earthquake.

Strong ground motion led to resonant phenomena in subsurface soils, to large strains and pore-pressure build-up; as a consequence, soil behavior became "hard-type", especially just near the fault planes. In these cases, strong nonlinearity of the soil response did not lead to noticeable de-amplification of motion on the surface.

Resonant amplification of seismic motion occurred in the upper soil layers at many sites during the three earthquakes, however, during the stronger Chi-Chi earthquake, resonant phenomena involved thicker soil layers, down to ~40–60 m, compared to ~10–15 m during the Kobe and Tottori earthquakes.

On the whole, along with a diversity of patterns of soil response depending on such factors as the structure of soil layers, mechanical properties of soils and their saturation with water, spectral composition and intensity of seismic motion, there is a similarity in the behavior of similar soils during different strong earthquakes, which indicates a possibility to forecast soil behavior in future strong earthquakes.

References

Aki, K., Irikura, K.: Characterization and mapping of earthquake shaking for seismic zonation. In: Proceeding 4th International Conference on Seismic Zonation, Aug. 25–29, Stanford, California, pp. 61–110 (1991)

Groshkov, A.L., Kalimullin, R.R., Shalashov, G.M.: Cubic nonlinear seismic effects, Doklady Acad. Sci. USSR **315**(1), 65–68 (1991) (in Russian)

Gushchin, V.V., Shalashov, G.M.: On the possibility of using nonlinear seismic effects in the problems of vibrational transmission of the Earth. Study of the Earth by non-explosive seismic sources. Nauka, Moscow 144–155 (1981) (in Russian)

Hardin, B.O., Drnevich, V.P.: Shear modulus and damping in soils: measurement and parameter effects. Proc. Am. Soc. Civil Eng., J. Soil Mech. Found. Div. **98**, 603–624 (1972)

Kadish, A., Johnson, P.A., Zinszner, B.: Evaluating hysteresis in earth materials under dynamic resonance. J. Geoph. Res. **101**, 25139–25147 (1996)

Landau, L.D., Lifshitz, E.M.: Mechanics. Pergamon Press (1960)

Landau, L.D., Lifshitz, E.M.: Theory of elasticity. Pergamon Press (1970)

Lee, C.-T., Cheng, C.-T., Liao, C.-W., Tsai, Y.-B.: Site classification of Taiwan free-field strong-motion stations. Bull. Seism. Soc. Am. **91**, 1283–1297 (2001)

Marmarelis, P.Z., Marmarelis, V.Z.: Analysis of Physiological Systems. The White-Noise Approach. Plenum Press, New York and London (1978)

Shalashov, G.M.: Cross-modulation of acoustic waves on cubic nonlinearity of solids. Acoust. J. **XXX**(3), 386–390 (1984) (in Russian)

Van Den Abeele, K.E.A., Johnson, P.A.: Elastic pulsed wave propagation in media with second- or higher-order nonlinearity. Part II. Simulation of experimental measurements on Berea sandstone. J. Acoust. Soc. Am. **99**, 3346–3352 (1996)

Zarembo, L.K., Krasilnikov, V.A.: Introduction to nonlinear acoustics. Nauka, Moscow (1966). (in Russian)

Zvolinsky, N.V.: Wave processes in inelastic media. Probl. Eng. Seismol. **23**, 4–19 (1982). (in Russian)

Chapter 6
Nonlinear Transformations of Seismic Waves in Soil Layers, High-order Spectra and Coherences as Characteristics of Nonlinearity

As shown in the previous chapters, the behavior of subsurface soft soils in the near-fault zones of strong earthquakes can be substantially nonlinear, and we may estimate the degree of nonlinearity in numerical experiments with models of soil behavior.

Nonlinear transformations of seismic waves propagating in subsurface soils—nonlinear media—include changes in their waveforms and spectra. These changes are often clearly seen in strong motion records.

Nonlinear effects in seismic wave fields can be considered as the results of changes of the medium properties caused by propagating waves and affecting both the propagation of the waves themselves (self-action) and the propagation of other seismic waves (interaction).

In this chapter, we discuss the evolution of seismic waves in a nonlinear medium, nonlinear wave processes accompanying the propagation of intense seismic waves, and quasi-stationary shape of the spectrum of seismic motion on the surface.

Nonlinearity of the soil response differently influences the ranges of low- and high-seismic frequencies, because it depends on the generation of combinational-frequency harmonics, mechanisms of decay instability regulated by Manley-Rowe relations, nonlinear hysteretic absorption and some other nonlinear mechanisms. As a result of the cumulative action of these mechanisms, spectrum of motion on the surface tends to take a quasi-stationary form $E(f) \sim f^{-k}$.

Nonlinear effects of modulation instability of seismic waves propagating in subsurface soils are studied in numerical experiments. Self-modulation and generation of sub-harmonics of the main (basic) frequencies are observed in intense seismic wave fields in cases of resonant amplification of seismic waves in soil layers; it precedes the stage of chaotization that occurs at further increase of the wave amplitudes.

Analysis of high-order spectra and coherences is described, which allows us to identify and estimate nonlinear components of soil response based on seismic records.

O. V. Pavlenko, *Seismic Waves in Soil Layers*,
https://doi.org/10.1007/978-3-031-52338-0_6

High-order spectra and coherencies (HOS and HOC) are used in system analysis to study the properties of nonlinear systems. In geophysics, mainly bispectrum and bicoherency are used; however, in order to detect nonlinear components in seismic signals, tricoherency and coherencies of higher odd orders should be more informative, because soils possess mainly odd-order nonlinearities.

6.1 Transformations of Seismic Waves in Soil Layers, the Quasi-Stationary Spectrum of Motion on the Surface

Chapter 5 describes numerical experiments, in which the propagation of monochromatic and noise-like seismic waves in soil layers is studied. Seismic wave propagation is accompanied by interactions of the waves, which can be described statistically.

Similar phenomena in the range of acoustic frequencies are studied in statistical nonlinear acoustics. For simple acoustic waves, statistical characteristics of nonlinear disturbances can be calculated accurately. A general expression is found for the correlation function of a random process in an arbitrary cross-section of a nonlinear medium, based on a two-dimensional characteristic function of an input signal. For a normal process, formulas are deduced that relate spectral density in the medium to the correlation function at the input (Rudenko and Soluyan 1975). If the input of a nonlinear system is a broadband noise, an abnormally low attenuation of high–frequency components is observed, which indicates pumping of the energy to the high frequency range; at the same time, the energy is also pumped to the low-frequency range of the spectrum. Thus, the spectrum of the broadband noise expands to both high– and low-frequency ranges, due to nonlinear interactions (Rudenko 1986).

Such transformations of the broadband spectra are attributed to the problem of acoustic turbulence. The most interesting is the equilibrium form of spectra, or the universal law of spectral decay at high frequencies. According to many researchers, in cases of strong nonlinearity (in the range of developed shock fronts) power spectra decrease as $\sim f^{-2}$, because the amplitudes of harmonic waves forming the gaps decrease as $\sim f^{-1}$ (Rudenko and Soluyan 1975). This dependence was obtained theoretically in the approximation of chaotic phases for a medium without dispersion in inertial (where sources and drains of energy are absent) frequency ranges (Kadomtsev and Karpman 1971).

In case, when the sources radiating waves in a nonlinear medium are scattered in space, the propagation of the waves is described by inhomogeneous nonlinear equations, the right parts of which are random functions, e.g., the Burgers equations or the equations for simple waves (Rudenko 1986). Calculating the correlation functions of noise-like waves to find approximate solutions of these equations and power spectra, we find that low–frequency components are emphasized in the quasi-stationary spectrum. This is usually associated with increasing the field scales due to merging of moving gaps. In addition, the nonlinearity forms high-frequency asymptotics related to the structure of the fronts of long-living shock waves, which has the form f^{-2} (Rudenko 1986). Sometimes, for small inverse Reynolds numbers x_p/x_d, i.e., for small ratios of characteristic nonlinear x_p and dissipative x_d wavelengths, the power asymptotic turns to the exponential one of the form $\exp(-\beta f)$.

As noted in Rudenko's review (1986), the emphasis on the low-frequency components is a general trend. Physically, this is because high-frequency dissipation cannot prevent accumulation of energy in low-frequency ranges. According to the Manley-Rowe relations, low-frequency modes do not transfer energy to high-frequency modes, but high-frequency modes can decay, transferring energy to low-frequency modes (Rabinovich and Trubetskov 1984).

Similar behavior patterns are established for seismic waves. In seismic wave fields, interactions of spectral components of the propagating waves lead to an avalanche-like multiplication of spectral lines, i.e., to nonlinear broadening of spectra and the appearance of a constant component. In time domain, strong nonlinear distortions manifest themselves in the waves becoming sawtooth.

When intense waves propagate in nonlinear media, the theory predicts the activation of nonlinear damping mechanisms that do not depend on the dissipative properties of the medium (Rudenko 1986); this is manifested in well-known effects of increase of damping with increasing seismic wave amplitudes (Chap. 1).

Nonlinear effects in subsurface soils lead to the rise of low-frequency components and smoothing of spectral peaks in motion on the surface; this is clearly seen in acceleration spectra of strong earthquakes. In records of seismic vertical arrays, the nonlinearity of soil response appears in different waveforms and spectra of motion on the surface and at depths.

Since intense seismic motion of any spectral composition at the input to soft soils tends to transform into motion on the surface with a spectrum of the form $E(f) \sim f^{-k}$ due to nonlinear wave distortions, information on the initial spectral contents is lost in subsurface soils in such cases.

Let us consider two numerical experiments showing transformations of spectra of seismic waves in subsurface soils. In the two experiments, the propagation of broadband and narrowband seismic noise signals of various intensities in a "simple" soil profile, described in (Joyner and Chen 1975) is studied. The soil profile is a 200 m-horizontally layered dry (no groundwater) alluvium with a constant density of 2.05 g/cm^3, with a smooth increase of seismic velocities with depth (Fig. 6.1), underlain by a crystalline foundation.

The behavior of soil layers is described by Ivan's rheological elastic–plastic model (Joyner and Chen 1975). The same normalized nonlinear stress–strain relations are used to describe the behavior of all the layers in the profile; the differences in the behavior of different layers are defined by the differences in maximum shear stresses $\tau_{\max}$ and shear moduli $G_{\max}$ in the layers. The corresponding hysteretic curves are shown in Fig. 1.2b (Chap. 1).

Figures 6.2 and 6.3 show the results of the numerical experiments: spectra of motion on the surface in cases of linear and nonlinear soil response to broadband and narrowband Gaussian white noise of increasing intensity. As seen from the figures,

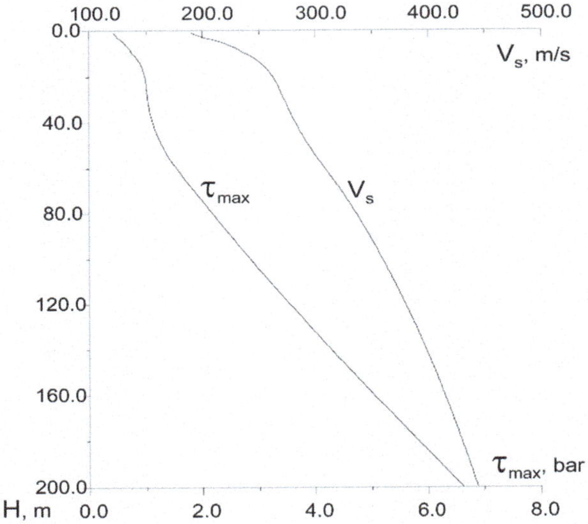

Fig. 6.1 Parameters of the "simple" soil profile: shear stress in failure τ_{max} and S-wave velocities, gradually increasing with depth

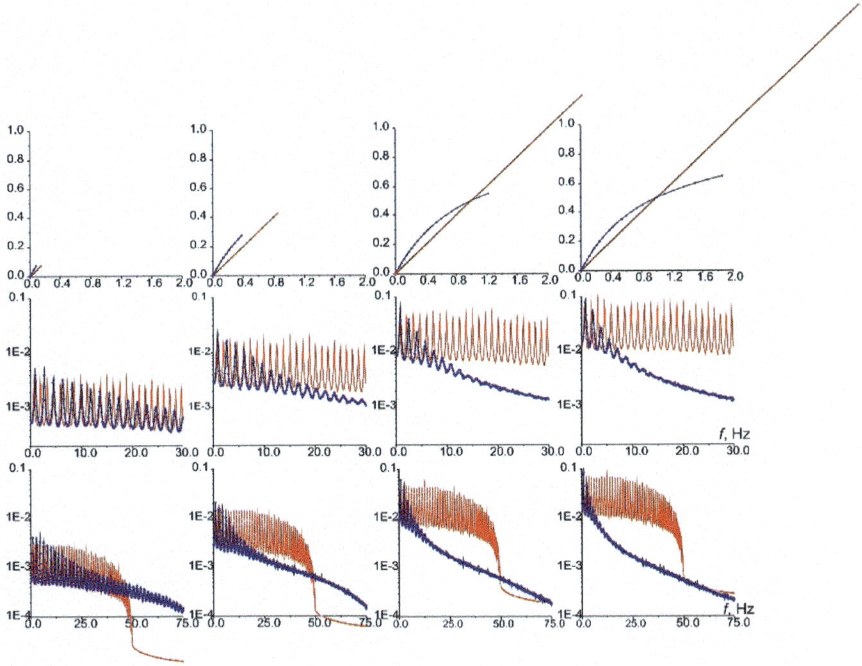

Fig. 6.2 Nonlinear transformations of spectra of a broadband seismic noise in the "simple" soil profile at various intensities of input motion. Upper row: the working intervals of linear and nonlinear stress–strain relations (only initial loading curves are shown); lower rows: spectra of motion on the surface in cases of linear (red) and nonlinear (blue) response of the soil profile

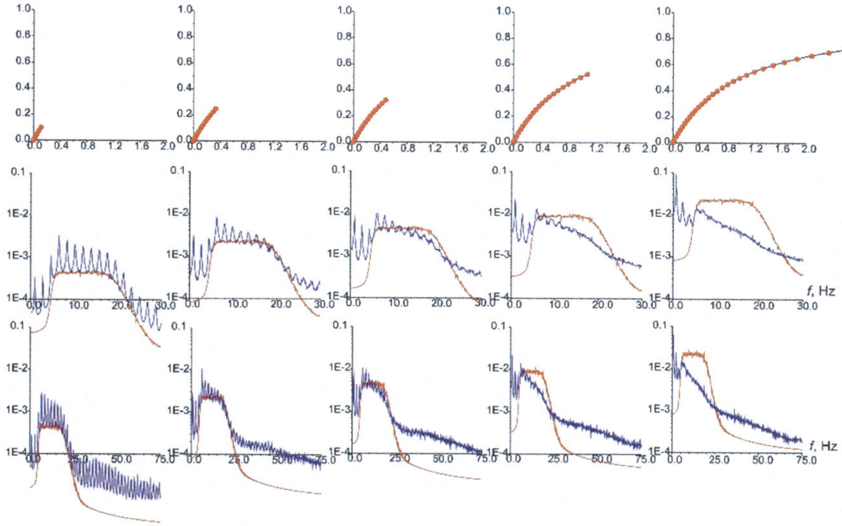

Fig. 6.3 Nonlinear transformations of spectra of a narrowband seismic noise in the "simple" soil profile at various intensities of input motion. Upper row: the working intervals of nonlinear stress–strain relations (only initial loading curves are shown); lower rows: spectra of input motion (red) and motion on the surface (blue) of the soil profile

in the case of a linear response, the spectral shape of motion on the surface is not changed with increasing intensity of input motion. In the case of a nonlinear response, spectra of motion on the surface gradually decrease at high frequencies and finally take the form $E(f) \sim f^{-k}$. The amplitudes of low-frequency components in linear and nonlinear cases are close to each other, but the amplitudes of the medium- and high-frequency components in the case of nonlinear response are lower, obviously due to the nonlinear damping.

In the case of a narrow–band input noise signal of rather high intensity, spectra of motion on the surface also tend to transform to the form $E(f) \sim f^{-k}$, and the energy of motion passes from the frequency band of the input signal to high- and low-frequency ranges.

Yu et al. (1992) obtained similar results. They performed numerical simulations of acceleration time histories of strong earthquakes and showed that nonlinearity of ground response weakens medium-frequency components, while low-frequency components remain the same, and high-frequency components slightly increase.

Thus, transformations of spectra of strong seismic motion to the form $E(f) \sim f^{-k}$ in subsurface soils occurs due to nonlinear (hysteretic) dissipation and redistribution of energy of seismic waves over the spectrum through seismic wave interactions. The increase in low-frequency components is consistent with the Manley-Rowe decay instability relations, stating that the energy transfer from low-frequency to high-frequency oscillations is difficult, but the decay of high-frequency oscillations and the transfer of their energy to low-frequency components is easy (Rabinovich and Trubetskov 1984).

6.2 Modulation Instability in Seismic Wave Fields

In Sect. 5.2, a numerical experiment is described, in which models of soil behavior during the 2000 Tottori earthquake at KiK-net sites were tested by harmonic signals (Fig. 5.8). During the propagation of monochromatic waves in subsurface soils, higher odd-order (3rd, 5th, etc.) harmonics of the main frequency were recorded. In cases of high-intensity motion, along with odd-order harmonics, even-order harmonics were also generated (2nd, 4th, 6th, etc.). Similar results were obtained in numerical experiments with other soil profiles.

Note that when testing the soil profile at TTRH02 site by input motion of high intensity, subharmonics of the fundamental frequency are also generated (Fig. 5.8). The conditions of subharmonic generation were studied in numerical experiments. Figure 6.4 shows the results of testing the soil profile at TTRH02 site by harmonic signals of increasing intensity with a frequency of f ~ 3 Hz.

For an input signal with the amplitude $V_0 = 1.0$ cm/s, the output was received with a frequency of 3 Hz, enriched with its even- and odd-order higher harmonics. Then the input amplitude was increased to 1.1 cm/s, the working intervals of the stress–strain relations also increased, and the point of inflection appeared. The output became amplitude-modulated: harmonics at 1, 5 (3 $\pm$ 2 Hz), 7, 11 (9 $\pm$ 2 Hz), 13, 17 (15 $\pm$ 2 Hz), 19 Hz, etc. appeared in its spectrum. Harmonics at frequencies 1 and 1.5 Hz can be treated as subharmonics of frequency 3 Hz. At a further increase in the input amplitude, the modulation kept, and then the output became noise-like (Fig. 6.4d).

The experiment was repeated with an input frequency $f = 6$ Hz, and a weak modulation effect was found for the input amplitudes V_0 ~ 3.0–5.0 cm/s. When the amplitude increased to 10.0 cm/s, the oscillations became chaotic. Thus, the effect clearly manifested at frequency $f = 3$ Hz virtually disappeared at frequency $f =$ 6 Hz.

The soil profile was tested by harmonics of various frequencies, and the results (Fig. 6.5) showed that at frequency $f = 3$ Hz, a resonant amplification of motion occurs at depths of ~ 9 m (obviously, there is a substantial decrease of the resonant frequency due to nonlinearity of the soil response: in a linear case, the resonant frequency would be ~ 5.8 Hz).

The nonlinearity of stress–strain relations leads to a noticeable decrease in the velocity of propagation of seismic waves with the resonant frequency f ~ 3 Hz, compared to other frequencies, i.e., to velocity dispersion. Thus, conditions for the development of self-modulation appear (Rabinovich and Trubetskov 1984).

The phenomenon of self-modulation of harmonic waves in nonlinear media with dispersion is studied quite well. Dispersion leads to the phenomenon that higher

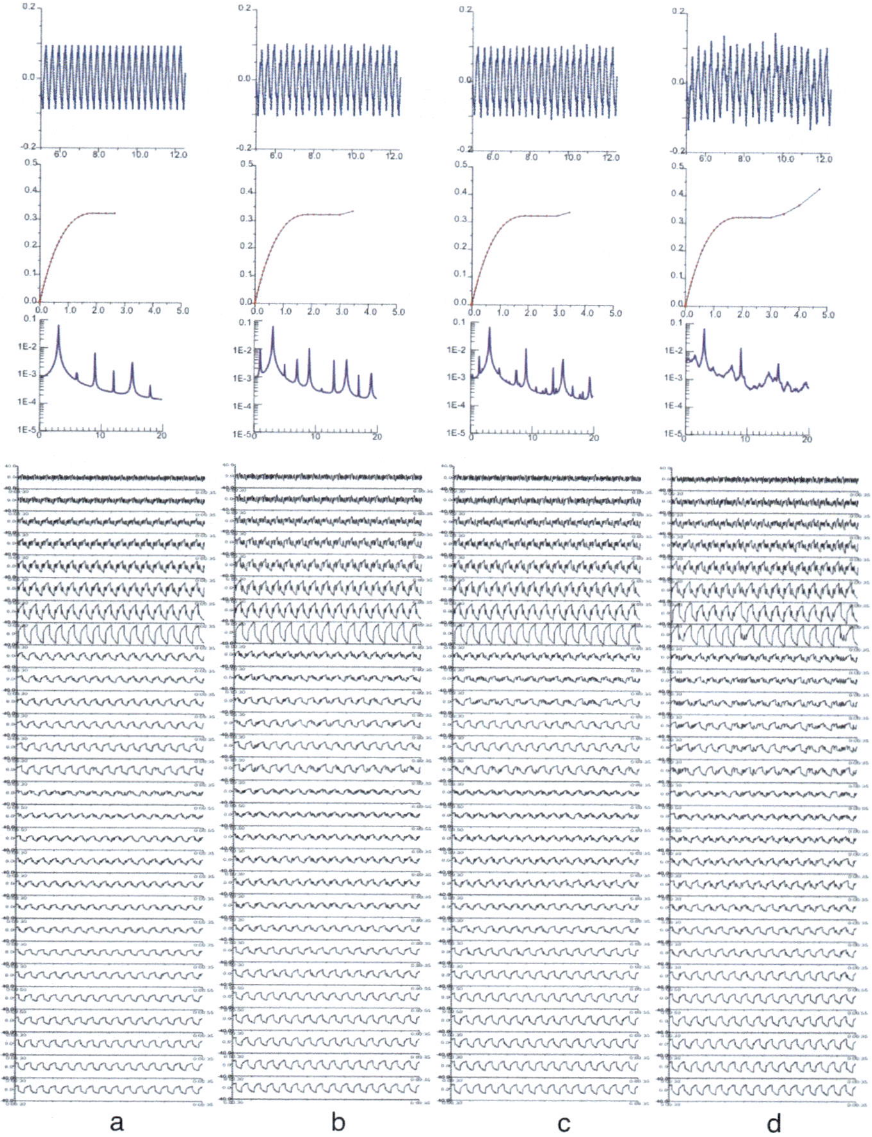

Fig. 6.4 Testing the soil profile at TTRH02 site by harmonic signals. From top to bottom: calculated motion on the surface, the corresponding working intervals of stress–strain relations (only initial loading curves are shown); spectra of motion on the surface; motion in the layers from the surface down to a depth of ~ 100 m (maximum amplitudes correspond to a depth of ~ 9-m, the bottom of the upper softer layer). The input signals are harmonics with a frequency of $f = 3$ Hz and the amplitudes $V_0 = 1.0$ cm/s (**a**), 1.1 cm/s (**b**), 1.2 cm/s (**c**) and 1.3 cm/s (**d**)

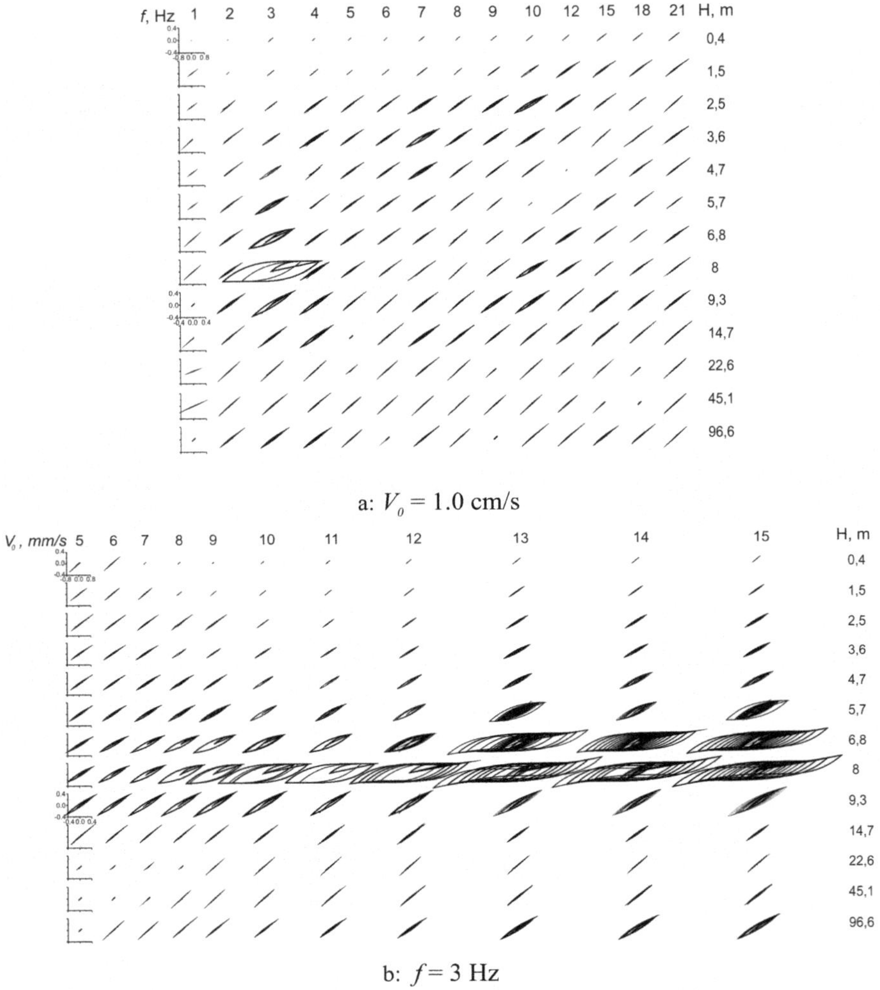

Fig. 6.5 Stresses and strains induced in the soil profile at TTRH02 site by testing harmonic signals: **a**—with amplitude $V_0 = 1.0$ cm/s and frequencies $f = 1$–21 Hz; **b**—with amplitudes $V_0 = 0.5$–1.5 cm/s and frequency $f = 3$ Hz

harmonics generated on the system's nonlinearity are not synchronized with the main-frequency waves and therefore are not increased, and the output becomes amplitude-modulated. The modulation results from the development of parametric instability, leading to the appearance of satellite waves with frequencies close to the carrier frequency. Modulation instability is possible only under a certain ratio of the signs of nonlinearity and dispersion of the group velocity (Lighthill's condition), providing the condition that the nonlinear detuning from the synchronism compensates linear de-synchronism from dispersion (Rabinovich and Trubetskov 1984).

Let's consider vertical distributions of stresses and strains shown in Fig. 6.5b. The waveforms and spectra of motion on the surface are determined, to a large extent, by the hysteretic curves in the upper 9 m, where the strains are maximal. When the amplitudes of the input increase to ~ 1.2 cm/s, spectra of motion on the surface become more complicated, and chaotic with a further increase of the amplitudes (Fig. 6.4). The chaotization stage corresponds to large working intervals of stress–strain relations and a large number of closed cycles of loading–unloading hysteretic curves. Expanding the working intervals beyond the point of inflection evidently contributes to the rapid complication and chaotization of motion.

Changes in the shapes of the hysteretic curves with increasing amplitudes of input motion correspond to changes in the trajectories of oscillations of the medium particles in phase space, i.e., changes in phase portraits of the system—bifurcations. As known (Rabinovich and Trubetskov 1984), the transition of a system to stochasticity through a series of period-doubling bifurcations is typical for dissipative nonlinear systems.

In the numerical experiments, tripling of the periods is observed (this periodicity is clearly seen in the layers with maximal strains at a depth of ~ 9 m in Fig. 6.4), which is obviously due to the predominance of cubic rather than quadratic nonlinearity in the soil response. The transition to stochastic motions occurs with increasing the amplitudes of input signals, i.e., with increasing the number of nonlinear interactions and combinational-frequency harmonics. Numerical experiments were repeated with other soil profiles, and for harmonics of resonant frequencies, self-modulation and generation of subharmonics were also observed.

Self–modulation and Fermi–Pasta-Ulam (FPU) recurrence phenomena in seismic wave fields were studied in field experiments by Dimitriu (1988). The spatial evolution of harmonic waves radiated by seismic vibrators in the frequency range 7–14.65 Hz was monitored by seismic profiles consisting of 8 vertical seismometers installed on the surface at distances of 10–70 m from the vibrator. Signals of various intensities were radiated, and the amplitudes of motion on the surface near the vibrator (at ~ 1 m) were V_0 ~ 0.015–0.6 cm/s.

At small amplitudes of the radiated signals, waveform recurrence phenomenon was observed, first described by Fermi, Pasta and Ulam.

Whenever a wave train was generated with a Mach number below some critical value ($\sim 10^{-5}$ at 1 m from the vibrator), it evolved recurrently. The Fourier spectra of the records show that, during the initial stage of evolution, at ~ 20 m from the vibrator, an instability developed, resulting in a dramatic growth of the second and third harmonics and their lower sideband components (whose appearance is attributed to ground nonlinearity), which took the energy of the main (carrier) frequency harmonic. As the instability effects reached their peak levels, the wave train seemed to be losing its coherence, often becoming strongly modulated (self-modulation). A balance of nonlinearity and dispersion effects occurred, and stable solitons interacting with each other were formed. However, due to the stability of their formation, only phase shifts appeared resulting from their interactions, the solitons decayed, and the wave returned to its initial state. With increasing the propagation distance, however, the energy gradually returned to the carrier frequency component, and the wave train demodulated, recovering its original (uniform and coherent) form (recurrence) at distances of ~ 30 m from the vibrator (Dimitriu 1988).

A complete return of the initial waveforms of the envelope and the carrier was observed, and a short distance (several carrier wavelengths) at which the return took place (Dimitriu 1988). A significant dependence of the observed phenomena on the frequency and a weaker dependence on the amplitude were noticed: the effect was intensified with increasing the amplitude of the signal and decreasing its frequency. For frequencies of 7–9 Hz, the spectra of propagating waves were sometimes dominated by the 2nd harmonic and its side frequencies, sometimes by the 3rd harmonic and its side frequencies. Only weak signs of instability were observed for the highest studied frequency of 14.65 Hz, while the clearest manifestations of instability were observed for frequency f ~ 10 Hz.

Dimitriu explained the observed phenomena from the viewpoint of the nonlinear theory of wave propagation. The nonlinear Schrodinger equation describes three main classes of phenomena observed during the propagation of quasi-harmonic waves in nonlinear media with dispersion, such as, modulation instability (or self-modulation), the existence of stationary envelope waves, including solitons, and periodically repeating in time and space recurrence of a weakly modulated wave, which approaches a periodic sequence of solitons in the process of its evolution. The Schrodinger equation has the form (Novikov et al. 1984):

$$iA_t + bA_{xx} + c|A|^2 A + dA = 0,$$

where A is the complex amplitude of the envelope of a propagating wave of the form exp*(i(kx–wt))*; the real constants $b = d^2\omega/dk^2$, c and d determine the magnitude and sign of dispersion, cubic nonlinearity and damping, respectively. The sign of the product bc determines the development of the process: if $bc < 0$, the wave is stable, and solitons are not formed; if $bc > 0$, instability develops, i.e., in case of an impulse input signal, the signal is splitting into several envelope solitons and an oscillating tail; in case of a harmonic input signal, the FPU recurrence phenomenon is observed. In Dimitriu's experiments, the minimal group velocity of the wave propagation corresponds to the frequency range f ~ 10–12 Hz, and the results of the experiment confirm this explanation.

In numerical experiments described by Figs. 6.4 and 6.5, the appearance of the velocity dispersion in the medium is associated with ground nonlinearity; the minimal wave propagation velocity also corresponds to the range of resonant frequencies, so the condition for modulation instability $bc > 0$ is fulfilled.

Thus, the propagation of intense harmonic seismic waves in soil layers is accompanied by generation of higher and combinational-frequency harmonics of the main frequency; this happens the faster the higher the degree of nonlinearity of the system. At a sufficiently high intensity of the input signal, an avalanche-like increase of the harmonics and their randomization occur, and the output spectrum becomes continuous. The stage of nonlinear modulation and subharmonic generation precedes the stage of chaotization.

In numerical experiments described above, soil profiles of fairly simple structures were studied, i.e., without seismic boundaries, with a smooth increase of seismic velocities with depth, or consisting of an upper soft and underlying dense layers. An example of a complex soil profile (a system with many degrees of freedom) is a soil profile at Port Island (PI) site with an upper liquefied layer; a numerical model of its behavior during the 1995 Kobe earthquake is described in Sect. 3.1. The soil profile at Port Island site includes five groups of layers with different mechanical properties: 0–13 m—artificial gravelly soil above the groundwater level, 13–17 m—the same below the groundwater level, 17–28 m—alluvial clays, 28–34 m—alluvial sand, deeper than 34 m—dense layers of gravels and clays.

Figure 6.6 shows the results of testing this soil profile by harmonic signals with frequency $f \sim 3$ Hz, which belongs to the range of resonance frequencies (~1–4 Hz) of the upper liquefied layer. With increasing the intensity of the input signal, nonlinear distortions of seismic waves increase, and spectra of motion on the surface are noticeably changed: first, odd-order higher harmonics of the main frequency appear (3rd, 5th, …), then, it's even-order higher harmonics (2nd, 4th, 6th, …) and subharmonics of the form $nf/3$ arise, and the vibrations are chaotized at an amplitude of the input of 0.008 m/s. At a further increase of the amplitude, a phenomenon similar to recurrence is observed, when the system periodically passes from the state of subharmonic generation with subsequent randomization to the state of generation of higher odd- and even-order harmonics, till the maximum amplitude of the input of ~ 0.525 m/s, apparently corresponding to the ultimate strength of the soil.

Table 6.1 illustrates an increase of the working intervals of stress–strain relations of the upper groups of soft layers (Fig. 3.1, Sect. 3.1): 1st, 2nd and 4th (soils in group 3 are dense clays) with increasing the amplitude of the input. The working intervals are measured in the number of points of the discrete stress–strain relations, as described in (Joyner and Chen 1975).

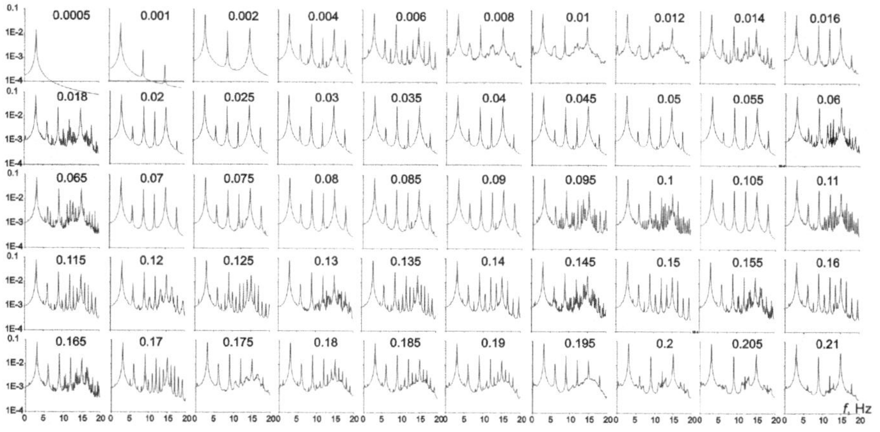

Fig. 6.6 Spectra of motion on the surface of the soil profile at Port Island site at the amplitudes of input harmonic signal $V_0 = 0.0005$–0.21 m/s and its frequency $f = 3$ Hz

Table. 6.1 Working intervals of stress–strain relations in the upper groups of layers of the soil profile at Port Island at different amplitudes of the input harmonic signal of frequency $f = 3$ Hz

V_O, m/s	0.0005	0.001	0.002	0.004	0.005	0.006	0.007
Group 1	3	4	5	6	6	6	7
Group 2	2	2	3	3	4	4	4
Group 3	3	3	4	5	5	6	6

Even at small amplitudes of the input motion, the soil response is substantially nonlinear. Liquefied soils in the upper layers are most sensitive to changes of the amplitudes of the input, and an increase of the working intervals of stress–strain relations in one of the groups of layers entails changes of the vertical distribution of stresses and strains throughout the soil thickness and changes in spectra of motion on the surface (Fig. 6.6).

When the amplitude of the input reaches ~ 0.006 m/s, subharmonics appear in spectra of motion on the surface; at a further increase of the amplitude, resonant amplification of the vibrations in soil layers occurs, while spectra of motion on the surface are qualitatively changed. Modulation occurs at resonant frequencies, and spectra shown in Fig. 6.6 illustrate a complicated behavior of a system with many degrees of freedom.

Subharmonics of the type $nf/3$ are generated (at the input amplitudes 0.004, 0.006, 0.014 m/s, etc.), as well as subharmonics of the type $nf/2$ (at the input amplitudes 0.018, 0.06, 0.095 m/s, etc.); in some cases, both at the same time. The types of the subharmonics and higher harmonics may be determined by the forms of stress–strain relations in soil layers: liquefied soils at Port Island possess significant even-order nonlinearities (Sect. 5.2), and only in this case, subharmonics of the type $nf/2$ appear in the output spectra. For other studied soil profiles, such subharmonics were observed in spectra of output motion in cases of high intensity of input signals.

For comparison, Fig. 6.7 shows spectra of motion on the surface of a simple soil profile (Fig. 6.1) tested by harmonic signals with amplitudes 0.001–0.69 m/s and frequencies $f = 1$ Hz (a) and 6 Hz (b).

As seen from the figure, when harmonic seismic signals propagate in soil layers, odd-order higher harmonics and subharmonics of the main frequency of the form $nf/3$ are generated; higher even-order harmonics appear in the output only in cases of sufficiently large amplitudes of the input signal. Frequency f ~ 1 Hz is resonant, and its modulation by a subharmonic 0.333 Hz appears at small amplitudes of the input, V_O ~ 0.05 m/s. At amplitudes V_O greater than 0.59 m/s, the vibrations become chaotic.

Frequency f ~ 6 Hz is not resonant, but a constructive interference of the vibrations occurs in individual layers, which probably determines the generation of subharmonics. Subharmonics of small amplitudes are generated; they modulate higher-order harmonics of frequency f ~ 6 Hz. At various amplitudes of the input signal, amplification occurs in layers at different depths. Slight signs of recurrence of spectral shapes are observed only for the harmonic at 6 Hz, which is evidently due to the

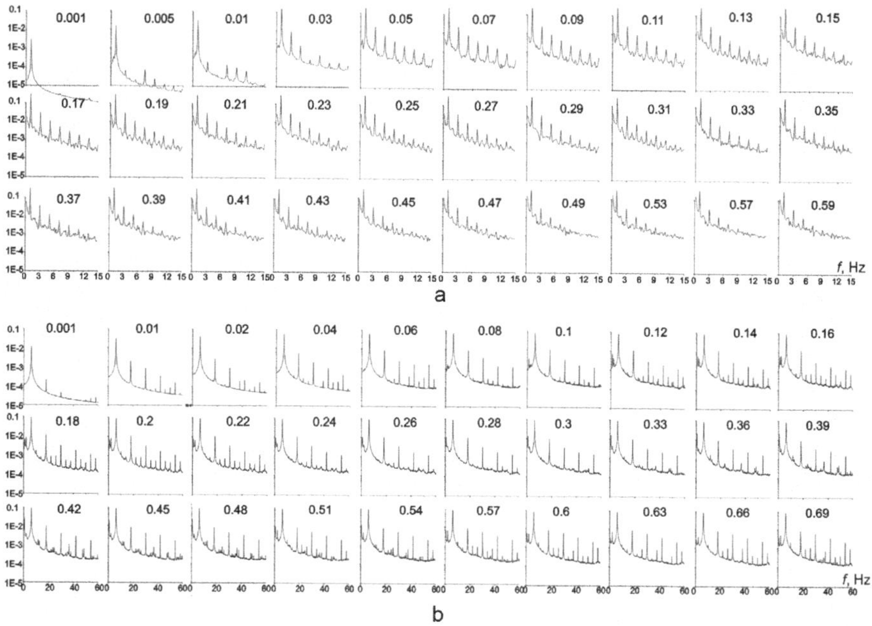

Fig. 6.7 Spectra of motion on the surface of a simple soil profile with a smooth increase of seismic velocities with depth (Fig. 6.1); the amplitudes of the input signal are $V_0 = 0.001–0.69$ m/s and frequencies $f = 1$ Hz (**a**) and $f = 6$ Hz (**b**)

features of the vertical distributions of stresses and strains in the soil layers. Chaotic vibrations are observed only for the resonant harmonic at 1 Hz, because we cannot obtain large enough strains at a frequency of 6 Hz.

6.3 High-order Spectra and Coherences as Identificators of Nonlinear-bounded Components in Seismic Wave Fields

The application of high–order spectra and coherences (HOS and HOC) in the analysis of time series has been developing since the 1950s. HOC are used in oceanography, geophysics and Earth sciences, in passive hydrolocation, biomedicine, communications, speech analysis, economics, plasma physics, and sunspot observations.

Three main points justifying the use of high-order spectra in signal processing are: extracting information hidden in time series about deviations from normality (Gaussian properties); estimating the phases of non-Gaussian parametric signals, and identifying and describing the properties of nonlinear mechanisms that generate time series (Nikias and Raghuveer 1987).

When we evaluate power spectra, the studied process is considered to be the result of the superposition of statistically uncorrelated harmonic components, and the distribution of energy among these components is estimated. In this case, only linear mechanisms determining the process are investigated; phase relations between the components are not considered. The information contained in the power spectrum is sufficient for a complete statistical description of a Gaussian random process with a known mean.

High-order spectra defined for a process by its high-order semi-invariants (cumulants) provide information about the deviations from normality and the presence of nonlinearities. Since all HOS (of orders > 2) are identically zero for a Gaussian process (Nikias and Raghuveer 1987), a nonzero high-order spectrum indicates a deviation from normality. Thus, in cases when the signal is a non–Gaussian stationary process and the additive noise is a Gaussian process, we get certain advantages in estimating the signal parameters based on HOS.

An important property of HOS is their ability to keep information about the phases of non-Gaussian parametric signals. Phase estimation in non-minimal-phase parametric processes plays a major role in deconvolution in geophysics, communication, and other fields of science.

Similar to the Volterra functionals, the estimation of HOS can play a key role in identifying and describing the types of nonlinearities of a system based on data at its output (Nikias and Raghuveer 1987).

Let's consider a linear time-invariant system with an input

$$x_k = \sum_m A_m \exp j(\omega_m k + \phi_m), \tag{6.3.1}$$

where $\{\phi_m\}$ are independent random variables uniformly distributed over the interval $[0, 2\pi]$. The output is described by the expression

$$y_k^{(1)} = \sum_m B_m \exp j(\omega_m k + \theta_m) \tag{6.3.2}$$

It can be easily shown that all semi-invariants $\{y_k^{(1)}\}$ of orders higher than two are equal to zero. Thus, zero high-order spectra of sequences $\{y_k^{(1)}\}$ will indicate that the output time series is generated exclusively by linear mechanisms.

If the same process $\{x_k\}$ is passed through a nonlinear system of a general type, for example, the N–th order Volterra processor,

$$\begin{aligned} z_k = &\sum_{h=0}^{M_1-1} h_1(i_1)x_{k-i_1} + \sum_{h=0}^{M_2-1}\sum_{h=0}^{M_2-1} h_{12}(i_1, i_2)x_{k-i_1}x_{k-i_2} + \cdots \\ &+ \sum_{i_1=0}^{M_N-1} \cdots \sum_{i_N=0}^{M_N-1} h_{12\ldots N}(i_1, i_2, \ldots, i_N)x_{k-i_1}\ldots x_{k-i_N} \end{aligned} \tag{6.3.3}$$

then in general, the output sequence $\{z_k\}$ is described by the expression

$$z_k = \sum_{\lambda=1}^{N} y_k^{(\lambda)}, \tag{6.3.4}$$

where $\{y_k^{(1)}\}$ is defined by the formula (6.3.2),

$$y_k^{(2)} = \sum_m \sum_n D_m D_n \exp j[(\omega_m + \omega_n)k + (\theta_m + \theta_n)], \tag{6.3.5}$$

$$y_k^{(3)} = \sum_m \sum_n \sum_e F_m F_n F_e \exp j[(\omega_m + \omega_n + \omega_e)k + (\theta_m + \theta_n + \theta_e)], \tag{6.3.6}$$

and so on. Nonzero bispectrum (trispectrum) of the sequence $\{z_k\}$ indicates the existence of a member $\{y_k^{(2)}\}$ ($\{y_k^{(3)}\}$) in relation (6.3.4) and consequently, the presence of quadratic (cubic) nonlinearity in the system. Thus, analysis of HOS of signals that have passed a nonlinear system allows us to determine the types of the system's nonlinearities and estimate them quantitatively (Nikias and Raghuveer 1987).

Similar to the Volterra series in time domain, in spectral domain, the response of the system can be represented by a set of spectra; the first is the power spectrum, then the bispectrum, trispectrum and other higher-order spectra (Marmarelis and Marmarelis 1978; Nikias and Raghuveer 1987). N-th order spectrum $C_N(\omega_1, \omega_2, \ldots, \omega_{N-1})$ of a real discrete stationary random process $x(k)$ is defined as the Fourier transform of its cumulant sequence of N-th order $c_N(\tau_1, \tau_2, \ldots \tau_{N-1})$, i.e.,

$$C_N(\omega_1, \omega_2, \ldots, \omega_{N-1}) = \sum_{\tau_1=-\infty}^{\infty} \cdots \sum_{\tau_{N-1}=-\infty}^{\infty} c_N(\tau_1, \tau_2, \ldots, \tau_{N-1}) \cdot \exp\{-j(\omega_1\tau_1 + \cdots + \omega_{N-1}\tau_{N-1})\} \tag{6.3.7}$$

The idea of spectral representation of a sequence of semi-invariants of the form (6.3.7) belongs to a Russian mathematician Andrey Kolmogorov.

The power spectrum, bispectrum and trispectrum are versions of N–th order spectra; the power spectrum ($N = 2$) is defined as

$$C_2(\omega_1) = \sum_{\tau_1=-\infty}^{\infty+} c_2(\tau_1) \exp\{-j(\omega_1\tau_1)\}, \tag{6.3.8}$$

bispectrum ($N = 3$)

$$C_3(\omega_1, \omega_2) = \sum_{m=-\infty}^{+\infty} \sum_{n=-\infty}^{+\infty} c_3(\tau_1, \tau_2) \exp\{-j(\omega_1\tau_1 + \omega_2\tau_2)\}, \tag{6.3.9}$$

trispectrum ($N = 4$)

$$C_4(\omega_1, \omega_2, \omega_3) = \sum_{\tau_1=-\infty}^{\infty} \sum_{\tau_2=-\infty}^{\infty} \sum_{\tau_3=-\infty}^{\infty} c_4(\tau_1, \tau_2, \tau_3) \exp\{-j(\omega_1\tau_1 + \omega_2\tau_2 + \omega_3\tau_3)\} \tag{6.3.10}$$

5th order spectrum ($N = 5$)

$$C_5(\omega_1, \omega_2, \omega_3, \omega_4) = \sum_{\tau_1=-\infty}^{\infty} \sum_{\tau_2=-\infty}^{\infty} \sum_{\tau_3=-\infty}^{\infty} \sum_{\tau_4=-\infty}^{\infty} c_5(\tau_1, \tau_2, \tau_3, \tau_4) \cdot \exp\{-j(\omega_1\tau_1 + \omega_2\tau_2 + \omega_3\tau_3 + \omega_4\tau_4)\} \tag{6.3.11}$$

6th order spectrum ($N = 6$)

$$C_6(\omega_1, \omega_2, \omega_3, \omega_4, \omega_5) = \sum_{\tau_1=-\infty}^{\infty} \sum_{\tau_2=-\infty}^{\infty} \sum_{\tau_3=-\infty}^{\infty} \sum_{\tau_4=-\infty}^{\infty} \sum_{\tau_5=-\infty}^{\infty} c_6(\tau_1, \tau_2, \tau_3, \tau_4, \tau_5) \cdot \exp\{-j(\omega_1\tau_1 + \omega_2\tau_2 + \omega_3\tau_3 + \omega_4\tau_4 + \omega_5\tau_5)\} \tag{6.3.12}$$

Here $c_2(\tau_1)$, $c_3(\tau_1, \tau_2)$, $c_4(\tau_1, \tau_2, \tau_3)$, $c_5(\tau_1, \tau_2, \tau_3, \tau_4)$, and $c_6(\tau_1, \tau_2, \tau_3, \tau_4, \tau_5)$ are sequences of cumulants of the second, third, fourth, fifth and sixth orders of time sequence $x(k)$, respectively.

For a real discrete stationary random process $x(k)$ with zero mean ($E\{x(k)\} = m_1 = 0$), cumulants of the 2nd and 3rd orders coincide with the moments of the 2nd and 3rd orders:

$$c_2(\tau_1) = m_2(\tau_1) = E\{x(k)x(k + \tau_1)\}, \tag{6.3.13}$$

$$c_3(\tau_1, \tau_2) = m_3(\tau_1, \tau_2) = E\{x(k)x(k + \tau_1)x(k + \tau_2)\}, \tag{6.3.14}$$

but this is not true for statistics of the 4th order and higher. Cumulants of the 4th and 5th orders are determined through the central moments (Malakhov 1978):

$$c_4(\tau_1, \tau_2, \tau_3) = m_4(\tau_1, \tau_2, \tau_3) - 3\{m_2(\tau_1)m_3(\tau_2)\}_s, \tag{6.3.15}$$

$$c_5(\tau_1, \tau_2, \tau_3, \tau_4) = m_5(\tau_1, \tau_2, \tau_3, \tau_4) - 10\{m_2(\tau_1)m_3(\tau_2, \tau_3)\}_s, \tag{6.3.16}$$

$$c_6(\tau_1, \tau_2, \tau_3, \tau_4, \tau_5) = m_6(\tau_1, \tau_2, \tau_3, \tau_4, \tau_5) - 15\{m_2(\tau_1)m_4(\tau_2, \tau_3, \tau_4)\}_s - 10\left\{m_3^3(\tau_1, \tau_2)\right\}_s + 30\left\{m_2^3(\tau_1)\right\}_s, \tag{6.3.17}$$

where $\{\}_s$ are symmetrization brackets (Malakhov 1978) constructed as follows:

$$3\{m_2(\tau_1) \cdot m_2(\tau_2)\}_s = m_2(\tau_1) \cdot m_2(\tau_2) + m_2(\tau_2) \cdot m_2(\tau_3) + m_2(\tau_1) \cdot m_2(\tau_3) \tag{6.3.18}$$

From the definition of HOS and the properties of high–order cumulants, it follows that HOS are generally complex, i.e. they can be represented as a module and a phase, and they are n-fold periodic with a period 2π (n is the order of HOS) and possess symmetry properties, so that knowledge of the spectrum in one of its symmetric domains (for example, bispectrum in the range $\omega_2 \geq 0$, $\omega_1 \geq \omega_2$, $\omega_1 + \omega_2 \leq \pi$) is enough for its full description (Nikias and Raghuveer 1987).

There are two main methods for estimating HOS: a conventional method of Fourier-analysis type and a parametric method based on autoregression (AR), sliding average (SA) and combined models (ARSA) (Nikias and Raghuveer 1987).

We used usual methods for estimating HOS based on fast Fourier transform that provide asymptotically unbiased and consistent estimates (Haubrich 1965).

High-order coherencies (HOC) are often more convenient in practical applications than HOS. They combine the concepts of HOS and power spectra and are defined as HOS normalized by power spectra. Similar to HOS, high-order coherencies possess symmetry properties.

Bicoherence is calculated by the formula (Haubrich 1965; Nikias and Raghuveer 1987):

$$r_3^2(\omega_1, \omega_2) = \frac{|C_3(\omega_1, \omega_2)|^2}{C_2(\omega_1)C_2(\omega_2)C_2(\omega_1 + \omega_2)} \tag{6.3.19}$$

Due to the symmetry properties, it is sufficient to calculate bicoherence in the octant $\omega_1 > \omega_2 > 0$ for its complete definition.

High-order coherences take values from 0 to 1 and can serve as the measures of phase coupling between the signal's spectral components. If the signal at the input of a system with quadratic (cubic) nonlinearity contains harmonics with frequencies ω_1 and ω_2, then the output signal contains both the original frequencies ω_1, ω_2, and the combinational frequencies $2\omega_1$, $2\omega_2$, $\omega_1 + \omega_2$, and $\omega_1 - \omega_2$ (in case of cubic nonlinearity $-3\omega_1$, $3\omega_2$, $\omega_1 + \omega_1 + \omega_2$, $\omega_1 + \omega_2 + \omega_2$, $\omega_1 + \omega_1 - \omega_2$ and others); the phases of the main and combinational harmonics being connected by certain relations called quadratic (cubic) phase coupling.

For two fixed frequencies ω_1 and ω_2, bicoherence $r_3^2(\omega_1, \omega_2)$ is equal to the part of power spectral density, released at frequency $\omega_1 + \omega_2$ that has a quadratic phase coupling with harmonics ω_1 and ω_2. Respectively, for three fixed frequencies ω_1, ω_2 and ω_3, tricoherence $r_4^2(\omega_1, \omega_2, \omega_3)$ is equal to the part of power spectral density released at frequency $\omega_1 + \omega_2 + \omega_3$ that has a cubic phase coupling with harmonics ω_1, ω_2 and ω_3.

If the values of high-order coherence $r_N^2(\omega_1, \omega_2, \ldots, \omega_{N-1})$ are close to 1, this indicates a high phase coupling, i.e., the combination frequency harmonic $\omega_1 + \omega_2 + \cdots + \omega_{N-1}$ may be a result of the interaction of harmonics of the main frequencies: $\omega_1, \omega_2, \ldots, \omega_{N-1}$; whereas, its low values (close to 0) indicate the independence of the harmonics $\omega_1, \omega_2, \ldots, \omega_{N-1}$ and $\omega_1 + \omega_2 + \cdots + \omega_{N-1}$.

Under the condition of spectral uniformity of nonlinear transformations, the diagonal values of high-order spectra represent the HOS values in other spectral ranges. This allows using one-dimensional HOS and HOC in many cases, instead of cumbersome multidimensional HOS, without significant loss of information.

Diagonal values of the 4th order coherence, i.e., tricoherence, are calculated by the formula

$$r_4^2(\omega) = \frac{|C_4(\omega)|^2}{C_2^3(\omega)C_2(3\omega)}, \tag{6.3.20}$$

diagonal values of the 5th order coherence, by the formula

$$r_5^2(\omega) = \frac{|C_5(\omega)|^2}{C_2^4(\omega)C_2(4\omega)}, \tag{6.3.21}$$

diagonal values of the 6th order coherence, by the formula

$$r_6^2(\omega) = \frac{|C_6(\omega)|^2}{C_2^5(\omega)C_2(5\omega)}. \tag{6.3.22}$$

For practical applications, we should know the variance of the estimates, or the 95% confidence interval of HOS estimates in cases when their true value is zero. Probability density functions for HOS are not found yet; however, we can approximately estimate the variance for the case when the time series *x(t)* is a stationary Gaussian random process.

To evaluate HOS or HOC, time sequence *x(k)* is divided into p equal non-overlapping intervals, over which averaging of the HOS or HOC estimates is performed. The estimates become smooth at sufficiently large p (Nikias and Raghuveer 1987).

Let each non-overlapping interval have a length of M points, and for the *jth* interval, coefficients of the Fourier expansion for the *kth* harmonic are a_{kj}, b_{kj}. Haubrich (1965) showed that, since coefficients a_{kj}, b_{kj} are independent for a stationary Gaussian process and normally distributed with zero means, and a variance $E({a^2}_{kj}) = E({b^2}_{kj}) = c^2$, the expected average of bicoherence is ~ *1/p*, and for large p values bicoherence has approximately χ^2-distribution with two degrees of freedom (being a sum of squares of two values, both approximately normally distributed). Thus, the average bicoherence for non–diagonal terms is *1/p* (for diagonal terms, it is *2/p*, due to the symmetry), and the 95% confidence interval for $r_3^2(\omega_1, \omega_2)$ is approximately *3/p* (Haubrich 1965).

Since higher-order coherences are constructed similar to bicoherence, it is easy to show that tricoherence, 5th and 6th order coherences also possess a χ^2-distribution with two degrees of freedom, and average values for diagonal terms are $(N-1)!/p$, where N is the order of the coherence, and the multiplier $(N-1)!$ appears due to the symmetry. Thus, the average values of HOS of a Gaussian process passed through a nonlinear system and enriched by combinational-frequency harmonics should be higher than $(N-1)!/p$. Increased bicoherence values indicate the presence of quadratic nonlinearity in the system and give its quantitative estimate. Increased values of tricoherence (5th or 6th order coherence) indicate the presence of cubic nonlinearity in the system (4th or 5th order nonlinearities) and give their quantitative estimates. Thus, we can determine the types and obtain quantitative estimates of the system's nonlinearities.

HOS have been applied in geophysics. Hasselman et al. (1963) used bispectral analysis to study nonlinear effects in low-frequency ocean waves. Haubrich (1965) investigated the presence of nonlinearly coupled vibrations in seismic noise and in coda-waves; he concluded that within the error limits, the estimates testify to the linearity of seismic noise and to high content of nonlinearly coupled waves in seismic coda, which is due to nonstationarity rather than nonlinearity. Zadro and Caputo (1968) used bispectral analysis of records of horizontal pendulums and a gravimeter to study the fine structure of free oscillations of the Earth excited by the Chilean and Alaskan earthquakes. The obtained results indicate nonlinear interactions of low-frequency oscillation modes.

Cartwright's research (Cartwright 1968) established the relationship between tides and random local variations of the sea level through bispectral analysis. Marmarelis and Marmarelis (1978) used high-order spectra for more efficient estimation of the Wiener kernels and functionals.

Here we calculate high-order spectra and coherencies of seismic noise passed through soil profiles in numerical experiments described in Chap. 5 and evaluate the informativeness of HOS in studying the nonlinearity of soil behavior.

Example 1 HOC of harmonic signals and the Gaussian white noise passed "simple" soil profile, i.e., horizontally layered sediments with a thickness of 200 m and density of 2.05 g/cm^3 were studied. Depth dependences of the shear stress in failure $\tau_{\max}$ and S-wave velocity V_S in the profile are shown in Fig. 6.1. HOC were calculated for time series of long duration, 65,000 points (~ 3 min at a sampling rate 340 Hz). Three cases were considered: the input signal was (a) harmonic, (b) a sum of two harmonics, and (c) the Gaussian white noise. Figure 6.8a–c show the results.

With increasing the intensity of input motion, the working intervals of stress–strain relations in soil layers increase, and nonlinear distortions of the propagating waves also increase, influencing the amplitudes and spectral content of motion on the surface. Changes of the spectral content of motion are illustrated by Fig. 6.8. If the input is harmonic, its nonlinear distortions are caused by the growth of the 3rd, 5th, 7th, … odd-order harmonics of the main frequency, which indicates odd-order nonlinearities of the soil.

The richer the spectrum of the input signal, the more complicated the spectrum of the response of a nonlinear system, as a result of wave interactions that are more diverse. Suppose the input signal is a sum of two harmonics, for example, at frequencies 5 Hz and 11 Hz (Fig. 6.8b); then odd-order higher harmonics at frequencies 15 Hz, 25 Hz, 35 Hz, 45 Hz, 33 Hz, 55 Hz, etc. are generated, as well as combinational-frequency harmonics at 1 Hz (11 − 5 − 5), 17 Hz (11 + 11–5), 21 Hz (5 + 5 + 11),

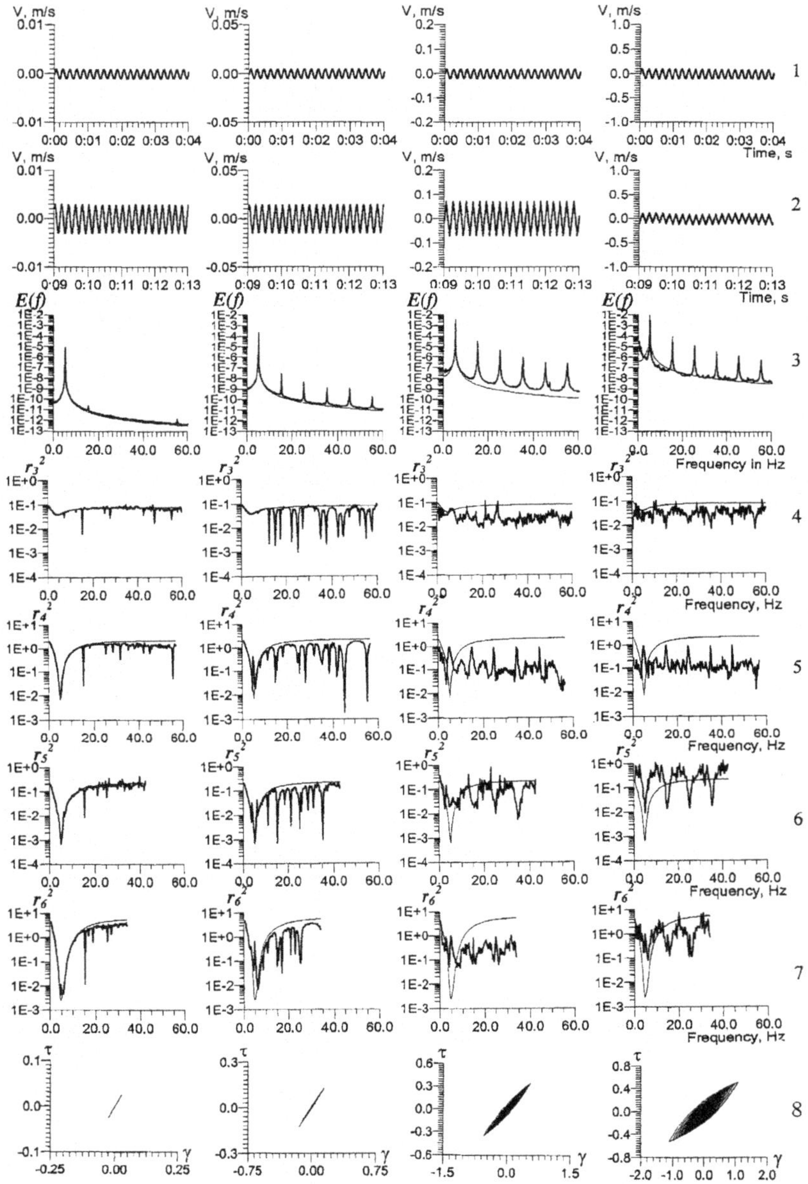

(a)

Fig. 6.8 Changes in shapes, spectra, and high-order coherences of seismic signals propagating in a nonlinear medium – soil layers. The input is: **a**—harmonic, **b**—a sum of two harmonics, **c**—the Gaussian noise in a wide frequency range. 1—input signals; 2—the response (motion on the surface); 3—power spectra of the input signals (thin lines) and the responses (thick lines); 4–7—one-dimensional bicoherence $r_3^2(x)$, tricoherence $r_4^2(x)$, fifth-order $r_5^2(x)$, and sixth-order $r_6^2(x)$ coherences of the input signals (thin lines) and the responses (thick lines); 8—the corresponding stress–strain diagrams (they depend on depth; maximum values are shown)

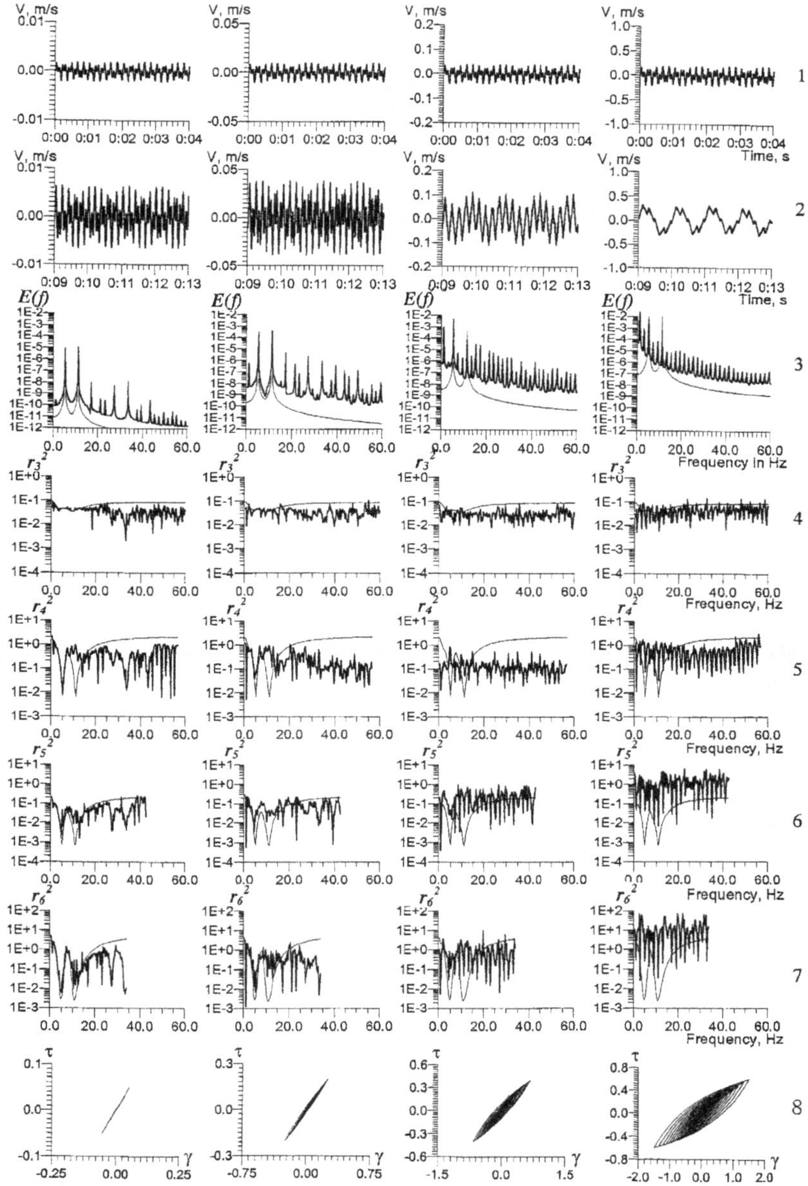

(b)

Fig. 6.8 (continued)

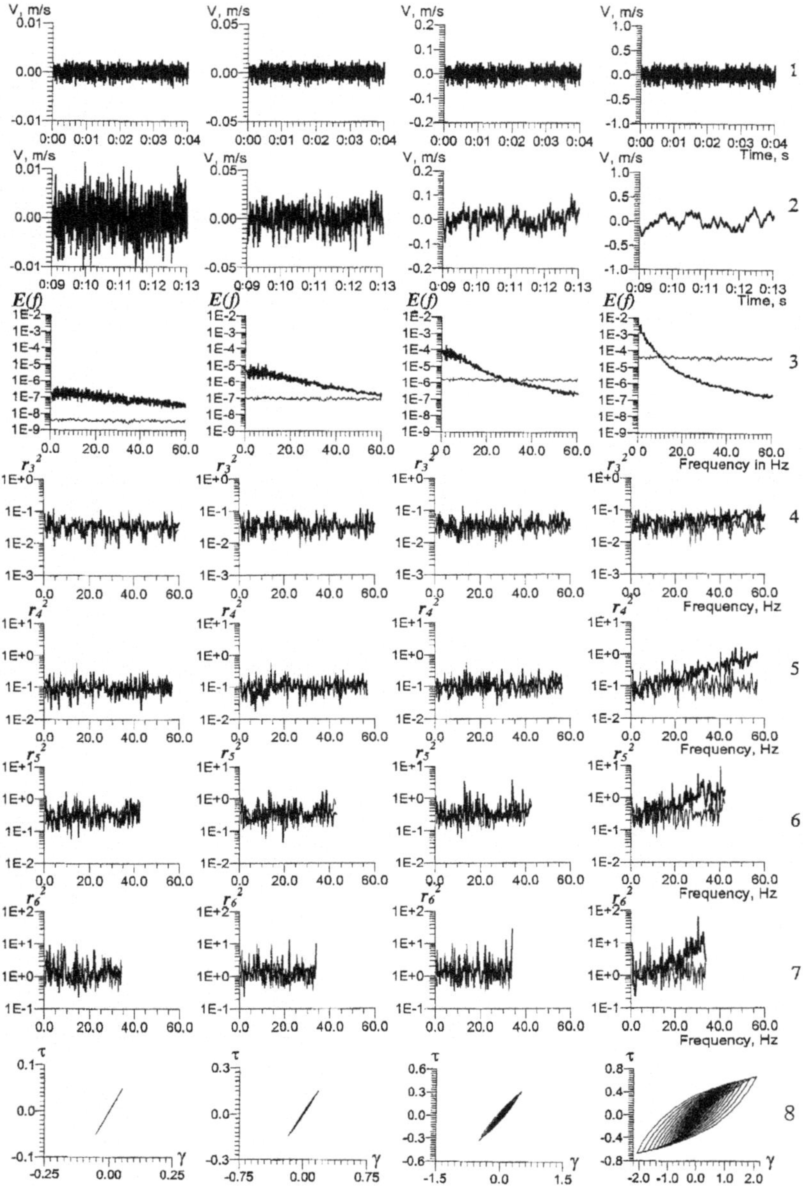

(c)

Fig. 6.8 (continued)

27 Hz (5 + 11 + 11), 17 Hz (11 + 11 – 5), 37 Hz (5 + 5 + 5 + 11 + 11), 43 Hz (5 + 5 + 11 + 11 + 11) and others. For odd-order nonlinearities, self-modulation effects are typical, i.e., generation of harmonics at the original frequencies, (5 + 5–5), (5–5 + 11) and others (Vinogradova et al. 1979).

The power spectrum and high-order spectra of the input and output quantitatively characterize nonlinear distortions of testing signals in the medium. Figure 6.8 shows bicoherence $r_3^2(\omega)$, tricoherence $r_4^2(\omega)$, and higher-order coherencies $r_5^2(\omega)$ and $r_6^2(\omega)$. As seen from the figure, an increase of the intensity of the input leads to the increase in tricoherence and 6th order coherence $r_6^2(\omega)$ of motion on the surface.

If the input signal is harmonic or a sum of harmonics (Fig. 6.8a, b), the increase of the coherences of the 4th and 6th orders at the frequencies of the input signals (5 Hz and 11 Hz) indicates generation of their 3rd and 5th higher harmonics. At large amplitudes of the inputs, the 4th and 6th order coherences are close to 1 (Fig. 6.8a, b). This means about 100% phase coupling between the main frequency harmonic and its 3rd and 5th harmonics; i.e., that the 3rd and 5th order harmonics are the results of nonlinear distortions of the main- frequency harmonic. Note high sensitivity of HOC (especially, tricoherence and the 6th order coherence) to the nonlinearity even at weak input motion (Fig. 6.8a).

If the amplitudes of input signals are large enough, their propagation in a nonlinear medium is accompanied by significant nonlinear distortions, when the number of combinational-frequency harmonics rapidly increases, and motion on the surface becomes noise–like (Fig. 6.8a, b). Power spectral density takes the form $E(f) \sim f^{-k}$, regardless of the spectral composition of the input signal, and the averages of squared HOC approach the value $(N-1)!/p$, which is characteristic of noise signals, and exceed it.

Table 6.2 shows the values of squared HOC, averaged within the spectral band, for the input Gaussian white noise of various intensities. The corresponding values $(N-1)!/p$ ($p = 55$) are given below for comparison.

From the data shown in the table and in Fig. 6.8b, we can see that at a sufficiently high intensity of the input noise, HOC exceed $(N-1)!/p$, especially at high frequencies (generation of higher harmonics is more effective at higher frequencies).

HOC values indicate the predominant contribution of cubic nonlinearity to the soil response, but the contributions of 5th and 6th order nonlinearities are also large.

Table. 6.2 Average values of squared High-Order Coherences of Motion on the Surface for input – the Gaussian white noise

The intensity of input motion, m/s	$\langle r_3^2(\omega)\rangle$	$\langle r_4^2(\omega)\rangle$	$\langle r_5^2(\omega)\rangle$	$\langle r_6^2(\omega)\rangle$
0.001	0.037	0.105	0.38	1.8
0.005	0.04	0.107	0.43	1.72
0.02	0.041	0.119	0.48	2.12
0.1	0.066	0.35	0.99	4.49
$(N-1)!/p$	0.036	0.11	0.43	2.1

As shown in Chap. 5, the types of nonlinearities of the soil response are determined by the shapes of stress–strain relations in soil layers, and at the maximum intensities of input signals achieved in the experiment, the functions describing the loading and unloading parts of stress–strain relations (Fig. 6.8c) can be represented by combinations of even and odd functions that are comparable in the amplitudes; they define the presence of even- and odd-order nonlinear components in the soil response. At lower intensities of input signals, stress–strain relations are predominantly odd functions, and nonlinear components of odd types predominate in the soil response.

Figure 6.8c indicates that sensitivity of HOC to weak nonlinearities of the soil response is quite low in case of noise input signals, so it is not easy to detect nonlinear components in real seismic noise. In order to detect and quantify nonlinear components in seismic noise, we can try (1) to analyze stationary noise records of long duration, to reduce variance; (2) to evaluate not only bicoherence, but also tricoherence, 5th and 6th order coherences; (3) to analyze many parameters: HOC, their average values in the whole spectral band and in the high frequency range, the numbers of values exceeding 95- and 99-percent confidence limits, deviations of the distributions of HOC squares from χ^2-distribution.

Example 2 2-D bicoherence of seismic noise recorded at soil and rock sites is calculated. Haubrich (1965) estimated bispectrum of seismic noise in the frequency range 0.005–0.5 Hz and did not find nonlinear components. As shown above, on one hand, soils possess nonlinearities of odd orders, so bispectrum and bicoherence can hardly show nonlinear components in seismic noise; on the other hand, noise-like signals are not too sensitive to the nonlinearities in the medium.

Bicoherences shown below are calculated in a wide frequency range; they give us a representation of the bispectrum and bicoherence and of the informativeness of their diagonal components. They also show the difference in bicoherence estimates for seismic noise recorded on soft soils and on dense crystalline rocks.

Bicoherences of seismic noise recorded at four sites with different ground conditions were studied. The records were obtained at:

- Geophysical Observatory Borovoe (Northern Kazakhstan) located on the periphery of the Kokchetav massif, on the outcrops of Precambrian rocks. Seismic equipment is installed on slightly cracked granites, and monolithic granites start at a depth of 6–10 m (hard rock). The area is aseismic, with a low level of natural and industrial seismic noise.
- Seismic station Dobrushka (Czech Republic) located in a quiet place in Eagle Mountains massif in northeastern Bohemia. The recording equipment is installed at the site where the upper 6 m are cretaceous deposits of sandstone–siltstone; below are granodiorites, smoothly turning to diorites, siliceous and biotite granites (rock).
- Geophysical test site Bystrovsky located on the shore of Novosibirsk reservoir, about 100 km to the south of Novosibirsk. The upper ~ 15 m are soft sandy-clay soils with V_S ~ 280 m/s; at depths of 15–40 m, fractured clay shales and limestones

are deposited with V_S ~ 2000 m/s. The bedrock of the basement (below ~ 50 m) are metamorphosed clay shales with V_S ~ 3500 m/s.
- Geophysical test site Uznozh located near Gomel city in Belarus. The upper layers down to ~ 150 m are sands, clays and chalk with V_S ~ 240–800 m/s. Below, at ~ 150–400 m, there is sandstone with V_S ~ 1200 m/s.

Table 6.3 provides information on the recording equipment, site conditions, analyzed records, and the obtained estimates. At Borovoe, Dobrushka and Bystrovsky sites, horizontal components of velocity time histories were processed. At Uznozh test site, seismic recording was carried out by a horizontal strainmeter with a base of 2 m and an optoelectronic displacement converter (accuracy ~ $2 \cdot 10^{-10}$ m, frequency band 0–100 Hz).

For a stationary Gaussian process, bicoherence is approximately χ^2-distributed with two degrees of freedom; its expected average is $r_3^2 \cong 1/p$ (~2/p for diagonal values) and with a probability of ~ 95%, $r_3{}^2 \leq 3/p$ (p is the number of analyzed time intervals).

Figures 6.9, 6.10, 6.11 and 6.12 present the spectra and bicoherences of seismic noise recorded at the four sites; the isolines are drawn with an interval of 1/p. The density of the isolines increases with increasing the softness of the subsurface soils: it is maximal at Uznozh test site.

Spectral components at frequencies 1, 1.23, 2, 2.08, 2.5, 3, 4, 5, 5.48, 6, 6.48, 7 and 9 Hz at Dobrushka station (Fig. 6.10) are evidently produced by industry. Bicoherence maxima at frequencies (2 Hz, 1 Hz), (3 Hz, 1 Hz), (5 Hz, 1 Hz), (5.48 Hz, 1 Hz), (3 Hz, 2 Hz), (4 Hz, 2 Hz), (4.16 Hz, 2.08 Hz) correspond to phase-coupled frequencies: 2 Hz, 1 Hz and 3 Hz; 3 Hz, 1 Hz and 4 Hz; 5 Hz, 1 Hz and 6 Hz, etc.

Maxima at frequencies (1 Hz, 1 Hz) and (2.5 Hz, 2.5 Hz) (Fig. 6.9) and (1 Hz, 1 Hz), (2 Hz, 2 Hz), (2.08 Hz, 2.08 Hz), (3 Hz, 3 Hz) (Fig. 6.10) indicate phase coupling of these harmonics with the double-frequency harmonics.

At Dobrushka site, two groups of harmonics are distinguished, apparently corresponding to different sources, such as, (1, 2, 3, 4, 5, 5.48, 6, 6.48, 7, 8 and 9 Hz) and (2.08, 4.16 and 6.24 Hz), because during the observations temporal variations

Table. 6.3 Deviations of the average bicoherence estimates of seismic noise from the bicoherence estimates of the Gaussian process

Site	Number of records	Equipment	Site conditions	p	δ_1 (%)	δ_2(%)
Borovoe	24, 20 min each	SM3-OS	Tunnel, 15 m	21	1.5	1.0
Dobrushka	250, 1 h each	IRIS/USGS	Tunnel, 8 m	69	1.4	0.95
Bystrovsky	4, 1 h each	SM3-OS	Surface	150	2.25	1.2
Uznozh	10, 3.5 min each	Optoelectronic strainmeter	Surface	30	9.0	1.8

δ_1, % is the excess of the average bicoherence over 1/p in percent, δ_2, % is the number of bicoherence values exceeding 3/p in percent of the total number of the calculated values

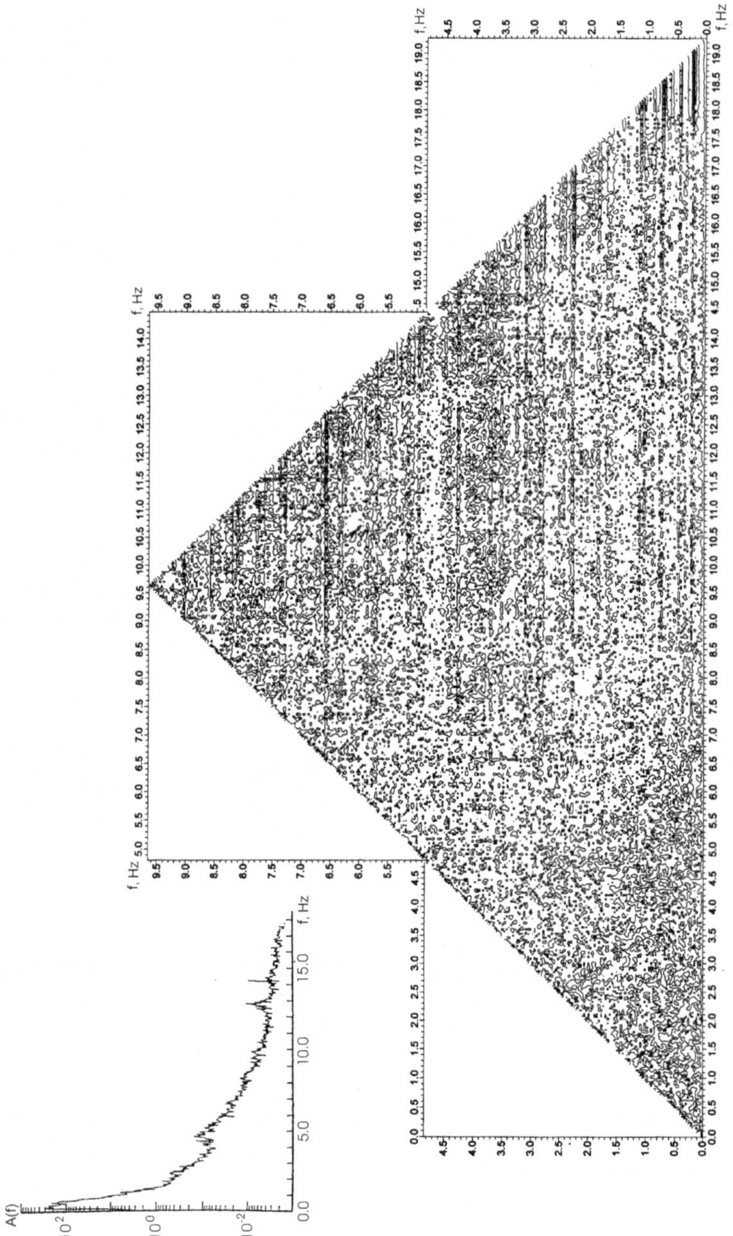

Fig. 6.9 The spectrum and bicoherence of seismic noise at Borovoe (Northern Kazakhstan)

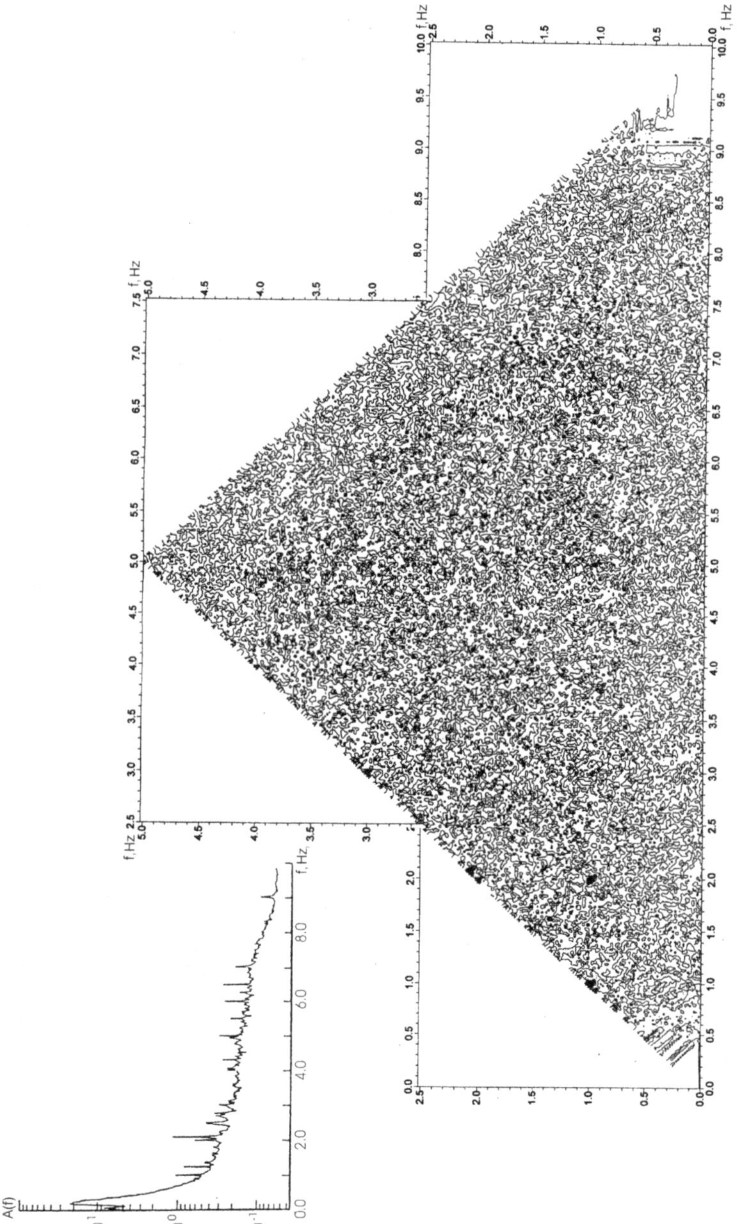

Fig. 6.10 The spectrum and bicoherence of seismic noise at Dobrushka station (Czech Republic)

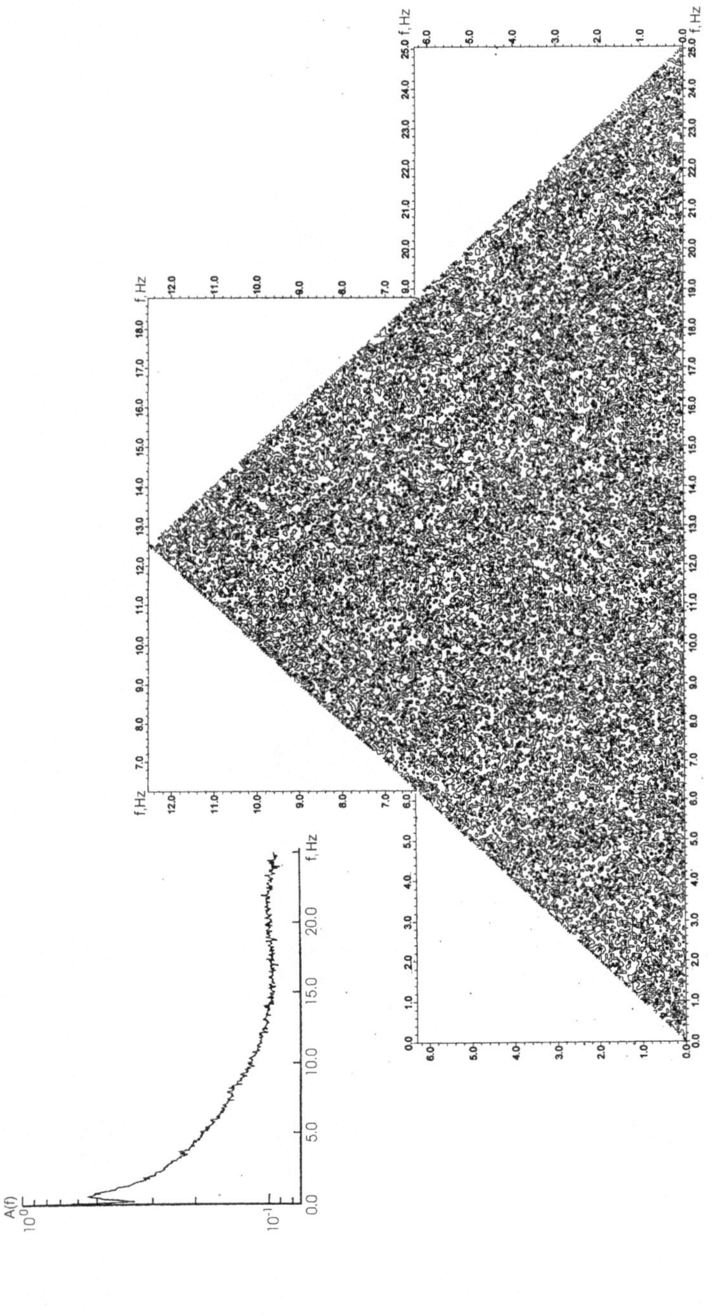

Fig. 6.11 The spectrum and bicoherence of seismic noise at Bystrovsky test site (Western Siberia)

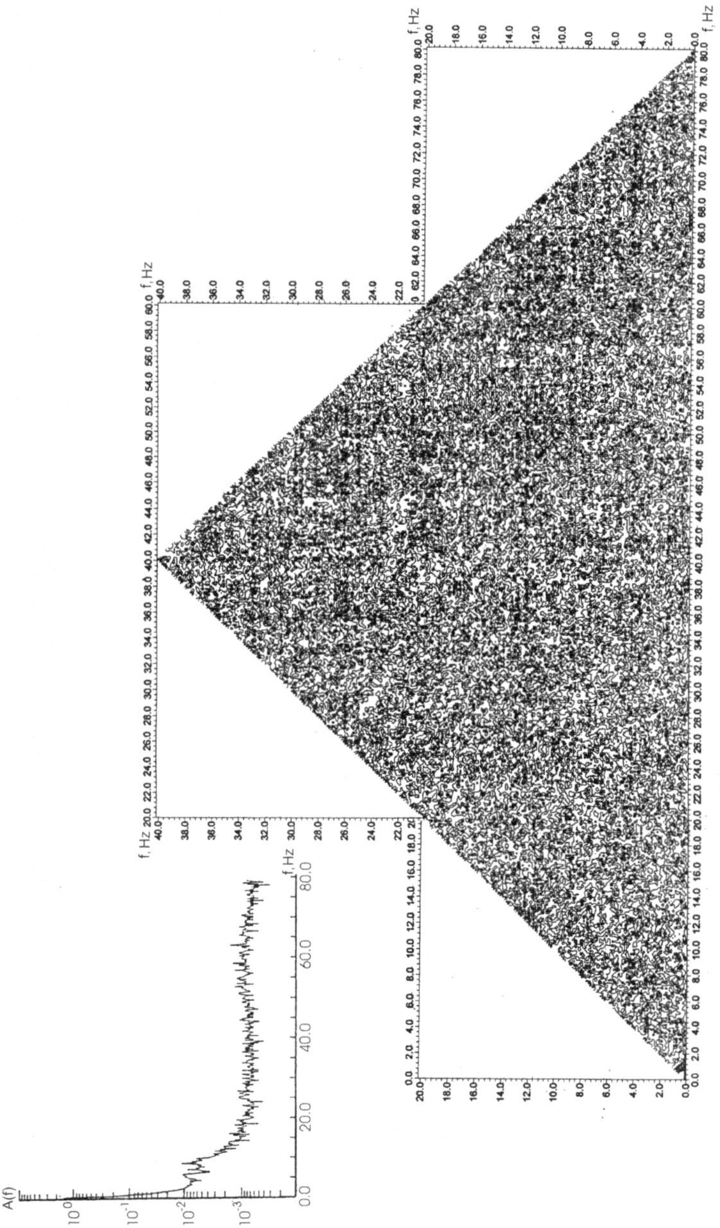

Fig. 6.12 The spectrum and bicoherence of seismic noise at Uznozh test site (Belarus)

of spectral amplitudes of each group correlate with each other, but do not correlate with amplitude variations of the other group. Apparently, harmonics of doubled and combinational frequencies are generated by industrial sources along with harmonics of the main frequencies, which is typical for industrial sources.

The estimates of bicoherence of seismic noise at different sites are based on records of different durations, and they were compared with the averages for the Gaussian process $1/p$ and a confidence limit of $3/p$. To calculate the average values of bicoherence, spectral ranges were selected that do not contain frequencies of storm microseisms and industrial harmonics. For all available records, the average bicoherence values and the numbers of values exceeding $3/p$ are calculated; the results are shown in Table 6.3.

For Borovoe and Dobrushka sites, the obtained averages coincide with $1/p$ within the error limits, and the value $3/p$ corresponds to approximately 99% confidence level. For Bystrovsky and Uznozh test sites, deviations of the average bicoherence from the values typical for the Gaussian white noise may be significant, but more records should be processed for the verification.

The variance of the average bicoherence estimates was evaluated empirically. For computer-generated random time series with the Gaussian distribution of amplitudes, the mean values, 95% confidence intervals and standard deviations of the mean values were calculated. The mean values of bicoherence estimates were very close to $1/p$, the 95% intervals were approximately equal to $\cong 2/p$, and the values $3/p$ defined the 99% confidence level.

Thus, seismic noise recorded at Borovoe and Dobrushka rock sites is virtually a Gaussian random process, but at soil sites Bystrovsky and Uznozh, it has deviations from the Gaussian process, which may be significant.

******

Thus, by means of numerical simulations and processing real records, we studied transformations of seismic waves in nonlinear media—subsurface soils.

The development of nonlinear wave effects can be traced by the evolution of harmonic seismic waves propagating in subsurface soils. At rather large amplitudes of the wave, when the soil behavior becomes nonlinear, the 3rd, 5th, 7th, etc. higher harmonics of the main frequency are sequentially generated; at a sufficiently high intensity of motion, the 2nd, 4th, 6th, etc. even-order harmonics are also generated. At a further increase of the amplitude of the propagating wave, nonlinear effects lead to the generation of new spectral components, the spectra become more complicated, and motion on the surface becomes chaotic.

The nonlinearity of soil response influences different frequency ranges, depending on the generation of high and combinational-frequency harmonics, mechanisms of decay instability regulated by Manley-Rowe relations, and nonlinear hysteretic damping. As a result of the combined action of these mechanisms, low-frequency components of seismic motion are not weakened, while the medium and high-frequency components are weakened (but may also be amplified if they were absent or low in the input), so that spectra of motion on the surface tend to take the quasi-stationary smoothed form $E(f) \sim f^{-k}$. Thus, nonlinear transformations of seismic waves in subsurface soils leads to losing information about spectral contents of the waves incident to soil layers.

At frequencies corresponding to the resonant frequencies of soil layers, nonlinearity of stress–strain relations leads to the dispersion of propagation velocities. As a result of the combined action of nonlinearity and dispersion, modulation instability phenomena can occur, i.e., self-modulation and generation of subharmonics of the main frequencies of the propagating waves. Mainly subharmonics of $nf/3$ type are generated, while subharmonics of $nf/2$ type are also generated in soils possessing both odd- and even-order nonlinearities.

High-order spectra and coherences allow us to identify and quantify the contents of nonlinearly bounded spectral components in seismic records; however, the records should be stationary and long-lasting; more information can be obtained, if not only bispectra (bicoherences) are evaluated, but also other high-order spectra (high-order coherences), such as, trispectra (tricoherences), spectra (coherences) of the 5th and 6th orders.

References

Cartwright, D.E.: A united analysis of tides and surges round North and East Britain. Philos. Trans. Roy. Soc. London A **263**, 1–55 (1968)

Dimitriu, P.P.: Self-modulation and recurrence phenomena in vibrator-induced steady-state sinusoidal ground vibrations. Phys. Earth Planet. Inter. **50**, 74–82 (1988)

Hasselman, K., Munk, W., MacDonald, G.: Bispectra of ocean waves. In: Rossenblatt, M. (ed.) Proceedings of Symposium Time Series Analysis. New York, Wiley, pp. 125–130 (1963)

Haubrich, R.A.: Earth Noise, 5 to 500 millicycles per second. J. Geoph. Res. **70**, 1415–1427 (1965)

Joyner, W.B., Chen, T.F.: Calculation of nonlinear ground response in earthquakes. Bull. Seism. Soc. Am. **65**(5), 1315–1336 (1975)

Kadomtsev, B.B., Karpman, V.I.: Nonlinear waves. Soviet Phys. Uspekhi. **103**(2), 193–232 (1971) (in Russian).

Malakhov, A.N.: Cumulant Analysis of Non-gaussian Random Processes and their Transformations. Sovetskoe Radio, Moscow (1978). (in Russian)

Marmarelis, P.Z., Marmarelis, V.Z.: Analysis of Physiological Systems. The White-Noise Approach. Plenum Press, New York and London (1978)

Nikias, C.L., Raghuveer, M.R.: Bispectral estimation: a digital signal processing framework. IEEE **75**, 869–891 (1987)

Novikov, S., Manakov, S.V., Pitaevsky, L.P., Zakharov, V.E.: Theory of Solitons. The Inverse Scattering Method. Springer New York, NY (1984)

Rabinovich, M.I., Trubetskov, D.I.: Introduction to the Theory of Vibrations and Waves. Nauka, Moscow (1984). (in Russian)

Rudenko, O.V., Soluyan, S.I.: Theoretical Foundations of Nonlinear Acoustics. Nauka, Moscow (1975). (in Russian)

Rudenko, O.V.: Interaction of intense noise waves. Uspekhi Fizicheskikh Nauk **149**(3), 413–447 (1986) (in Russian)

Vinogradova, M.B., Rudenko, O.V., Sukhorukov, A.P.: Teorija voln. Nauka, Moscow (1979) (in Russian)

Yu, G., Anderson, J.G., Siddharthan, R.V.: On the characteristics of nonlinear soil response. Bull. Seism. Soc. Am. **83**, 218–244 (1992)
Zadro, B., Caputo, M.: Spectral, Bispectral analysis and Q of the free oscillations of the Earth. Suppl. Nuovo Cimento V **I**(1), 67–81 (1968)

Chapter 7
Seismic Microzonation Methods Accounting for Nonlinear Soil Response

The features of soil behavior in the near-fault zones of large earthquakes are discussed that should be accounted for in seismic microzonation, i.e., in evaluation of soil response in future strong earthquakes.

We describe (1) the development of seismic zonation maps in Russia, USA, and Japan, (2) regional databases for ground motion estimation, (3) significant parameters of soil profiles, (4) the behavior of soils in the near-fault zones, (5) nonlinear damping mechanisms that attenuate motion at high and medium frequencies and only weakly affect low frequencies, (6) changes of elastic moduli of soft and dense soils during strong motion, (7) the influence of saturation of soils with underground water on characteristics of motion on the surface, (8) resonant and nonlinear phenomena in soil layers, (9) the influence of directivity effects in extended sources of large earthquakes on soil response and generation of high PGA in the near-fault zones.

Considering these features, we conclude that seismic microzonation should be based on the calculation of acceleration time histories of future strong earthquakes, accounting for the soil response, as suggested by Aki and Irikura (1991). We should compute time series for given source-receiver pairs using existing methods and our current knowledge on the earthquake sources, propagation paths, and site conditions. Only complete data on the composition and physical parameters of soil layers, combined with adequate methods for calculation of soil response provide us with reliable estimates of the parameters of motion on the surface.

Seismic zonation maps in Russia, USA, and Japan

During seismic microzonation we assess the influence of local geotechnical conditions on the parameters (intensity, waveforms, spectrum, etc.) of motion on the surface during a strong earthquake. These parameters define stability or destruction of buildings at soil sites during strong earthquakes.

Accounting for local site effects in strong ground motion helps us to reduce damage from natural disasters, and seismologists are looking for reliable and not expensive ways to do this.

O. V. Pavlenko, *Seismic Waves in Soil Layers*,
https://doi.org/10.1007/978-3-031-52338-0_7

In Russia, seismic microzonation is performed at the final stage of seismic zonation studies, after developing Seismic Zonation and Detailed Seismic Zonation maps that provide seismic hazard estimates at so-called average soil conditions, approximately corresponding to Class C sites in Eurocode.

The first official seismic zonation map of the former USSR territory was compiled in 1937 in Seismological Institute of the Academy of Sciences of the Soviet Union. That map and three following generations of seismic zonation maps (all of them created in Schmidt Institute of Physics of the Earth in Moscow) were incorporated in seismic building codes in 1957, 1968, and 1978. The maps were constantly upgraded with accumulation of new strong motion data (Ulomov et al. 1999).

Three probabilistic Seismic Zonation maps are currently in use in Russia, such as, OSR-2015-A, OSR-2015-B, and OSR-2015-C; they represent the modifications of the previous maps OSR–97-A, OSR-97-B, and OSR-97-C, adjusted for the latest strong earthquakes. The maps show the estimates of seismic intensity (on average soils) in balls of MSK-64 scale expected with 10% (map A), 5% (map B), and 1% (map C) probability of exceedance in 50 years (Ulomov et al. 1999).

The maps were compiled by a new methodology using a database of seismological, geological and geophysical information of the territory of Northern Eurasia, with lineament-domain-focal (LDF) models of seismic source zones and the assessment of seismic effect on the surface (Ulomov 1997; Ulomov et al. 1999; Shumilina et al. 2000). Seismic intensity was calculated by statistical modelling of earthquake catalogs based on the information of long-term characteristics of seismicity in the regions.

In the new methodology, the approaches of Riznichenko (1966), Keilis-Borok with the colleagues (Molchan et al. 1970) and Cornell (1968) were developed. The limitations of these approaches, like using short series of seismic observations to describe long-term characteristics of the seismic regime, linearity of the earthquake recurrence law in its extrapolation to maximum magnitudes, representation of earthquake sources as point sources, fixing them at certain depths, using magnitudes M_{LH} as energy characteristics, neglecting scattering of ball values for given magnitude and distance, etc.—were overcome. The maps were calculated using regional dependencies of ball (of seismic intensity) on the magnitude and distance for extended seismic sources (Shumilina et al. 2000).

The experience of the USA and Japan in the estimation and mapping of predicted parameters of strong ground motion is analyzed by Aki and Irikura (1991). In these countries, seismic zoning maps describe seismic motion at the bedrock.

The development of national USA maps for seismic zonation is reviewed by Algermissen (1983). Since the 1980s, two national maps were used: one is included in the Uniform Building Code (UBC 1979), and the other is described in the Applied Technology Council report (ATC 1978). The UBC map developed by Algermissen (1969) with subsequent modifications divides the United States into 5 zones (from 0 to 4) based on the maximum Modified Mercalli intensity observed historically in each zone. The ATC report was aimed at the development of nationally applicable seismic design provisions and included maps of "effective peak accelerations" and "effective peak velocity" that could be used to calculate lateral forces. These maps are based on the estimates of the expected maximum accelerations on rock in a 50-year period with a 10% probability of exceedance (Algermissen and Perkins 1976). Such maps have been compiled for ground acceleration and ground velocity for exposure times of 10, 50 and 250 years.

The estimates of the probability of exceedance are based on the assumption that earthquake occurrence follows Poisson's law in time and the Guttenberg-Richter magnitude-frequency law and regional empirical attenuation relations are used. This method of seismic hazard analysis is classified as Type III single model PSHA (Probabilistic Seismic Hazard Analysis) method in the report by National Research Council (NRC 1988). More sophisticated PSHA method, including the information on the characteristics of Quaternary faults, such as, their locations, segment lengths and slip rates has also been described in the report (NRC 1988). Mapping of peak accelerations and velocities based on such geological data was performed by Wesnousky (1986) for California.

In Japan, seismic zonation maps have a longer history than in the USA; the first maps were compiled in 1951 by Kawasumi (1951). Numerous maps of peak ground acceleration and velocity for various periods of earthquake recurrence from 50 to 200 years have been compiled and published in Japan, but almost all the comments and papers on the subject are written in Japanese. Most maps are based on the historic data going back to the period of 1300 years, interpreted using various statistical models of earthquake occurrence, and some attempts to include the data on active faults have also been carried out (Wesnousky et al. 1984).

Development of regional databases for ground motion estimation, significant parameters of soil profiles

In addition to zonation maps for bedrock motion, microzonation maps for selected areas are being developed in the USA and Japan to include the effects of local geology. Most existing maps are based on the relations between the site amplification factor, *S*-wave velocity, density, geologic age, texture of sediments, and other measurable geological and geotechnical parameters. In Japan the maps are often divided into meshes of sizes 500 m × 500 m or 1 km × 1 km (Shima 1978). In the USA the maps show zones of different site effects (Fumal and Tinsley 1985).

To successfully predict parameters of strong ground motion in future earthquakes, accumulation of various seismic, geological and geotechnical data at the studied areas is required. Macroseismic data are desirable, for example, on the intensity of motion and building damage at the sites, where seismic, geological and geotechnical measurements were carried out (Aki and Irikura 1991). Conventional geologic maps provide only a rough differentiation of young unconsolidated deposits; however, they can be used to identify areas of possible soil liquefaction by superimposing them on the maps of the groundwater level.

Various examples from the practice of seismological research indicate the importance of numerical simulations in seismic microzonation studies, in order to take it beyond the empirical prediction based on surface geology (Sanchez-Sesma 1987; Aki 1988). Numerical simulation methods have become a powerful tool for understanding the mechanisms of soil response during strong earthquakes. The profiling parameters that are necessary for calculations are *S*-wave velocities, shear stress in failure, densities and attenuation in soil layers.

Formulas for estimating *S*-wave velocity and density based on geological and geotechnical data have been developed, for example, in Russia (Nazarov 1969), Japan (Ohta and Goto 1978), etc. The main geological data are the age of the sediment (Holocene, Pleistocene, Tertiary, etc.) and the soil texture, classified into clay, silt, sand, or gravel.

Shear wave velocity is broadly recognized as a single most important parameter affecting site amplification. The relationships between *S*-wave velocity and the geological age have been developed for Los Angeles region by Fumal and Tinsley (1985). Classification of surface geology has also been compiled for microzonation of Tokyo, with a cell size of 1 × 1 km (Shima 1978). The data show that the best site (in *S*-wave velocities) in Tokyo is comparable to the worst site in Los Angeles. However, in the vicinity of San Francisco Bay there are areas comparable to Tokyo, with bay mud ($V_S \sim$ 55–115 m/s) and artificial fill ($V_S \sim$ 159–222 m/s) (Fumal 1978). The worst known urban area is in the Mexico City, where *S*–wave velocity in lake sediments reach 30 m/s.

Analyzing geotechnical data, Rogers et al. (1985) have found for Los Angeles and San Francisco that the most significant factor controlling site amplification is void ratio, which strongly correlates (inversely) with *S*-wave velocity. In addition to the void ratio and *S*-wave velocity, the thickness of unconsolidated sediments and the depth to the bedrock are significant parameters controlling amplification at periods 0.2–0.5 s ($f \sim$ 2–5 Hz). At periods > 0.5 s (f < 2 Hz), the depth to the bedrock and the thickness of Quaternary sediments are controlling factors (Rogers et al. 1985).

Goto et al. (1982) showed that geotechnical parameters calculated from standard penetration test (*SPT-N*) closely relate to the site amplification. The *SPT-N* value was used in Ohta-Goto formula (1978) and in the predictions of nonlinear site effect (Sugito and Kameda 1990).

Coefficients of local amplification for weak motion can be determined from coda waves of weak local earthquakes (Phillips and Aki 1986). In areas where no records were obtained, artificial seismic sources can be used (Borcherdt 1970; Rogers et al. 1985; King et al. 1990). For long-period (3–6 s) waves, the amplification can be estimated from the background noise (Kagami et al. 1982, 1986).

In addition to records of strong earthquakes that occurred in the region in the past, the regional database for ground motion estimation should include detailed descriptions of seismic intensity and quantitative estimates of building damage (Ohta et al. 1979; Iglesias 1989).

Geotechnical databases used to predict local site amplification usually include borehole measurements of *P*- and *S*-wave velocities, densities and *SPT-N* values. Attenuation parameter h or Q-factor ($Q = 1/2\,h$) are also used.

Reliable measurements of Q (or h) parameter require vertical arrays of seismographs in the boreholes. The performed studies revealed that generally, Q increases with frequency approximately as the power law $Q = Q_0 f^{\alpha}$, where α varies from 0 to 1. Sakai et al. (1989) found from borehole array records of local earthquakes with magnitudes $M \sim 6$ that in the frequency range 0–10 Hz, $Q = 6.47 f^{\,0.35}$ at Nakano site (fine sands and clays) and $Q = 1.92f^{0.8}$ at Miyagino site (clays and gravels). Kobayashi et al. (1989) concluded that Q can be expressed by the formula $Q = V_S f/20$ for a silty-clay site and $Q = V_S f/60$ for a sandy-gravel site; V_S is expressed in m/s. A similar result was obtained from deep borehole data in Kanto area: $Q = 4f$–$5f$ in Holocene sediments and $Q = 50f$ at frequencies 0.5–3.2 Hz in Quaternary and Tertiary deposits in the entire sedimentary layer (Kinoshita et al. 1986).

On the other hand, Seale and Archuleta (1989) found from records of local earthquakes (including records of earthquakes with magnitudes $M = 5.8$ and 6.4) obtained in a 166-m borehole in Mammoth Lake area that $Q = 10$, regardless of frequency up to ~ 10 Hz in a surface layer composed of glacial moraine. Joyner et al. (1976) based on records of the 186-m borehole array on the shore of the San Francisco Bay found that $Q = 16$ for bay mud, independently of frequency. The so-called κ-effect found by Anderson and Hough (1984) also implies the presence of a subsurface layer with a frequency-independent low Q. Frequency-independent Q ~ 9–10 for filled land in Tokyo were reported by Shima et al. (1985). At higher frequencies ~ 30–90 Hz, the following frequency-independent Q values were obtained: $Q = 8$ (Pleistocene sand), $Q = 20$ (Holocene silt), $Q = 6.5$ (Tertiary mudstone), $Q = 5$ (loam) (Kudo and Shima 1970).

Development of seismic microzonation methods accounting for nonlinear soil response

According to Aki and Irikura (1991), in the 1990s, the most urgent need to improve seismic zonation methods was the clear understanding of nonlinear effects at soil sites that may be more pervasive than seismologists thought earlier. Another important task was to develop links with the community of users of earth science information. The user community is diverse, and the required parameters of ground motion are also diverse. It is impractical to prepare general multi-purpose zonation maps to meet all their needs.

They suggested the following procedure. Since any ground motion parameter can be extracted from the acceleration time series, we shall compute the time series for given source–receiver pairs using the state-of-the-art methods based on our current knowledge on the earthquake source, propagation path and recording site conditions. We then extract the ground motion parameters requested by the customers and attach the probability of the occurrence of the particular earthquake to the parameters. The procedure is repeated for all relevant source–receiver pairs, and the results are synthesized into a site-specific, or a map view of the parameter for a given probability of exceedance in the usual manner of the probabilistic seismic hazard analysis (PSHA). However, for planning purposes maps depicting potentially dangerous areas will always be useful; for example, seismic intensity maps, maps of landslide areas, maps of liquefaction potential, etc. These maps allow us to identify zones where special studies of possible disasters associated with seismic hazard should be undertaken (Aki and Irikura 1991).

To date, linear mechanisms of seismic wave transformations in soil layers (amplification due the transition of waves to surface layers with lower seismic velocities and densities and resonant phenomena in the upper layers) are rather well studied and accounted for in the practice of engineering seismology. At the same time, nonlinear soil response during strong earthquakes is not always well predicted.

The response of soft soils can be considered as linear in weak seismic motion, and in conditions of strong motion, the degree of nonlinearity of the soil response depends on the intensity of motion.

Since microzonation is usually performed for territories subjected to strong earthquakes, the adequate consideration of the nonlinearity of soil response is necessary.

As shown in Chap. 3, the nonlinearity of soil response leads to changes in spectral content of seismic motion, sometimes significant, and in amplification of seismic waves in soil layers. In conditions of strong motion, nonlinear damping occurs (in addition to linear damping) that weakens high and medium frequency components and virtually does not influence low-frequency components. If low-frequency components were weak or absent in the input motion, they may be amplified or emerge in the output motion on the surface.

Changes in spectra of seismic motion associated with the nonlinearity of soil response appear in decreasing resonant frequencies of soil layers and transformations of spectra of motion on the surface to the form $E(f) \sim f^{-k}$. Nonlinearity of soil response reduces amplification of seismic waves in soil layers, compared to the linear case. In water-saturated soils, nonlinear damping sometimes does not compensate "linear" amplification, so that in the whole seismic waves are amplified by soil layers even during strong motion.

The influence of the nonlinearity of soil response on the amplitudes and spectra of intense seismic waves propagating in soil layers is shown in Fig. 7.1.

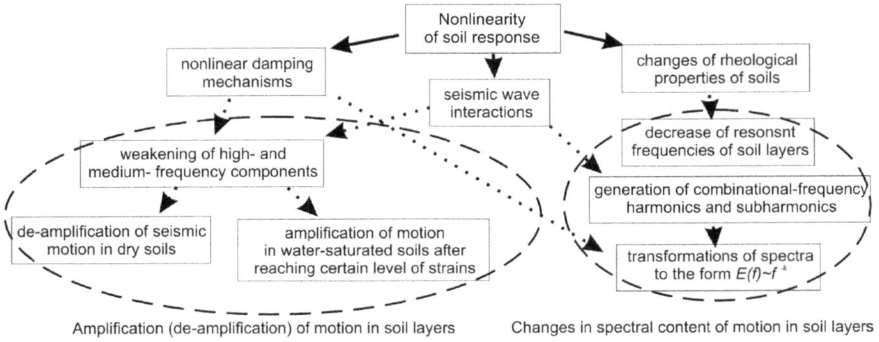

Fig. 7.1 The influence of nonlinearity of soil response on the amplitudes and spectra of intense seismic waves propagating in soil layers

The degree of nonlinearity of soil response depends on the soil characteristics, such as, soil composition, depth, saturation with water, density, etc., and on the intensity and spectral content of seismic motion. Soil characteristics vary from site to site; therefore, various proposed relations between parameters of motion on the surface at soil sites and the earthquake magnitudes and epicentral distances obviously have a large scattering. Such relations should be constructed for certain types of soil conditions.

The constructed models of soil behavior in strong earthquakes (Chap. 3) indicate that nonlinearity of soil behavior is stronger at the bottoms of soft subsurface layers, i.e., at their boundaries with the underlying dense strata (usually in the upper ~ 5–40 m), where the highest strains occur; this should be taken into account in earthquake-resistant design and construction.

Thus, the correct approach in seismic microzonation is calculation of acceleration time histories of possible future earthquakes accounting for nonlinearity of soil response. The acceleration time histories are then used to estimate amplitudes, frequency content and duration of the strong motion.

Acceleration time histories for a given earthquake defined by its magnitude, source location and depth, are usually calculated at a given site on rock. The well-known calculation methods include the stochastic method (Boore 2003) and its modifications, accounting for a finite-fault source (Beresnev and Atkinson 1998), parameters of crack propagation, radiation directivity; empirical Green's functions method (Hartzell 1978; Irikura 1983) and the combinations of the stochastic method and the empirical Green's functions method (Hutchings 1994; Wald et al. 1996; Dreger and Kaverina 2000; Gusev and Pavlov 2009, etc.). The accuracy of the calculation programs is verified by testing them with real earthquake records.

The method developed by A. A. Gusev and V. M. Pavlov (Institute of Volcanology and Seismology of Russian Academy of Sciences, Petropavlovsk-Kamchatsky) has shown a rather high accuracy. In this method, the source of a strong earthquake is described by a set of Green's functions, seismic radiation is assumed stochastic, and seismic waves propagate to the receivers in a layered weakly inelastic medium. The source radiation model is adjusted to a specific seismological situation, statistical probabilistic estimates of ground motion parameters are obtained, and time series corresponding to these estimates are generated. The far-field spectrum adjusts in its high-frequency part to the average regional spectrum.

At the second step of calculations, the acceleration time histories on rock are recalculated to the surfaces of given soil profiles. The computer programs should adequately account for the mechanisms of seismic wave transformations in soil layers, and they use different models of soil behavior. Nonlinear analysis provides more accurate estimates than equivalent linear analysis, however tests on real earthquake records indicate that there is no program that works well in all situations.

As shown in Chap. 3, conventional programs for calculation of soil response in strong motion sometimes underestimate the degree of nonlinearity of soil behavior, neglect differences in the behavior of separate soil layers and changes in the rheological properties of the upper softer layers during strong motion.

Thus, in the calculations of soil response, computer codes based on nonlinear analysis are preferred, modified in such a way that they can account for (1) various types of nonlinear soil behavior (soft-type, hard-type, liquefaction) depending on the composition of the soil, its depth and saturation with water; (2) possible changes in the rheological properties of the soil during strong motion (soil softening at the beginning and recovering at the end of strong motion, liquefaction and slow recovering, etc.).

The assumed models of nonlinear soil behavior in strong ground motion are important parts of the computer codes. The results obtained so far indicate that normalized nonlinear stress–strain relations obtained in laboratory experiments by Hardin and Drnevich (1972) well describe the behavior of soils under moderate dynamic loadings. In conditions of strong motion and high nonlinearity, these relations can be used to describe the behavior of denser soils at depths of ~ 20–25 m and more (depending on the earthquake magnitude and the proximity to the source), whereas the behavior of soft surface soils is usually described by stress–strain relations that are significantly more nonlinear.

Based on the analysis of records of the 1995 Kobe, 2000 Tottori and 1999 Chi-Chi earthquakes (M_w ~ 6.7–7.6), the areas of strong nonlinearity of soil response can be estimated, where changes of the rheological properties of soils were observed.

During the Kobe and Tottori earthquakes ($M_w = 6.7$–6.8, fault lengths L ~ 40 km), strong manifestations of soil nonlinearity (the content of nonlinear components in soil response ~ 40% and more) were observed at ~ 7–8 km from the fault (~ 1/4 of the fault length L). At distances ~ 15–20 km (~ 1/2 of the fault length L) soil nonlinearity was weaker (the content of nonlinear components in soil response ~ 10–12%). During the Chi-Chi earthquake ($M_w = 7.6$, fault length L ~ 110 km), strong soil nonlinearity was located within ~ 20–25 km from the fault (~ 1/4 of the fault length L). Within this zone, shear moduli in the upper ~ 30 m of soft soils reduced by ~ 40–50% and more from the initial values, and the content of nonlinear components in the soil response was ~ 30–80% of the intensity of the response. Within this zone, strong motion caused changes in the rheology of the upper layers, and spectra of motion on the surface were close to $E(f) \sim f^{-k}$. Thus, the area of strong nonlinearity of soil response, where changes of rheological properties of soft soils occur, is located within ~ 1/4L from the fault.

During the three earthquakes, at the majority of soil sites located within ~ 1/4L from the fault, soft soils showed hard-type nonlinearity. Strong motions induced resonant phenomena in surface water-saturated layers leading to large strains and pore pressure build-up. As a result, stress–strain relations harden, and the intensity of motion on the surface increased. Resonant phenomena in the upper softer layers were widely spread during the three earthquakes, however, during the Chi-Chi earthquake, they involved deeper layers, down to ~ 40–60 m.

As shown in Chap. 4, records of stronger earthquakes, such as, 2003 Tokachi-oki earthquake (M_w ~ 8.3) and 2011 Tohoku earthquake (M_w ~ 9.0), also show hard-type soil behavior and high PGA exceeding 1 g in the near-fault zones.

During large earthquakes with extended sources (M_w ~ 8.0 and higher), the effects of source directivity play an important role in soil behavior and distribution of PGA in the near-fault zones. In the sources of such earthquakes, the cracks can propagate at high speeds along lengthy smoothed parts of the fault planes, so that seismic waves radiated by the crack tips during certain time periods arrive at some remote stations in shorter periods or almost simultaneously, and shock waves are generated. At these stations, soils experience additional compression, harden, and effectively amplify strong motion, and strong motion records at soil sites possess short duration, high PGA exceeding 1 g, and high frequencies $f \sim 4–6$ Hz.

During the 2003 Tokachi-oki earthquake ($M_w \sim 8.3$, fault plane size ~ 140 km by 160 km), such effects were observed at epicentral distances of ~ 180–200 km. During the 2011 Tohoku earthquake ($M_w \sim 9.0$, fault plane size ~ 200 km by 500 km), they were observed at epicentral distances of ~ 270–310 km. The lengths of the smoothed sections on the faults, where the crack tips moved at high speeds (~ 4.0–4.3 km/s) and superposition of seismic waves occurred, were ~ 100–120 km.

Obviously, we may expect similar effects during future large earthquakes with magnitudes $M_w > 8.0$.

In the whole, along with various manifestations of soil response, depending on the composition, mechanical properties, water saturation of soil layers, as well as spectral content and intensity of incident seismic motion, there is a similarity in the behavior of similar soils during different earthquakes, indicating a possibility to predict soil behavior in future strong earthquakes.

References

Aki, K., Irikura, K.: Characterization and mapping of earthquake shaking for seismic zonation. In: Proceedings of 4th International Conference on Seismic Zonation, Aug 25–29, Stanford, California, vol. 1, pp. 61–110 (1991)

Aki, K.: Local site effect of ground motion, earthquake engineering and soil dynamics, II: recent advances in ground-motion evaluation. In: Lawrence Von Thun, J. (ed.) American Society of Civil Engineers Geotechnical Special Publions, vol. 20, pp. 103–155 (1988)

Algermissen, S.T., Perkins, D.M.: A probabilistic estimate of maximum acceleration in rock in the continuous United States. U.S. Geol. Surv. Open-File Report 76–416 (1976)

Algermissen, S.T.: Seismic risk studies in the United States. In: Proceedings of 4th World Conference on Earthquake Engineering, Santiago, Chile, vol. 1, pp. 14–27 (1969)

Algermissen, S.T.: An introduction to the seismicity of the United States. Earthq. Eng. Res. Inst. (1983)

Anderson, J.G., Hough, S.E.: A model for the shape of the Fourier amplitude spectrum of acceleration at high frequencies. Bull. Seism. Soc. Am. **74**, 1969–1993 (1984)

Beresnev, I.A., Atkinson, G.M.: FINSIM—a FORTRAN program for simulating stochastic acceleration time histories from finite faults. Seism. Res. Lett. **69**(1), 27–32 (1998)

Boore, D.M.: Simulation of ground motion using the stochastic method. Pure Appl. Geoph. **160**, 635–676 (2003)

Borcherdt, R.D.: Effects of local geology on ground motionnear San-Francisco Bay. Bull. Seism. Soc. Am. **60**, 29–61, 2213 (1970)

Cornell, C.A.: Engineering seismic risk analysis. Bull. Seism. Soc. Am. **58**(5), 1583–1606 (1968)

Dreger, D.S., Kaverina, A.: Seismic remote sensing for the earthquake source process and near-source strong shaking: a case study of the October 16, 1999 Hector Mine earthquake. Geoph. Res. Lett. **27**, 1941–1944 (2000)

Fumal, T.E., Tinsley, J.C.: Mapping shear-wave velocities of near-surface geologic materials. In: Evaluating Earthquake Hazards in the Los Angeles region. U.S.G.S. Professional Paper 1360, pp. 127–149 (1985)

Fumal, T.E.: Correlations between seismic wave velocities and physical properties of geologic materials in the San Francisco Bay region, California. U.S. Geol. Surv. Open-File Rep. 78–1067 (1978)

Goto, H., Kameda, H., Sugito, M.: Use of N-value profilas for estimation of site dependent earthquake motions. Jpn. Soc. Civil Eng. **317**, 69–78 (1982)

Gusev, A.A., Pavlov, V.M.: Broadband simulation of earthquake ground motion by a spectrum-matching, multiple-pulse technique. Earthq. Spectra **25**(2), 257–276 (2009)

Hardin, B.O., Drnevich, V.P.: Shear modulus and damping in soils: measurement and parameter effects. Proc. Am. Soc. Civil Eng. J. Soil Mech. Found. Div. **98**, 603–624 (1972)

Hartzell, S.H.: Earthquake aftershocks as Green's functions. Geoph. Res. Lett. **5**(1), 1–4 (1978)

Hutchings, L.: Kinematic earthquake models and synthesized ground motion using empirical Green's functions. Bull. Seiem. Soc. Am. **84**(4), 1028–1050 (1994)

Iglesias, J.: The Mexico earthquake of September 19, 1985—Seismic zoning of Mexico City after the 1985 earthquake. Earthq. Spectra **5**, 257–271 (1989)

Irikura, K.: Semi-empirical estimation of strong ground motions during large earthquakes. Bull. Disaster Prevention Res. Inst. (kyoto Univ.) **32**, 63–104 (1983)

Joyner, W.B., Warrick, R.E., Oliver, A.A.: Analysis of seismograms from a downhole array in sediments near San Francisco Bay. Bull. Seism. Soc. Am. **66**, 937–958 (1976)

Kagami, H., Okada, S., Shiono, K., Oner, M., Dravinski, M., Mal, A.K.: Observation of 1- to 5-second microtremors and their application to earthquake engineering, Part III: a two-dimensional study of site effects in the San Fernando valley. Bull. Seism. Soc. Am. **76**, 1801–1812 (1986)

Kagami, H., Duke, C.M., Liang, G.C., Ohta, Y.: Observation of 1 to 5-second microtremorsand their application to earthquake engineering. Part II. Evaluation of site effects upon seismic wave amplification due to extremely deep soil deposits. Bull. Seism. Soc. Am. **72**, 987–998 (1982)

Kawasumi, T.: Measures of earthquake danger and expectancy of maximum intensity throughout Japan as inferred from the seismic activity in historical times. Bull. Earthq. Res. Inst. Tokyo Univ. **29**, 469–482 (1951)

King, K.W., Tarr, A.C., Carver, D.L., Williams, R.A., Worley, D.M.: Seismic ground-response studies in Olympia, Washington, and vicinity. Bull. Seism. Soc. Am. **80**, 1057–1078 (1990)

Kinoshita, H., Mikoshiba, T., Hoshino, T.: Estimation of the average amplification characteristics of a sedimentary layer for short period S-waves. Zisin J. Seism. Soc. Jpn. **2**(39), 67–80 (1986)

Kudo, K., Shima, E.: Attenuation of shear waves in soil. Bull. Earthq. Res. Inst. **48**, 145–158 (1970)

Molchan, G.M., Keilis-Borok, V.I., Vilkovich, G.V.: Seismicity and principal seismic effects. Geoph. J. Roy. Astr. Soc. **21**(3), 323–335 (1970)

Nazarov, G.N.: Methodological Guidelines for Complex Seismogeological and Engineering-Geological Studies. VIA Publishing House, Moscow (1969)

Ohta, Y., Goto, N.: Empirical shear wave velocity equations in terms of characteristic soil indexes. Earthq. Eng. Struct. Dyn. **6**, 167–187 (1978)

Ohta, Y., Goto, N., Ohashi, H.: A questionnaire survey for estimating seismic intensities. Bull. Faculty Eng. Hokkaido Univ. **92**, 241–250 (1979)

Phillips, W.S., Aki, K.: Site amplification of coda waves from local earthquakes in central California. Bull. Seism. Soc. Am. **76**, 627–648 (1986)

Riznichenko, Y.V.: Raschet sotryasaemosti tochek zemnoi poverkhnosti ot zemletryasenii v okruzhayushchei oblasti (Computation of tremors at points on the Earth's surface from earthquakes in the surrounding regions). Izv. AN SSSR, Fizika Zemli **5**, 16–32 (1966) (in Russian)

Rogers, A.M., Tinsley, J.C., Borcherdt, R.D.: Predicting relative ground response. Evaluating earthquake hazards in the Los Angeles region—an earth-science perspective. U.S. Geol. Surv. Prof. Pap. **1360**, 221–247 (1985)

Sakai, S., Ichikawa, M., Yoshimura, N.: Evaluation of amplification characteristics of body wave in ground surface layers. In: Proceedings of National Symposium on Effects of Surface Geology on Seismic Motion, pp 19–24, (1989)

Sanchez-Sesma, F.J.: Site effects in strong ground motion. Soil Dyn. Earthq. Eng. **6**, 124–132 (1987)

Seale, S.H., Archuleta, R.J.: Site amplification and attenuation of strong ground motion. Bull. Seism. Soc. Am. **79**, 1673–1695 (1989)

Shima, E., Yanagisawa, M., Kohketsu, K., Zama, S., Hoshino, T.: Behavior of seismic waves in the newly developed cut-down and fill-in grounds. Bull. Earthq. Res. Inst. Tokyo Univ. **60**, 179–199 (1985)

Shima, E.: Seismic micro-zoning map of Tokyo. In: 2nd International Conference on Microzonation pp. 433–443 (1978)

Shumilina, L.S., Gusev, A.A., Pavlov, V.M.: An improved technique for determination of seismic hazard. J. Earthq. Predict. Res. **8**(1), 104–110 (2000)

Sugito, M., Kameda, H.: Non-linear soil amplification model with verification by vertical strong motion array records. In: 4th U.S. NCEE, pp. 1–10 (1990)

Ulomov, V.I., et al.: Seismic hazard of northern Eurasia. Annali Geofis **42**, 1023–1038 (1999)

Ulomov, V.I.: On the identification and seismological parameterization of earthquake source zones. The Caucasus and adjacent area. Historical and Prehistorical Earthquakes in the Caucasus. NATO ASI. Ser. 2. Environment, vol. 28. ILP Publication n. 333. Kluwer Academic Publishers Dordrecht Boston London, pp. 503–522 (1997)

Wald, D.J., Heaton, T.H., Hudnut, K.W.: The slip history of the 1994 Northridge, California, earthquake determined from strong-motion, teleseismic, GPS and leveling data. Bull. Seism. Soc. Am. **86**, s49–s70 (1996)

Wesnousky, S.G., Scholtz, C.H., Shimazaki, K., Matsuda, T.: Integration of geological and seismological data for the analysis of seismic hazard: a case study of Japan. Bull. Seism. Soc. Am. **74**, 687–708 (1984)

Wesnousky, S.G.: Earthquakes, quaternary faults and seismic hazards in California. J. Geoph. Res. **91**, 12587–12631 (1986)

The manufacturer's authorised representative in the EU is Springer Nature Customer Service Centre GmbH, Europaplatz 3, 69115 Heidelberg, Germany. If you have any concerns regarding our products, please contact ProductSafety@springernature.com

Printed and bound by CPI Group (UK) Ltd, Croydon, CR0 4YY
01/10/2026
02225965-0012